Advances in

ATOMIC, MOLECULAR, AND OPTICAL PHYSICS

VOLUME 55

Editors

ENNIO ARIMONDO
University of Pisa
Pisa, Italy

PAUL R. BERMAN
University of Michigan
Ann Arbor, Michigan

CHUN C. LIN
University of Wisconsin
Madison, Wisconsin

Editorial Board

C. JOACHAIN
Université Libre de Bruxelles
Brussels, Belgium

M. GAVRILA
F.O.M. Insituut voor Atoom- en Molecuulfysica
Amsterdam, The Netherlands

M. INOKUTI
Argonne National Laboratory
Argonne, Illinois

ADVANCES IN
ATOMIC, MOLECULAR, AND OPTICAL PHYSICS

Edited by

E. Arimondo
PHYSICS DEPARTMENT
UNIVERSITY OF PISA
PISA, ITALY

P.R. Berman
PHYSICS DEPARTMENT
UNIVERSITY OF MICHIGAN
ANN ARBOR, MI, USA

and

C.C. Lin
DEPARTMENT OF PHYSICS
UNIVERSITY OF WISCONSIN
MADISON, WI, USA

Volume 55

AMSTERDAM • BOSTON • HEIDELBERG • LONDON • NEW YORK
OXFORD • PARIS • SAN DIEGO • SAN FRANCISCO • SINGAPORE
SYDNEY • TOKYO

Academic Press is an imprint of Elsevier

Academic Press is an imprint of Elsevier
84 Theobald's Road, London WC1X 8RR, UK
Radarweg 29, PO Box 211, 1000 AE Amsterdam, The Netherlands
Linacre House, Jordan Hill, Oxford OX2 8DP, UK
30 Corporate Drive, Suite 400, Burlington, MA 01803, USA
525 B Street, Suite 1900, San Diego, CA 92101-4495, USA

First edition 2008

Copyright © 2008 Elsevier Inc. All rights reserved

No part of this publication may be reproduced, stored in a retrieval system or transmitted in any form or by any means electronic, mechanical, photocopying, recording or otherwise without the prior written permission of the publisher

Permissions may be sought directly from Elsevier's Science & Technology Rights Department in Oxford, UK: phone (+44) (0) 1865 843830; fax (+44) (0) 1865 853333; email: permissions@elsevier.com. Alternatively you can submit your request online by visiting the Elsevier web site at http://elsevier.com/locate/permissions, and selecting: *Obtaining permission to use Elsevier material*

Notice
No responsibility is assumed by the publisher for any injury and/or damage to persons or property as a matter of products liability, negligence or otherwise, or from any use or operation of any methods, products, instructions or ideas contained in the material herein. Because of rapid advances in the medical sciences, in particular, independent verification of diagnoses and drug dosages should be made

ISBN-13: 978-0-12-373710-6

ISSN: 1049-250X

For information on all Academic Press publications
visit our website at books.elsevier.com

Printed and bound in USA

08 09 10 11 12   10 9 8 7 6 5 4 3 2 1

Working together to grow
libraries in developing countries

www.elsevier.com | www.bookaid.org | www.sabre.org

ELSEVIER   BOOK AID International   Sabre Foundation

Contents

CONTRIBUTORS ix

Direct Frequency Comb Spectroscopy

Matthew C. Stowe, Michael J. Thorpe, Avi Pe'er, Jun Ye, Jason E. Stalnaker, Vladislav Gerginov and Scott A. Diddams

1. Introduction and Historical Background 2
2. Comb Control and Detection 5
3. Direct Frequency Comb Spectroscopy 12
4. Multi-Frequency Parallel Spectroscopy 30
5. Coherent Control Applications 38
6. Future Outlook 45
7. Concluding Remarks 52
8. Acknowledgements 52
9. References 53

Collisions, Correlations, and Integrability in Atom Waveguides

Vladimir A. Yurovsky, Maxim Olshanii and David S. Weiss

1. Introduction 62
2. Effective 1D World 64
3. Bethe Ansatz and beyond 85
4. Ground State Properties of Short-Range-Interacting 1D Bosons: Known Results and Their Experimental Verification 103
5. What Is Special about Physics in 1D 113
6. Summary and Outlook 125
7. Acknowledgements 126
8. Appendix: Some Useful Properties of the Hurwitz Zeta Function 126
9. References 127

MOTRIMS: Magneto–Optical Trap Recoil Ion Momentum Spectroscopy

Brett D. DePaola, Reinhard Morgenstern and Nils Andersen

1. Introduction to MOTRIMS 140
2. Relative Total Electron Transfer Cross Sections 147
3. Case Studies in Total Electron Transfer Collisions 149

4.	Case Studies of Differential Electron Transfer Cross Sections	161
5.	Probing Excitation Dynamics	172
6.	Future Applications	180
7.	Concluding Comments	184
8.	Acknowledgements	185
9.	References	185

All-Order Methods for Relativistic Atomic Structure Calculations

Marianna S. Safronova and Walter R. Johnson

1.	Introduction and Overview	192
2.	Relativistic Many-Body Perturbation Theory	194
3.	Relativistic SD All-Order Method	198
4.	Motivation for Further Development of the All-Order Method	206
5.	Recent Developments in the Calculations of Monovalent Systems: Non-Linear Terms and Triple Excitations	209
6.	Many-Particle Systems	213
7.	Applications of High-Precision Calculations	219
8.	Conclusion	230
9.	Acknowledgements	231
10.	References	231

B-Splines in Variational Atomic Structure Calculations

Charlotte Froese Fischer

1.	Introduction	236
2.	The Hartree–Fock Approximation	238
3.	Multiconfiguration Hartree–Fock Approximation	249
4.	B-Spline Theory	251
5.	B-Spline Methods for the Many-Electron Hartree–Fock Problem	261
6.	B-Spline MCHF Equations	275
7.	Conclusion	288
8.	Acknowledgements	289
9.	References	289

Electron–Ion Collisions: Fundamental Processes in the Focus of Applied Research

Alfred Müller

1.	Introduction	294
2.	Basics of Electron–Ion Collisions	298
3.	Experimental Access to Data	319

4. Overview of Experimental Results on Free-Electron–Ion Collisions . . 355
5. Conclusions 397
6. Acknowledgements 399
7. References 399

Robust Probabilistic Quantum Information Processing with Atoms, Photons, and Atomic Ensembles

Luming Duan and Christopher R. Monroe

1. Introduction 420
2. Quantum Communication with Atomic Ensembles 421
3. Quantum State Engineering with Realistic Linear Optics 434
4. Quantum Computation through Probabilistic Atom–Photon Operations 442
5. Summary 459
6. Acknowledgements 460
7. References 460

INDEX 465
CONTENTS OF VOLUMES IN THIS SERIAL 473

CONTRIBUTORS

Numbers in parentheses indicate the pages on which the author's contributions begin.

MATTHEW C. STOWE (1), JILA, National Institute of Standards and Technology and University of Colorado, Department of Physics, University of Colorado, Boulder, Colorado 80309-0440, USA

MICHAEL J. THORPE (1), JILA, National Institute of Standards and Technology and University of Colorado, Department of Physics, University of Colorado, Boulder, Colorado 80309-0440, USA

AVI PE'ER (1), JILA, National Institute of Standards and Technology and University of Colorado, Department of Physics, University of Colorado, Boulder, Colorado 80309-0440, USA

JUN YE (1), JILA, National Institute of Standards and Technology and University of Colorado, Department of Physics, University of Colorado, Boulder, Colorado 80309-0440, USA

JASON E. STALNAKER (1), Time and Frequency Division, National Institute of Standards and Technology, Boulder, Colorado 80305, USA

VLADISLAV GERGINOV (1), Time and Frequency Division, National Institute of Standards and Technology, Boulder, Colorado 80305, USA

SCOTT A. DIDDAMS (1), Time and Frequency Division, National Institute of Standards and Technology, Boulder, Colorado 80305, USA

VLADIMIR A. YUROVSKY (61), School of Chemistry, Tel Aviv University, Tel Aviv 69978, Israel

MAXIM OLSHANII (61), Department of Physics & Astronomy, University of Southern California, Los Angeles, CA 90089, USA

DAVID S. WEISS (61), Department of Physics, Pennsylvania State University, University Park, Pennsylvania 16802-6300, USA

B.D. DEPAOLA (139), Department of Physics, Kansas State University, Manhattan, KS 66506, USA

R. MORGENSTERN (139), KVI, Atomic Physics, Rijksuniversiteit Groningen - Zernikelaan 25, NL-9747 AA Groningen, The Netherlands

N. ANDERSEN (139), Niels Bohr Institute, University of Copenhagen, Universitetsparken 5, DK-2100 Copenhagen, Denmark

M.S. SAFRONOVA (191), Department of Physics and Astronomy, 223 Sharp Lab, University of Delaware, Newark, Delaware 19716, USA

W.R. JOHNSON (191), Department of Physics, University of Notre Dame, Notre Dame, IN 46556, USA

CHARLOTTE FROESE FISCHER (235), Department of Electrical Engineering and Computer Science, Vanderbilt University, Box 1679 B, Nashville, Tennessee, 37235, USA

ALFRED MÜLLER (293), Institut für Atom- und Molekülphysik, Justus-Liebig-Universität Giessen, Leihgesterner Weg 217, D-36392 Giessen, Germany

L.-M. DUAN (419), FOCUS, MCTP, and Department of Physics, University of Michigan, Ann Arbor, MI 48109-1040, USA

C. MONROE (419), FOCUS, MCTP, and Department of Physics, University of Michigan, Ann Arbor, MI 48109-1040, USA

DIRECT FREQUENCY COMB SPECTROSCOPY

MATTHEW C. STOWE[1], MICHAEL J. THORPE[1], AVI PE'ER[1], JUN YE[1], JASON E. STALNAKER[2], VLADISLAV GERGINOV[2] and SCOTT A. DIDDAMS[2]

[1] JILA, National Institute of Standards and Technology and University of Colorado, Department of Physics, University of Colorado, Boulder, Colorado 80309-0440, USA
[2] Time and Frequency Division, National Institute of Standards and Technology, Boulder, Colorado 80305, USA

1. Introduction and Historical Background 2
2. Comb Control and Detection 5
   2.1. Comb Degrees of Freedom 5
   2.2. Control of Comb via an Optical Cavity 8
3. Direct Frequency Comb Spectroscopy 12
   3.1. Theoretical Treatments 13
   3.2. Single-Photon DFCS 15
   3.3. Multi-Photon DFCS 22
   3.4. Short Wavelength DFCS 29
4. Multi-Frequency Parallel Spectroscopy 30
   4.1. Cavity Enhanced DFCS 32
5. Coherent Control Applications 38
   5.1. High Resolution Coherent Control 39
   5.2. Extension of DFCS to Strong Field Coherent Control 40
6. Future Outlook 45
   6.1. Atomic Clock Applications 45
   6.2. XUV Comb Development 46
   6.3. Future VUV Comb Spectroscopy 49
   6.4. Optical Frequency Synthesis and Waveform Generation 51
7. Concluding Remarks 52
8. Acknowledgements 52
9. References 53

Abstract

Besides serving as a frequency counter or clockwork, an optical frequency comb can be used directly for spectroscopy, thus the name Direct Frequency Comb Spectroscopy (DFCS). Precise phase coherence among successive ultrashort pulses allow one to explore both fast dynamics in the time domain and high-resolution structural information in the frequency domain. Coherent accumulation of weak pulses can lead to strong field effects. Combined with spectral manipulation, high-resolution quantum control can be implemented. The large number of frequency comb components also provide a massive set of parallel detection channels to gather spectroscopic information. In this chapter we provide a detailed review of some of the current applications that exploit these unique features, and discuss several future directions of DFCS.

1. Introduction and Historical Background

At the end of the 1990s the mode-locked femtosecond laser and its associated optical frequency comb made an unanticipated entry into the field of precision optical frequency metrology (Udem et al., 1999a, 1999b; Diddams et al., 2000; Jones et al., 2000). It is difficult to overstate the impact of this new technology on frequency metrology research (Hall, 2006; Hänsch, 2006). In simple terms, the femtosecond laser frequency comb transformed optical frequency standards into optical clocks by providing a straightforward means to count optical cycles (Diddams et al., 2001; Ye et al., 2001). Just as important, a stabilized frequency comb permits the easy comparison of optical standards separated by hundreds of terahertz (Diddams et al., 2001; Stenger et al., 2002), offers new tools for frequency transfer (Holman et al., 2004; Foreman et al., 2007), introduces new approaches to length metrology (Ye, 2004; Minoshima and Matsumoto, 2000), and provides the carrier–envelope control required for phase-sensitive nonlinear optics and attosecond physics (Paulus et al., 2001; Baltuška et al., 2003).

Beyond its enabling advances in optical frequency metrology, a stabilized optical frequency comb can also be a versatile spectroscopic tool, providing excellent accuracy, high spectral purity, and at the same time broad spectral coverage. In fact, the concept of using the frequency comb from a mode-locked laser for high resolution spectroscopy precedes the more recent comb-related excitement by several decades. The 1970s saw the emergence of the theory and experiments containing the roots of the ideas that will be discussed in detail in this review. While the pulses were longer (nano- or picosecond compared with femtosecond) and the frequency control was rudimentary by today's standards, many of the early experiments already highlighted the advantages of a mode-locked laser frequency comb for multi-photon spectroscopy and measuring optical frequency differences.

Considering our present-day understanding of the frequency comb, it is not surprising that these early experiments emerged as either time- or frequency-domain interpretations of the interaction of a series of coherent pulses with an atomic system. From the time-domain perspective, two early experiments were cast as variations of Ramsey spectroscopy applied to two-photon transitions (Salour and Cohen-Tannoudji, 1977; Teets et al., 1977). In both cases, linewidths less than the pulse bandwidth were observed due to the coherence

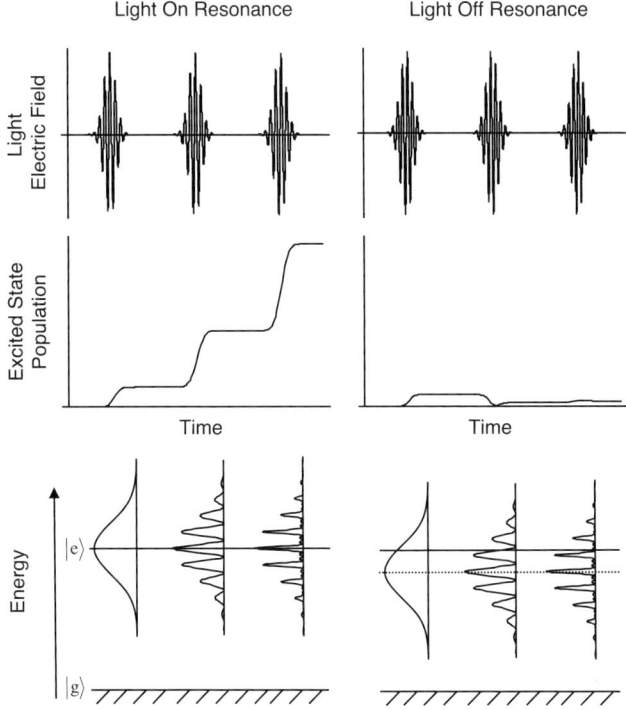

FIG. 1. Time- and frequency-domain interpretations of the interaction of a series of short pulses with an atomic system. The bandwidth of the pulse is much broader than the linewidth of the resonance. When the light is on resonance, the contribution to the atomic polarization adds constructively from pulse to pulse, and population accumulates in the excited state (left column). In the frequency domain, this corresponds to one of the comb modes being on resonance with the transition as shown in the lower left frame. Note that the frequency comb is not defined for a single pulse, but builds up from the interference of several pulses, as shown in the lower plots. In the off resonance case, subsequent pulses arrive out of phase with the evolving polarization and population is moved back to the ground state (right column). This corresponds to the case of a comb mode not overlapping with the transition frequency.

between the temporally-separated pulses as illustrated in Fig. 1. Early experiments that approached spectroscopy from the frequency domain focused on the advantage the frequency comb brings to multi-photon transitions in which pairs of comb modes are summed or subtracted to match the atomic resonance. Two examples are shown in Fig. 2. The case of the lambda transition shown in Fig. 2(a) can be viewed as Raman spectroscopy in which multiple comb components contribute to the resonant transfer of population between the two ground states under the requirement that the difference in energy between the ground

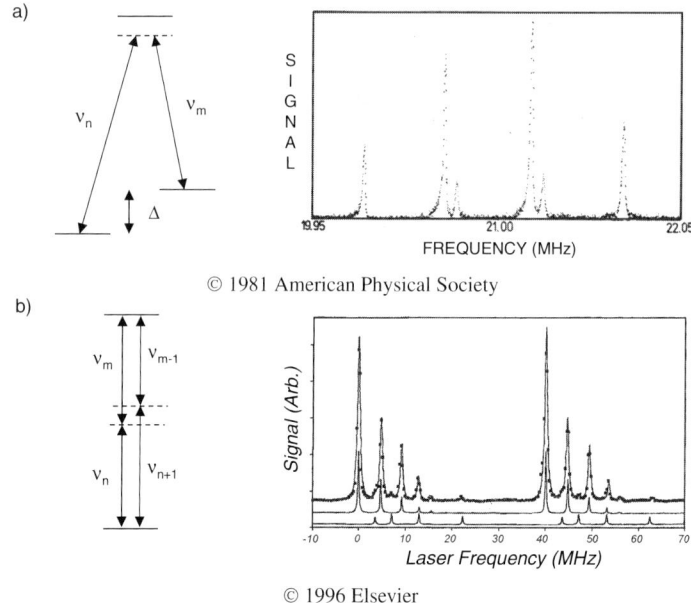

FIG. 2. Two-photon comb spectroscopy in (a) lambda and (b) ladder systems. In the lambda system resonances are observed when the difference in two comb mode frequencies (or an integer multiply thereof) matches the lower level splitting. An example from the work of Mlynek et al. (1981) is shown where the pulse repetition rate is varied. In this case, the six different resonances result from different hyperfine and magnetic sub-levels of the sodium D_1 transition. In the case of the more conventional two-photon transition (b), an example from the work of Snadden et al. (1996) in Rb shows the resonances due to the hyperfine structure in the 5S–5D transition. Because of the periodic structure of the frequency comb, the data repeat every 40 MHz (1/2 the comb spacing). [Figure (a) reprinted with permission from Mlynek et al. (1981). Figure (b) reprinted with permission from Snadden et al. (1996).]

states corresponds to a harmonic of the comb mode spacing (Mlynek et al., 1981; Fukuda et al., 1981; Harde and Burggraf, 1982). Related approaches have proved to be of interest in the area of coherent population trapping (Diddams et al., 1998; Brattke et al., 1998). In Fig. 2(b) one sees the classic two-photon transition in which multiple pairs of comb modes sum together to reach the excited state (Eckstein et al., 1978; Baklanov and Chebotayev, 1977; Hänsch and Wong, 1980; Snadden et al., 1996). This basic approach has been extended to experiments in saturated absorption and polarization spectroscopy (Couillaud et al., 1980; Ferguson et al., 1979). In contrast to the case of the lambda transition, this spectroscopy is sensitive not only to the mode spacing but also the absolute frequency of the comb modes—a point that we will return to later.

2. Comb Control and Detection

In this section we provide a brief review of optical frequency comb structure and the means for its stabilization. More detailed reviews can be found in Cundiff et al. (2001), Cundiff and Ye (2003), Ye et al. (2003), Ye and Cundiff (2005). Passive optical cavities will also be discussed here, which will serve three roles: frequency comb stabilization; buildup of intracavity field strength; and enhancement of spectroscopic detection sensitivity.

2.1. COMB DEGREES OF FREEDOM

The frequency spectrum of an ideal mode-locked laser is generally composed of hundreds of thousands of equidistant modes spaced by the repetition rate of the laser, thus the name optical frequency comb. In practice, however, a mode-locked laser exhibits noise on the optical phase and repetition rate of the emitted pulses. In this section we briefly review the relevant degrees of freedom governing the laser frequency spectrum and the common methods of stabilization.

There are only two degrees of freedom necessary to uniquely specify the frequency of every comb mode (Reichert et al., 1999). The first is the repetition rate of the laser, f_r, defined as the inverse of the inter-pulse period. As can be seen in Fig. 3 the comb mode spacing is exactly f_r. Secondly, the rate change of optical phase from pulse to pulse must be specified. Referring to Fig. 3 the carrier–envelope offset phase, ϕ_{ce}, is defined as the relative phase between the peak of the pulse envelope and the underlying electric field. In general the ϕ_{ce} changes between pulses, it is this change, $\Delta\phi_{ce}$, which determines the offset frequency, f_o, of the comb modes as shown in Fig. 3. Using these definitions the frequency of the mode indexed by n is given by $\nu_n = nf_r + f_o$, with $f_o = f_r \frac{\Delta\phi_{ce}}{2\pi}$. For a laser with a repetition rate between 1 to 100 MHz and center wavelength of 800 nm the mode index n is of order 10^5 to 10^6. In general, f_r and f_o can be independently measured and then controlled via two servo transducers acting on the cavity length and the dispersion of the mode-locked laser.

The laser repetition rate f_r is generally directly measured on a fast, multi-GHz bandwidth, photodiode. This produces a series of radio frequency (RF) beats at multiples of f_r which can be used for phase locking to an RF reference such as a crystal oscillator with low short-term noise. To achieve superior phase sensitivity, one of the high harmonics of f_r is used for the phase lock, for example the tenth harmonic of a 100 MHz repetition rate laser can be locked to a 1 GHz reference oscillator. The beat phase between the measured f_r harmonic and the reference is filtered and amplified for use as an error signal to servo the laser cavity length via a piezo-electric transducer (PZT) mounted output coupler. Because an optical frequency is related to the repetition rate by the mode number, any noise on

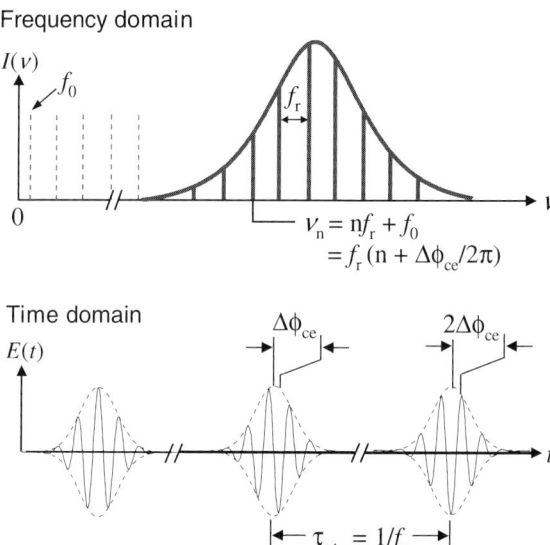

FIG. 3. (Top panel) Frequency-domain picture of the femtosecond pulse train indicating the comb modes under a Gaussian spectral envelope. The modes are equally spaced by f_r and offset from exact f_r harmonics by f_o. The expression for the frequency of the nth mode is given. (Bottom panel) Time-domain picture of the corresponding infinite periodic pulse train. Note that ϕ_{ce} of the first pulse is zero then evolves to a nonzero value by the second pulse, thus $\Delta\phi_{ce} \neq 0$ so $f_o \neq 0$.

the repetition rate is multiplied by 10^5–10^6 for typical mode numbers. Thus, the short term stability of the reference oscillator plays a crucial role in the resulting comb mode linewidth. With some of the best available RF sources one can obtain fractional instabilities on the order of 10^{-13} in one second. In fact, the ultimate performance in frequency comb stabilization is via an optical approach: direct comparisons between frequency comb modes against optical references such as a single frequency continuous-wave (cw) laser (Diddams et al., 2000; Ye et al., 2000, 2001) or a high-finesse optical cavity (Jones and Diels, 2001; Jones et al., 2004). The direct optical approach permits the transfer of superior stability optical standards to the frequency comb, yielding comb instabilities of order 10^{-15} in one second and Hz-level linewidths (Bartels et al., 2004a; Ludlow et al., 2006).

Direct detection and stabilization of the offset frequency, f_o, is somewhat more complicated than f_r. The typical approach for measuring f_o (Reichert et al., 1999; Telle et al., 1999) requires an octave spanning pulse bandwidth. If the pulse bandwidth spans one octave then it contains two frequencies, $\nu_1 = n_1 f_r + f_o$ and $\nu_2 = n_2 f_r + f_o$, where $n_2 = 2n_1$. The second harmonic of ν_1 is $2\nu_1 = 2n_1 f_r + 2f_o$, which upon forming a heterodyne beat with ν_2 yields a beat note at the desired f_o frequency. This technique is referred to as self-referencing because it forms the beat note entirely from the laser output without any separate reference frequencies. It is now also possible to obtain an octave bandwidth directly from a suitably constructed Ti:sapphire laser (Ell et al., 2001; Fortier et al., 2006a, 2003; Mücke et al., 2005). A common technique for generating the required bandwidth is via four-wave mixing in an extra-cavity highly nonlinear fiber. Other techniques for measuring f_o require less than an octave of bandwidth. For example, one needs only 2/3 of an octave bandwidth by using a heterodyne beat between two sets of adjacent modes, one formed by doubling the high-frequency portion of the original comb spectrum and the other from tripling the low-frequency portion (Telle et al., 1999; Ramond et al., 2002).

It was the development of dispersion modified micro-structured fiber (Ranka et al., 2000; Dudley et al., 2006) that enabled the first experimental demonstration of self-referencing (Jones et al., 2000; Apolonski et al., 2000). Micro-structure fiber, also called photonic crystal fiber, exhibits two important features for efficiently producing an octave spanning bandwidth. Micro-structure fiber gets its name from the regular pattern of air holes that extend the length of the fiber and surrounding the core, forming a photonic band gap to confine the light. By tuning the waveguide dispersion of the fiber the net dispersion can be made zero near the desired input pulse wavelength. Typically such a fiber has a core diameter of about 1.7 µm for a zero dispersion point at 800 nm. Confining the pulse to such a small area can generate extremely large peak intensities, which due to the zero dispersion property may persist through the fiber with minimal pulse stretching. This combination of high intensity over a long interaction length allows for efficient four-wave mixing. In practice transform limited pulses with energies of order nanojoules can be used with approximately 10 cm of fiber to generate the necessary octave bandwidth.

To understand how to servo the laser and thus control f_o we must first understand the origin of the phase, ϕ_{ce}. Recall that ϕ_{ce} is defined between the pulse envelope, which travels at the group velocity v_g, and the underlying electric field, which travels at the phase velocity v_p. In the laser cavity the difference of v_g and v_p results in a generally nonzero $\Delta\phi_{ce}$ given by

$$\Delta\phi_{ce} = l_c \omega_c \left(\frac{1}{v_g} - \frac{1}{v_p} \right),$$

where l_c is the cavity round trip length and ω_c is the carrier frequency. With this knowledge in mind it is clear that tuning the dispersion of the cavity will also change f_o. This was first accomplished by tilting the end mirror of a linear cavity in which the spectrum is spatially dispersed via the intracavity prisms, the resulting wavelength-dependent extra path length changes the group velocity. Experimentally this can be done by mounting the end mirror after the intracavity prism pair on a split PZT that is driven such that the mirror tilts in response to the error signal (Reichert et al., 1999; Cundiff et al., 2001). Alternatively, one can change the dispersion of the laser cavity via control of the pump power (Holman et al., 2003a). This approach can achieve a typical servo bandwidth of the order of 100 kHz. Once f_o is stabilized, f_r can be controlled using an error signal generated from a heterodyne beat between a cw optical frequency standard and one of the neighboring comb modes. This stabilization scheme comprises the basic gearwork for an optical atomic clock.

2.2. Control of Comb via an Optical Cavity

A passive, high-finesse, and low-dispersion optical cavity is useful in many aspects to aid in high precision DFCS, as will be discussed in the following sections. A broad bandwidth optical frequency comb can be efficiently coupled into such a cavity, enabling its use as a frequency reference as well as enhancement of power and interaction length. Indeed, understanding the intricate interactions between a train of ultrashort pulses and passive optical cavities, along with the development of the capability to efficiently couple and coherently store ultrashort pulses of light inside a high-finesse optical cavity, has been an area of focused research (Jones and Ye, 2002). Several important studies have ensued. Intracavity storage and amplification of ultrashort pulses in the femtosecond regime require precise control of the reflected spectral phase of the resonator mirrors as well as the optical loss of the resonator. While the reflected phase and group delay of the mirrors only change the effective length of the resonator, the group delay dispersion (GDD) and higher-order derivatives of the group delay with respect to wavelength affect the pulse shape. The net cavity GDD over the bandwidth of the pulse needs to be minimized in order to maintain the shape of the resonant pulse and allow for the coherent addition of energy from subsequent pulses (Thorpe et al., 2005; Schliesser et al., 2006).

Direct stabilization of a frequency comb to a high-finesse optical cavity benefits from a low cavity dispersion that leads to a large spectral range over which comb and cavity modes can overlap precisely (Jones and Diels, 2001; Jones et al., 2004). A simplified schematic for cavity stabilization of femtosecond pulses is shown in Fig. 4. The pulse train is mode-matched to the passive high-finesse cavity after passing through an electro–optical phase modulator (EOM). Both f_r and f_o need proper adjustments to optimize frequency overlap between the comb components and the corresponding cavity modes. The light reflected from the cavity is spectrally dispersed with a grating and error signals from the cavity are obtained at different spectral regions resonant with the cavity (ω_i) using the standard Pound–Drever–Hall RF sideband technique (Drever et al., 1983).

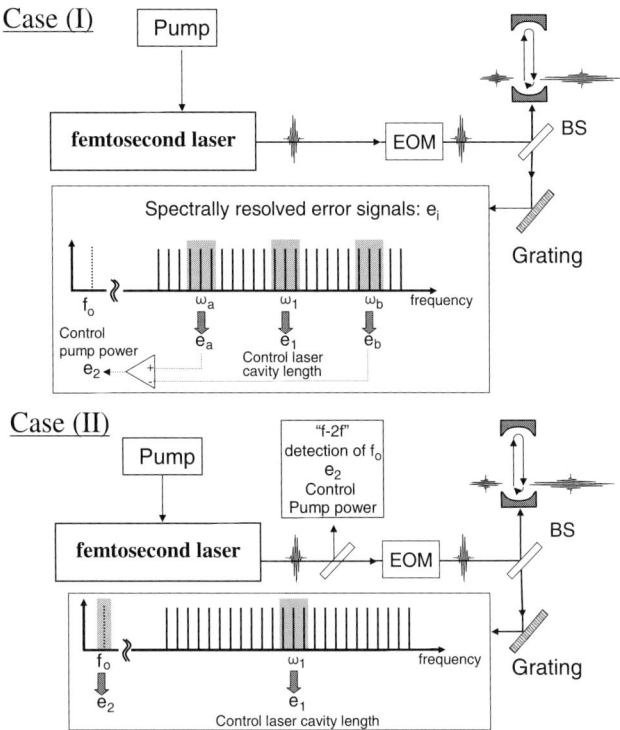

FIG. 4. Stabilization of a train of ultrashort pulses to a passive reference cavity. In scheme (I) "all-cavity locking", both degrees of freedom of the fs frequency comb are detected with reference to the cavity and subsequently stabilized to the cavity. In scheme (II), an independent detection and stabilization of f_o is implemented using an octave-spanning comb bandwidth and an f-to-$2f$ interferometer. f_r is then stabilized via the cavity-generated error signal between a collection of comb components versus a corresponding set of cavity modes. In both panels, shaded regions indicate spectral positions at which the femtosecond comb is stabilized. e_i: error signal recovered at spectral region ω_i; BS: beam splitter; EOM: electro–optical modulator.

The error signals are then used to lock the two degrees of freedom of the frequency comb relative to the cavity. Two distinct stabilization schemes have been compared in detail, leading to the in-depth understandings of the optimum conditions for cavity stabilization of a frequency comb (Jones et al., 2004). In the first scheme, both error signals for comb stabilization are recovered from the cavity reference. The first error signal (e_1) is obtained by comparing a collection of comb components (usually thousands of them) at the center of the spectrum (ω_1) against a corresponding set of cavity modes. This error signal can then be used to stabilize the length of the femtosecond laser cavity. To completely fix the comb, a second error signal (e_2) is necessary, it is derived from a comparison of two errors obtained at two spectral regions located symmetrically around the center, as shown in Fig. 4. The second error signal is used to control the laser cavity dispersion. In scheme (II), the second error signal is derived from a standard f-to-$2f$ self-referencing spectrometer, which corresponds to a direct detection and stabilization of f_o. The value of f_o is chosen to allow the maximum number of frequency comb modes to be matched to the cavity.

The stability of the frequency comb can be determined in both the optical and the radio frequency domains. It has been verified that the second scheme performed better than the first one due to the effective long "lever arm" (Jones et al., 2004). In principle, one could improve the performance of the first scheme by using a larger spectral separation between ω_a and ω_b. However, the ability to simultaneously optimize error signals in different spectral regions is necessarily limited by the parabolic variation of the cavity free-spectral-range (FSR) frequency, characteristic of the simple quarter-wave stack dielectric mirrors employed. By employing lower dispersion mirrors, the performance of the first scheme approaches that of the second. Using an independent, stable cw laser as a reference, measurement of the linewidth and stability of the cavity-stabilized optical frequency comb reveal superior performance comparable to that achieved by cavity-stabilized cw lasers.

The resulting frequency/phase stability between the frequency comb and the cavity modes demonstrates a fully coherent process of intracavity pulse buildup and storage. Similar to the state-of-art stabilization of cw lasers, a cavity-stabilized ultrafast laser demonstrates superior short-term stability of both f_r and f_o. The improved stability is beneficial for frequency-domain applications, where the relative phase or "chirp" between comb components is unimportant (e.g. optical frequency metrology), as well as time-domain applications where the pulse shape and/or duration is vital, such as in nonlinear optical interactions. These applications include optical frequency measurement, carrier–envelope phase stabilization, all-optical atomic clocks, optical frequency synthesizer, and coherent pulse synthesis.

To further improve the quality of femtosecond enhancement cavities, a femtosecond comb-based measurement protocol has been developed to precisely characterize mirror loss and dispersion (Thorpe et al., 2005; Schliesser et al., 2006). This technical capability has facilitated production of large bandwidth, low-loss and low-dispersion mirrors. In addition, nonlinear responses of intracavity optical elements have been studied, demonstrating their limitation on power scalability (Moll et al., 2005). This study has led to the design of novel cavity geometries to overcome this limitation (Moll et al., 2006). In short, nearly three-orders of power enhancement inside a femtosecond buildup cavity within a spectral bandwidth of 30 nm has been achieved, resulting in an intracavity pulse train that (1) is completely phase coherent to the original comb from the oscillator, (2) has the original laser's repetition rate (100 MHz), (3) has a pulse peak energy exceeding 5 µJ (average power > 500 W), intracavity peak intensity approaching 10^{14} W/cm^2, and (4) is under 60 fs pulse duration. We also note that this enhancement cavity approach is compatible with a number of femtosecond laser systems, including mode-locked Ti:sapphire and fiber lasers.

The use of power-enhanced intracavity pulses for high harmonic generation will be discussed in Section 6.2. This is partly motivated by the fact that coherent optical spectroscopy has now led to the recovery of a record-high quality factor ($Q > 2.4 \times 10^{14}$) for a doubly "forbidden" natural resonance observed in a large ensemble of ultracold Sr atoms trapped in an optical lattice (Boyd et al., 2006). This unprecedented spectral resolving power impacts fields ranging from precision frequency metrology to quantum optics and quantum information science. Ultrastable lasers, together with optical frequency combs, can now maintain optical phase coherence beyond 1 s and transfer this stability across hundreds of terahertz (Bartels et al., 2004a; Ludlow et al., 2006). As it becomes increasingly challenging to maintain phase coherence beyond multiple seconds, it is natural to look beyond the visible domain and consider speeding up the "wheel of precision measurement" to higher carrier frequencies. Two related experimental directions are currently being pursued to address this vision. One is the generation of phase coherent, high repetition-rate frequency combs in the VUV (50–200 nm) spectral domain (Jones et al., 2005; Gohle et al., 2005). In parallel, direct frequency comb spectroscopy (Marian et al., 2004) is being explored to ready ourselves for quantum optics and precision spectroscopy once phase coherent sources (necessarily in the form of a train of pulses) become available in VUV. In addition to the power enhancement aspect, a femtosecond cavity effectively increases the interaction length between matter and light, allowing direct frequency comb spectroscopy to measure weak linear or nonlinear atomic and molecular absorption with dramatically increased sensitivity and across a vast spectral bandwidth (Thorpe et al., 2006, 2007).

The coherently enhanced pulse train stored in the cavity can also be switched out using a cavity-dumping element (Bragg cell), resulting in phase-coherent, amplified pulses at a reduced repetition frequency (Potma et al., 2003; Vidne et al., 2003; Jones and Ye, 2004). The net cavity group delay dispersion over the bandwidth of the pulse is minimized to maintain the shape of the resonant pulse. This coherent pulse-stacking technique has been applied to both picosecond (Potma et al., 2003) and femtosecond pulses (Jones and Ye, 2004), demonstrating amplifications >500. An important application of these advanced pulse-control technologies is in the field of nonlinear optical spectroscopy and nanoscale imaging. Using two tightly synchronized picosecond lasers, one can achieve a significant improvement in experimental sensitivity and spatial resolutions for vibrational imaging based on coherent anti-Stokes Raman spectroscopy (CARS) for acquisition of chemically selective maps of biological samples (Potma et al., 2002,

2004). The technologies of pulse synchronization and coherent pulse stacking fit well with this task of combining spectroscopy and microscopy.

3. Direct Frequency Comb Spectroscopy

Femtosecond lasers have been used to reveal ultrafast dynamics and provide wide bandwidth spectroscopic information in atomic, molecular, and condensed matter systems (Zewail, 2000; Jonas, 2003). The advent of precision femtosecond optical combs brings a new set of tools for precision atomic and molecular spectroscopy. For example, ultrafast lasers are now being used not only for time-resolved spectroscopy on fast dynamics, but also for precision spectroscopy of electronic transitions. Indeed, coherent control of dynamics and precision measurement are merging into a joint venture (Stowe et al., 2006). The ability to preserve optical phase coherence and superior spectral resolution over a wide spectral bandwidth permits detailed and quantitative studies of atomic and molecular structure and dynamics. The spectral analysis can be performed over a broad wavelength range, allowing precise investigations of minute changes in atomic and molecular structure over a large dynamic range (Chen and Ye, 2003). For example, absolute frequency measurement of vibration-overtone transitions and other related resonances (such as hyperfine splitting) can reveal precise information about the molecular potential-energy surface and relevant perturbation effects (Chen et al., 2004, 2005).

With precise control of both time- and frequency-domain properties of a pulse train, these two features can be combined into a single spectroscopic study, as demonstrated in the following sections. The broad frequency comb bandwidth available for modern comb spectroscopy measurements distinguishes them from the earlier picosecond-pulse based experiments, leading to truly united time–frequency spectroscopy for broad-bandwidth spectroscopy (Marian et al., 2004). Precision spectroscopy of global atomic structure is achieved with a direct use of a single, phase-stabilized femtosecond optical comb. The pulsed nature of excitation allows real-time monitoring and control capabilities for both optical and quantum coherent interactions and state transfer. It is a synthesis of the fields of precision spectroscopy and coherent control: at short time scales the coherent accumulation and population transfer is monitored, at long times all the information pertinent to the atomic level structure is measured at a resolution limited only by the natural atomic linewidth, with a spectral coverage spanning hundreds of terahertz. This powerful combination of frequency-domain precision and time-domain dynamics represents a new paradigm for spectroscopy.

The precision and control of both the frequency and time domain achievable with femtosecond frequency combs introduce many new features to spectroscopy. In particular the spectroscopic resolution and precision are not compromised by the use of the inherently spectrally broad ultrafast pulses. This is due to the long coherence time of the phase-stabilized femtosecond pulse train. Also, no prior knowledge of atomic transition frequencies are essential for this technique to work. The results presented in this chapter demonstrate the revolutionary advantage of direct frequency comb spectroscopy (DFCS), i.e., the comb frequencies may be absolutely referenced, for example to a cesium atomic clock, enabling absolutely referenced precision spectroscopy over a bandwidth of several tens of nanometers (Marian et al., 2004, 2005; Mbele et al., in preparation). Furthermore, spectral phase manipulation can now be combined with the time-domain optical phase coherence to enable quantum coherent control at resolutions limited only by the natural linewidth (Stowe et al., 2006). Indeed, the field of coherent control of atomic and molecular systems has seen advances incorporating high power femtosecond laser sources and pulse shaping technology. As will be discussed further, this has allowed for demonstrations of robust coherent population transfer via adiabatic passage techniques, coherent control of two-photon absorption, resolution enhancement of coherent anti-Stokes Raman scattering, and progress toward cold atom photoassociation.

3.1. THEORETICAL TREATMENTS

The theoretical work for modeling the interaction of pulsed optical fields with atoms has been explored in several directions. Most treatments focus primarily on the interaction of broad-bandwidth pulses with two or three level atoms and neglect propagation effects and the motional degrees of freedom of the atom. We briefly present an overview of some work done concerning effects not directly addressed in this review. Perhaps the most fundamental question is how does a few cycle femtosecond pulse interact with a two-level atom. The works of Genkin (1998), Casperson (1998) address this question and the limits of common approximations such as the rotating wave approximation. Concerning pulse propagation, work has been done illustrating pulse shaping effects and the pulse area theorem applied to ultrashort pulses (Delagnes et al., 2003; Ziolkowski et al., 1995). Analytic expressions for the excitation of a two-level atom by a train of pulses have been derived in Vitanov and Knight (1995), Greenland (1983).

For much of the experimental work reviewed in this chapter the relevant physics is the single and multi-photon absorption in multi-level atoms or molecules from an optical pulse train of moderate intensity. The root concept behind direct frequency spectroscopy can be found in Ramsey spectroscopy, see Fig. 1, in which only two narrow band pulses excite the atom. Basically, the first broadband pulse excites a small fraction of population to all allowed transitions within the pulse bandwidth. The second and subsequent pulses interact with the coherences generated by the first pulse and, depending on the relative phase of the pulse with the coherence, population is either excited or de-excited. We refer to the coherent interaction of a series of pulses with an atomic or molecular transition as coherent accumulation. This is the key physical concept behind the signal size and resolution enabled by DFCS. To excite only one of many states in the bandwidth of the pulse, f_o and f_r may be chosen to always be in phase with only one coherence. In the frequency-domain picture this is simply to say one comb mode is resonant and all others are detuned. To properly treat multi-photon absorption the sum of comb modes must be considered, for example in a two-photon transition there may be hundreds of thousands of resonant mode pairs. Due to the fact that more than one pair of modes may excite a two-photon transition the spectral phase of each mode, and intermediate state detunings, must be considered to account for quantum interference effects.

Two models are used for the majority of the theoretical predictions for the experiments presented in Sections 3.2 and 3.3. The first is a perturbative model suitable for two-photon absorption of transform limited pulses in multi-level atoms, for example the many hyperfine levels of the 5S, 5P, and 5D states of rubidium 87. This technique, presented by Felinto et al. (2004), uses the impulsive excitation approximation such that the spectral content and phase of the pulse are not included. It provides a computationally fast method to predict the complicated optical pumping effects present between the many comb modes and electronic states. Other experiments require either a non-perturbative treatment or the capability of modeling spectral phase effects for multi-photon absorption. In this case one can directly numerically solve the Louiville equation for the density matrix of the system for a three-level model under arbitrary shaped pulse excitation. The advantage of this approach is that it correctly predicts pulse shaping effects on multi-photon transitions and the non-perturbative effect of power broadening of atomic transitions relevant to correctly modeling parts of Sections 3.3 and 5.

In many experiments it is possible to treat only those comb modes that are resonant with a transition. The simplest case is when the comb is tuned to be resonant with only one single-photon transition and the nearest mode for all other transitions is detuned several linewidths. Similarly, for a resonantly enhanced two-photon transition it may be reasonable to analytically model the interaction using only the two resonant modes. The two-photon transition amplitude, C_{gf}, useful for calculating most results in Section 3.3 is,

$$C_{gf} \propto \frac{E_n E_m}{i(\omega_{gf} - (m+n)2\pi f_r - 4\pi f_o) + \pi \gamma_f} \\ \times \left(\frac{1}{i(\omega_{gi} - 2\pi(nf_r + f_o)) + \pi \gamma_i} \right. \\ \left. + \frac{1}{i(\omega_{gi} - 2\pi(mf_r + f_o)) + \pi \gamma_i} \right) \tag{1}$$

where $E_{n,m}$ are the electric fields of the nth and mth modes of the comb, $\gamma_{i(f)}$ is the intermediate (final) state scattering rate, $\omega_{gi(gf)}$ is ground to intermediate (final) state transition frequency. In general, the total transition amplitude is the sum over all modes of the comb and intermediate states. The full equation with comparison to time-domain techniques may be found in Yoon et al. (2000).

In molecular systems, where the number of individual states covered by the bandwidth of the pulses may be too large to handle by direct solution in state space, one can use techniques developed for the modeling of wave-packet dynamics, such as the split-operator method. In these techniques, one directly solves the time-dependent Schrödinger equation with a set of electronic potentials coupled by the time-dependent fields of the pulses, see Garraway and Suominen (1995) for a comprehensive review. The theoretical results shown in Section 5.2 were obtained with such a model. Like in the case of atomic excitation, the coherent accumulation was modeled by simulating a single femtosecond pulse interaction followed by free evolution of the molecular coherences between pulses, this procedure is then repeated for every pulse.

3.2. SINGLE-PHOTON DFCS

Of the many applications directly using the output of a femtosecond laser frequency comb (FLFC) for spectroscopy, single-photon spectroscopy is the easiest to realize experimentally. A single frequency component of the broad comb spectrum is used to selectively excite an atomic or molecular transition. The spectra are recorded by measuring the excited-state fluorescence while changing the laser's repetition rate f_r or carrier–envelope offset frequency f_o. The time-domain properties of the laser determine only the frequency and width of the selected component, but the requirements for pulse overlapping or preserving the phase relation across multiple comb components are relaxed. The femtosecond laser frequency comb can be viewed simply as a multimode laser with narrow-linewidth modes at known optical frequencies. As a result, all methods of cw laser spectroscopy can be applied with the advantage of absolute frequency calibration determined by the knowledge of f_r and f_o. Additionally, the combs have spectral components where no tunable cw laser sources exist, and they cover a broad spectral interval allowing spectroscopy of different atomic or molecular systems to be done with the same laser system. In particular the high peak intensity attainable with femtosecond pulses allows for efficient nonlinear conversion to spectral regions with difficult or no accessibility by traditional cw laser techniques, see Section 3.4 for a specific example.

Several disadvantages of using FLFC for single-photon spectroscopy have to be mentioned. The total laser intensity is distributed among $\sim 10^5$ comb modes, resulting in a per mode intensity typically below 1 µW. Consequently spectroscopy must be performed with very low laser power levels. This limits the sub-Doppler spectroscopy experiments to atomic beams or laser cooled systems, where the Doppler width has already been reduced geometrically or using laser cooling (Gerginov et al., 2005; Marian et al., 2004). Saturated absorption is possible, but in order to reach significant saturation, tight laser beam focusing is required which leads to time of flight broadening and other systematic effects. A possible solution to the low intensity problem is amplification of the comb's output, as shown in Section 3.2.3.

When the repetition rate is swept in order to measure the atomic transition spectrum, the same difficulties as when using multimode lasers are encountered. Due to interleaving, all spectral features will be present in a frequency window determined by the separation of the comb components at the optical frequency of interest. The identification of the spectral lines becomes problematic. The presence of many other comb components, even if filtered by an interference filter or using techniques described in the sections below, leads to systematic effects such as AC Stark shifts. It is difficult to directly measure the intensity of any single comb mode, although it may be inferred from the laser's spectral shape, power, and mode separation. The method of measuring the excited population in the following experiments is either directly via fluorescence or population transfer with a cw probe laser. This is preferable because in transmission, all components present in the comb spectrum contribute to the background.

Determining the absolute frequency of a transition measured by DFCS requires the determination of the resonant comb mode index, n. There are generally two ways to determine the index. In some applications the frequency may be already known to within half the comb mode spacing. If this is the case then it is trivial to unambiguously determine n and calculate the absolute transition frequency using $\nu_n = nf_r + f_o$. The second more general technique is to measure the linecenter with two or more different f_r. In principle, that is in the absence of uncertainty corresponding to the f_o and f_r of the measured linecenter, the idea is straightforward. Measure the linecenter with $\nu_n = nf_r + f_o$ and $\nu_n = n(f_r + \Delta f_r) + (f_o + \Delta f_o)$, where Δf_r is small enough such that the resonant mode remains the same. From these measurements it is clear $n = -\Delta f_o/\Delta f_r$. In practice there is measurement uncertainty and Δf_r must be made larger, in general the resonant comb mode will change but Δn may be determined. Cases may also arise in which two different atomic transitions are spaced by approximately f_r, these lines would overlap and it may be necessary to choose an appropriate f_r range, or filter the pulse spectrum, to reduce the congestion of resonant transitions.

3.2.1. Single-Photon Spectroscopy of Rubidium

The first DFCS experiment we discuss measured the 5S to 5P single-photon transitions in laser cooled ^{87}Rb atoms (Marian et al., 2005). This simple application

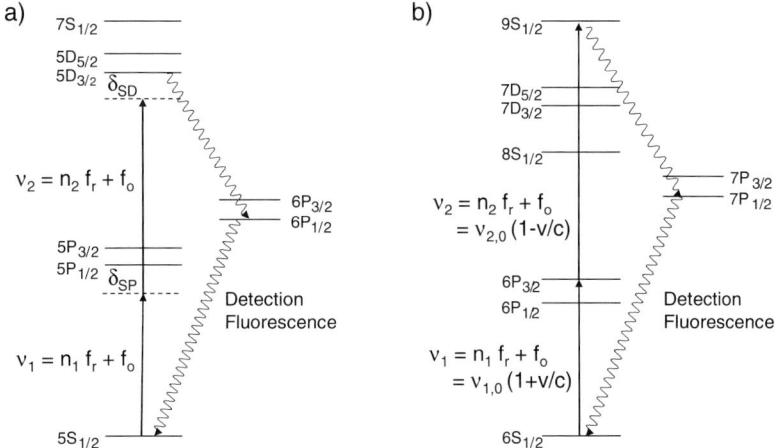

FIG. 5. Relevant energy levels studied in experiments in (a) Rb and (b) Cs. The two-photon transitions reached by two modes of the comb are discussed in Section 3.3. Although not shown the comb resolution allows the excitation of specific hyperfine levels.

is a clear demonstration of how DFCS can be used to determine the absolute frequency of any allowed transition within the laser bandwidth, see Fig. 5(a). A mode-locked Ti:sapphire laser operated at 100 MHz repetition rate, outputting 20 fs pulses, with a center wavelength of 780 nm was used to excite a sample of ^{87}Rb atoms in a magneto–optical trap (MOT). A crystal oscillator with low short-term instability was used to lock f_r resulting in a comb mode linewidth of 330 kHz at 1 ms integration time. To absolutely reference the frequencies of the comb modes f_r was measured on a cesium referenced counter and steered appropriately. To shutter the fs probe pulses a Pockels cell with an 8 ns rise time was used. In brief, the experimental cycle generally used was as follows: the MOT was loaded for 7.8 ms, the magnetic field was turned off and the atoms were held in optical molasses for 2 ms, then all MOT related optical fields were extinguished before probing the atoms for 200 µs. Weakly focusing the fs pulses into the MOT to a diameter of about 130 µm results in an on axis average intensity of 0.8 mW/cm^2. Due to the fact that the 5P fluorescence is at a wavelength present in the pulse spectrum the photomultiplier tube used to measure the fluorescence signal is turned off during the excitation time to avoid background counts. Similarly the fs laser was shuttered during the PMT operation time. Within the 200 µs probing window the atoms were repeatedly excited and then PMT was used to count the 5P fluorescence signal immediately after excitation. The f_r and f_o of the comb were chosen such that only the $5S_{1/2}F = 2$ to $5P_{3/2}F = 3$ transition is near resonance, all other transitions are several linewidths detuned. Figure 6(b) shows the measured lineshape recovered by scanning f_o and translat-

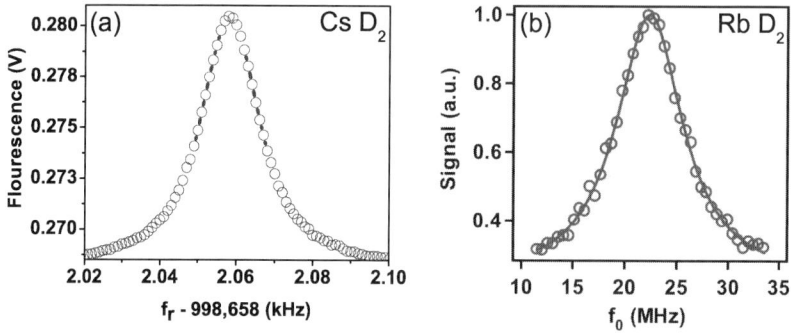

FIG. 6. (a) ^{133}Cs D$_2$ $F = 4 \rightarrow F = 5$ line scan versus f_r with $f_o = 120$ MHz and 1.5 nW/mode. (b) ^{87}Rb D$_2$ $F = 2 \rightarrow F = 3$ line scan versus f_o with f_r fixed.

ing the resonant comb mode over the atomic transition. To minimize any Doppler shift of the linecenter versus time, from repeated scattering via $5P_{3/2}F = 3$, the pulses were counter-propagated through the atoms balancing the radiation pressure. However, there is still residual radiation pressure that causes the linecenter to shift versus probing time. Taking this into account the linecenter versus time is measured and extrapolated to zero probing time to recover a center frequency of $\nu = 384, 228, 115, 271(87)$ kHz.

The broad bandwidth of a Ti:sapphire frequency comb has also recently been used to study velocity selective excitation of a room temperature cell of Rb atoms, see the work of Aumiler et al. (2005), Ban et al. (2006). In these experiments the Doppler broadening at room temperature is significantly larger than the intermode spacing of the frequency comb used for excitation. This leads to a velocity selective excitation and optical pumping of both the ground and excited 5P state hyperfine levels. By measuring the absorption profile versus frequency with a separate cw laser several transparency windows are observed corresponding to different velocity classes of the thermal Rb ensemble. Similar experiments have been performed that used significantly narrower bandwidth mode-locked lasers to demonstrate electromagnetic induced transparency (EIT) with a comb (Sautenkov et al., 2005; Arissian and Diels, 2006). The idea is very similar to the work discussed in Section 1. By tuning the inter-pulse period or f_r to a harmonic of the ground state hyperfine splitting the atoms can be driven into a dark state. Future experiments in this direction may lead to the ability to directly lock f_r of the comb to the narrow transparency window achievable with EIT.

3.2.2. Single-Photon Spectroscopy of Cesium

Another experiment demonstrating single-photon spectroscopy in ^{133}Cs is done using a thermal atomic beam. It uses a self-referenced Ti:sapphire femtosecond laser frequency comb with a 1 GHz repetition rate (Ramond et al., 2002; Bartels et al., 2004a; Gerginov et al., 2005). Its repetition rate, f_r, and the carrier–envelope offset frequency, f_o, of the femtosecond laser are phase-locked to frequency synthesizers referenced to a stable hydrogen maser, which is calibrated by a cesium atomic fountain clock (Heavner et al., 2005; Ramond et al., 2002; Morgner et al., 2001). The fractional frequency instability of the comb teeth is determined by that of the hydrogen maser, given by $\sim 2 \times 10^{-13} \tau^{-1/2}$, where τ is the integration time measured in seconds. A single-mode fiber is used to deliver part of the comb's output to the highly-collimated atomic beam (Gerginov et al., 2004). The optical power is stabilized using the zero diffraction order of an acousto–optical modulator (AOM), placed before the optical fiber. It should be noted that for single-photon absorption spectroscopy the spectral phase of the femtosecond pulses does not effect the signal strength, therefore pulse stretching in, for example, the fiber or AOM is of no concern. However, in multi-photon absorption the spectral phase may be of importance and will be addressed in Section 5.

To excite the $6s\,^2S_{1/2} \to 6p\,^2P_{1/2}$ (D$_1$) and $6s\,^2S_{1/2} \to 6p\,^2P_{3/2}$ (D$_2$) lines in Cs, the comb spectrum was filtered using interference filters centered at 900 and 850 nm, respectively. The Cs atomic beam is collimated to a rectangular profile. The output of the single-mode fiber is collimated and intersects the thermal atomic beam at a right angle after the nozzle and above a large-area photodetector.

Around 895 nm, the FLFC excites four different components of the transition $6s\,^2S_{1/2}\,(F_g = 3, 4) \to 6p\,^2P_{1/2}(F_e = 3, 4)$. Due to the presence of a comb tooth every 1 GHz, the fluorescence signals also repeat every 1 GHz change in optical frequency (interleaving), corresponding to a change in f_r of ~ 3 kHz.

The FLFC spectrum, narrowed in the vicinity of 850 nm, excites six components of the transition $6s\,^2S_{1/2}(F_g = 3, 4) \to 6p\,^2P_{3/2}(F_e = 2, 3, 4, 5)$. The fluorescence versus femtosecond laser repetition rate for one of the Cs D$_2$ lines is shown in Fig. 6(a).

In this measurement, previous knowledge of the optical frequencies were used to identify each spectral component. Due to the large repetition rate of the femtosecond laser (1 GHz), such identification is possible knowing the optical frequencies with a precision of several tens of megahertz, significantly better than $f_r/2$. As previously mentioned using different repetition rates of the femtosecond laser to measure the same spectra, identification without previous knowledge of optical frequencies is possible (Halzworth et al., 2001; Ma et al., 2003; Marian et al., 2005).

Due to reduced signal-to-noise ratio in the FLFC experiment, the AC Stark shift of the optical frequency was not measured. The first-order Doppler shift was canceled with the technique described in detail in Gerginov et al. (2004, 2006). Due to the reduced signal-to-noise ratio, this cancellation is limited by the statistical uncertainty in the optical frequency determination, and is probably the reason for the small offset of the FLFC data compared with the one obtained with the CW laser. From the data presented in Gerginov et al. (2005), it is clear that the statistical and fit uncertainties approach the ones obtained using a cw laser.

3.2.3. Excitation of Forbidden Transitions

An extension of the ideas and techniques of single-photon spectroscopy described above to the study of forbidden transitions requires additional considerations. In particular, the difficulties encountered due to the small power per optical mode are heightened. Two possible solutions to this problem are the use of cold atoms, which allow for long interrogation times, and the amplification of the comb. In addition to the power considerations, the study of forbidden transitions also requires the optical comb modes to be narrower than what is needed for the study of dipole-allowed transitions. Thus, the stabilization of the comb's repetition rate directly to an RF source is insufficient and greater stability can be achieved by stabilizing a single optical mode to a highly stable laser, in addition to the usual stabilization of the carrier–envelope offset frequency. These extensions were demonstrated in the context of studying the narrow $4s^2\,^1S_0 \to 4s4p\,^3P_1$ transition in calcium (Ca) (Fortier et al., 2006b).

The experimental setup for the experiment of Fortier and co-workers is shown in Fig. 7. The $4s^2\,^1S_0 \to 4s4p\,^3P_1$ transition in Ca is forbidden in the spin–orbit

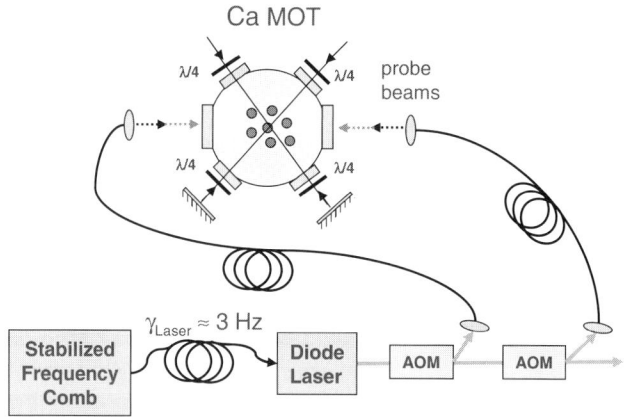

FIG. 7. Experimental depiction for high-precision optical spectroscopy of cold Ca atoms using a femtosecond laser frequency comb. Acousto–optical modulators (AOM) switched light between two optical fibers that deliver counter-propagating pulses to the Ca magneto–optical trap (MOT). An f-to-$2f$ self-referencing technique allows phase stabilization of the laser offset frequency, f_o. Phase stabilization of the laser repetition rate, f_r, is obtained by phase locking the comb to a fiber laser at 1068 nm that is stabilized to a high-finesse optical cavity. Measurement of the laser repetition with respect to a hydrogen-maser provided absolute frequency calibration in the measurements.

approximation and has a natural linewidth of 374 Hz (Degenhardt et al., 2005), and this transition has been studied extensively as a optical frequency standard (Degenhardt et al., 2005; Oates et al., 2000). In this experiment, the frequency comb was derived from an octave-spanning mode-locked Ti:sapphire laser with a ≈1 GHz repetition rate (Fortier et al., 2006a). Due to the transmission of the output coupler, the output power of the laser peaked around the 657 nm wavelength needed to excite the $4s^2\,{}^1S_0 \rightarrow 4s4p\,{}^3P_1$ transition. The power per mode at this wavelength was ≈1 μW. With this type of power and the small dipole matrix element of the transition, one has Rabi frequencies of ∼10 kHz. In order to achieve sizable population transfer, it is necessary to have interrogation times of ∼10 μs. These types of interrogation times can be achieved with moderately cooled atoms with temperatures of a few millikelvin.

Because the frequency comb was in a different laboratory from the Ca atoms it was necessary to fiber couple the comb light. In addition, the light was passed through an AOM to allow for switching of the light. In order to overcome the losses introduced by these processes, the frequency comb light was amplified by using the comb light to injection lock a diode laser (Cruz et al., 2006). Two regimes of amplification were observed depending on the current applied to the slave diode. With the current of the slave diode near the lasing threshold, the ≈0.5 nm of injected light was amplified more or less uniformly, providing an amplification factor of ∼10. This amplification compensated for the losses described above and allowed for saturation spectroscopy of the transition and observation of the photon-recoil splitting (Hall et al., 1976; Oates et al., 2005), see Fig. 8(a). With the current of the slave diode significantly above the lasing threshold, the amplification was dominated by specific optical modes. By fine adjustment of the diode current and temperature, it was possible to preferentially amplify the optical mode resonant with the transition. In this regime, an amplification of ∼2000 was observed. This amplification was sufficient to observe the Ramsey–Bordé fringes (Bordé et al., 1984) and achieve spectral resolutions of 1 kHz, see Fig. 8(b).

In addition to the restrictions imposed by the low optical powers available, the study of forbidden transitions requires that the individual optical modes be narrower than is necessary for the study of allowed transitions. If the repetition rate is stabilized directly to an RF source, any noise present in the RF source is multiplied by the optical mode number. This places stringent requirements on the short-term stability of the RF source if the comb is to be used to study narrow transitions. This problem can be circumvented by instead stabilizing a single comb mode, with mode number n, to a stable cw laser of frequency ν_{cw}. The frequency of a given comb mode, m, is then given by

$$\nu_m = \frac{m}{n}(\nu_{cw} + f_b) + f_o\left(1 - \frac{m}{n}\right), \qquad (2)$$

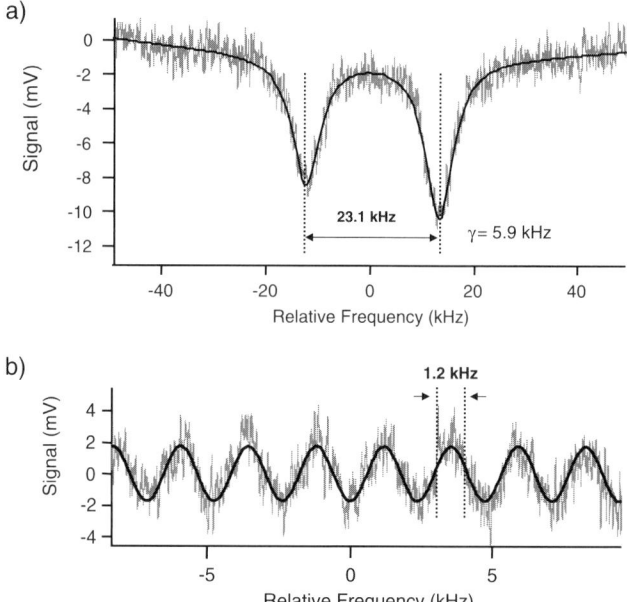

FIG. 8. (a) Saturation absorption dip observed on the Doppler profile of the Ca clock transition using two counter-propagating pulses. The double peak is the photon-recoil doublet. (b) Time-resolved optical Bordé–Ramsey fringes.

where f_b is the heterodyne beat signal between the comb mode n and the cw laser. The frequency ν_m can be scanned by changing the frequency of the heterodyne beat note f_b while the frequency comb is stabilized to light from a cw fiber laser at 1068 nm. The cw fiber laser was stabilized by doubling a portion of the light and referencing it to a ultrastable optical cavity (Young et al., 1999). The residual linewidth of the light at 1068 nm was $\sim$1 Hz. This stability was transferred to the entire bandwidth of the comb, providing a sufficiently narrow linewidth for the spectroscopy of the $4s^2\,^1S_0 \to 4s4p\,^3P_1$ transition.

3.3. MULTI-PHOTON DFCS

Direct frequency comb spectroscopy is not only useful for single-photon transitions but also multi-photon transitions. Two general cases exist in the context of two-photon transitions, those with an intermediate resonance and those in which the pulse spectrum is far detuned from an intermediate resonance. In the case where one mode is resonant with the intermediate state, the system can often be understood by considering only two modes. However, when there is no intermediate resonance in the spectrum it should be noted that if one pair of comb modes is two-photon resonant, then all modes will form a resonant pair due to the equal spacing of modes.

3.3.1. Multi-Photon DFCS of Rubidium

The following experiment demonstrates the versatility of DFCS for measuring two-photon transitions. The absolute transition frequencies of several hyperfine levels in the 5D and 7S manifolds were measured, via resonant enhanced transitions through an intermediate 5P hyperfine level (Marian et al., 2004, 2005). Primary focus was placed on the $5S_{1/2}F = 2 \to 5P_{3/2}F = 3 \to 5D_{5/2}F = 4$ transition because it is the strongest transition in the comb bandwidth. This is due to the fact it has the highest dipole moments, and the $5P_{3/2}F = 3$ level may only decay to $5S_{1/2}F = 2$, avoiding excessive optical pumping to off resonant levels. The setup is essentially the same as the one used for measuring single-photon transitions in laser cooled ^{87}Rb described in Section 3.2.1. One difference is that the population of the 5D or 7S states was determined by monitoring the fluorescence at 420 nm resulting from the cascade decay of the 6P state to the 5S state. There are a total of 14 transitions between the 5S and 5D states alone, this large number of possible transitions makes for an excellent DFCS demonstration using different comb modes from one laser for spectroscopy.

Figure 9 shows the measured signal, proportional to the 5D population, versus a scan of f_r by only 26 Hz while keeping f_o fixed. Within this one scan it is possible to identify 28 lines. The optical frequency for the two-photon transition is of the order 770 THz, this corresponds to a f_r harmonic of 7.7×10^6, dividing $f_r \cong 100$ MHz by this value gives a value of $\Delta f_r = 13$ Hz before the next comb mode becomes resonant with any given transition. The single-photon transition being at approximately $\frac{1}{2} \times 770$ THz repeats after every 26 Hz change of f_r. Identified in Fig. 9 are the single-photon resonant and detuned lines, thus 28 lines for 14 possible transitions, although in this range of f_r some lines overlap. Several transition frequencies for both one- and two-photon absorption are provided in Table I.

As an explicit experimental demonstration of the resolution and signal enhancement of multi-pulse coherent accumulation, the $5S_{1/2}F = 2 \to 5P_{3/2}F = 3 \to 5D_{5/2}F = 4$ transition is measured versus a controllable number of pulses (Stowe et al., 2006). A Pockels with a 8 ns rise time is used as a pulse picker and triggered in phase with f_r such that any number of pulses can be used to excite the atoms. Figure 10 shows the resulting $5D_{5/2}F = 4$ population and lineshape versus pulse number, rescaled to match the theoretical predictions. Perhaps the clearest example of coherent accumulation, shown in the inset of Fig. 10, is the quadratic scaling of population at short times, or pulse number. This is because the total accumulated pulse area prior to atomic decoherence is the sum of the

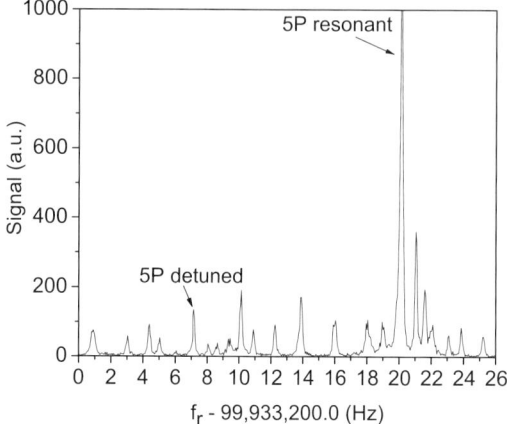

FIG. 9. Full spectrum of all allowed two-photon transitions from $5S \to 5D$ of ^{87}Rb. This spectrum was measured in only a 26 Hz scan of f_r. Different modes of the comb become resonant with transitions at various values of f_r allowing for large spectral coverage. Two peaks are identified in the figure both corresponding to the $5S_{1/2}F = 2 \to 5P_{3/2}F = 3 \to 5D_{5/2}F = 4$ transition. The larger of the two peaks is exactly resonant with both the single- and two-photon transitions. The weaker line, although two-photon resonant, is off resonant with any intermediate states.

Table I
Rubidium two-photon transition frequencies measured by DFCS

Measured transition (from $5S_{1/2}F = 2$)	Measured frequency (kHz)
$5D_{5/2}F = 2$	770,569,184,527.9 (49.3)
$5D_{5/2}F = 3$	770,569,161,560.5 (11.1)
$5D_{5/2}F = 4$	770,569,132,748.8 (16.8)
$5D_{3/2}F = 3$	770,480,275,633.7 (12.7)
$5D_{3/2}F = 2$	770,480,231,393.9 (38.1)
$5P_{3/2}F = 3$	384,228,115,309.0 (63.0)
$5P_{1/2}F = 2$	377,105,206,938.7 (179.0)

individual pulse areas. In analogy with driving a two-level system with a resonant cw laser the population initially scales as the square of time, or equivalently as the square of accumulated area. At longer times, or larger number of pulses, the decoherence due to the natural lifetime of the 5D state begins to limit the coherent accumulation until eventually steady state is reached. In this experiment the resonant comb modes are strong enough to cause substantial power broadening of the measured linewidth. As can be seen in Fig. 10 the linewidth clearly becomes narrower as more pulses are coherently accumulated, eventually reaching

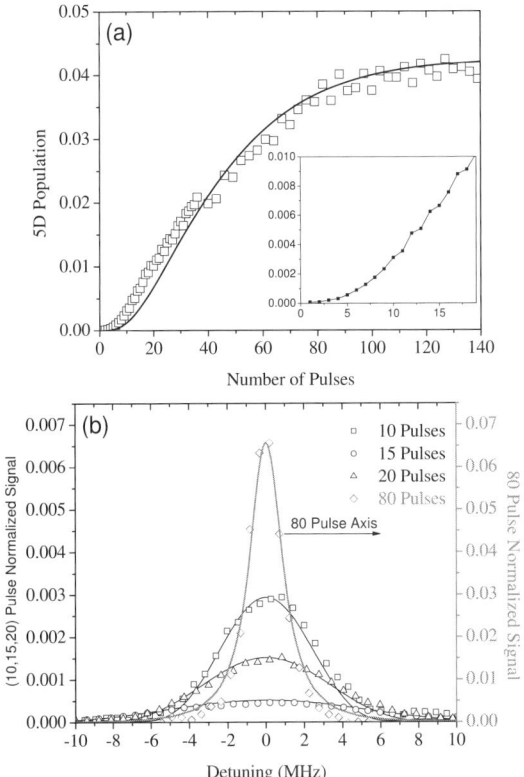

FIG. 10. Shown in (a) is the excited state population versus pulse number. Notice two features in particular, for short times the scaling is quadratic versus pulse number (see inset), and the population reaches steady state at approximately 80 pulses. Scanning f_o recovers the lineshape versus pulse number shown in (b). The power broadened asymptotic linewidth of 2.2 MHz is reached at approximately 80 pulses.

the asymptotic power broadened linewidth of 2.2 MHz at 80 pulses. In principle the resolution obtainable by DFCS is only limited by decoherence mechanisms, typically due to the atomic transition because modern fs comb linewidths can be reduced to the 1 Hz level.

To determine the linecenter with high accuracy it is necessary to conduct a higher resolution scan of the line of interest. Several sources of systematic error must be carefully addressed to determine the unperturbed linecenter of interest. For all measurements the fs pulses are counter-propagated through the same set of atoms within the MOT to balance the radiation pressure. Provided the intensities of the two counter-propagating beams are well matched the net momentum transfer along the probing direction via photon scattering, primarily from the 5P

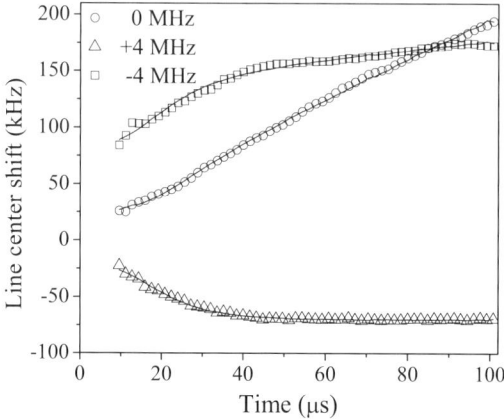

FIG. 11. The measured linecenter shift of the $5D_{5/2}F = 4$ level versus time relative to an arbitrary reference value. For times below approximately 40 μs the fs comb power is increasing as the LCD shutter used in this experiment turns on, after 40 μs the power is constant. Three cases are presented for different intermediate state detunings, all are nominally on resonance with $5D_{5/2}F = 4$. Two effects may be seen in this data, first the Stark shift is approximately equal and opposite about the 0 MHz detuning result for the ±4 MHz detuning cases. Secondly, notice when the fs comb power is constant the Stark shift is constant (red and blue detunings). However, the on resonance case undergoes significant accumulative Doppler shift due to radiation pressure from the residual intensity and alignment mismatch between the counter-propagating probe beams.

states, is significantly reduced. Repeated scattering also heats the atomic ensemble, broadening the linewidth. Similarly a residual net intensity imbalance of the probe beams can accelerate the atoms resulting in a Doppler shift of the measured linecenters. The linecenter results reported in this review are based on measuring and fitting the linecenter versus time then extrapolating to zero interrogation time. One interesting result is that for resonantly enhanced two-photon transitions the comb mode connecting the $5S_{1/2}F = 2 \to 5P_{3/2}F = 3$ transition can be either resonant, blue detuned, or red detuned. By detuning the comb from exact intermediate resonance the heating rate is greatly reduced (Marian et al., 2004).

Another systematic error commonly encountered in two-photon transitions is the AC Stark shift due to off-resonant intermediate states. Again in this case focus is placed on the $5S_{1/2}F = 2 \to 5P_{3/2}F = 3 \to 5D_{5/2}F = 4$ transition which is somewhat simplified because transitions to the final state are only allowed via the $5P_{3/2}F = 3$ intermediate state due to hyperfine selection rules. A measurement of the $5D_{5/2}F = 4$ linecenter versus time, shown in Fig. 11, illustrates the positive or negative Stark shift as a function of detuning from the intermediate state. This emphasizes the importance of being exactly on resonance with the intermediate state when conducting highly accurate spectroscopy, but also that if the comb is made exactly resonant the Stark shift is reduced. Furthermore it should be mentioned that, due to the regular spacing of modes, when one mode is resonant with the intermediate state then for every mode detuned to the red of the intermediate resonance there is one to the blue side with an equal detuning. To within limits due to the spectral shape of the pulse the Stark shifts from off resonant modes tends to cancel out, regardless of the spectral phases of the individual modes.

3.3.2. Multi-Photon DFCS of Cesium

As discussed in Section 3.3, two-photon spectroscopy is greatly enhanced in the case where the two-photons are resonant with an intermediate state. While this enhancement can be accomplished using the two degrees of frequency control available with the comb as discussed above, it is also possible to utilize the motion of the atoms to eliminate the need to fine tune both the carrier–envelope offset frequency and the repetition rate. This Doppler-induced enhancement was demonstrated in the work of Mbele et al. (in preparation).

In that work, two-photon spectroscopy was performed on four different excited states in Cs. A schematic of the experimental setup is shown in Fig. 12(a). The Cs was contained in a vapor cell at room temperature, resulting in a velocity distribution with an average velocity of $v = 190$ m/s. The output of the frequency comb was split and the two beams were counter-propagated though the vapor cell. The frequency comb was derived from an octave-spanning mode-locked Ti:sapphire laser with a ≈ 1 GHz repetition rate (Fortier et al., 2006a). The carrier–envelope offset frequency was stabilized using the f-to-$2f$ technique described in Section 2.1. The repetition rate was stabilized to a synthesizer referenced to an H maser. The repetition rate of the synthesizer was scanned and fluorescence from the $7p\,^2P \rightarrow 6s\,^2S_{1/2}$ transition was detected with a photomultiplier tube, see Fig. 5(b). The resulting spectrum is shown in the upper plot of Fig. 12(b). The different peaks are a result of different hyperfine transitions between the ground state and four excited states: the $8s\,^2S_{1/2}$, $9s\,^2S_{1/2}$, $7d\,^2D_{3/2}$, and $7d\,^2D_{5/2}$. By filtering the different optical beams, it was possible to isolate and identify the peaks and make precision measurements of the two-photon transition frequencies.

Each of the peaks is a result of an enhancement of a single pair of comb modes resonant with both stages of the excitation, similar to the enhancement observed by Weber and Sansonetti (1987). Because of the large velocity distribution, there are atoms which are Doppler shifted into resonance with the first stage of the excitation regardless of the repetition rate and carrier–envelope offset frequency. For specific repetition rates, these Doppler shifted atoms will also be resonant with the second stage of the excitation. By considering the velocity groups resonant with each stage of the excitation it can be shown that both of the two comb modes are resonant when the repetition rate is

$$f_r = \frac{2v_{1,0} v_{2,0} - f_o(v_{1,0} + v_{2,0})}{n_1 v_{2,0} + n_2 v_{1,0}}, \tag{3}$$

were $v_{1,0}$ and $v_{2,0}$ are the Doppler-free frequencies of the first and second stages of excitation, respectively, of the two-photon transition. This eliminates the need for independent adjustment of the carrier–envelope offset frequency in order to achieve the double resonance condition. Because the velocity class that is excited depends on the frequency of both stages of the two-photon excitation, the spectrum also provides information concerning the intermediate state frequency.

The velocity selective enhancement selects out a single pair of comb modes. Consequently, one can consider the excitation of a given state as equivalent to the excitation by two cw lasers. The middle trace in Fig. 12(b) shows the

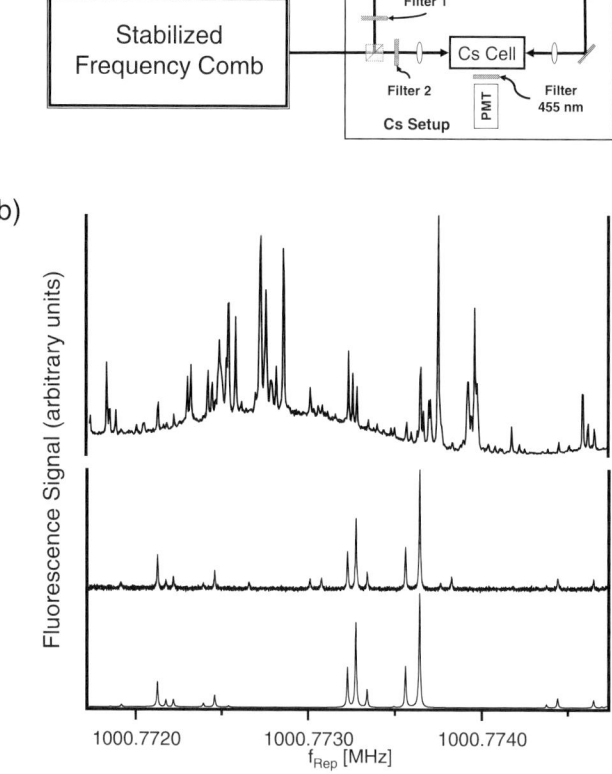

FIG. 12. Two-photon excitation of Cs states. (a) shows a block diagram of the experiment. The top trace in (b) shows the spectrum obtained with an unfiltered comb. Four different excited states are present. The middle trace in (b) shows the spectrum obtained by filtering the comb so that only the $6s\,^2S_{1/2} \rightarrow 6p\,^2P_{3/2} \rightarrow 8s\,^2S_{1/2}$ transition contributes. The bottom trace shows the calculated spectrum.

experimental spectrum acquired by filtering the two arms to observe only the $6s\,^2S_{1/2} \rightarrow 9s\,^2S_{1/2}$ transition through the $6p\,^2P_{3/2}$ intermediate state. Below it is the calculated spectrum generated by integrating the two-photon formula, Eq. (1), over the velocity distribution. It is possible to determine the frequency of the transitions, as well as the hyperfine constants of the states, by calculating spectra with different transition frequencies and hyperfine constants and matching the calculations to the experimental spectra. In the experiment of Mbele et al. (in preparation), the frequencies and hyperfine coefficients of the states were determined with an accuracy of 50–200 kHz.

3.4. Short Wavelength DFCS

One of the most promising future applications of DFCS is to truly take advantage of the phase coherence of the comb combined with the efficient nonlinear conversion enabled by high peak intensities. Already some experiments have been conducted by frequency mixing the fs pulses to shorter wavelengths where cw lasers are not easily accessible. There are two experiments conducted to date utilizing multi-pulse quantum interference for spectroscopy in the VUV spectrum. The first demonstration of this technique used two-photon absorption at 212 nm in krypton by pairs of phase coherent pulses (Witte et al., 2005). The most recent

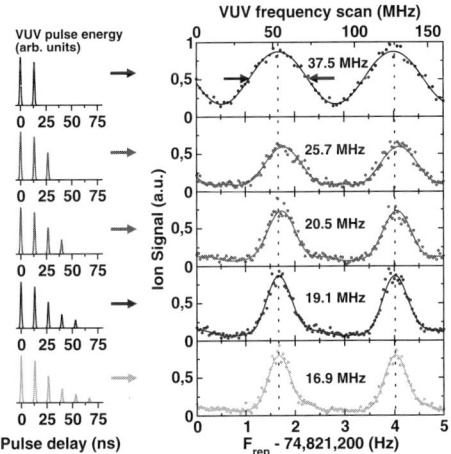

Fig. 13. Measured xenon ion yield versus f_r for 2, 3, 4, 5, 6 pulses. The strength of the pulses decreases versus pulse number as the amplifier inversion decays. These results emphasize several aspects of DFCS. The signal linewidth reduces versus the number of pulses used, note it is sinusoidal for two pulses similar to Ramsey spectroscopy. Secondly, the single resonant line repeats periodically due to the comb structure of the laser. [Reprinted with permission from Zinkstok et al. (2006).]

© 2006 Elsevier

example demonstrates multi-pulse DFCS of the $5p^6\,^1S_0 \rightarrow 5p^5\,(^2P_{3/2})5d[\frac{1}{2}]_1$ transition in xenon at 125 nm, see Zinkstok et al. (2006). Pulses from a Rb-atomic-clock-referenced Ti:sapphire frequency comb are injected into a multipass Ti:sapphire amplifier that outputs pulses of about 25 µJ centered at 750 nm with a 0.7 nm spectral width. From two to six pulses may be injected into the amplifier then frequency doubled in BBO and subsequently tripled in either oxygen or krypton. The resulting pulses are phase coherent and have an energy of 50 fJ. These pulses are used to coherently excite xenon atoms in an atomic beam of <10 MHz Doppler width. The excited atoms are then ionized by a second pulse and the ions are detected. Using this technique the authors were able to scan f_r and recover a transition linewidth as narrow as 7.5 MHz FWHM. Shown in Fig. 13 is the ion signal in arbitrary units versus the f_r detuning from the mean value for excitation by two to six VUV pulses.

4. Multi-Frequency Parallel Spectroscopy

In most DFCS experiments, the atomic system under study has acted as the high-resolution spectral discriminator that effectively selects an individual frequency comb element (or groups of comb elements) out of a greater number of elements that passed through the sample. The interaction of the optical frequency comb with the atomic system is then detected by the fluorescence or ionization from an excited state. A different approach involves the use of a high-resolution spectrometer to spectrally resolve individual comb elements in a parallel architecture. This allows a multi-channel detector to subsequently measure the amplitude (and potentially the relative phase—with appropriate reference) of the comb elements.

The challenge of resolving individual comb modes arises from the combined density and bandwidth of a typical optical frequency comb. A Fabry–Perot etalon has sufficient finesse to easily resolve the individual modes of a frequency comb with mode spacing below 100 MHz; however, the free spectral range of such an etalon does not typically exceed 100 GHz, which limits the bandwidth over which it can be used. On the other hand, grating-based spectrometers of reasonable size and cost can cover hundreds of THz, but the typical resolution does not exceed 10 GHz, which is insufficient to isolate individual modes of most femtosecond frequency combs. Such conflicting requirements of high resolution over large bandwidths are not necessarily new to the field of optics, and one solution involves using two spectrally dispersive elements (e.g. an etalon and a grating) along orthogonal spatial dimensions (Jenkins and White, 1976). An efficient implementation of this idea, originally directed toward separating densely-spaced optical communications channels (Xiao and Weiner, 2004; Wang et al., 2005) employs a diffraction grating orthogonal to a virtually-imaged phased array (VIPA) disperser (Shirasaki, 1996), which is essentially a variation

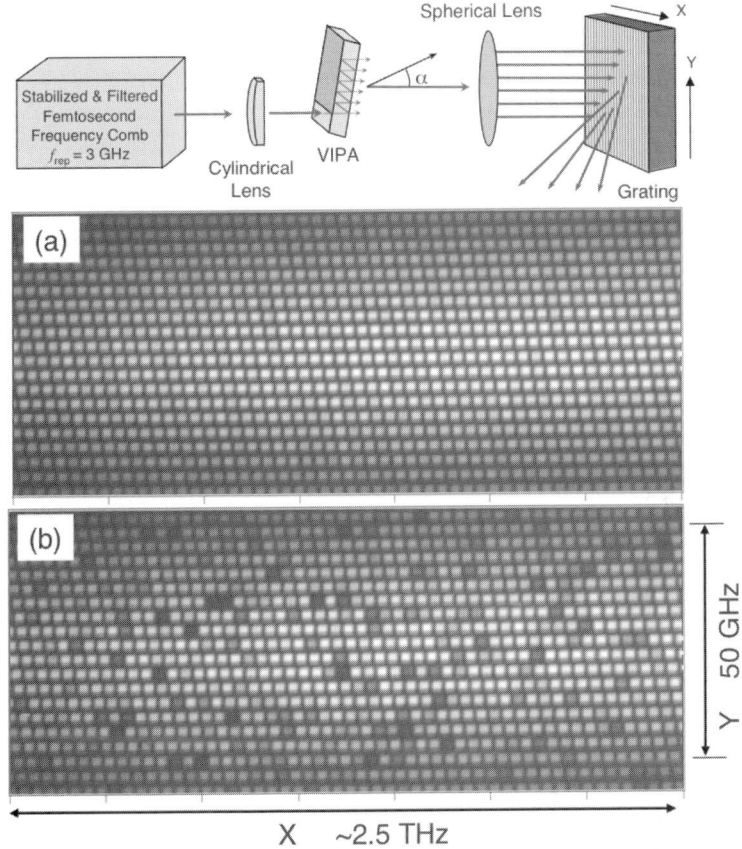

FIG. 14. A high resolution virtually-imaged phased array (VIPA) disperser is used in combination with a lower resolution diffraction grating to spatially resolve the filtered frequency comb of a frequency-stabilized, broadband Ti:sapphire femtosecond laser. (a) The spectrometer output in the region of 633 nm, as measured on a CCD element, consists of a 2-d array of the frequency comb modes, where each dot of this image represents an individual mode. Within a column (Y), which is tilted by the grating dispersion, the dots are separated by the mode spacing (3 GHz in this case). Within each row (X), the dots are separated by the VIPA free spectral range (FSR 50 GHz in this case). (b) Same frame as above, but now with iodine cell present. Note that some modes are missing or noticeably attenuated due to the presence of the iodine.

on the well-known plane-parallel etalon. In this case, the grating should provide spectral resolution better than that of the VIPA's free spectral range (FSR).

A high-resolution spectrometer designed for resolving individual comb modes is diagrammed in Fig. 14 (Diddams et al., 2006). The VIPA is a plane-parallel solid glass etalon, where the input beam is focused to a line and injected at an angle through an uncoated entrance window on the etalon's front face. The remainder of the front face is coated with a high-reflective dielectric, while the back face has a dielectric coating with 96% reflectivity. The multiple reflections within the VIPA etalon interfere in such a way that the exiting beam has its different frequencies emerging at different angles. In this case, the VIPA has a free spectral range of 50 GHz, determined by its thickness and index of refraction. The 2400 line/mm grating, used at a large angle of incidence and oriented with its dispersive axis orthogonal to that of the VIPA, provides 20 GHz spectral resolution that is sufficient to separate the various orders of the VIPA etalon.

The output of the VIPA/grating spectrometer is imaged onto a CCD camera, resulting in the array of dots, representing the individual comb modes, see Fig. 14(a). In this example, a frequency comb with elements spaced by 3 GHz was created by filtering the 1 GHz comb produced by a broad bandwidth Ti:sapphire laser. Approximately 2200 individual modes have been resolved within a 6.5 THz bandwidth captured in a few millisecond exposure on the CCD, for clarity the bandwidth is restricted to 2.5 THz in Fig. 14. In the vertical direction of this image, the data repeats every 50 GHz (at the VIPA FSR), and adjacent columns are also separated by the FSR of the VIPA. Full details on the indexing and counting of the numerous modes, as well as the frequency calibration are described in a recent publication (Diddams et al., 2007). The repetitive nature of the data is evident in Fig. 14(b), which is an image acquired with the iodine vapor cell inserted in the beam path. The cell is at room temperature and multi-passed to yield an equivalent length of 2 m. As clearly seen, numerous modes which coincide with absorption features of iodine are attenuated. Given the particular phase-locked values of f_o and f_r, the absolute frequency of each mode (or equivalently each CCD pixel) can be determined, thus providing a rapid means of identifying atomic or molecular species. While the VIPA spectrometer resolution is presently 1.2 GHz (at 475 THz), spectroscopic features down to the linewidth of the comb lines can be distinguished. Scanning the repetition rate or offset frequency of the laser enables one to scan out the full optical spectrum with a resolution suitable for the system under study (Diddams et al., 2007).

4.1. Cavity Enhanced DFCS

With every optical comb component efficiently coupled into a respective high-finesse cavity mode, one can establish a network of parallel channels for ultrasensitive detection of molecular dynamics and trace analysis. This configuration provides an ideal spectroscopic paradigm suitable for the next generation of atomic and molecular measurements. The approach presents simultaneously the following attractive characteristics: (i) A large spectral bandwidth allowing for the observation of global energy level structure of many different atomic and molecular species; (ii) High spectral resolution for the identification and quantitative analysis of individual spectral features; (iii) High sensitivity for detection of trace amounts of atoms or molecules and for recovery of weak spectral features; and (iv) A fast spectral acquisition time, which takes advantage of high sensitivity, for the study of dynamics.

The development of cavity-enhanced direct frequency comb spectroscopy utilizing a broad bandwidth optical frequency comb coherently coupled to a high-finesse optical cavity has been demonstrated (Thorpe et al., 2006, 2007). Hundreds of thousands of optical comb components, each coupled into a specific cavity mode, collectively provide sensitive intracavity absorption information simultaneously across a 100 nm bandwidth in the visible and near IR spectral region. By placing various atomic and molecular species inside the cavity, real-time, quantitative measurements of the trace presence, transition strengths and linewidths, and population redistributions due to collisions and temperature changes had been demonstrated. This novel capability to sensitively and quantitatively monitor multi-species molecular spectra over a large optical bandwidth in real-time provides a new spectroscopic paradigm for studying molecular vibrational dynamics, chemical reactions, and trace analysis. The continuing development of state-of-the-art laser sources in the infrared spectral regions, possibly even covering the important 3 μm area, will further improve the system's sensitivity.

The use of cavity enhancement allows for the measurement of optical absorption with sensitivities that are many orders of magnitude higher than what can be achieved with single pass absorption measurements. Cavity enhancement is most easily understood as an increased interaction length between the probe light and an intracavity absorber. Once coupled into the cavity, a photon will on average traverse the cavity a number of times that is proportional to the inverse of the mirror losses before decaying out of the cavity. Since the mirror losses can be made very small (as low as 10 parts per million) the enhancement in the light/matter interaction length can be more than 100,000. Because of its high sensitivity, cavity enhancement is useful when investigating atomic and molecular systems that have weak oscillator strengths, or in cases where only a trace amount of the target atom or molecule is present.

In the mid to late eighties the effect of cavity enhancement led to the development of cavity ring-down spectroscopy (CRDS), see for example Anderson et al. (1984), O'Keefe and Deacon (1988). CRDS has the useful feature that the laser intensity is removed from the measurement of optical absorption. Instead, the beam incident on the cavity is turned off and the rate of photon decay out of the cavity is measured. In this way, laser intensity fluctuations and noise associated with coupling the laser to the optical cavity do not affect the absorption measurement. By the mid to late nineties, CRDS was being used in a wide variety of applications and had achieved fantastic detection sensitivities, for example 3×10^{-12}/cm at 1 s integration time (Ye and Hall, 2000) or using the more complicated but sensitive technique, NICE-OHMS, 2×10^{-14}/cm at 1 s is possible (Ye et al., 1998).

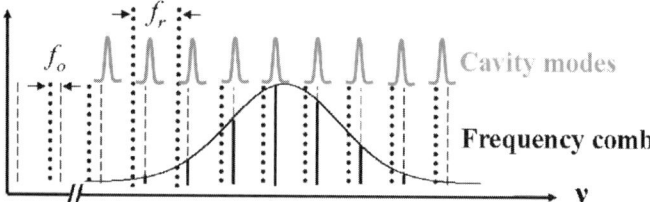

FIG. 15. When the frequencies of an optical frequency comb are made to overlap with the resonant modes of a high-finesse optical cavity, >90% of the broadband light can be coupled into the cavity. Nonuniform cavity mode spacing caused by dispersion limits the spectral bandwidth that can be coupled into the cavity.

Over the past decade, applications that require large spectral bandwidths such as the simultaneous detection of many different molecular species have led to the development of cavity enhanced systems based on mode-locked lasers (Crosson et al., 1999; Gherman et al., 2004; Thorpe et al., 2006). Using mode-locked lasers, broadband light can be coupled into a high-finesse optical cavity with very high efficiency, see Fig. 15. As a result, low average powers can be used to perform spectroscopy over a large spectral bandwidth. Also, the high peak intensities produced by ultrashort pulses are convenient for frequency conversion and spectral broadening such that modern mode-locked lasers can access any spectral region from the extreme ultraviolet to the far infrared and provide large spectral bandwidths for investigation and detection of a wide variety of atomic and molecular species (Dudley et al., 2006; Jones et al., 2005). In addition, massively parallel detection schemes have enabled the acquisition of broadband and high resolution atomic and molecular spectra in sub-millisecond timescales (Scherer et al., 2001; Diddams et al., 2007).

Efficient coupling of an optical frequency comb into a high-finesse optical cavity requires independent control of both the repetition frequency (f_r) and carrier offset frequency (f_o) of the comb. Using these controls the comb modes can be appropriately spaced and shifted to overlap the cavity modes (Jones and Ye, 2002). The bandwidth of the incident comb that can be coupled into the cavity is a function of the cavity dispersion and finesse. Cavity dispersion makes the resonant modes of the cavity have a nonuniform spacing as shown in Fig. 15 (Thorpe et al., 2005). Specifically, the frequency-dependent free spectral range (FSR) of the cavity is given by:

$$\text{FSR}(\omega) = \frac{c}{2L + c\frac{\partial \phi}{\partial \omega}\big|_{\omega_o}}. \qquad (4)$$

Here c is the speed of light, L is the cavity length, and ϕ is the frequency-dependent phase shift due to reflections from cavity mirrors and the intracavity

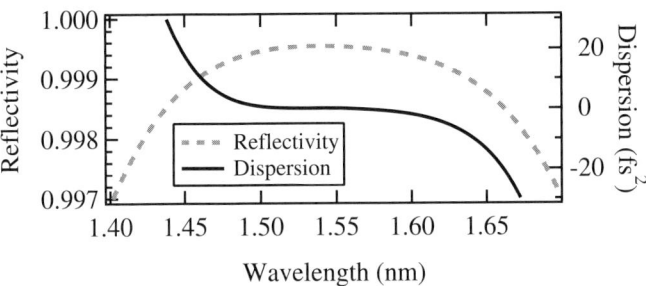

FIG. 16. The reflectivity and dispersion curves of Bragg reflectors constructed from alternating layers of silicon dioxide and niobium oxide centered at 1550 nm.

medium. The cavity finesse determines the linewidth of the cavity modes according to the relationship $\Delta \nu = \text{FSR}/\mathcal{F}$ where $\mathcal{F}$ is the cavity finesse and $\Delta \nu$ is the cavity linewidth. For a high-finesse optical cavity constructed from two identical mirrors the finesse can be approximated by $\mathcal{F} \approx \pi/(1-R)$ where R is the reflectivity of the cavity mirrors. The spectral bandwidth that can be coupled into the optical cavity can be found by evaluating the walk-off integral:

$$\Delta \nu = \frac{\text{FSR}(\omega)}{\mathcal{F}} = \frac{1}{\pi} \int_0^{\omega_{1/2}} \left(\frac{\text{FSR}(\omega)}{f_r} - 1 \right) d\omega. \tag{5}$$

Here the spectral bandwidth coupled into the cavity is $2\omega_{1/2}$. The walk-off integral illustrates the tradeoff between the cavity finesse and the coupled spectral bandwidth. As the finesse increases (leading to increased sensitivity) the bandwidth coupled into the cavity decreases. The typical dispersion and reflectivity curves for Bragg reflectors used to construct broadband enhancement cavities are shown in Fig. 16. If the comb is locked to the cavity modes, coupled bandwidths of 30 to 60 nm at 1550 nm (limited by the dispersion and reflectivity of the mirrors) are typical. For broader bandwidth measurements, the comb can be dithered over the cavity modes. Using this method, the measurement bandwidth is only limited by reflectivity of the cavity mirrors (typically 200 nm at 1550 nm), however, the measurement rate or duty cycle is reduced by the dither.

Using techniques of spectral broadening, optical frequency combs can have a very broad spectrum allowing a single laser source to perform spectroscopy over more than an octave of spectrum. A modest example of a spectrum that can be generated using highly nonlinear fiber is shown in Fig. 17. Because of this, it is typically the cavity mirrors that limit the spectral bandwidth of cavity enhanced measurement systems. However, the continued development of broad-band, high reflectivity, and low-loss mirror coatings as well as the recent development of

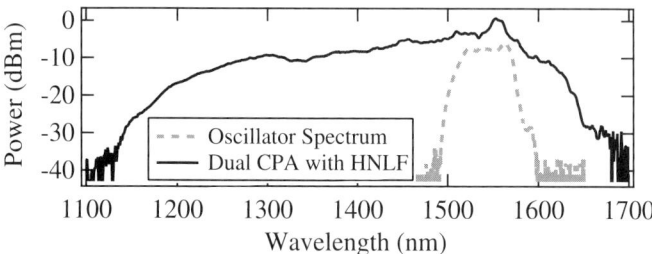

FIG. 17. The spectrum of an Er^{+3} mode-locked fiber laser before and after amplification and spectral broadening via highly nonlinear fiber. Both spectra were recorded with 1 nm resolution.

broad-band, high-finesse prism cavities will relax this limitation in future systems.

A number of parallel detection techniques have been developed to simultaneously read-out the many parallel channels that are available in cavity transmission. The simplest form of parallel detection uses a direct absorption measurement. Here, the light transmitted from the cavity is spectrally dispersed and the intensity is recorded at different spectral regions with a CCD or a diode array. The transmitted beam can be spectrally resolved with either a grating spectrometer producing a 1-dimensional dispersion pattern or with a virtually-imaged phased array (VIPA) spectrometer that provides a higher resolution and more compact 2-dimensional dispersion pattern (Diddams et al., 2007). To simultaneously record multiple ring-down signals requires some more complicated techniques. One clever way to do this involves the use of a rotating mirror to convert the time-domain information contained in the ring-down signal into a spatial position of the beam on a CCD chip (Scherer et al., 2001).

A typical frequency comb based cavity ring-down system shown in Fig. 18 consists of a mode-locked laser, a high-finesse optical cavity, a broadband detection scheme, and feedback to maintain the comb and cavity in resonance.

The inset in Fig. 18 describes the basic requirements for coupling the comb to the cavity. First, the FSR of the cavity must be equal to the laser f_r. Next, f_o must be set such that the comb frequencies are shifted to overlap the cavity modes. Once light is built-up into the cavity, an optical switch is used to turn off the light incident on the cavity at which time the photons in each cavity mode decay out of the cavity with a characteristic time constant that depends on the cavity length and the total cavity losses,

$$\tau = \frac{1}{\text{FSR}(\omega)\text{Loss}} = \frac{1}{\text{FSR}(\omega)(1 - R_1 R_2 e^{-2\alpha(\omega)L})}. \quad (6)$$

Here, R_1 and R_2 are the reflectivities of the cavity mirrors, L is the cavity length, and $\alpha(\omega)$ is the frequency-dependent intracavity absorption. The light that decays

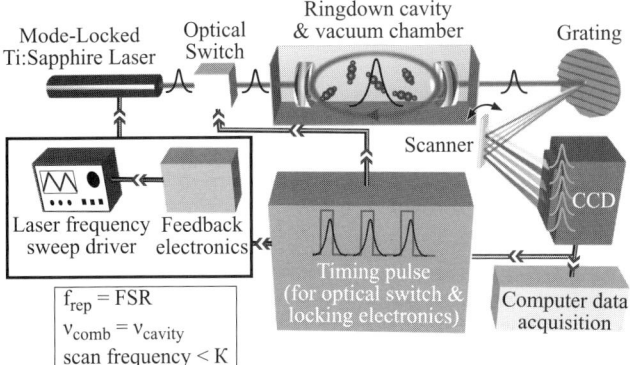

FIG. 18. Schematic of broadband cavity ring-down spectroscopy. The inset box lists the basic requirements for coupling the frequency comb to the cavity and the maximum rate that cavity enhanced measurements can be made, where $\kappa = \frac{1}{\tau}$.

from the cavity is spectrally dispersed with a grating. The decay rate is measured by sweeping the dispersed beam across a CCD chip with a rotating mirror (Scherer et al., 2001). In this system, the comb is dithered over the cavity modes such that the cavity builds-up and rings-down twice per modulation cycle. The feedback scheme that keeps the comb resonant with the cavity modes typically consists of two low speed servos. The first servo has a bandwidth of 10 Hz and maintains the cavity transmission peak at the center of the f_r dither by actuating f_r. The second servo actuates f_o via the laser pump power or by rotating a dispersive element inside the laser cavity to maximize the height of the cavity transmission peak.

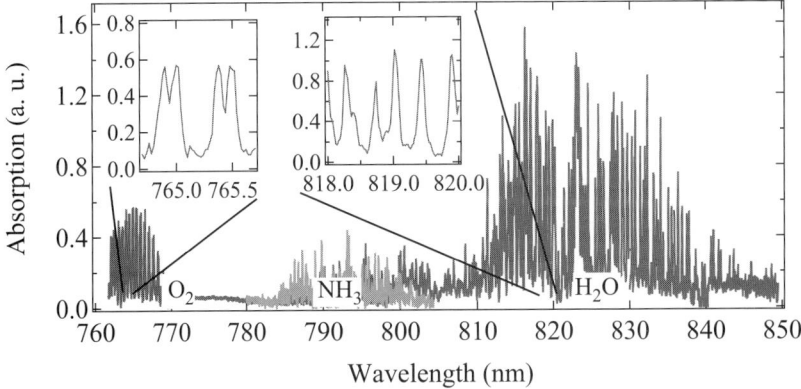

FIG. 19. The spectra of O_2, NH_3 and C_2H_2 near 800 nm.

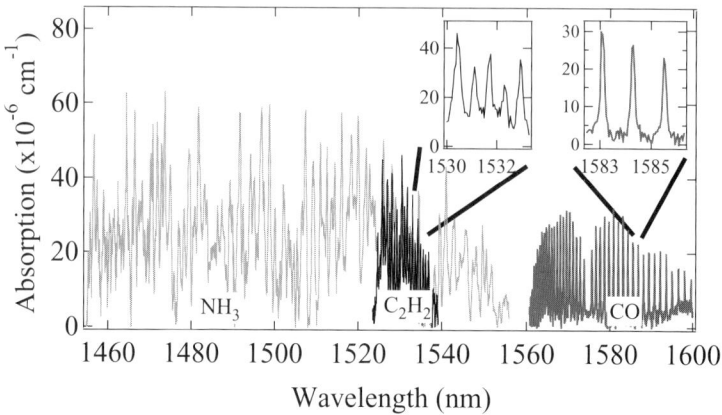

FIG. 20. The spectra of CO, NH$_3$ and C$_2$H$_2$ near 1550 nm.

Recently, two systems have been constructed to demonstrate CRDS based on an optical frequency comb (Thorpe et al., 2006, 2007). The first of these systems was based on a mode-locked Ti:sapphire laser with a 100 nm spectral bandwidth centered at 800 nm. Figure 19 shows the spectra of O$_2$, NH$_3$, and H$_2$O recorded with the Ti:sapphire comb-based CRDS system. The ring-down cavity used for these measurements was a two mirror linear cavity with mirror reflectivity $R = 0.9993$ and a FSR of 200 MHz.

The second frequency comb CRDS system was based on a mode-locked erbium-doped fiber laser. Figure 20 shows the spectra of NH$_3$, C$_2$H$_2$, and CO recorded with the erbium fiber laser CRDS system. The ring-down cavity used for these measurements was a two mirror linear cavity with mirror reflectivity $R = 0.9996$ and a FSR of 100 MHz.

5. Coherent Control Applications

Direct frequency comb spectroscopy has also been applied to conduct coherent control experiments at vastly enhanced spectral resolution. A basic feature of coherent control is the manipulation of the final state of a system by controlling quantum interference routes. In contemporary coherent quantum control, short pulses are commonly used due to their broad spectrum and known, stable spectral phase. Shaping the spectrum and the spectral phase of the pulse allows control over the interference between different quantum paths, steering an interaction toward a desired outcome. The main improvement offered by the frequency comb is that it allows for the coherent interaction time to be much longer than the duration of a single (shaped) pulse. Due to the inter-pulse coherence enabled by a

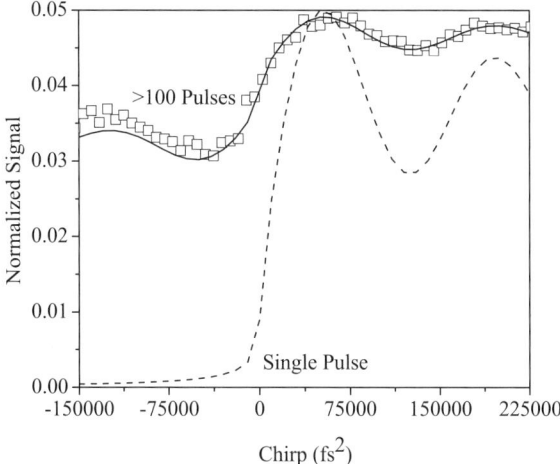

FIG. 21. The measured $5D_{5/2} F = 4$ ^{87}Rb fluorescence is shown (squares) versus chirp for steady state excitation. The solid line indicates the theory prediction at steady-state and the dashed line is for single-pulse excitation.

phase-stabilized frequency comb, it is possible to use the entire train for a coherent control process.

5.1. High Resolution Coherent Control

Some of the pioneering work on coherent control introduced either linear frequency chirp or discrete phase steps in a pulse spectrum to manipulate two-photon absorption in Alkali atoms such as ^{87}Rb, see for example Balling et al. (1994), Chatel et al. (2003), Meshulach and Silberberg (1998). These experiments were conducted with only one pulse, therefore the resolution was significantly lower than allowed by DFCS. The experiment presented in Stowe et al. (2006) extends some of the pioneering work on coherent control to high resolution by introducing linear frequency chirp to a coherent pulse train, or in other words adjusting the phase of the comb modes used in DFCS. In this experiment a grating based pulse stretcher/compressor was used to introduce up to ± 225000 fs^2 of dispersion to the pulses before they excite the $5S_{1/2} F = 2 \to 5P_{3/2} F = 3 \to 5D_{5/2} F = 4$ transition in ^{87}Rb. Figure 21 shows the measured 5D population versus the pulse chirp. Particularly interesting is that the difference in signal between positive and negative chirp decreases significantly between single and multi-pulse experiments. In the single pulse case negative chirp excites essentially no population due to the fact the wavelength connecting the $5S_{1/2} F = 2 \to 5P_{3/2} F = 3$ transition arrives after the $5P_{3/2} F = 3 \to 5D_{5/2} F = 4$ wavelength, essentially the time ordering is wrong to complete a two-photon transition via the intermediate state. However, for multiple pulsed excitation the coherence between the atomic states excited by the first pulse remains intact for the second pulse, which can complete the two-photon transition. This is a clear example of an important difference between single and multiple shaped pulse excitation of an atomic or molecular system exhibiting a coherence time longer than the inter-pulse period. Several previous papers (Balling et al., 1994; Chatel et al., 2004, 2003) have demonstrated and modeled the oscillation of the excited 5D population versus chirp. The period of this oscillation is inversely proportional to the detuning of the intermediate state from half the two-photon transition frequency, for example 1 THz for $5P_{3/2}$ and 8 THz for $5P_{1/2}$. Unlike previous single pulse coherent control experiments the high resolution enabled by the comb can be used to control the population transfer to a single excited hyperfine level, $5D_{5/2}F = 4$. Furthermore, due to hyperfine selection rules, this level may only be reached from a single intermediate state, $5P_{3/2}F = 3$. For this reason the measurement shown in Fig. 21 and the observed oscillation period versus chirp can be attributed to only one intermediate state, as opposed to all the states within the 5P manifold. This demonstrates that DFCS can extend some classic single pulse coherent control experiments, limited to THz order resolution, to natural linewidth limited resolutions. Coherent control experiments utilizing DFCS can now truly control the population transfer to a single quantum state.

5.2. Extension of DFCS to Strong Field Coherent Control

While analysis of coherent quantum control is relatively simple in the weak field perturbative domain (Shapiro and Brumer, 1986, 2003a, 2003b; Tannor and Rice, 1985; Rice and Zhao, 2000), extension to strong fields is not straightforward. Analytic models exist only for simple cases (Vitanov et al., 2001; Gaubatz et al., 1990; Dudovich et al., 2005; Lee et al., 2004) and solutions are often found by numerical optimizations (Rabitz et al., 2000). Here, the concept of coherent accumulation of a train of pulses also proves very useful. While at first, the need to consider the effect of the whole pulse train seems to further complicate the analysis, it can in fact considerably simplify the design. Specifically, since each pulse in the train achieves only a fraction of the overall population transfer, one can exploit analytic perturbative models to design "ideal" weak pulses, relying on the coherent accumulation of all pulses to achieve the high overall efficiency. This avoids the complication of strong field design while gaining the high spectral selectivity offered by the frequency comb.

As a model problem to illustrate the concept of *strong-field* control with the comb, we discuss a narrow-band Raman transition between molecular vibrational levels from a single vibrational level embedded in a dense environment of other

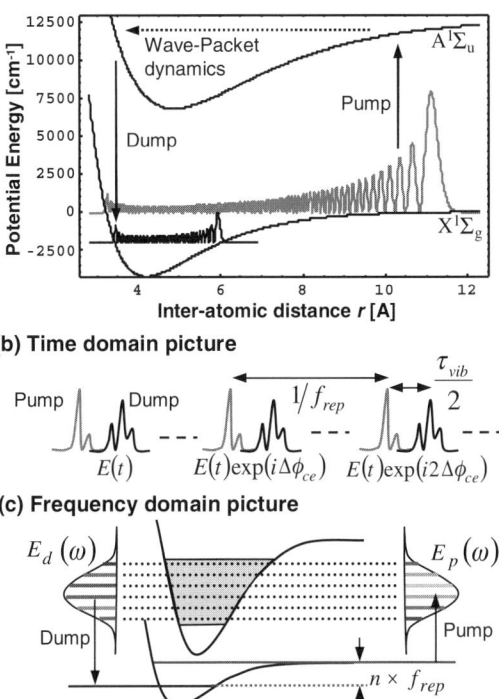

FIG. 22. Basic Raman control scheme. (a) Typical electronic potentials and vibrational levels (here, Morse potential fits of Rb$_2$). Population is transferred from the input vibrational level near the dissociation limit ($v' = 130$) to a deeply bound level ($v = 45$), mediated via a broadband wave-packet in the excited electronic potential. (b) Time-domain picture. A train of phase coherent pump–dump pulse pairs interacts with the molecule. Pulse pairs are shaped to achieve efficient population transfer. The intrapair time is half the vibration time (τ_{vib}) of the intermediate wave-packet and the inter-pair time is the repetition time of the source $1/f_r$. (c) Frequency-domain picture. A tooth-to-tooth match between the pump and the dump frequency combs, with f_r equal to a sub-harmonic of the net Raman energy difference, locks the relative phase between pulse pairs to the free evolving Raman phase.

levels near the dissociation limit, to a single deeply bound vibrational level, as shown in Fig. 22, Pe'er et al. (2007). Each pump–dump pulse pair in the coherent train of shaped pulse pairs is weak, i.e., it transfers only a small fraction of the input population to the target state and only coherent accumulation of many pulses enables a high overall transfer efficiency. In the time domain, each pump pulse excites a wave-packet that starts to oscillate in the excited electronic potential. After half a vibration, this wave-packet reaches the inner turning point where the dump pulse drives it to the target state. Since population appears on the excited potential only for half a vibration, this scheme eliminates spontaneous emission losses. In order to enable coherent accumulation of population at the target state, the temporal phase difference between pulse pairs should match the phase of the free evolving Raman coherence. In the frequency domain, the combs of the pump and the dump pulses must overlap tooth to tooth and the repetition rate f_r must match a sub-harmonic of the Raman energy difference, as illustrated in Fig. 22(c).

The coherent accumulation process gradually transfers the population from the input to the target state. A weak pump pulse can indeed excite a small fraction of population from the input state, but the very first dump pulse must have a "pulse area" near π to drive all this population to the *empty* target state. The second dump pulse, however should have an "area" of only $\pi/2$, because now the excited population is about equal to the population already in the target state from the previous dump, just like in a Ramsey experiment. Similarly, as the input state is depleted, the pump "area" should slowly increase to excite the same population every time, reaching $\sim \pi$ for the very last pump pulse. In general, the fraction of population excited (or dumped) by a pulse of "area" A is $\sin^2(A/2)$. As population is accumulated in the target state and depleted from the input state, the dump "area" of the nth pulse $A_d[n]$ should decrease and the pump "area" $A_p[n]$ increase according to

$$\sin^2\left(\frac{A_d[n]}{2}\right) = \frac{1}{n}, \quad \sin^2\left(\frac{A_p[n]}{2}\right) = \frac{1}{N-n+1}, \tag{7}$$

for each pulse pair to transfer the same fraction of population (total of N pulses). Another equally valid solution is to fix the pump area and vary the dump according to the evolving population distribution. In general, for any given pump series, a dump series can be matched according to the nth fraction of excited population. Clearly the very first dump pulses and the last pump pulses are of areas near π and cannot be considered weak, but for a large N, the majority of the population is transferred by the accumulative effect of all pulses, which are mostly weak. Consequently, even if the first (last) pulses do not have the required area or are complicated by multi-photon effects, the overall process can remain efficient. Experimentally, current intensity modulators can easily meet the required pulse area variation for standard pulse repetition times of 1–10 ns.

For an efficient pump–dump process it is required that the wave-packet $|\psi_p\rangle$, excited by the pump from the input state $|i\rangle$ and propagated for half a vibration, will overlap perfectly with the wave-packet $|\psi_d^r\rangle$ that would have been excited from the target state $|t\rangle$, by the time-reversed dump. For weak pulses we can express these two wave-packets, using first order perturbation theory, as

$$|\psi_p\rangle \propto \sum_\omega E_p(\omega) e^{i\phi_D(\omega)} |\omega\rangle \langle \omega | d_{el} | i \rangle,$$

$$|\psi_d^r\rangle \propto \sum_\omega E_d^r(\omega)|\omega\rangle\langle\omega|d_{el}|t\rangle, \tag{8}$$

where $|\omega\rangle$ denotes the vibrational states in the excited potential using the detuning ω from the pulse carrier frequency as a vibrational index, $F_p(\omega) = \langle\omega|d_{el}|i\rangle$ and $F_d(\omega) = \langle\omega|d_{el}|t\rangle$ are the pump and the dump transition dipole matrix elements, which under the Condon approximation are proportional to the Franck–Condon factors $\langle\omega|i\rangle$, $\langle\omega|t\rangle$. d_{el} is the electronic transition dipole moment and $E_p(E_d^r)$ is the spectral amplitude of the pump (time-reversed dump) field. $\phi_D(\omega)$ is the spectral phase acquired by the wave-packet between the pulses, which reflects both the delay of half a vibration and the dispersion of the wave-packet as it oscillates in the anharmonic excited potential. For a given molecular potential, both the dipole matrix elements and $\phi_D(\omega)$ can be easily calculated.

As a result, perfect overlap of the two wave-packets can be achieved by shaping the pump field according to the dump dipole matrix elements and vice versa:

$$E_p(\omega) \propto F_d(\omega)A(\omega), \qquad E_d(\omega) \propto F_p(\omega)A(\omega)\exp[i\phi_D(\omega)], \tag{9}$$

where A is an arbitrary spectral amplitude, common to both fields. Intuitively, this spectral shaping avoids pumping of what cannot be dumped (due to a node in the dump dipole matrix elements), and vice versa.

Within the perturbative discussion relevant to Eq. (9), the common spectral amplitude $A(\omega)$ is completely arbitrary. Yet, it is desirable to minimize the total number of pulses for both practical reasons (accumulated phase noise) and fundamental ones (the total interaction time is limited by the coherence time of the input state), so one would probably prefer to use the strongest pulses that can still qualify as "weak". The limiting power level is where two-photon (Raman) processes by a single pump or dump pulse become pronounced, and here common shaping will have an important effect. Specifically, in many cases of molecular dynamics, positively chirped (red to blue) excitation pulses can strongly suppress Raman processes that adversely affect the input wave-packet during the pulse (Bardeen et al., 1995; Cao et al., 1998; Malinovsky and Krause, 2001), leaving the excitation, although strong, essentially one-photon. Thus, chirping the pulses can improve considerably the dumping efficiency for the first dump pulse that is necessarily strong because it dumps to an empty target state. Common chirping can also help to resolve vibrational structure around the target state deep in the molecular well, yet it cannot resolve rotational/hyperfine structure, nor the dense environment around the input state. This is where the coherent accumulation proves powerful.

Representative simulation results of the full coherent accumulation process are shown in Fig. 23 for the interaction of a train of 40 pulse-pairs with the molecule. The pump pulses depleted >90% of the input population, and when the inter-pulse phase was tuned to the Raman condition, 95% of this population reached the target

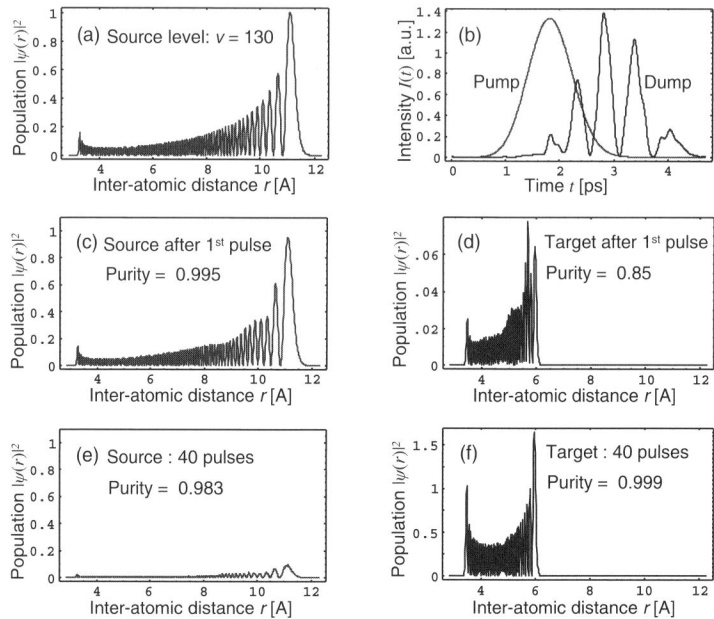

FIG. 23. Simulation results for the coherent accumulation process. The pump pulse power was fixed, exciting about 5.7% of the input population each time, and the dump area was varied to match with the accumulation progress. The pulses were ~10 nm in bandwidth, chirped out to ~1.5 ps by dispersion of 500,000 fs^2. (a) the input state population density, (b) the intensity temporal profile of the pulses (after strong chirp). (c) and (d) are the input and target wave-packet population densities after one pulse-pair, and (e) and (f) are the corresponding results after 40 pulses.

state. The purifying nature of the coherent accumulation process is demonstrated by the obtained final wave-packet—practically a single state.

The coherent accumulation proves to be quite robust against intensity fluctuations. Scaling the intensity of the pump or dump pulse train (or both) by a factor of two in simulation leaves the total transfer efficiency constant within a few percent. The exact variation of dump pulse area according to the accumulated population is also not critical. Even if the dump area is kept constant, the transfer efficiency is >50% over a range of factor of two in intensity. The reason for this robustness is that even when not all population is dumped, the anharmonicity of the excited potential scrambles the lingering wave-packet between pulses, such that the interference with the next pumped wave-packet is neither constructive, nor destructive, but averages out (with background population decaying at $1/f_r$ time scales). Therefore, pulse area errors do not coherently add from pulse to pulse.

To conclude, the presented scheme forms a unique and powerful combination of frequency-domain control (comb) and time-domain control (molecular dynamics). As such, it enables performance of coherent control tasks with both high efficiency and unprecedented spectral resolution. Due to the use of weak pulses the process can be analyzed within a perturbative model, thus opening an analytic path to strong field problems that were so far accessible only by numerical optimizations.

6. Future Outlook

6.1. ATOMIC CLOCK APPLICATIONS

A general trend throughout the history of precision timekeeping has been to employ oscillators with an ever-increasing frequency, since a clock operating at a higher frequency provides greater precision through the finer sub-division of a unit of time. Following this trend over the past one hundred years, one finds the advance from pendulum clocks, to crystal quartz oscillators, to microwave atomic clocks, and now most recently, atomic clocks operating at optical and ultraviolet frequencies (Diddams et al., 2001; Ye et al., 2001). It is likely that this progression will continue in the future, and DFCS could play an important role in the push to still higher frequencies since the high peak powers of the associated femtosecond pulses provide an efficient means to generate coherent radiation at high frequencies. Some of the exciting new techniques for producing frequency combs in the extreme ultraviolet regime will be discussed in detail in the following section, but first we present a scheme of employing DFCS to control the frequency comb of a femtosecond laser, and thereby produce optical and microwave frequencies referenced directly to an atomic transition (Gerginov et al., 2005).

If the optical frequency $\nu_n = f_o + nf_r$ of a certain comb component can probe, and then be directly stabilized to an atomic transition, the repetition rate instability will be given by $\delta f_r/f_r = (1/n)\delta\nu_n/\nu_n$, where $\delta\nu_n/\nu_n$ is the optical frequency instability. In this way, an optical clock is realized without the use of cw lasers. The stability of the optical atomic reference is mapped onto the stability of f_r, which in turn can be measured and used as a frequency reference. Such an experiment has been realized by using the setup described in Section 3.2.2. The $F = 3 \rightarrow F = 2$ transition of the Cs D$_2$ line was excited using DFCS and the subsequent fluorescence was detected. By modulating f_r the fluorescence signal and its derivative shown in Fig. 24 were measured. The error signal was then used to steer f_r such that the specific optical mode ($n = 352,210$) is frequency-locked to the center of the $F_g = 3 - F_e = 2$ transition. Now atomically-stabilized, the stability of f_r was measured, and a fractional frequency instability plot are shown in Fig. 25. The plot shows that instability of better than 10^{-11} can be easily achieved in 100 s integration time. While this example takes advantage of convenient, yet relatively broad near-infrared transition in cesium, the basic approach is one that could be applied to much higher (and narrower) frequency atomic transitions.

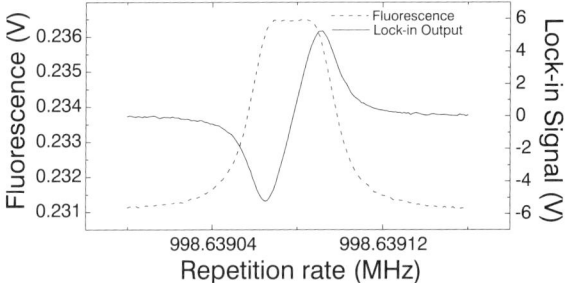

FIG. 24. Error signal used to lock the femtosecond laser repetition rate. A specific FLFC component excites $F_g = 3 - F_e = 2$ transition of the Cs D$_2$ line. The repetition rate was modulated at rate of 27 Hz with 15 Hz depth. Phase detection was used with 5 mV sensitivity and 1 s time constant.

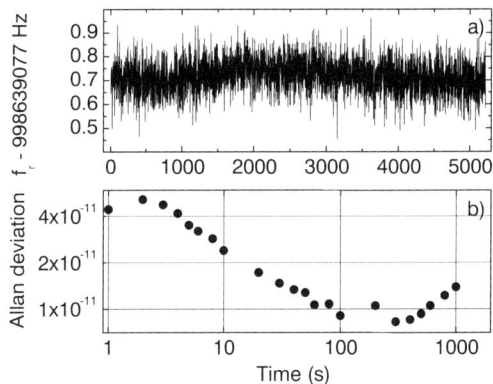

FIG. 25. Femtosecond laser repetition rate measured with $n = 352,210$ optical component locked to $F_g = 2 \to F_e = 2$ transition. The Allan variance calculated from the data is also shown.

6.2. XUV COMB DEVELOPMENT

While some of the powerful applications of the DFCS technique have been discussed in the previous sections, it is anticipated that its most powerful application is yet to come. Many scientific discoveries and technological advances are awaiting in the landscape of precision measurements and quantum control in the extreme ultraviolet (XUV) spectral region. With cw lasers unlikely to develop significantly into the XUV regime, DFCS is a uniquely attractive option for performing high resolution spectroscopy and quantum optics in the vast spectral region spanning from VUV to XUV.

To extend the coherent frequency comb structure and related precision measurement capabilities into the deep UV, recent research has focused on the demon-

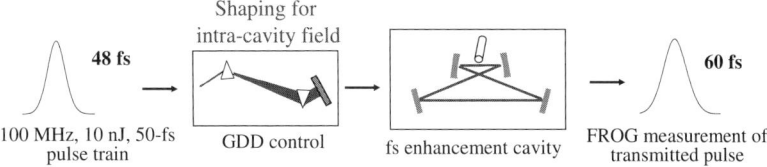

FIG. 26. Schematic setup of intracavity high-harmonic generation. The femtosecond pulse train, after proper dispersion compensation, is incident and stabilized to a high-finesse cavity of a bow-tie geometry, enhancing the pulse energy by nearly three orders of magnitude while maintaining a high repetition frequency inside the cavity. A gas target at the cavity focus enables phase-coherent high-harmonic generation, resulting in a phase-stable frequency comb in the VUV spectral region. Measurement of the pulse shape transmitted through the cavity permits optimization of both the cavity lock and dispersion.

stration of high-harmonic generation (HHG) at a 100 MHz repetition rate enabled by a femtosecond enhancement cavity (Jones et al., 2005; Gohle et al., 2005), utilizing the technology discussed in Section 2.2. The process of HHG provides a coherent source of vacuum-ultraviolet to soft x-ray radiation (Corkum, 1993; Lewenstein et al., 1994). Traditionally HHG has only been demonstrated with high-energy, low repetition rate amplified laser systems to provide the peak intensities needed for ionization of a gas target. The small conversion efficiency of the process, combined with the low repetition rate of amplified laser systems, results in low average powers in the XUV generation. Furthermore, the use of these sources as precision spectroscopic tools is limited, as the original laser frequency comb structure is lost in the HHG process. By use of a femtosecond laser coupled to a passive optical cavity, coherent frequency combs in the XUV spectral region are generated via high-harmonics of the laser without any active amplification or decimation of the repetition frequency. Using this technique it's possible to significantly improve the average power conversion efficiency and reduce the system cost and size, while dramatically improving the spectral resolution. Since little of the fundamental pulse energy is converted, a femtosecond enhancement cavity is ideally suited for HHG as the driving pulse is continually recycled after each pass through the gas target. The development of a frequency comb in the XUV and its extreme spectral resolution will likely enable revolutions in precision measurement, quantum control, and ultrafast science, similar to those in the visible and IR regimes.

Figure 26 depicts a simple scheme of intracavity HHG approach. A standard mode-locked femtosecond Ti:sapphire laser with f_r of 100 MHz, 50-fs pulse duration, and 8 nJ pulse energy is used. The pulse train from the laser passes through a prism-based compressor before impinging upon the passive optical cavity. To couple the pulse train from the mode-locked laser into the cavity, both f_o and f_r of the laser are adjusted such that the optical comb components are maxi-

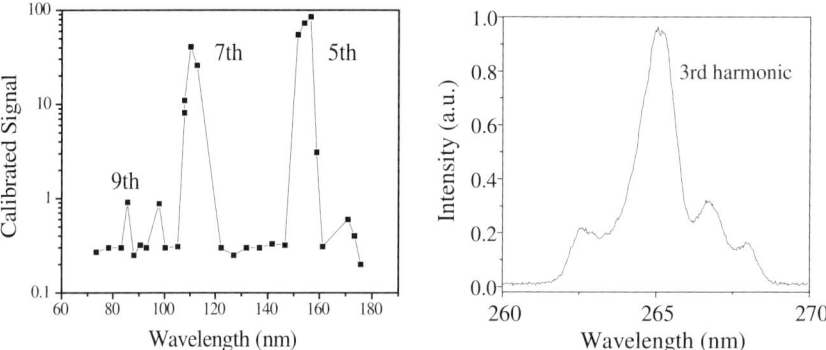

FIG. 27. The output spectrum of the 100-MHz intracavity high harmonic generation.

mally aligned to a set of resonant cavity modes. The enhancement cavity builds up the pulse energy from 8 nJ to 4.8 µJ, while maintaining the original pulse width and f_r. The peak intracavity intensity of $>3\times 10^{13}$ W/cm^2 is obtained at the intracavity focus. The gas target of Xe atoms is placed slightly after the cavity focus to maximize the HHG light, reducing absorption of the generated optical harmonics. The typical gas pressure at the interaction region is >10 Torr, with <1 mTorr of background pressure for the evacuated chamber where the fs enhancement cavity is located.

To couple the HHG light out of the cavity, a thin sapphire plate is placed at Brewster's angle (for the fundamental comb at near infrared) inside the cavity. This additional optical element, while lowering the cavity finesse only slightly (<5%), requires a negative group-delay-dispersion coated mirror to compensate for the additional material dispersion. The Fresnel reflection coefficient of the intracavity plate rises towards shorter wavelengths in the EUV, with a maximum reflectivity of ~10% between 30 to 80 nm. The HHG light is then diffracted with a MgF-coated aluminum grating and the resultant spectrum spanning from the 3rd to the 9th harmonics is shown in Fig. 27. With a similar system, the Garching group has observed up to the 15th harmonic (Gohle et al., 2005). The average power of the 3rd harmonic light generated inside the cavity reaches more than 10 µW. The single-shot efficiency ($\approx 10^{-8}$) of high harmonic generation using this technique is comparable to traditional amplifier-based systems at similar intensity levels. This demonstrates the dramatic increase in high-harmonic average power that can be accessed using a high repetition rate.

While the use of the intracavity Brewster plate does not impose a serious limitation on the pulse bandwidth under the intracavity dispersion compensation, its nonlinear response does limit the power scalability of the system (Moll et al., 2005). To solve this problem, a novel enhancement cavity configuration that will allows the use of more powerful lasers was designed. One of the focusing cavity mirrors has a 200 μm hole drilled in the middle to couple out the high-harmonics. By using a higher order cavity mode such as TEM01, it is still possible to build up sufficient peak power inside the cavity for HHG, despite the hole in one focusing mirror. The generated VUV comb, however, leaks out of the mirror hole due to the significantly smaller diffraction angles enjoyed by the shorter wavelength light beam (Moll et al., 2006). This cavity geometry will allow a larger buildup power inside the cavity without any intracavity optics. Using this geometry, more powerful laser systems can be explored, such as high energy (>25 nJ pulse energy), chirped-pulse Ti:sapphire oscillators developed by Apolonski and Krausz (Naumov et al., 2005) and power-scalable Yb fiber lasers developed by Hartl of IMRA America, Inc.

High precision measurement of phase/frequency fluctuations in the high-harmonic generation process has also been performed. Two sets of the frequency combs at 266 nm that represent the 3rd harmonic of the fundamental IR comb are brought together for beat detection. One comb is generated by the Xe gas via the HHG process while a second comb is produced by a bulk BBO non-linear crystal. The RF spectrum of the beat note in Fig. 28 shows a clear presence of the comb structure in the UV. The resolution bandwidth-limited 1 Hz beat signal as shown in Fig. 28 demonstrates that the full temporal coherence of the original near-IR comb has been precisely transferred to high harmonics.

6.3. FUTURE VUV COMB SPECTROSCOPY

Another motivation for short wavelength spectroscopy is that the simplest atoms, hydrogen and helium, have their first electronic transitions in the VUV spectral range. As previously emphasized, the extremely large peak intensities achievable with femtosecond pulses enables significantly more efficient frequency mixing to generate shorter wavelengths. One of the most interesting future applications of DFCS to extend the realm of coherent spectroscopy of atoms and molecules to shorter wavelengths. Frequency combs also improve the resolution limit past that of current dye laser amplifier techniques used for measuring VUV transitions. One and two-photon cw-laser spectroscopy of the lowest lying electronic transitions in these simplest of atoms have provided direct checks of quantum electrodynamics (Hänsch, 2006; Eikema et al., 1997; de Beauvoir et al., 2000; Bergeson et al., 2000). For example, the Lamb shift of ground state hydrogen and helium has been measured with traditional cw techniques. Another reason for studying these atoms is that the nucleus is simpler in structure and better known than heavy elements. This knowledge of the nucleus allows for better measurements of the Rydberg constant. Currently research is under way in several groups to conduct DFCS of helium to improve measurements of both the ground state Lamb shift

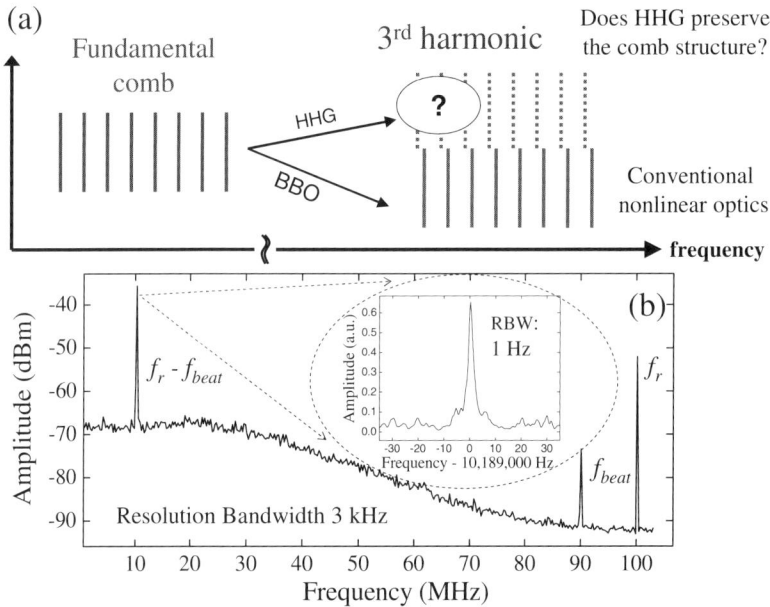

FIG. 28. Coherent heterodyne beat signal is detected between the HHG in Xe gas and bound optical nonlinearities in BBO. These two frequency combs (both at 3rd harmonic) spectrally overlap and provide the optical heterodyne beat signal at an offset radio frequency introduced by an acousto-optical modulator placed in one arm of the interferometer. The two corresponding pulses trains are overlapped in time. The coherent beat signal and repetition frequency detection are shown in (b), demonstrating the HHG comb is phase coherent with respect to its parent comb from the laser, with the coherence limited only by the observation time. The linewidth shown in the inset of (b) is resolution bandwidth (RBW) limited at 1 Hz.

and the Rydberg constant over current cw-laser based measurements. Figure 29 indicates the wavelengths necessary for some one and two-photon transitions of interest in helium. Hydrogen-like, singly charged helium ions is also under current investigation by the group in Garching led by T. Hänsch and T. Udem. One aspect of the regularly spaced mode structure of the comb, important for two-photon transitions, is that if any pair of modes is two-photon resonant then all modes can form a resonant pair. If there is no intermediate state within the bandwidth of the comb, transform limited pulses allow for all of the transition amplitudes driven by the corresponding comb modes to add constructively to the two-photon transition. This is an attractive feature of combs for multi-photon spectroscopy because the entire spectral power contributes to exciting the transition. Another application closely related to spectroscopy at short wavelengths is the laser cooling of atoms at these wavelengths with a frequency comb (Kielpinski, 2006).

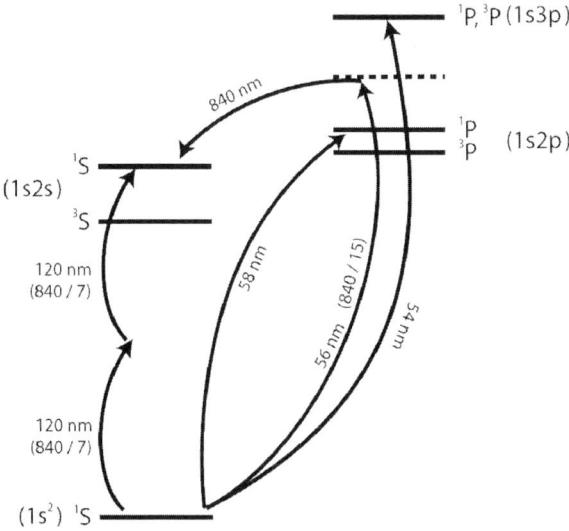

FIG. 29. Several possible interesting excitation paths of helium and corresponding wavelengths based on harmonics of a 840 nm Ti:sapphire laser.

6.4. OPTICAL FREQUENCY SYNTHESIS AND WAVEFORM GENERATION

Femtosecond comb technology has made dramatic advances in time-domain experiments as it has in optical frequency metrology. Stabilization of the "absolute" carrier–envelope phase at a level of tens of milliradians has been demonstrated and maintained over minutes, laying the groundwork for electric field synthesis. The capability of precisely controlling pulse timing and the carrier–envelope phase allows one to manipulate pulses using novel techniques and achieve unprecedented levels of flexibility and precision. For example, comb spectra emitted by independent mode-locked laser sources can now be coherently stitched together (Ye et al., 2002), allowing DFCS to be performed at distinctly different spectral regions while maintaining optical coherence. The related technology development will even allow DFCS to be performed remotely, with optical fiber networks delivering precise frequency comb structure with minimum perturbation (Holman et al., 2004; Hudson et al., 2006).

To establish phase coherence among independent mode-locked lasers, it is necessary to first achieve synchronization among these lasers and reduce the remaining timing jitter below the optical oscillation period (Ma et al., 2001; Shelton et al., 2002a; Schibli et al., 2003). Detecting timing jitter should be carried out at a high harmonic of f_r to attain much-enhanced detection sensitivity. The second step in achieving phase locking of separate femtosecond lasers requires effective stabilization of the phase difference between the two optical carrier waves (Shelton et al., 2001, 2002b; Holman et al., 2003b; Bartels et al., 2004b; Kobayashi et al., 2005). Phase locking demands that the spectral combs of individual lasers be maintained exactly coincident in the region of spectral overlap so that the two sets of combs form a continuous and phase-coherent entity. A coherent heterodyne-beat signal between the frequency combs can be used to yield information about the difference in the offset frequencies, which can then be controlled, resulting in two pulse trains with nearly identical phase evolution.

Phase-coherent femtosecond combs have been extended to the mid-infrared spectral region (Foreman et al., 2003, 2005; Zimmermann et al., 2004). Being able to combine the characteristics of two or more pulsed lasers working at different wavelengths provides a flexible approach to coherent control and it is particularly important for difficult to reach spectral regions. Two independent, pulse-synchronized, and phase-locked mode-locked Ti:sapphire lasers can be used in a difference-frequency generation (DFG) experiment to produce phase coherent frequency combs in the mid-IR region, with the wavelength spanning a few to tens of microns. The capability of expanding the spectral coverage of frequency combs while maintain phase coherence is an important ingredient for the goal of optical arbitrary waveform synthesizer. The other important component is precise spectral shaping, including both amplitude and phase. Indeed, an optical arbitrary waveform synthesizer will be the ultimate tool for DFCS, permitting precision spectroscopy and high-resolution quantum control of matter across a vast range of energy scale.

7. Concluding Remarks

Once wide-bandwidth, precisely stabilized optical frequency combs became available in laboratories, direct frequency comb spectroscopy has made rapid progress and found many different applications. Its future impact remains promising for a number of scientific and technological fields, including precision spectroscopy, high-resolution quantum control, arbitrary optical waveform synthesis, long-distance communications, and extreme nonlinear physics. Development of frequency combs at various spectral regions, with the capability of controlling both the time-domain coherence and spectral-domain phase, will no doubt make DFCS an attractive tool to study many atomic, molecular, and condensed matter systems.

8. Acknowledgements

We sincerely thank many of our colleagues and coworkers who have made critical contributions to the work reported in this review. In particular, R.J. Jones led the initial effort and, along with K.D. Moll, T.R. Schibli, D. Yost, and D. Hudson, carried out a large portion of the femtosecond enhancement cavity work. A. Marian, D. Felinto, and J. Lawall are important contributors to the initial development of broadband DFCS work. A. Bartels, T. Fortier, L. Hollberg, Y. Le Coq, V. Mbele, C. Oates, D. Ortega, and C. Tanner are acknowledged for their invaluable contributions to the Ca and Cs spectroscopy, and the VIPA-based spectrometer. We benefited from useful discussions with E. Eyler, J.L. Hall, S.T. Cundiff, F. Cruz and A. Weiner. Low-loss, low-dispersion, and wide-bandwidth mirrors are provided by R. Lalezari of Advanced Thin Films. We also thank A. Fernandez, A. Apolonski, F. Krausz, I. Hartl, and M. Fermann for providing high energy femtosecond oscillators. Finally, we gratefully acknowledge funding support from NIST, AFOSR, ONR, NSF, and NASA. A. Pe'er thanks the Fulbright foundation for financial support. J.E. Stalnaker gratefully acknowledges the support of the National Research Council. Finally, we thank T. Schibli and T. Fortier for their many useful comments on the manuscript.

9. References

Anderson, D.Z., Frisch, J.C., Masser, C.S. (1984). Mirror reflectometer based on optical cavity decay time. *Appl. Opt.* **23**, 1238–1245.
Apolonski, A., Poppe, A., Tempea, G., Spielmann, C., Udem, T., Holzwarth, R., Hänsch, T.W., Krausz, F. (2000). Controlling the phase evolution of few-cycle light pulses. *Phys. Rev. Lett.* **85**, 740–743.
Arissian, L., Diels, J.-C. (2006). Repetition rate spectroscopy of the dark line resonance in Rubidium. *Opt. Comm.* **264**, 169–173.
Aumiler, D., Ban, T., Skenderović, H., Pichler, G. (2005). Velocity selective optical pumping of Rb hyperfine lines induced by a train of femtosecond pulses. *Phys. Rev. Lett.* **95**. 233001.
Baklanov, Y.V., Chebotayev, V.P. (1977). Narrow resonances of two-photon absorption of supernarrow pulses in a gas. *Appl. Phys.* **12**, 97–99.
Balling, P., Maas, D.J., Noordam, L.D. (1994). Interference in climbing a quantum ladder system with frequency-chirped pulses. *Phys. Rev. A* **50**, 4276–4285.
Ban, T., Aumiler, D., Skenderovic, H., Pichler, G. (2006). Mapping of the optical frequency comb to the atom-velocity comb. *Phys. Rev. A* **73**. 043407-1-10.
Baltuška, A., Udem, T., Uiberacker, M., Hentschel, M., Goulielmakis, E., Gohle, C., Holzwarth, R., Yakovlev, V.S., Scrinzi, A., Hänsch, T.W., Krausz, F. (2003). Attosecond control of electronic processes by intense light fields. *Nature* **421**, 611–615.
Bardeen, C.J., Wang, Q., Shank, C.V. (1995). Selective excitation of vibrational wave-packet motion using chirped pulses. *Phys. Rev. Lett.* **75**, 3410–3413.
Bartels, A., Oates, C.W., Hollberg, L., Diddams, S.A. (2004a). Stabilization of femtosecond laser frequency combs with subhertz residual linewidths. *Opt. Lett.* **29**, 1081–1083.
Bartels, A., Newbury, N.R., Thomann, I., Hollberg, L., Diddams, S.A. (2004b). Broadband phase-coherent optical frequency synthesis with actively linked Ti:sapphire and Cr:forsterite femtosecond lasers. *Opt. Lett.* **29**, 403–405.
Bergeson, S.D., Baldwin, K.G.H., Lucatorto, T.B., McIlrath, T.J., Cheng, C.H., Eyler, E.E. (2000). Doppler-free two-photon spectroscopy in the vacuum ultraviolet: Helium 1^1S–2^1S transition. *J. Opt. Soc. Am. B* **17**, 1599–1606.

Bordé, C.J., Salomon, Ch., Avrillier, S., Van Lerberghe, A., Bréant, Ch., Bassi, D., Scoles, G. (1984). Optical Ramsey fringes with traveling waves. *Phys. Rev. A* **30**, 1836–1848.

Boyd, M.M., Zelevinsky, T., Ludlow, A.D., Foreman, S.M., Blatt, S., Ido, T., Ye, J. (2006). Optical atomic coherence at the 1-second time scale. *Science* **314**, 1430–1433.

Brattke, S., Kallmann, U., Hartmann, W.-D. (1998). Coherent dark states of Rubidium 87 in a buffer gas using pulsed laser light. *Eur. Phys. J. D* **3**, 159–161.

Cao, J., Bardeen, C.J., Wilson, K.R. (1998). Molecular "π pulse" for total inversion of electronic state population. *Phys. Rev. Lett.* **80**, 1406–1409.

Casperson, L.W. (1998). Few-cycle pulses in two-level media. *Phys. Rev. A* **57**, 609–621.

Chatel, B., Degert, J., Stock, S., Girard, B. (2003). Competition between sequential and direct paths in a two-photon transition. *Phys. Rev. A* **68**. 041402-1-4.

Chatel, B., Degert, J., Girard, B. (2004). Role of quadratic and cubic spectral phases in ladder climbing with ultrashort pulses. *Phys. Rev. A* **70**. 053414-1-10.

Chen, L.-S., Ye, J. (2003). Extensive, high-resolution measurement of hyperfine interactions: Precise investigations of molecular potentials and wave functions. *Chem. Phys. Lett.* **381**, 777–783.

Chen, L.-S., Cheng, W.Y., Ye, J. (2004). Hyperfine interactions and perturbation effects in the $B0_u^+(3\Pi_u)$ state of $^{127}I_2$. *J. Opt. Soc. Am. B* **21**, 820–832.

Chen, L.-S., de Jong, W.A., Ye, J. (2005). Characterization of the molecular iodine electronic wave functions and potential energy curves through hyperfine interactions in $B0_u^+(3\Pi_u)$ state. *J. Opt. Soc. Am. B* **22**, 951–961.

Corkum, P.B. (1993). Plasma perspective on strong-field multiphoton ionization. *Phys. Rev. Lett.* **71**, 1994–1997.

Couillaud, B., Ducasse, A., Sarger, L., Boscher, D. (1980). High-resolution saturated absorption spectroscopy with coherent trains of short light pulses. *Appl. Phys. Lett.* **36**, 407–409.

Crosson, E.R., Haar, P., Marcus, G.A., Schwettman, H.A., Paldus, B.A., Spence, T.G., Zare, R.N. (1999). Pulse-stacked cavity ring-down spectroscopy. *Rev. Sci. Instrum.* **70**, 4–10.

Cruz, F.C., Stowe, M.C., Ye, J. (2006). Tapered semiconductor amplifiers for optical frequency combs in the near infrared. *Opt. Lett.* **31**, 1337–1339.

Cundiff, S.T., Ye, J. (2003). Colloquium: Femtosecond optical frequency combs. *Rev. Mod. Phys.* **75**, 325–342.

Cundiff, S.T., Ye, J., Hall, J.L. (2001). Optical frequency synthesis based on mode-locked lasers. *Rev. Sci. Instrum.* **72**, 3749–3771.

de Beauvoir, B., Schwob, C., Acef, O., Jozefowski, L., Hilico, L., Nez, F., Julien, L., Clairon, A., Biraben, F. (2000). Metrology of the hydrogen and deuterium atoms: Determination of the Rydberg constant and Lamb shifts. *Eur. Phys. J. D* **12**, 61–93.

Degenhardt, C., Stoeher, H., Lisdat, C., Wilpers, G., Schnatz, H., Lipphardt, B., Nazarova, T., Pottie, P.-E., Sterr, U., Helmcke, J., Riehle, F. (2005). Calcium optical frequency standard with ultracold atoms: Approaching 10^{-15} relative uncertainty. *Phys. Rev. A* **72**. 62111-1-7.

Delagnes, J.C., Blanchet, V., Bouchene, M.A. (2003). Role of the radiated field in the propagation of an ultra-short chirped pulse. *Opt. Comm.* **227**, 125–131.

Diddams, S.A., Diels, J.-C., Atherton, B.W. (1998). Differential intracavity phase spectroscopy and its application to a three-level system in samarium. *Phys. Rev. A* **58**, 2252–2264.

Diddams, S.A., Jones, D.J., Ye, J., Cundiff, S.T., Hall, J.L., Ranka, J.K., Windeler, R.S., Holzwarth, R., Udem, T., Hänsch, T.W. (2000). Direct link between microwave and optical frequencies with a 300 THz femtosecond laser comb. *Phys. Rev. Lett.* **84**, 5102–5105.

Diddams, S.A., Udem, T., Bergquist, J.C., Curtis, E.A., Drullinger, R.E., Hollberg, L., Itano, W.M., Lee, W.D., Oates, C.W., Vogel, K.R., Wineland, D.J. (2001). An optical clock based on a single trapped $^{199}Hg^+$ ion. *Science* **293**, 825–828.

Diddams, S., Hollberg, L., Mbele, V. (2006). High-resolution spectral fingerprinting with a stabilized femtosecond laser frequency comb. In: *15th International Conference on Ultrafast Phenomena, Technical Digest (CD)*. Optical Society of America. Paper WA3.

Diddams, S., Hollberg, L., Mbele, V. (2007). Molecular fingerprinting with the resolved modes of a femtosecond laser frequency comb. *Nature* **445**, 627–630.

Drever, R.W.P., Hall, J.L., Kowalski, F.V., Hough, J., Ford, G.M., Munley, A.J., Ward, H. (1983). Laser phase and frequency stabilization using an optical resonator. *Appl. Phys. B* **31**, 97–105.

Dudley, J.M., Genty, G., Coen, S. (2006). Supercontinuum generation in photonic crystal fiber. *Rev. Mod. Phys.* **78**, 1135–1184.

Dudovich, N., Polack, T., Pe'er, A., Silberberg, Y. (2005). Simple route to strong field control. *Phys. Rev. Lett.* **94**. 083002-1-4.

Eckstein, J.N., Ferguson, A.I., Hänsch, T.W. (1978). High-resolution 2-photon spectroscopy with picosecond light-pulses. *Phys. Rev. Lett.* **40**, 847–850.

Eikema, K.S.E., Ubachs, W., Vassen, W., Hogervorst, W. (1997). Lamb shift measurement in the 1 S_1 ground state of Helium. *Phys. Rev. A* **55**, 1866–1884.

Ell, R., Morgner, U., Kartner, F.X., Fujimoto, J.G., Ippen, E.P., Scheuer, V., Angelow, G., Tschudi, T., Lederer, M.J., Boiko, A., Luther-Davies, B. (2001). Generation of 5-fs pulses and octave-spanning spectra directly from a Ti:sapphire laser. *Opt. Lett.* **26**, 373–375.

Felinto, D., Acioli, L.H., Vianna, S.S. (2004). Accumulative effects in the coherence of three-level atoms excited by femtosecond-laser frequency combs. *Phys. Rev. A* **70**. 043403-1-10.

Ferguson, A.I., Eckstein, J.N., Hänsch, T.W. (1979). Polarization spectroscopy with ultrashort light pulses. *Appl. Phys.* **18**, 257–260.

Foreman, S.M., Jones, D.J., Ye, J. (2003). Flexible and rapidly configurable femtosecond pulse generation in the mid-IR. *Opt. Lett.* **28**, 370–372.

Foreman, S.M., Marian, A., Ye, J., Petrukhin, E.A., Gubin, M.A., Mücke, O.D., Wong, F.N.C., Ippen, E.P., Kärtner, F.X. (2005). Demonstration of a HeNe/CH_4-based optical molecular clock. *Opt. Lett.* **30**, 570–572.

Foreman, S.M., Holman, K.W., Hudson, D.D., Jones, D.J., Ye, J. (2007). Remote transfer of ultrastable frequency references via fiber networks. *Rev. Sci. Instrum.* **78**. 021101-1-25.

Fortier, T.M., Jones, D.J., Cundiff, S.T. (2003). Phase stabilization of an octave-spanning Ti:sapphire laser. *Opt. Lett.* **28**, 2198–2200.

Fortier, T.M., Bartels, A., Diddams, S.A. (2006a). Octave-spanning Ti:sapphire laser with a repetition rate >1 GHz for optical frequency measurements and comparisons. *Opt. Lett.* **31**, 1011–1013.

Fortier, T.M., Le Coq, Y., Stalnaker, J.E., Ortega, D., Diddams, S.A., Oates, C.W., Hollberg, L. (2006b). Kilohertz-resolution spectroscopy of cold atoms with an optical frequency comb. *Phys. Rev. Lett.* **97**. 163905-1-4.

Fukuda, Y., Hayashi, J., Kondo, K., Hashi, T. (1981). Synchronized quantum beat spectroscopy using periodic impact excitations with CW mode-locked laser pulses. *Opt. Comm.* **38**, 357–360.

Garraway, B.M., Suominen, K. (1995). Wave-packet dynamics: new physics and chemistry in femto-time. *Rep. Prog. Phys.* **58**, 365–419.

Gaubatz, U., Rudecki, P., Schiemann, S., Bergmann, K. (1990). Population transfer between molecular vibrational levels by stimulated Raman scattering with partially overlapping laser fields. A new concept and experimental results. *J. Chem. Phys.* **92**, 5363–5376.

Genkin, G.M. (1998). Rabi frequency and nonlinearity of a two-level atom for an ultrashort optical pulse. *Phys. Rev. A* **58**, 758–760.

Gerginov, V., Tanner, C.E., Diddams, S., Bartels, A., Hollberg, L. (2004). Optical frequency measurements of $6s\,^2S_{1/2} \to 6p\,^2P_{3/2}$ transition in a ^{133}Cs atomic beam using a femtosecond laser frequency comb. *Phys. Rev. A* **70**. 42505-1-8.

Gerginov, V., Tanner, C.E., Diddams, S.A., Bartels, A., Hollberg, L. (2005). High-resolution spectroscopy with a femtosecond laser frequency comb. *Opt. Lett.* **30**, 1734–1736.

Gerginov, V., Calkins, K., Tanner, C.E., McFerran, J.J., Diddams, S., Bartels, A., Hollberg, L. (2006). Optical frequency measurements of $6s\,^2S_{1/2} \to 6p\,^2P_{1/2}(D_1)$ transitions in ^{133}Cs and their impact on the fine-structure constant. *Phys. Rev. A* **73**. 32504-1-10.

Gherman, T., Eslami, E., Romanini, D., Kassi, S., Vial, J.-C., Sadeghi, N. (2004). High sensitivity broad-band mode-locked cavity-enhanced absorption spectroscopy: Measurement of Ar*($^3P_2$) atom and N_2^+ ion densities. *J. Phys. D* **37**, 2408–2415.

Gohle, C., Udem, T., Herrmann, M., Rauschenberger, J., Holzwarth, R., Schuessler, H.A., Krausz, F., Hänsch, T.W. (2005). A frequency comb in the extreme ultraviolet. *Nature* **436**, 234–237.

Greenland, P.T. (1983). Resonances excited by a train of delta function pulses. *J. Phys. B: At. Mol. Phys.* **16**, 2515–2521.

Hall, J.L. (2006). Nobel lecture: Defining and measuring optical frequencies. *Rev. Mod. Phys.* **78**, 1279–1295.

Hall, J.L., Bordé, C.J., Uehara, K. (1976). Direct optical resolution of the recoil effect using saturated absorption spectroscopy. *Phys. Rev. Lett.* **37**, 1339–1342.

Halzwarth, R., Nevsky, A.Yu., Zimmermann, M., Udem, Th., Hänsch, T.W., Von Zanthier, J., Walther, H., Knight, J.C., Wadsworth, W.J., Russell, P.St.J., Skvortsov, M.N., Bagayev, S.N. (2001). Absolute frequency measurement of iodine lines with a femtosecond optical synthesizer. *Appl. Phys. B* **73**, 269–271.

Hänsch, T.W. (2006). Nobel lecture: Passion for precision. *Rev. Mod. Phys.* **78**, 1297–1309.

Hänsch, T.W., Wong, N.C. (1980). Two-photon spectroscopy with an FM laser. *Metrologia* **16**, 101–104.

Harde, H., Burggraf, H. (1982). Ultrahigh-resolution coherence spectroscopy by means of periodic excitation with picosecond pulses. *Opt. Comm.* **40**, 441–445.

Heavner, T.P., Jefferts, S.R., Donley, E.A., Shirley, J.H., Parker, T.E. (2005). NIST-F1: Recent evaluations and accuracy improvements. *Metrologia* **42**, 411–422.

Holman, K.W., Jones, R.J., Marian, A., Cundiff, S.T., Ye, J. (2003a). Intensity-related dynamics of femtosecond frequency combs. *Opt. Lett.* **28**, 851–853.

Holman, K.W., Jones, D.J., Ye, J., Ippen, E.P. (2003b). Orthogonal control of the frequency comb dynamics of a mode-locked laser diode. *Opt. Lett.* **28**, 2405–2407.

Holman, K.W., Jones, D.J., Hudson, D.D., Ye, J. (2004). Precise frequency transfer through a fiber network by use of 1.5 µm mode-locked sources. *Opt. Lett.* **29**, 1554–1556.

Hudson, D.D., Foreman, S.M., Cundiff, S.T., Ye, J. (2006). Synchronization of mode-locked femtosecond lasers through a fiber link. *Opt. Lett.* **31**, 1951–1953.

Jenkins, F.A., White, H.E. (1976). "Fundamentals of Optics", 4th edition. McGraw Hill, New York.

Jonas, D.M. (2003). Two-dimensional femtosecond spectroscopy. *Annual Rev. Phys. Chem.* **54**, 425–463.

Jones, R.J., Diels, J.-C. (2001). Stabilization of femtosecond lasers for optical frequency metrology and direct optical to radio frequency synthesis. *Phys. Rev. Lett.* **86**, 3288–3291.

Jones, R.J., Ye, J. (2002). Femtosecond pulse amplification by coherent addition in a passive optical cavity. *Opt. Lett.* **27**, 1848–1850.

Jones, R.J., Ye, J. (2004). High-repetition-rate coherent femtosecond pulse amplification with an external passive optical cavity. *Opt. Lett.* **29**, 2812–2814.

Jones, D.J., Diddams, S.A., Ranka, J.K., Stentz, A., Windeler, R.S., Hall, J.L., Cundiff, S.T. (2000). Carrier–envelope phase control of femtosecond mode-locked lasers and direct optical frequency synthesis. *Science* **288**, 635–639.

Jones, R.J., Thomann, I., Ye, J. (2004). Precision stabilization of femtosecond lasers to high-finesse optical cavities. *Phys. Rev. A* **69**. 051803-1-4.

Jones, R.J., Moll, K.D., Thorpe, M.J., Ye, J. (2005). Phase-coherent frequency combs in the vacuum ultraviolet via high-harmonic generation inside a femtosecond enhancement cavity. *Phys. Rev. Lett.* **94**. 193201-1-4.

Kielpinski, D. (2006). Laser cooling of atoms and molecules with ultrafast pulses. *Phys. Rev. A* **73**. 063407-1-6.

Kobayashi, Y., Yoshitomi, D., Kakehata, M., Takada, H., Torizuka, K. (2005). Long-term optical phase locking between femtosecond Ti:sapphire and Cr:forsterite lasers. *Opt. Lett.* **30**, 2496–2498.

Lee, K.F., Villeneuve, D.M., Corkum, P.B., Shapiro, E.A. (2004). Phase control of rotational wave packets and quantum information. *Phys. Rev. Lett.* **93**. 233601-1-4.

Lewenstein, M., Balcou, P., Ivanov, M.Y., Lhuillier, A., Corkum, P.B. (1994). Theory of high-harmonic generation by low-frequency laser fields. *Phys. Rev. A* **49**, 2117–2132.

Ludlow, A.D., Boyd, M.M., Zelevinsky, T., Foreman, S.M., Blatt, S., Notcutt, M., Ido, T., Ye, J. (2006). Systematic study of the ^{87}Sr clock transition in an optical lattice. *Phys. Rev. Lett.* **96**. 033003-1-4.

Ma, L.S., Shelton, R.K., Kapteyn, H.C., Murnane, M.M., Ye, J. (2001). Sub-10-femtosecond active synchronization of two passively mode-locked Ti:sapphire oscillators. *Phys. Rev. A* **64**. 021802 -1-4.

Ma, L.-S., Zucco, M., Picard, S., Robertsson, L., Windeler, R.S. (2003). A new method to determine the absolute mode number of a mode-locked femtosecond-laser comb used for absolute optical frequency measurements. *IEEE Journal of Selected Topics in Quantum Electronics* **9**, 1066–1071.

Malinovsky, V.S., Krause, J.L. (2001). Efficiency and robustness of coherent population transfer with intense, chirped laser pulses. *Phys. Rev. A* **63**. 043415-1-8.

Marian, A., Stowe, M.C., Lawall, J.R., Felinto, D., Ye, J. (2004). United time-frequency spectroscopy for dynamics and global structure. *Science* **306**, 2063–2068.

Marian, A., Stowe, M.C., Felinto, D., Ye, J. (2005). Direct frequency comb measurements of absolute optical frequencies and population transfer dynamics. *Phys. Rev. Lett.* **95**. 023001-1-4.

Mbele, V., Stalnaker, J.E., Gerginov, V., Diddams, S.A., Tanner, C.E., Hollberg, L. (in preparation).

Meshulach, D., Silberberg, Y. (1998). Coherent quantum control of two-photon transitions by a femtosecond laser pulse. *Nature* **396**, 239–242.

Minoshima, K., Matsumoto, H. (2000). High-accuracy measurement of 240-m distance in an optical tunnel by use of a compact femtosecond laser. *Appl. Opt.* **39**, 5512–5517.

Mlynek, J., Lange, W., Harde, H., Burggraf, H. (1981). High-resolution coherence spectroscopy using pulse trains. *Phys. Rev. A* **24**, 1099–1102.

Moll, K.D., Jones, R.J., Ye, J. (2005). Nonlinear dynamics inside femtosecond enhancement cavities. *Opt. Express* **13**, 1672–1678.

Moll, K.D., Jones, R.J., Ye, J. (2006). Output coupling methods for cavity-based high-harmonic generation. *Opt. Express* **14**, 8189–8197.

Morgner, U., Ell, R., Metzler, G., Schibli, T.R., Kärtner, F.X., Fujimoto, J.G., Haus, H.A., Ippen, E.P. (2001). Nonlinear optics with phase-controlled pulses in the sub-two-cycle regime. *Phys. Rev. Lett.* **86**, 5462–5465.

Mücke, O.D., Ell, R., Winter, A., Kim, J.W., Birge, J.R., Matos, L., Kartner, F.X. (2005). Self-referenced 200 MHz octave-spanning Ti:sapphire laser with 50 attosecond carrier–envelope phase jitter. *Opt. Exp.* **13**, 5163–5169.

Naumov, S., Fernandez, A., Graf, R., Dombi, P., Krausz, F., Apolonski, A. (2005). Approaching the microjoule frontier with femtosecond laser oscillators. *New J. Phys.* **7**, 216.

Oates, C.W., Curtis, E.A., Hollberg, L. (2000). Improved short-term stability of optical frequency standards: Approaching 1 Hz in 1 s with the Ca standard at 657 nm. *Opt. Lett.* **25**, 1603–1605.

Oates, C.W., Wilpers, G., Hollberg, L. (2005). Observation of large atomic-recoil-induced asymmetries in cold atom spectroscopy. *Phys. Rev. A* **71**, 23404-1-6.

O'Keefe, A., Deacon, D.A.G. (1988). Cavity ringdown optical spectrometer for absorption-measurements using pulsed laser sources. *Rev. Sci. Instrum.* **59**, 2544–2551.

Paulus, G.G., Grasbon, F., Walther, H., Villoresi, P., Nisoli, M., Stagira, S., Priori, E., De Silvestri, S. (2001). Absolute-phase phenomena in photoionization with few-cycle laser pulses. *Nature* **414**, 182–184.

Pe'er, A., Shapiro, E.A., Stowe, M.C., Shapiro, M., Ye, J. (2007). Precise control of molecular dynamics with a femtosecond frequency comb—A weak field route to strong field coherent control. *Phys. Rev. Lett.* **98**. 113004-1-4.

Potma, E.O., Jones, D.J., Cheng, J.X., Xie, X.S., Ye, J. (2002). High-sensitivity coherent anti-Stokes Raman scattering microscopy with two tightly synchronized picosecond lasers. *Opt. Lett.* **27**, 1168–1170.

Potma, E.O., Evans, C., Xie, X.S., Jones, R.J., Ye, J. (2003). High-repetition rate picosecond pulse amplification with an external passive optical cavity. *Opt. Lett.* **28**, 1835–1837.

Potma, E.O., Xie, X.S., Muntean, L., Preusser, J., Jones, D.J., Ye, J., Leone, S.R., Hinsberg, W.D., Schade, W. (2004). Chemical imaging of photoresists with coherent anti-Stokes Raman scattering (CARS) microscopy. *J. Phys. Chem. B* **108**, 1296–1301.

Rabitz, H., Vivie-Riedle, R., Motzkus, M., Kompa, K. (2000). Whither the future of controlling quantum phenomena? *Science* **288**, 824–828.

Ramond, T.M., Diddams, S.A., Hollberg, L. (2002). Phase-coherent link from optical to microwave frequencies by means of the broadband continuum from a 1-GHz Ti:sapphire femtosecond oscillator. *Opt. Lett.* **27**, 1842–1844.

Ranka, J.K., Windeler, R.S., Stentz, A.J. (2000). Visible continuum generation in air-silica microstructure optical fibers with anomalous dispersion at 800 nm. *Opt. Lett.* **25**, 25–27.

Reichert, J., Holzwarth, R., Udem, T., Hänsch, T.W. (1999). Measuring the frequency of light with mode-locked lasers. *Opt. Comm.* **172**, 59–68.

Rice, S.A., Zhao, M. (2000). "Optical Control of Molecular Dynamics". John Wiley & Sons, New York.

Salour, M.M., Cohen-Tannoudji, C. (1977). Observation of Ramsey's interference fringes in the profile of Doppler-free two-photon resonances. *Phys. Rev. Lett.* **38**, 757–760.

Sautenkov, V.A., Rostovtsev, Y.V., Ye, C.Y., Welch, G.R., Kocharovskaya, O., Scully, M.O. (2005). Electromagnetically induced transparency in rubidium vapor prepared by a comb of short optical pulses. *Phys. Rev. A* **71**. 063804-1-4.

Scherer, J.J., Paul, J.B., Jiao, H., O'Keefe, A. (2001). Broadband ringdown spectral photography. *Appl. Opt.* **40**, 6725–6732.

Schibli, T.R., Kim, J., Kuzucu, O., Gopinath, J.T., Tandon, S.N., Petrich, G.S., Kolodziejski, L.A., Fujimoto, J.G., Ippen, E.P., Kärtner, F.X. (2003). Attosecond active synchronization of passively mode-locked lasers by balanced cross correlation. *Opt. Lett.* **28**, 947–949.

Schliesser, A., Gohle, C., Udem, T., Hänsch, T.W. (2006). Complete characterization of a broadband high-finesse cavity using an optical frequency comb. *Opt. Exp.* **14**, 5975–5983.

Shapiro, M., Brumer, P. (1986). Laser control of product quantum state populations in unimolecular reactions. *J. Chem. Phys.* **84**, 4103–4104.

Shapiro, M., Brumer, P. (2003a). "Principles of the Quantum Control of Molecular Processes". John Wiley & Sons, Hoboken.

Shapiro, M., Brumer, P. (2003b). Coherent control of molecular dynamics. *Rep. Prog. Phys.* **66**, 859–942.

Shelton, R.K., Ma, L.S., Kapteyn, H.C., Murnane, M.M., Hall, J.L., Ye, J. (2001). Phase-coherent optical pulse synthesis from separate femtosecond lasers. *Science* **293**, 1286–1289.

Shelton, R.K., Foreman, S.M., Ma, L.S., Hall, J.L., Kapteyn, H.C., Murnane, M.M., Notcutt, M., Ye, J. (2002a). Subfemtosecond timing jitter between two independent, actively synchronized, mode-locked lasers. *Opt. Lett.* **27**, 312–314.

Shelton, R.K., Ma, L.S., Kapteyn, H.C., Murnane, M.M., Hall, J.L., Ye, J. (2002b). Active synchronization and carrier phase locking of two separate mode-locked femtosecond lasers. *J. Mod. Opt.* **49**, 401–409.

Shirasaki, M. (1996). Large angular dispersion by a virtually imaged phased array and its application to a wavelength demultiplexer. *Opt. Lett.* **21**, 366–368.

Snadden, M.J., Bell, A.S., Riis, E., Ferguson, A.I. (1996). Two-photon spectroscopy of laser-cooled Rb using a mode-locked laser. *Opt. Comm.* **125**, 70–76.

Stenger, J., Schnatz, H., Tamm, C., Telle, H.R. (2002). Ultraprecise measurement of optical frequency ratios. *Phys. Rev. Lett.* **88**. 073601-1-4.

Stowe, M.C., Cruz, F.C., Marian, A., Ye, J. (2006). High resolution atomic coherent control via spectral phase manipulation of an optical frequency comb. *Phys. Rev. Lett.* **96**. 153001-1-4.

Tannor, D.J., Rice, S.A. (1985). Control of selectivity of chemical reaction via control of wave packet evolution. *J. Chem. Phys.* **83**, 5013–5018.

Teets, R., Eckstein, J., Hänsch, T.W. (1977). Coherent 2-Photon excitation by multiple light pulses. *Phys. Rev. Lett.* **38**, 760–764.

Telle, H.R., Steinmeyer, G., Dunlop, A.E., Stenger, J., Sutter, D.H., Keller, U. (1999). Carrier–envelope offset phase control: A novel concept for absolute optical frequency measurement and ultrashort pulse generation. *App. Phys. B* **69**, 327–332.

Thorpe, M.J., Jones, R.J., Moll, K.D., Ye, J., Lalezari, R. (2005). Precise measurements of optical cavity dispersion and mirror coating properties via femtosecond combs. *Opt. Express* **13**, 882–888.

Thorpe, M.J., Moll, K.D., Jones, R.J., Safdi, B., Ye, J. (2006). Broadband cavity ringdown spectroscopy for sensitive and rapid molecular detection. *Science* **311**, 1595–1599.

Thorpe, M.J., Hudson, D.D., Moll, K.D., Lasri, J., Ye, J. (2007). Cavity-ringdown molecular spectroscopy based on an optical frequency comb at 1.45–1.65 μm. *Opt. Lett.* **32**, 307–309.

Udem, T., Reichert, J., Holzwarth, R., Hänsch, T.W. (1999a). Absolute optical frequency measurement of the Cesium D-1 line with a mode-locked laser. *Phys. Rev. Lett.* **82**, 3568–3571.

Udem, T., Reichert, J., Holzwarth, R., Hänsch, T.W. (1999b). Accurate measurement of large optical frequency differences with a mode-locked laser. *Opt. Lett.* **24**, 881–883.

Vidne, Y., Rosenbluh, M., Hänsch, T.W. (2003). Pulse picking by phase-coherent additive pulse generation in an external cavity. *Opt. Lett.* **28**, 2396–2398.

Vitanov, N.V., Knight, P.L. (1995). Coherent excitation of a two-state system by a train of short pulses. *Phys. Rev. A* **52**, 2245–2261.

Vitanov, N.V., Fleischhauer, M., Shore, B.W., Bergmann, K. (2001). Coherent manipulation of atoms and molecules by sequential laser pulses. *Adv. At. Mol. Opt. Phys.* **46**, 55–190.

Wang, S.X., Xiao, S., Weiner, A.M. (2005). Broadband, high spectral resolution 2-D wavelength-parallel polarimeter for dense WDM systems. *Opt. Express* **13**, 9374–9380.

Weber, K.-H., Sansonetti, C.J. (1987). Accurate energies of nS, nP, nD, nF, and nG levels of neutral cesium. *Phys. Rev. A* **35**, 4650–4660.

Witte, S., Zinkstok, R.T., Ubachs, W., Hogervorst, W., Eikema, K.S.E. (2005). Deep-ultraviolet quantum interference metrology with ultrashort laser pulses. *Science* **307**, 400–403.

Xiao, S., Weiner, A.M. (2004). 2-D wavelength demultiplexer with potential for $\geqslant 1000$ channels in the C-band. *Opt. Express* **12**, 2895–2901.

Ye, J. (2004). Absolute measurement of a long, arbitrary distance to less than an optical fringe. *Opt. Lett.* **29**, 1153–1155.

Ye, J., Cundiff, S.T. (Eds.) (2005). *Femtosecond Optical Frequency Comb Technology: Principle, Operation, and Application*. Springer, New York.

Ye, J., Hall, J.L. (2000). Cavity ringdown heterodyne spectroscopy: High sensitivity with microwatt light power. *Phys. Rev. A* **61**. 061802-1-4(R).

Ye, J., Ma, L.S., Hall, J.L. (1998). Ultrasensitive detections in atomic and molecular physics – Demonstration in molecular overtone spectroscopy. *J. Opt. Soc. Am. B* **15**, 6–15.

Ye, J., Hall, J.L., Diddams, S.A. (2000). Precision phase control of an ultrawide-bandwidth femtosecond laser: A network of ultrastable frequency marks across the visible spectrum. *Opt. Lett.* **25**, 1675–1677.

Ye, J., Ma, L.S., Hall, J.L. (2001). Molecular iodine clock. *Phys. Rev. Lett.* **87**, 270801-1-4.

Ye, J., Cundiff, S.T., Foreman, S., Fortier, T.M., Hall, J.L., Holman, K.W., Jones, D.J., Jost, J.D., Kapteyn, H.C., Leeuwen, K.A.H.V., Ma, L.-S., Murnane, M.M., Peng, J.L., Shelton, R.K. (2002). Phase-coherent synthesis of optical frequencies and waveforms. *Appl. Phys. B—Lasers & Opt.* **74**, S27–S34.

Ye, J., Schnatz, H., Hollberg, L.W. (2003). Optical frequency combs: From frequency metrology to optical phase control. *IEEE J. Sel. Top. Quant. Electron.* **9**, 1041–1058.

Yoon, T.H., Marian, A., Hall, J.L., Ye, J. (2000). Phase-coherent multilevel two-photon transitions in cold Rb atoms: Ultra-high resolution spectroscopy via frequency-stabilized femtosecond laser. *Phys. Rev. A* **63**, 011402-1-4.

Young, B.C., Cruz, F.C., Itano, W.M., Bergquist, J.C. (1999). Visible lasers with subhertz linewidths. *Phys. Rev. Lett.* **82**, 3799–3802.

Zewail, A.H. (2000). Femtochemistry: Atomic-scale dynamics of the chemical bond. *J. Phys. Chem. A* **104**, 5660–5694.

Zimmermann, M., Gohle, C., Holzwarth, R., Udem, T., Hänsch, T.W. (2004). Optical clockwork with an offset-free difference-frequency comb: Accuracy of sum- and difference-frequency generation. *Opt. Lett.* **29**, 310–312.

Zinkstok, R.Th., Witte, S., Ubachs, W., Hogervorst, W., Eikema, K.S.E. (2006). Frequency comb laser spectroscopy in the vacuum-ultraviolet region. *Phys. Rev. A* **73**. 061801-1-4.

Ziolkowski, R.W., Arnold, J.M., Gogny, D.M. (1995). Ultrafast pulse interactions with two-level atoms. *Phys. Rev. A* **52**, 3082–3094.

COLLISIONS, CORRELATIONS, AND INTEGRABILITY IN ATOM WAVEGUIDES

VLADIMIR A. YUROVSKY[1], MAXIM OLSHANII[2] and DAVID S. WEISS[3]

[1] School of Chemistry, Tel Aviv University, Tel Aviv 69978, Israel
[2] Department of Physics & Astronomy, University of Southern California, Los Angeles, CA 90089, USA
[3] Department of Physics, Pennsylvania State University, University Park, Pennsylvania 16802-6300, USA

1. Introduction 62
2. Effective 1D World 64
   2.1. Atom Waveguides 64
   2.2. Feshbach Resonance Interactions 68
   2.3. s-Wave Scattering 71
   2.4. p-Wave Scattering 81
3. Bethe Ansatz and beyond 85
   3.1. Many-Body 1D Problems 85
   3.2. Tonks–Girardeau Gas 86
   3.3. Bethe Ansatz 89
   3.4. Gross–Pitaevskii and Bogoliubov Methods 98
4. Ground State Properties of Short-Range-Interacting 1D Bosons: Known Results and Their Experimental Verification 103
   4.1. Thermodynamic Limit 103
   4.2. Correlation Functions 104
   4.3. Further Extensions 108
   4.4. Experimental Studies of 1D Quasicondensates 108
   4.5. Experimental Tests of the Lieb–Liniger Equation of State and Correlation Functions . 109
5. What Is Special about Physics in 1D 113
   5.1. Effects of Integrability and Non-Integrability 113
   5.2. Solitons 122
6. Summary and Outlook 125
7. Acknowledgements 126
8. Appendix: Some Useful Properties of the Hurwitz Zeta Function 126
9. References 127

Abstract

Elongated atom traps can confine ultracold gases in the quasi-one-dimensional regime. We review both the theory of these atom waveguides and their experimental realizations, with emphasis on the collisions of waveguide-bound particles. Under certain conditions, quasi-one-dimensional gases are well described by integrable one-dimensional many-body models with exact quantum solutions. We review the thermodynamic and correlation properties of one such model that has been experimentally realized, that of Lieb and Liniger. We describe the necessary criteria for integrability and its observable effects. We also consider ways to lift integrability, along with some observable effects of non-integrability.

1. Introduction

Since Bethe obtained exact solutions for the one-dimensional (1D) spin-1/2 Heisenberg model (Bethe, 1931), 1D systems of interacting particles have been widely studied theoretically, acting as something of a testing ground for the study of many-body physics. 1D many-body systems were also the subject of some of the first computer simulations (Fermi et al., 1955). Although experimental 1D condensed matter systems have been developed, most notably consisting of electrons in wires, these systems remain coupled to other degrees of freedom. They exhibit characteristic 1D phenomena, like spin–charge separation (Voit, 1995), but they do not realize one of the central features that has made 1D systems so theoretically popular, integrability. Ultracold atomic gases can be made to couple extremely weakly to the outside world. They can be trapped in perfectly well-known potentials, and their mutual interactions are very well understood. Accordingly, they have emerged as model systems for the study of a wide range of many-body phenomena, including Bose–Einstein condensation (BEC), degenerate Fermi gases, and the BCS state, formed by correlated pairs of fermions (see reviews by Parkins and Walls, 1998; Ketterle et al., 1999; Dalfovo et al., 1999; Courteille et al., 2001; Leggett, 2001; Yukalov, 2004; Regal and Jin, 2006; Bloch et al., 2007; the book by Pitaevskii and Stringari, 2003, and references therein). Ultracold atoms have also been used to make nearly perfect model 1D systems.

Atom waveguides are potentials that tightly confine a particle's motion in two transverse directions, leaving it free or nearly free in the third, axial, direction. Gases in these waveguides can be made cold enough that only the ground transverse mode is significantly populated. Particle motion in this regime is quasi-1D. Even with no real transverse excitations, virtual excitation of transverse waveguide modes must be taken into account to derive effective 1D interaction potentials (Olshanii, 1998). The motion can also involve a finite number of channels associated with transverse excitation. We are concerned in this review with both quasi-1D and idealized strictly-1D systems.

Effective 1D interaction potentials arise from underlying 3D interactions and depend on how tightly the atoms are trapped transversely, so they can be experimentally controlled in several ways. The 3D interaction strength can be changed using a Feshbach resonance (see Timmermans et al., 1999), which appears when the energy of a colliding pair in the open channel lies in the vicinity of a bound state in a closed channel. In a waveguide, the related phenomenon of confinement-induced resonance (CIR) occurs (Bergeman et al., 2003), where the excited transverse modes play the role of closed channels. Via a CIR, particles in an atom waveguide can form broad, quasi-1D, two-body bound states (Bergeman et al., 2003; Moritz et al., 2005), similar to broad Feshbach molecules (see Köhler et al., 2006).

Some, but not all, 1D many-body systems are integrable, and some of these long theoretically studied integrable models have now been experimentally realized. In the case of hard-core interactions, theoretical solutions can be obtained by mapping the interacting Bose gas onto an ideal Fermi gas (Girardeau, 1960) and vice versa. The corresponding quantum state is a strongly-interacting, Tonks–Girardeau (TG) gas. Many systems composed of particles with equal masses and interactions have Bethe-ansatz solutions. Such solutions exist for non-diffractive models with finite-range and zero-range interactions, and they include the well-known Lieb–Liniger, McGuire, Yang–Gaudin, and Calogero–Sutherland models. Even if a system consists of bosons, if it is non-diffractive then its ground state is not a BEC, but a Fermi sea, which is the phenomenon of fermionization of bosons in 1D. Integrability allows a system to be analytically solved, but it also leads to unique observable consequences. These include the existence of solitons—which are stable isolated waves, the suppression of thermalization, and the prohibition of chemical reactions (atom–molecule transformations).

1D systems do not, in general, have the long-range order that characterizes a BEC. However, they can exist in a quasicondensate regime (Petrov et al., 2000), where the 1D system can be divided into finite-length condensate segments, which have mutually uncorrelated phases. Other 1D and quasi-1D systems regimes exist, such as the true BEC regime and the 3D quasicondensate regime (Petrov et al., 2001; Menotti and Stringari, 2002; Bouchoule et al., 2007).

Quantum degenerate gases can be characterized by their correlation functions. For instance, the long-range order of a BEC corresponds to the non-zero long-range limit of a one-particle density matrix. Two- and three-body correlation functions can be directly related to inelastic collisions.

Integrability is not a universal feature of 1D models. For instance, both a CIR and a Feshbach resonance break integrability. Integrability can also be lifted by coupling several parallel waveguides, or by having effectively finite-range 1D interactions. Observable effects of non-integrability include thermalization and atom–molecule transformations. In some circumstances, non-integrability can also suppress inelastic processes.

Some features of quantum-degenerate 1D gases have analogies in non-linear optics. This is especially true for solitons, which have been extensively studied (see books by Kivshar and Agrawal, 2003; Malomed, 2006) both in Kerr media with cubic non-linearities, which are counterparts of atomic gases, and in media with quadratic nonlinearities, which are analogous to Feshbach-resonance interactions.

The properties of 1D quantum-degenerate gases can be also modified by applying a periodic axial potential. The addition of this potential gives rise to a host of phenomena, including the existence of a 1D Mott-insulator phase. The present review does not encompass systems with periodic axial potentials, which were the subject of a recent review by Morsch and Oberthaler (2006) and many other recent publications (Krämer et al., 2003; Orso et al., 2005; Rey et al., 2004, 2005a, 2005b, 2006, 2007; Gea-Banacloche et al., 2006).

Section 2 of the present review starts with an overview of two-body interactions in atom waveguides. This includes a survey of how atom waveguides are made in the laboratory, and discussions of Feshbach resonances, effective 1D interactions of even and odd parity (which correspond to s- and p-scattering in 3D), bound states and scattering amplitudes. Section 3 presents theoretical methods for many-body 1D problems. Major attention is focused on Fermi–Bose mapping, on the Bethe-ansatz method and on the limits of its applicability in the presence of resonant interactions, as well as on mean-field and Bogoliubov methods. A survey of theoretical results for Lieb–Liniger correlation functions is presented in Section 4, followed by a survey of experimental results for 1D Bose gases. Section 5 discusses thermalization, solitons and other observable effects of integrability and non-integrability. Throughout, we use a system of units in which Planck's constant is $\hbar = 1$.

2. Effective 1D World

2.1. ATOM WAVEGUIDES

2.1.1. Experimental Realization

1D atom trapping requires that all energies related to atomic motion and interactions be much smaller than the energy needed to excite the atom vibrationally in the transverse directions. So while the wavefunction of an atom is three dimensional (3D), it only has dynamics in one of them. To separate out the transverse trapping energy scale, it is necessary, but not sufficient, to create traps that are prolate and highly anisotropic. This has been accomplished for isolated traps using either magnetic or dipole forces. Optical lattices have been used to create arrays of 1D traps. We will here give an overview of the various options for making anisotropic atom traps.

Highly anisotropic magnetic traps can be made when the fields are not axially symmetric, which is also a requirement for a static field trap to have a non-zero field minimum. Such is the case for the Ioffe–Pritchard trap (Pritchard, 1983) and its variants, where the longitudinal and transverse spring constants can be separately adjusted. A magnetic waveguide formed above the surface of a current-carrying chip (Hänsel et al., 2001; Folman et al., 2002; Boyd et al., 2006) can have a very weak longitudinal spring constant along most of its length with steep potential walls near its ends. Focused laser beams provide a naturally anisotropic atom trap, although the more anisotropic they are, the weaker the transverse confinement (Görlitz et al., 2001). Although anisotropy does not ensure that a system is 1D, the 1D regime has been reached using each of the above types of traps. Experiments that have been performed include the formation of dark and bright solitons (Burger et al., 1999; Khaykovich et al., 2002; Strecker et al., 2002), studies of 1D Fermi–Bose mixtures (Schreck et al., 2001), and studies of phase and density fluctuations and correlations (Richard et al., 2003; Clement et al., 2005; Hugbart et al., 2005; Esteve et al., 2006).

Flat-bottomed light guides have been demonstrated (Meyrath et al., 2005), as have magnetic atom guides around a loop (Gupta et al., 2005; Arnold et al., 2006). These traps, which in principle avoid longitudinal trapping altogether, have not yet been used to reach the 1D regime.

Optical lattices (Jessen and Deutsch, 1996; Morsch and Oberthaler, 2006), because of the very short transverse length scale of each tube, can reach more decisively into the 1D regime. Three or four laser beams propagating in a plane can produce an array of anisotropic 1D traps. The most common arrangement is two orthogonal standing waves of slightly different frequencies. As long as the beat period is much shorter than other trapping time scales, the interference between the two standing waves washes out, and the lattice is insensitive to the relative phases between them (Winoto et al., 1999). This arrangement yields a square lattice of tubes. As long as the lattice is made deep enough that tunneling times exceed the time of the experiment, the tubes act as independent 1D systems.

The sign of the detuning with respect to the dominant atomic resonances significantly affects the anisotropy that can be obtained in an optical lattice. In a red-detuned lattice, where atoms are attracted to regions of high intensity, increasing the transverse trapping also increases the axial trapping. The central tubes made by a 2D lattice with wavevector k, beam waist w_z, and peak depth U_0 for each standing wave, have an axial spring constant given by

$$\kappa_{zr} = \frac{8U_0}{w_z^2}. \tag{2.1}$$

In a blue-detuned lattice, where atoms are repelled from regions of high intensity, atoms experience only the ac Stark shift associated with the transverse zero point energy. The light therefore acts as a weak anti-trap in the axial direction, with a (negative) spring constant,

$$\kappa_{zb} = -\kappa_{zr}\sqrt{\frac{k^2}{32U_0 m}}, \tag{2.2}$$

which has a much smaller magnitude than κ_{zr} for deep lattices. To make an array of trapped 1D gases, a blue-detuned lattice requires an additional axial trap, either magnetic or dipole. The axial trapping can thus be independently adjusted, allowing for higher trap anisotropy.

Optical lattices have been used in experiments to explore phase coherence (Greiner et al., 2001), to study reduced three-body association (Tolra et al., 2004), to observe the TG gas (Kinoshita et al., 2004), to study dipole oscillation damping (Fertig et al., 2005), to investigate pair correlations (Kinoshita et al., 2005), to observe confinement-induced quasi-1D molecules (Moritz et al., 2005), to study p-wave interactions (Gunter et al., 2005), and to investigate the effect of integrability on a system out of equilibrium (Kinoshita et al., 2006).

2.1.2. General Theoretical Description

Consider a system of two interacting particles in a 3D harmonic potential. It is described by the Hamiltonian

$$\hat{H} = \sum_{j=1}^{2}\left[-\frac{1}{2m_j}\nabla_j^2 + m_j \frac{(\omega_x^{(j)} x_j)^2 + (\omega_y^{(j)} y_j)^2 + (\omega_z^{(j)} z_j)^2}{2}\right] + V(\mathbf{r}_1 - \mathbf{r}_2), \tag{2.3}$$

where m_j are the particle masses, $\mathbf{r}_j$ are their radius-vectors with components x_j, y_j, and z_j, $\omega_{x,y,z}^{(j)}$ are trap oscillation frequencies, and $V(\mathbf{r}_1 - \mathbf{r}_2)$ is the interaction potential. Whenever the harmonic potentials for the two particles are equal, i.e.,

$$m_1\big(\omega_l^{(1)}\big)^2 = m_2\big(\omega_l^{(2)}\big)^2 \quad \text{for } l = x, y, z,$$

the Hamiltonian can be reexpressed as the sum of a center-of-mass term, $\hat{H}_{CM}$, and a relative motion term,

$$\hat{H} = \hat{H}_{CM} - \frac{1}{2\mu}\nabla_r^2 + \mu\frac{\omega_x^2 x^2 + \omega_y^2 y^2 + \omega_z^2 z^2}{2} + V(\mathbf{r}), \tag{2.4}$$

where $\mathbf{r} = \mathbf{r}_1 - \mathbf{r}_2$ is the relative radius-vector, $\mu = m_1 m_2/(m_1 + m_2)$ is the reduced mass, and

$$\omega_l = \sqrt{\frac{m_1}{2\mu}}\omega_l^{(1)}$$

are the relative motion frequencies. In the case of identical particles ($m_1 = m_2$ and $\omega_l^{(1)} = \omega_l^{(2)}$) the frequencies are the same as the single-particle frequencies. As the center-of-mass Hamiltonian is independent of the interparticle interactions, only the relative motion will be considered below.

In a harmonic waveguide the particle motion is restricted in two directions, x and y, and is free along the waveguide axis z, such that $\omega_x = \omega_y = \omega_\perp$ and $\omega_z = 0$. The wavefunction of non-interacting particles ($V \equiv 0$) can be represented as the product of a harmonic oscillator transverse wavefunction and a plane wave for axial motion.

Interacting particles can be simply described by the relative wavefunction $\psi(\mathbf{r})$ in a similar form,

$$\psi(\mathbf{r}) = \psi_{00}(z)\langle x, y|00\rangle, \tag{2.5}$$

where $|nm_z\rangle$ are eigenstates of the transverse Hamiltonian

$$\hat{H}_\perp |nm_z\rangle = (2n + |m_z| + 1)\omega_\perp |nm_z\rangle, \tag{2.6}$$

$$\hat{H}_\perp = -\frac{1}{2\mu}\left(\frac{\partial^2}{\partial\rho^2} + \frac{1}{\rho}\frac{\partial}{\partial\rho} + \frac{1}{\rho^2}\frac{\partial^2}{\partial\theta^2}\right) + \frac{\mu}{2}\omega_\perp^2\rho^2, \tag{2.7}$$

$\rho = \sqrt{x^2 + y^2}$, θ is the angular coordinate in the xy plane, n is the vibrational transverse quantum number, and m_z is the axial component of the angular momentum. Projection of the Schrödinger equation

$$\left[-\frac{1}{2\mu}\frac{\partial^2}{\partial z^2} + \hat{H}_\perp + V(\mathbf{r})\right]\psi(\mathbf{r}) = E\psi(\mathbf{r}) \tag{2.8}$$

onto the ground transverse mode $|00\rangle$ leads to a 1D Schrödinger equation with the potential $V_{1D}(z) = \langle 00|V|00\rangle$. In 3D mean field theory one often describes the effective interaction with the potential (Dalfovo et al., 1999; Pitaevskii and Stringari, 2003),

$$V(\mathbf{r}) = \frac{2\pi}{\mu}a_{el}\delta(\mathbf{r}), \tag{2.9}$$

where a_{el} is the elastic scattering length. This leads to

$$V_{1D}(z) = 2\omega_\perp a_{el}\delta(z). \tag{2.10}$$

There are, however, two problems with this derivation. First, the potential (2.9) has non-zero matrix elements $\langle n0|V|00\rangle$, so there will be virtual excitation of transverse modes. Second, a 3D δ-function with a finite interaction strength has a divergence that prevents it from correctly describing two-body scattering (see Kokkelmans et al., 2002; Yurovsky, 2005). As a result Eq. (2.10) misses important properties of interparticle interactions in confined geometries. These defects are eliminated in the derivation by Olshanii (1998) (see also Moore et al., 2004; Girardeau et al., 2004). In Section 2.3.1, we will present a generalization of that derivation that includes Feshbach resonant interactions (Yurovsky, 2005, 2006). To set up that derivation, in the next section we will introduce a formalism to describe Feshbach resonances.

2.2. Feshbach Resonance Interactions

Section 2.1.2 shows how the properties of interacting particles in atom waveguides depend on the elastic scattering length. The elastic scattering length can be tuned with a Feshbach resonance (see Timmermans et al., 1999) as in experiments by Moritz et al. (2005), where quasi-1D molecules were observed. A Feshbach resonance appears in multichannel two-body scattering when the collision energy in an open channel is near the energy of a bound (molecular) state in a closed channel. The open and closed channels can be coupled by resonant optical fields (Fedichev et al., 1996; Bohn and Julienne, 1997) or by hyperfine interactions (Tiesinga et al., 1992, 1993), using the Zeeman effect to control the energy detuning. It is important to note that a Feshbach resonance does not simply change the elastic scattering length, rather, it does so with a strong dependence on the collision energy.

Low-energy multichannel scattering can be conveniently analyzed using the M-matrix formalism (Ross and Shaw, 1961; Shaw and Ross, 1962). Let us introduce a set of state vectors $|j\rangle$ describing the internal states of a colliding pair in channel j, including relative motion with angular momentum l_j. Interactions that induce transitions between the channels are localized in some internal region. Outside this region, a set of relative wavefunctions can be expressed as

$$|\psi_k(r)\rangle = \sum_j \left[p_j^{-l_j} M_{jk} j_{l_j}(p_j r) - \delta_{kj} p_k^{l_k+1} n_{l_k}(p_k r) \right] |j\rangle, \qquad (2.11)$$

where r is the interparticle distance, $p_j = \sqrt{2\mu(E - D_j)}$ is the relative momentum in channel j with the asymptotic interaction energy D_j, and E is the collision energy. The spherical Bessel and Neumann functions are defined so that they have the asymptotics (Abramowitz and Stegun, 1964)

$$j_l(x) \underset{x \to \infty}{\sim} \sin(x - l\pi/2)/x, \qquad n_l(x) \underset{x \to \infty}{\sim} -\cos(x - l\pi/2)/x. \qquad (2.12)$$

A scattering state wavefunction with an incoming wave in a single input channel $|i\rangle$ can be expanded as

$$|\psi_i^+(r)\rangle = \sum_k f_{ki} p_k^{-l_k} |\psi_k(r)\rangle, \qquad (2.13)$$

where the scattering amplitudes f_{ki} are related to the M matrix by

$$\sum_k \left(p_j^{-l_j} M_{jk} p_k^{-l_k} - i p_j \delta_{jk} \right) f_{ki} = \delta_{ji} \vartheta(E - D_j). \qquad (2.14)$$

An advantage of the M matrix is that it has a simple effective-range expansion (Ross and Shaw, 1961; Shaw and Ross, 1962)

$$M_{jk}(E) \approx M_{jk}(0) + \delta_{jk} C_j R_0^{-2l_j+1} p_j^2, \qquad (2.15)$$

where the parameters $M_{jk}(0)$ and C_j are related to potentials in the internal interaction region with range R_0, and the dimensionless coefficients C_j are of order unity.

Although Eq. (2.15) diverges when the interaction range approaches zero for $l_j \neq 0$, the wavefunction can also be obtained by solving the free-space Schrödinger equation

$$\left[-\frac{1}{2\mu}\frac{d^2}{dr^2} - \frac{1}{\mu r}\frac{d}{dr} + \frac{l_j(l_j+1)}{2\mu r^2} + D_j - E\right]\langle j|\psi(r)\rangle = 0, \quad (2.16)$$

satisfying the following boundary conditions:

$$\frac{d^{2l_j+1}}{dr^{2l_j+1}}\left[r^{l_j+1}\langle j|\psi(r)\rangle\right]_{r=0}$$
$$= 2^{l_j} l_j! \sum_k \frac{M_{jk}}{(2l_k-1)!!}\left[r^{l_k+1}\langle k|\psi(r)\rangle\right]_{r=0}. \quad (2.17)$$

Here $|\psi(r)\rangle$ is an arbitrary linear combination of $|\psi_k(r)\rangle$, including $|\psi_i^+(r)\rangle$. Equation (2.17) is a generalization of the single-channel s-wave Bethe–Peierls boundary condition (Bethe and Peierls, 1935)

$$\frac{d}{dr}\left[r\psi(r)\right]_{r=0} = -\frac{1}{a_{el}}\left[r\psi(r)\right]_{r=0}, \quad (2.18)$$

which is equivalent to the Fermi–Huang pseudopotential (Fermi, 1934, 1936; Huang and Yang, 1957; Huang, 1987), acting on wavefunctions as

$$\hat{V}_{FH}(\mathbf{r})\psi(\mathbf{r}) = \frac{2\pi}{\mu}a_{el}\delta(\mathbf{r})\frac{\partial}{\partial r}\left[r\psi(\mathbf{r})\right]_{r=0}. \quad (2.19)$$

Multichannel s-wave boundary conditions (Demkov and Ostrovskii, 1988; Kartavtsev and Macek, 2002; Kim and Zubarev, 2003; Yurovsky and Band, 2007) are a special case of Eq. (2.17). Higher-partial-wave pseudopotentials (Kanjilal and Blume, 2004; Stock et al., 2005; Derevianko, 2005; Idziaszek and Calarco, 2006; Pricoupenko, 2006) are equivalent to single-channel boundary conditions (Macek and Sternberg, 2006), which are a special case of Eq. (2.17) as well.

The state $|\psi_i^+(r)\rangle$ can involve closed channels with $D_j > E$ and imaginary p_j. In this case the scattering amplitudes f_{jk}, which involve the closed channels, are determined by Eq. (2.14) so that the closed-channel components contain only decaying exponentials, $\exp(-|p_j|r)$.

Consider now the case of two channels, the closed one with $D_c > E$, and the open one with $D_o = 0$. The closed channel supports a bound resonant state with energy E_{Fesh} whenever $(-1)^{l_c} M_{cc}(E_{\text{Fesh}}) < 0$. Equation (2.14) leads to the following expression for the open-channel scattering amplitude,

$$f_{oo}(p_o) = \frac{p_o^{2l_o}}{M_{\text{eff}}(E) - ip_o^{2l_o+1}}, \quad (2.20)$$

where

$$M_{\text{eff}}(E) = M_{oo}(E) - \frac{M_{oc}M_{co}}{M_{cc} + (-1)^{l_c}|p_c|^{2l_c+1}}. \tag{2.21}$$

In the two-channel case the closed-channel component can be eliminated from Eq. (2.17), leading to a single-channel boundary condition,

$$\frac{d^{2l_o+1}}{dr^{2l_o+1}}\left[r^{l_o+1}\langle o|\psi(r)\rangle\right]_{r=0} = 2^{l_o}l_o!\frac{M_{\text{eff}}(E)}{(2l_o-1)!!}\left[r^{l_o+1}\langle o|\psi(r)\rangle\right]_{r=0}. \tag{2.22}$$

Whenever the closed-channel interaction strength $|M_{cc}|$ is large enough, $M_{\text{eff}}(E)$ can be represented in the conventional resonant form

$$M_{\text{eff}}(E) = M_{oo}(E) + M_{oo}(0)\frac{\mu_{oc}\Delta}{\mu_{oc}(B - B_0 - \Delta) - E}, \tag{2.23}$$

where Δ is the phenomenological resonance strength, μ_{oc} is the difference between the magnetic momenta of the closed and open channels, and B_0 is the resonant value of the magnetic field B. The resonant-state energy $E_{\text{Fesh}} = \mu_{oc}(B - B_0 - \Delta)$. Equations (2.23) and (2.20) demonstrate that f_{oo} depends resonantly on both the detuning $B - B_0$ and the collision energy E.

For s-wave scattering from a zero-range potential, when $M_{oo}(E) = M_{oo}(0) = -1/a_{el}$, Eq. (2.23) reduces to the expression for the effective scattering length $a_{\text{eff}} = -1/M_{\text{eff}}(E)$ in Yurovsky (2005),

$$a_{\text{eff}}(E) = a_{el}\left[1 + \frac{\mu_{oc}\Delta}{E - \mu_{oc}(B - B_0)}\right]. \tag{2.24}$$

(The conventional resonant scattering length $a_{\text{eff}}(0)$ has a singularity at $B = B_0$, which explains the choice of resonance position.) The boundary condition (2.22) then becomes the Bethe–Peierls boundary condition, which is equivalent to the Fermi–Huang pseudopotential (2.19) with a_{el} replaced by $a_{\text{eff}}(E)$,

$$\hat{V}_{\text{eff}}(\mathbf{r})\psi(\mathbf{r}) = \frac{2\pi}{\mu}a_{\text{eff}}(E)\delta(\mathbf{r})\frac{\partial}{\partial r}\left[r\psi(\mathbf{r})\right]_{r=0}. \tag{2.25}$$

If there exist channels below the input channel o, i.e., if one of the colliding particles is in an excited internal state, the collision can lead to deactivation due to short-range interactions (Yurovsky and Band, 2007). When the deactivation energy is high, this process is determined by the short-range behavior of the scattering state wavefunction (2.13) in the two-channel approximation, which involves only the input and closed channels. Its limit at the origin

$$\langle o|\psi_o^+(r)\rangle \underset{r\to 0}{\sim} -\frac{1}{r}\sqrt{\frac{\mu}{p_o}}a_{\text{eff}}(E) \tag{2.26}$$

increases near the resonance. The deactivation rate coefficient for slow collisions shows resonant behavior,

$$K_{\text{free}} = \frac{4\pi S}{\mu} |a_{\text{eff}}(E)|^2, \tag{2.27}$$

where the factor S accounts for the channel coupling (see Yurovsky and Band, 2007). There is no dependence on collision energy here except for that of $a_{\text{eff}}(E)$. This dependence is only substantial close to the resonance.

2.3. s-WAVE SCATTERING

2.3.1. s-Wave Interaction in Atom Waveguides

Consider the scattering in an atom waveguide [described by Eq. (2.8) with $V \equiv \hat{V}_{\text{eff}}$ given by Eq. (2.25)] of two particles that are initially in the relative transverse state n, $m_z = 0$ with relative energy E. We can expand the relative wavefunction in terms of the transverse modes and plane waves

$$\psi(\mathbf{r}) = (2\pi)^{-1/2} \sum_{n=0}^{\infty} \sum_{m_z} \int_{-\infty}^{\infty} dq \, \tilde{\psi}_{nm_z}(q) e^{iqz} |nm_z\rangle. \tag{2.28}$$

The matrix element of the Fermi–Huang pseudopotential (2.25)

$$\langle n'm_z | \hat{V}_{\text{eff}}(\mathbf{r}) | \psi(\mathbf{r}) \rangle = \frac{2\pi}{\mu} a_{\text{eff}}(E) \langle n'm_z | 0 \rangle \delta(z) \frac{\partial}{\partial r} [r\psi(\mathbf{r})]_{r=0}$$

$$= \frac{2}{\mu} \sqrt{\pi} \frac{a_{\text{eff}}(E)}{a_\perp} \delta_{0m_z} \delta(z) \eta_n(E), \tag{2.29}$$

where

$$a_\perp = (\mu \omega_\perp)^{-1/2} \tag{2.30}$$

is the transverse harmonic oscillator length and

$$\eta_n(E) = \lim_{r \to 0} \frac{\partial}{\partial r} [r\psi(\mathbf{r})], \tag{2.31}$$

generally depends on the initial state parameters n and E. Because $\langle n'm_z|0\rangle = \delta_{0m_z}/(\sqrt{\pi}a_\perp)$, the transverse state is non-zero at the origin only if $m_z = 0$. Therefore only cylindrically symmetric states ($m_z = 0$) are involved in the scattering. Equations (2.8), (2.28), and (2.29) lead to the following set of equations for the wavefunction components:

$$\left[E - (2n' + |m_z| + 1)\omega_\perp - \frac{q^2}{2\mu} \right] \tilde{\psi}_{n'm_z}(q)$$

$$= \frac{\sqrt{2}}{\mu} \frac{a_{\text{eff}}(E)}{a_\perp} \delta_{0m_z} \eta_n(E). \tag{2.32}$$

The $m_z = 0$ components can be expressed in terms of the 1D multichannel T matrix

$$\tilde{\psi}_{n'0}(q) = \delta(q - p_n)\delta_{nn'} + \frac{2\mu}{p_{n'}^2 - q^2 + i0} \langle n'q|T(p_0)|np_n\rangle, \tag{2.33}$$

where

$$p_n = \sqrt{2\mu[E - (2n+1)\omega_\perp]} \tag{2.34}$$

is the relative axial momentum in the channel that corresponds to the transverse mode $|n0\rangle$. Substitution of Eq. (2.33) into Eq. (2.32) leads to

$$\langle n'q|T(p_0)|np_n\rangle = \frac{\sqrt{2}}{\mu} \frac{a_{\text{eff}}(E)}{a_\perp} \eta_n(E), \tag{2.35}$$

which is independent of the final state n' and its momentum q.

The wavefunction (2.28) can then be expressed as

$$\psi(\mathbf{r}) = (2\pi)^{-1/2}|n0\rangle \exp(ip_n z)$$
$$- 2\sqrt{\pi} i \frac{a_{\text{eff}}(E)}{a_\perp} \eta_n(E) \sum_{n'=0}^{\infty} \frac{|n'0\rangle}{p_{n'}} \exp(ip_{n'}|z|). \tag{2.36}$$

The sum over n' diverges as $r \to 0$. The divergent part can be extracted using the identity (see Moore et al., 2004)

$$\lim_{r \to 0} \left[\sum_{n=0}^{\infty} \frac{|n0\rangle}{p_n} \exp(ip_n|z|) + \frac{i}{2\sqrt{\pi}} \frac{a_\perp}{r} \right] = -\frac{i}{2\sqrt{\pi}} \zeta\left(\frac{1}{2}, \frac{1}{2} - \frac{E}{2\omega_\perp}\right) \tag{2.37}$$

where $\zeta(\nu, \alpha)$ is the Hurwitz zeta function [see Eq. (8.1)]. The wavefunction (2.36) can then be separated into regular and irregular parts, $\psi(\mathbf{r}) = \psi_{\text{reg}}(\mathbf{r}) - \eta_n(E)a_{\text{eff}}(E)/r$. Substitution of this expression into Eq. (2.31) gives an equation in η

$$\eta_n(E) = \lim_{r \to 0} \frac{\partial}{\partial r}\left[r\psi_{\text{reg}}(\mathbf{r}) - \eta_n(E)a_{\text{eff}}(E)\right] = \psi_{\text{reg}}(0)$$
$$= \frac{1}{\sqrt{2\pi}a_\perp} - \frac{a_{\text{eff}}(E)}{a_\perp} \eta_n(E)\zeta\left(\frac{1}{2}, -\left(\frac{a_\perp p_0}{2}\right)^2\right). \tag{2.38}$$

Finally, we get the elements of the T matrix,

$$\langle n'q|T(p_0)|np_n\rangle \equiv \frac{1}{2\pi} T_{\text{conf}}(p_0)$$

$$= \frac{1}{\pi \mu a_\perp} \left[a_\perp a_{\text{eff}}^{-1}(E) + \zeta\left(\frac{1}{2}, -\left(\frac{a_\perp p_0}{2}\right)^2\right) \right]^{-1}. \quad (2.39)$$

These are independent of n and n'. Since $\eta_n(E)$ depends on ψ [see Eq. (2.31)] and, therefore, on T, Eq. (2.35) is a form of the Lippmann–Schwinger equation for the T matrix. It can be continued outside the energy shell, demonstrating that the T matrix is independent of the final momentum. By symmetry, it is also independent of the initial momentum and only depends on the collision energy, which is related to p_0. The independence of T matrix on the initial and final states is a common feature of zero-range spherically-symmetric interactions. The total energy in Eq. (2.39), $E = p_0^2/(2\mu) + \omega_\perp$, includes the energy of the ground transverse mode, and leads to a shift of the Feshbach resonance.

The validity of the present derivation is based on the assumption of the hermiticity of multichannel scattering in a *harmonic waveguide* with a zero-range interaction described by the boundary conditions (2.22). Hermiticity has only been proved for multichannel scattering in *free space* (Kim and Zubarev, 2003). However, Eq. (2.39) was also derived in Yurovsky (2005) using a renormalization procedure for multichannel δ-function interactions, and in that approach hermiticity is not in doubt.

2.3.2. Relation between 1D and 3D Scattering Parameters

Whenever $\omega_\perp < E < 3\omega_\perp$, or $p_0 < 2\sqrt{\mu\omega_\perp}$, only the ground transverse mode ($n = 0$) channel is open, and the T matrix (2.39) can be approximated by a solution of the 1D two-channel problem (Lipkin, 1973; Yurovsky, 2005, 2006) described by the coupled equations

$$E_c \varphi_0(z) = -\frac{1}{2\mu}\frac{d^2\varphi_0}{dz^2} + \left[U_a \varphi_0(0) + \sqrt{2}U_{am}^* \varphi_1^{am}\right]\delta(z),$$
$$E_c \varphi_1^{am} = D_{1D}\varphi_1^{am} + \sqrt{2}U_{am}\varphi_0(0). \quad (2.40)$$

Here $E_c = p_0^2/(2\mu) = E - \omega_\perp$ is the 1D collision energy, $\varphi_0(z)$ is the 1D open channel wavefunction and φ_1^{am} is the amplitude for the system to be in the 1D closed channel. This model approximates the 1D closed-channel bound state to be infinitesimal in size with infinite binding energy (Yurovsky, 2005). The non-resonant interaction strength U_a, the channel coupling U_{am}, and the detuning D_{1D} can be related to the 3D scattering and waveguide parameters in the following way (Yurovsky, 2006). The two-channel problem (2.40) exactly reduces to the single-channel problem, described by the Schrödinger equation

$$E_c \varphi(z) = -\frac{1}{2\mu}\frac{d^2\varphi}{dz^2} + U_{\text{eff}}(E_c)\delta(z)\varphi(0), \quad (2.41)$$

where

$$U_{\text{eff}}(E_c) = U_a + \frac{2|U_{am}|^2}{E_c - D_{1D}}. \tag{2.42}$$

The T matrix for this problem can be expressed as (Lipkin, 1973; Yurovsky, 2005)

$$\langle p'|T_{1D}^+(p_0)|p\rangle = \frac{1}{2\pi}\left[U_{\text{eff}}^{-1}(E_c) + \frac{i\mu}{p_0}\right]^{-1}. \tag{2.43}$$

It is independent of the initial p and final p' momenta and depends only on the p_0-dependent collision energy. The open-channel wavefunction can be expressed as

$$\varphi(z) = \exp(ip_0 z) + f_{1D}^+(p_0)\exp(ip_0|z|) \tag{2.44}$$

in terms of the 1D scattering amplitude

$$f_{1D}^+(p_0) = -\frac{2\pi i\mu}{p_0}\langle p_0|T_{1D}^+(p_0)|p_0\rangle. \tag{2.45}$$

The T matrix (2.39) is exactly reproduced by Eq. (2.43) with

$$U_{\text{eff}}(E_c) = \left[\frac{1}{2\omega_\perp a_{\text{eff}}(E_c+\omega_\perp)} + \frac{\mu a_\perp}{2}\zeta\left(\frac{1}{2}, -\frac{E_c}{2\omega_\perp}\right) - \frac{i\mu}{p_0}\right]^{-1}. \tag{2.46}$$

This interaction strength can be approximated by Eq. (2.42) with

$$U_a = \frac{2a_{el}\omega_\perp}{\beta}, \qquad \beta = 1 - C\frac{a_{el}}{a_\perp} - C'\frac{a_{el}}{a_\perp}\left(\mu_{oc}\frac{B-B_0-\Delta}{2\omega_\perp} - \frac{1}{2}\right),$$

$$D_{1D} = \frac{(1-Ca_{el}/a_\perp)[\mu_{oc}(B-B_0-\Delta)-\omega_\perp] + \mu_{oc}\Delta}{\beta},$$

$$|U_{am}|^2 = a_{el}\frac{2\omega_\perp\mu_{oc}\Delta + C'[\mu_{oc}(B-B_0-\Delta)-\omega_\perp]^2 a_{el}/a_\perp}{2\beta^2} \tag{2.47}$$

[three terms of the expansion (8.5) are used in this derivation]. This approximation is applicable for a wide range of parameters (see Yurovsky, 2006). If $a_\perp \neq Ca_{el}$ it can be applied to slow collisions, as long as $E_c \ll 2\omega_\perp a_\perp/a_{el}$.

In the absence of the Feshbach resonance ($\Delta = 0$), a substitution of two terms of the expansion (8.5) into Eq. (2.46) leads to the interaction strength

$$U_{\text{conf}} = \frac{2\omega_\perp a_{el}}{1 - Ca_{el}/a_\perp}, \tag{2.48}$$

first derived by Olshanii (1998). This expression has a denominator, unlike Eq. (2.10). The pole at $a_\perp = Ca_{el}$ is the confinement-induced resonance (CIR).

The interaction strength (2.46) can also be approximated by Eq. (2.42) for a weak resonance, $\Delta \ll |B-B_0-\Delta-\omega_\perp/\mu_{oc}|$. In this case, the 1D parameters can be estimated as

$$U_a \approx 0, \quad |U_{am}|^2 \approx \frac{2\omega_\perp}{C'\mu a_\perp}, \quad D_{1D} \approx -\frac{2\omega_\perp}{C'}\left(\frac{a_\perp}{a_{el}} - C\right). \quad (2.49)$$

The interaction strength (2.42) will be energy-dependent in this case as well. However, the detuning D_{1D} now substantially exceeds the transverse frequency $\omega_\perp$. Therefore, in the quasi-1D regime, whenever the energy is less then $\omega_\perp$, the energy-dependence is very weak.

2.3.3. Bound States

Resonances are generally related to poles of the T matrix, which correspond to bound states if they are located on the positive imaginary axis. In the present case, the T matrix (2.39) has such poles and the corresponding binding energies $E_b = -E_c$ are determined by the transcendental equation

$$a_{\text{eff}}(\omega_\perp - E_b)\zeta\left(\frac{1}{2}, \frac{E_b}{2\omega_\perp}\right) + a_\perp = 0. \quad (2.50)$$

Weakly bound states with $E_b \ll 2\omega_\perp$ become quasi-1D, corresponding to poles of the 1D T matrix (2.43). Their binding energies $E_b = \kappa^2/(2\mu)$, where $\kappa = -ip_0$ are positive solutions of the cubic equation

$$\kappa^3 + \mu U_a \kappa^2 + 2\mu D_{1D}\kappa + 2\mu^2 D_{1D}U_a - 4\mu^2|U_{am}|^2 = 0, \quad (2.51)$$

and the 1D parameters are expressed by Eq. (2.47).

In the absence of the Feshbach resonance, when $a_{\text{eff}} \equiv a_{el}$, Eq. (2.50) turns into an equation by Bergeman et al. (2003). In this case, for deeply bound states with $E_b \gg 2\omega_\perp$ and for $0 < a_{el} \ll a_\perp$, the first term of the expansion (8.6) leads to $E_b \approx (2\mu a_{el}^2)^{-1}$, which is the same as the free space result. However, there exists a bound state in the confined system even when $a_{el} < 0$. Here the binding energy can be approximately expressed as $E_b \approx 2\mu(\omega_\perp a_{el})^2$ for $|a_{el}| \ll a_\perp$, as in the 1D system with the potential (2.10). Such bound states have been observed with fermions experimentally (Moritz et al., 2005; see Fig. 1).

The CIR can be understood as a *multi-channel* Feshbach resonance (Bergeman et al., 2003) where the excited transverse modes [$|n0\rangle$ with $n > 0$ and imaginary p_n, as $E < 3\omega_\perp$, see Eq. (2.34)] serve as closed channels. The excited modes form a bound resonant state with energy $3\omega_\perp - E_b$, where E_b is the solution of Eq. (2.50). For $\Delta = 0$, the singularity in Eq. (2.48) appears at $a_\perp = Ca_{el}$, where the resonant state energy crosses the continuum threshold $\omega_\perp$.

The CIR can also be approximated by a *two-channel* Feshbach resonance, since 1D parameters (2.49) are independent of the Feshbach resonance parameters and are therefore applicable to the case of non-resonant 3D scattering. The location of the CIR is again determined by the condition $a_\perp = Ca_{el}$. The infinite set of closed

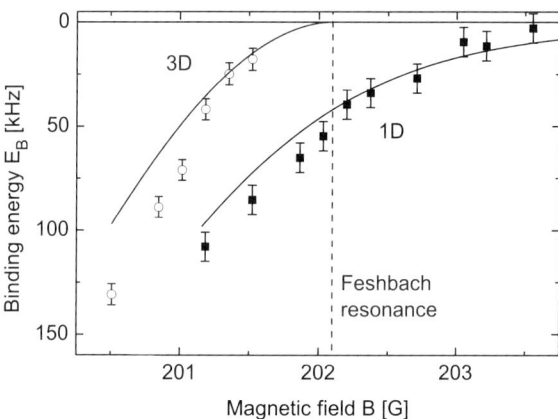

FIG. 1. Experimental observation quasi-1D molecules. At fields higher than the Feshbach resonance molecules do not form in free space, but in waveguides they do, a direct result of the CIR. [Reprinted figure with permission from Moritz et al. (2005).]

© 2005 American Physical Society

channels, which corresponds to all transverse excitations in the exact description (Bergeman et al., 2003), is approximated here by a single channel supporting an infinitely-bound state.

In the case of a resonant 3D interaction, the bound states of two atoms in an atom waveguide are superpositions of the closed and open channels of the Feshbach resonance. The open-channel component is a superposition of all transverse modes. The size of the closed-channel component is negligible compared to $a_\perp$. It is therefore unaffected by the waveguide, and does not need to be expanded in terms of the transverse modes. Its contribution can be expressed as

$$W_c = \int d^3r \left|\langle c|\psi(r)\rangle\right|^2 = \frac{2\pi \mu_{oc} \Delta}{\mu a_{el}[E - \mu_{oc}(B - B_0 - \Delta)]^2} (a_{\text{eff}} \eta)^2, \quad (2.52)$$

where η is defined by Eq. (2.31). In the case of a bound state with binding energy E_b, defined by Eq. (2.50), the contribution of the nth transverse mode can be expressed as

$$W_n = \frac{4\pi}{|p_n|^3 a_\perp^2} (a_{\text{eff}} \eta)^2. \quad (2.53)$$

Thus the relative contribution of the closed channel is (see also Yurovsky, 2006)

$$\frac{W_c}{W_0} = \frac{|p_0|^3 a_\perp^2 \mu_{oc} \Delta}{2\mu a_{el}[E - \mu_{oc}(B - B_0 - \Delta)]^2} \quad (2.54)$$

and the total contribution of the excited transverse modes is

$$\frac{1}{W_0} \sum_{n=1}^{\infty} W_n = \left(\frac{E_b}{2\omega_\perp}\right)^{3/2} \zeta\left(\frac{3}{2}, \frac{E_b}{2\omega_\perp}\right) - 1. \quad (2.55)$$

The 1D closed channel incorporates contributions from both the 3D closed channel and the excited waveguide modes. The ratio of the contributions of the 1D closed and open channels, written respectively as

$$W_c^{1D} = |\varphi_1^{am}|^2, \qquad W_o^{1D} = \int_{-\infty}^{\infty} |\varphi_0(z)|^2 \, dz, \quad (2.56)$$

can be expressed as (Yurovsky, 2006)

$$\frac{W_c^{1D}}{W_o^{1D}} = \frac{2|U_{am}|^2 \sqrt{2\mu E_b}}{(E_b + D_{1D})^2}, \qquad W_o^{1D} = \frac{E_b + D_{1D}}{3E_b + U_a\kappa + D_{1D}}, \quad (2.57)$$

where E_b and κ are expressed by Eq. (2.51).

Two particles with a Feshbach resonance interaction in an atomic waveguide can have several bound states (see Fig. 2). Far below the resonance, at $B_0 - B \gg \max(\Delta, \omega_\perp/\mu_{oc})$, a quasi-3D bound state, dominated by the closed channel, has the binding energy $E_b \approx \mu(B_0 - B)$. The structures of other bound states depend on a_{el} and $a_\perp$. Far from the resonance it can be either a quasi-3D bound state, with binding energy $E_b \approx 1/(2\mu a_{el}^2)$ (at $a_{el} < 0$ only), consisting mostly of the excited waveguide modes of the open channel, or a quasi-1D bound state with $E_b \approx \mu U_a^2/2$ (for $U_a < 0$ only), consisting mostly of the open-channel ground mode. One of the bound states vanishes at the threshold $B = B_0 + \Delta + \omega_\perp/\mu_{oc}$, where Eq. (2.50) has a solution $E_b = 0$. Thus the weak bound state exists even above the resonance (see Figs. 1 and 2), where the resonant scattering length is negative, $a_{\text{eff}}(0) < 0$, in contrast to the free space case with the threshold at $B = B_0$, where $a_{\text{eff}}(0) = 0$. Near the threshold, the major contribution to the weak bound state is from the ground transverse mode of the open channel. Further below the threshold the bound state composition depends on the resonance type (see Góral et al., 2004; Chin, 2005; Yurovsky, 2006). Parameters of a closed-channel dominated resonance [see Fig. 2(a)] will satisfy the condition $\mu a_{el}^2 \mu_{oc} \Delta < 0.2$. Otherwise, the resonance will be open-channel dominated, with the largest contribution coming from the excited transverse modes of the open channel [see Fig. 2(b)]. In the last case, the energy-dependence of a_{eff} can be neglected and the binding energy can be calculated within a "single-channel" model as a solution of Eq. (2.50) where $a_{\text{eff}}(E)$ is replaced by the resonant scattering length $a_{\text{eff}}(0)$, as in Moritz et al. (2005). However, this model predicts a non-physical singularity just above the threshold [see Fig. 2(b)], and is applicable only in a small region below the

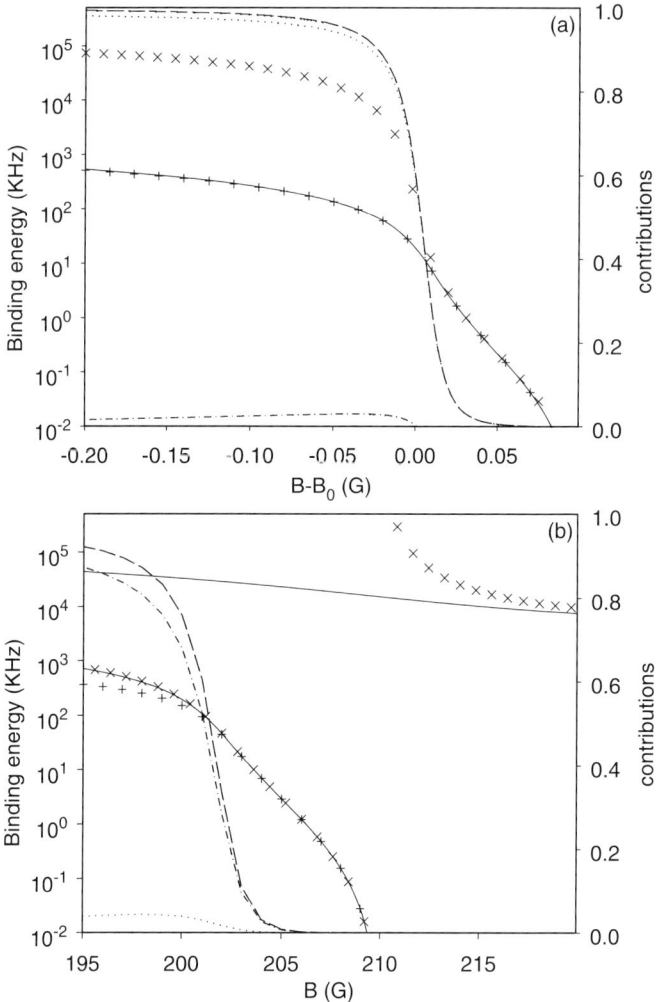

FIG. 2. The binding energy E_b calculated as a function of the external magnetic field for the 543 G resonance in ^{6}Li with $\omega_\perp = 200 \times 2\pi$ KHz [part (a)] and for the 202 G resonance in K with $\omega_\perp = 69 \times 2\pi$ KHz [part (b)]. The solid line, pluses, and crosses represent, respectively, solutions of the exact equation (2.50), the 1D approximation (2.51), and the open-channel model. The contributions of the closed channel and excited modes of the open channel are plotted by dot-dashed and dotted lines, respectively. The dashed line represents the contribution of the 1D closed channel. [This is a composite of figures from Yurovsky (2006).]

threshold in the case of a closed-channel dominated resonance [see Fig. 2(a)]. The 1D approximation [see Eq. (2.51)] can be applied to weak bound states for both resonance types.

2.3.4. Collisions

We will now consider collisions of two particles in the transverse mode $|n0\rangle$ with a relative axial momentum p_n. The collision can result in either transmission or reflection into a transverse mode $|n'0\rangle$. According to Eqs. (2.28) and (2.33), the corresponding probabilities can be expressed as

$$P^{\text{tr}}_{n'n} = \left| \delta_{n'n} - i\frac{\mu}{p_n} T_{\text{conf}}(p_0) \vartheta(E_c - 2n'\omega_\perp) \right|^2, \quad (2.58)$$

$$P^{\text{ref}}_{n'n} = \left| \frac{\mu}{p_n} T_{\text{conf}}(p_0) \vartheta(E_c - 2n'\omega_\perp) \right|^2 \quad (2.59)$$

in terms of the T matrix T_{conf}, (2.39) which depends on the single parameter $p_0 = \sqrt{2\mu E_c}$, where the total collision energy $E_c = p_n^2/(2\mu) + 2n\omega_\perp$ includes the transverse excitation energy. The Hurwitz zeta function $\zeta(1/2, -E_c/(2\omega_\perp))$ in Eq. (2.39) has poles at $E_c = 2n\omega_\perp$ [see Eq. (8.1)], and therefore $T_{\text{conf}}(p_0)$ crosses zero at a new transverse channel threshold. Let us now consider the validity of various approximations (Yurovsky, 2005). The 1D approximation is expressed by Eq. (2.43) with 1D parameters given by Eq. (2.47). It is valid at low collision energies (see Fig. 3). Substituting the leading term in the expansion (8.7) into Eq. (2.39) leads to a 3D approximation $T_{3D}(p_0)$, which is proportional to f_{00} of Eq. (2.20) when $l_o = 0$. It captures the average behavior at high collision energies, but it does not reproduce the jumps at the transverse excitation thresholds.

If one of the colliding particles is in an excited internal state, the collision can lead to its deactivation due to short-range interactions (Yurovsky and Band, 2007). Like in the free-space case [see Eq. (2.26)], the deactivation rate is determined by the limit at the origin of the scattering-state wavefunction (2.36)

$$\psi(\mathbf{r}) \underset{r \to 0}{\sim} -\frac{1}{r} \frac{1}{2\omega_\perp} \sqrt{\frac{\mu}{p_0}} T_{\text{conf}}(p_0). \quad (2.60)$$

The ratio of this limit to the free-space one (2.26) can be approximately expressed for slow collisions and $a_{el} \ll a_\perp$, using Eqs. (2.39) and (2.45), as the 1D wavefunction (2.44), $T_{\text{conf}}(p_0)/[2\omega_\perp a_{\text{eff}}(E)] \approx 1 + f^+_{1D} = \varphi(0)$. This ratio is also equal to the 1D elastic transmission amplitude [see Eq. (2.44)]. The ratio of the deactivation rate coefficient in a waveguide, K_{conf}, to its value in free space, K_{free} [see Eq. (2.27)], can then be approximated as

$$\frac{K_{\text{conf}}}{K_{\text{free}}} \approx P^{\text{tr}}_{00}. \quad (2.61)$$

At low collision energies the transmission probability, $P^{\text{tr}}_{00} \approx 1 - P^{\text{ref}}_{00}$, tends to zero (see Fig. 3), and two particles can not approach together. This leads to

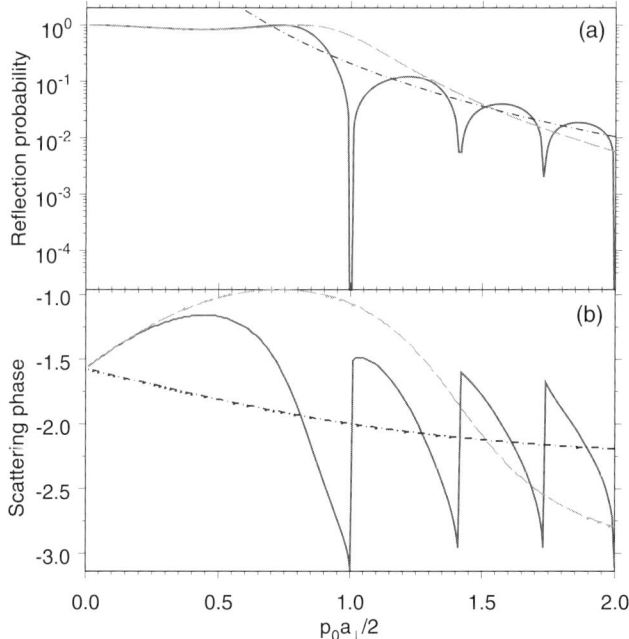

FIG. 3. The reflection probability $P_{n'n}^{\text{ref}}$ [(a), see Eq. (2.59)] and the scattering phase $\arg[T_{\text{conf}}(p_0)] - \pi/2$ (b) calculated as functions of the dimensionless collision momentum with the parameter values $a_{el}/a_\perp = 0.1$, $B = B_0 + \omega_\perp/\mu_{oc}$, and $\Delta = 20\omega_\perp/\mu_{oc}$. The solid, dashed, and dot-dashed lines are related, respectively, to the exact expression (2.39), to the 1D approximation (2.43), and to the 3D one. [The figure is similar to Fig. 1 in Yurovsky (2005), modified slightly using the relations between 1D and 3D parameters in Eq. (2.47).]

$\varphi(0) \to 0$, as is the case for fermions and hard-core interacting bosons (see Section 3.2), and to the suppression of deactivation. The deactivation rate,

$$K_{\text{conf}} \approx \pi a_\perp^4 p_0^2 S/\mu, \qquad (2.62)$$

will be then proportional to the collision energy and independent of a_{eff} (Yurovsky and Band, 2007), losing the resonant enhancement. Suppression of inelastic collisions was predicted in Gangardt and Shlyapnikov (2003b) for indistinguishable bosons within the many-body Lieb–Liniger model. However, it is mostly a two-body effect [see also Eq. (4.23) and the associated discussion] on the s-wave collisions of arbitrary particles, and holds true even when the Lieb–Liniger model is inapplicable.

2.3.5. Further Extensions

All the above considerations are related to zero-range interactions and harmonic waveguide potentials. In the case of a rectangular waveguide potential, analyzed by Kim et al. (2005), the binding energy nearly coincides with the harmonic potential results. Different trapping potentials for two particles, including non-harmonic potentials, were analyzed in Peano et al. (2005b, 2005a), and can lead to additional CIRs. An effect of finite interaction range has been considered by Bergeman et al. (2003) and by Naidon et al. (2007). As demonstrated in Naidon et al. (2007), the results of numerical calculations are well-reproduced by Eq. (2.48) with a_{el} replaced by $(a_{el}^{-1} - C_0 R_0 p_0^2)^{-1}$, where $C_0 R_0$ is an effective radius. Finite-range effects have been similarly incorporated into an energy-dependent length in a two-body problem with 3D confinement (Blume and Greene, 2002; Bolda et al., 2002, 2003). These results can be explained within the M matrix formalism. Indeed, outside the interaction region, the wavefunction is a solution of the free-space Schrödinger equation with the boundary condition (2.22). According to Eqs. (2.15) and (2.23), the coefficients $M_{\text{eff}}(E)$ can be expressed as

$$M_{\text{eff}}(E) = -a_{\text{eff}}^{-1}(E) + C_0 R_0 p_0^2. \tag{2.63}$$

This allows the effects of a Feshbach resonance, confinement, and finite range interactions to all be accounted for at the same time.

The energy dependence can alternatively be reproduced, in a limited range of collision energies, by an energy-independent finite-range potential (Girardeau et al., 2004).

2.4. p-WAVE SCATTERING

According to Eqs. (2.20), (2.15) and (2.23), the amplitude of elastic p-wave scattering ($l_o = 1$) can be expressed as

$$f_p \equiv f_{oo} = -\frac{p_o^2}{V_p^{-1}(E) + i p_o^3}, \tag{2.64}$$

where the energy-dependent scattering volume

$$V_p^{-1}(E) = -M_{oo}(0) - C_1 R_0^{-1} p_o^2 - M_{oo}(0) \frac{\mu_{oc} \Delta}{\mu_{oc}(B - B_0 - \Delta) - E} \tag{2.65}$$

can include the effects of the finite range interaction and the Feshbach resonance. A p-wave scattering length, a_p, can be defined as $V_p(0) = -M_{\text{eff}}^{-1}(0) = a_p^3$ (Girardeau and Olshanii, 2004).

Unlike for s-wave scattering, higher partial wave amplitudes can only be parameterized by a single parameter (the corresponding scattering length) to zeroth order in p_o. The first correction depends on the second parameter (R_0) [in the s-wave case the first correction is the universal term $i p_o^{2l_o+1}$ in the denominator of Eq. (2.20)]. This property reflects the non-universality of higher partial wave scattering.

Another peculiarity associated with higher partial waves is the divergence of the scattering amplitudes in the zero-range limit, as $R_0 \to \infty$. However, for a finite value of R_0 at low collision energies, whenever $|C_1|R_0^{-1}p_o^2 \ll |a_p|^{-3}$, the wavefunction outside the interaction region can be obtained using a boundary condition of the form of Eq. (2.22) with $M_{\text{eff}}(E) = V_p^{-1}(E)$. The pseudopotential considered in Kanjilal and Blume (2004), Stock et al. (2005), Derevianko (2005), Idziaszek and Calarco (2006), Pricoupenko (2006) is equivalent to this boundary condition.

p-Waves dominate the low-energy scattering of spin-polarized fermions since s-wave scattering is forbidden by spin statistics. p-Wave scattering in harmonic waveguides has been considered by Granger and Blume (2004) using non-orthogonal frame transformations. When a single transverse channel is open, the open-channel component of the relative wavefunction can be expressed as $\psi(\mathbf{r}) = \varphi^-(z)|00\rangle$. Since two spin-polarized fermions cannot occupy an even parity state, the axial motion can be described by the 1D odd-parity principal-value wavefunction

$$\varphi^-(z) = \sin p_0 z + K_{1D}^- \frac{z}{|z|} \cos p_0 z, \tag{2.66}$$

which is expressed in terms of the 1D K matrix (Granger and Blume, 2004)

$$K_{1D}^- = -\frac{6V_p p_0}{a_\perp^2}\left[1 - 12\frac{V_p}{a_\perp^3}\zeta\left(-\frac{1}{2}, 1 - \left(\frac{p_0 a_\perp}{2}\right)^2\right)\right]^{-1}. \tag{2.67}$$

The wavefunction (2.66) can be obtained as an odd-parity solution of the 1D free-space Schrödinger equation

$$-\frac{1}{2\mu}\frac{d^2\varphi}{dz^2} = \frac{p_0^2}{2\mu}\varphi \tag{2.68}$$

with the contact boundary condition (Girardeau and Olshanii, 2004)

$$\varphi(0+0) - \varphi(0-0) = \mu U_{1D}^-[\varphi'(0+0) + \varphi'(0-0)], \tag{2.69}$$

where $U_{1D}^- = K_{1D}^-/(\mu p_0)$. At low collision energies, when $p_0 a_\perp \ll 1$, the Hurwitz zeta function here can be expressed in the form of Eq. (8.3), leading to (Girardeau and Olshanii, 2004)

$$U_{1D}^- = -6V_p\omega_\perp\left[1 + 3\frac{V_p}{\pi a_\perp^3}\zeta\left(\frac{3}{2}\right)\right]^{-1}. \tag{2.70}$$

The boundary condition (2.69) is equivalent to the δ' interaction pseudopotential (Girardeau and Olshanii, 2004)

$$\hat{v}_{1D}^- = U_{1D}^-\delta'(z)\hat{\partial}_\pm, \quad \hat{\partial}_\pm\varphi(z) = \frac{1}{2}[\varphi'(0+0) + \varphi'(0-0)]. \tag{2.71}$$

Its matrix element is

$$\langle \varphi_1 | v_{1D}^- | \varphi_2 \rangle = -\frac{1}{2} U_{1D}^- [\varphi_1'(0)]^* [\varphi_2'(0+0) + \varphi_2'(0-0)]. \tag{2.72}$$

The even part of φ satisfies the convenient projection property $\hat{v}_{1D}^- \varphi^+ = 0$. In the momentum representation the pseudopotential can be expressed as

$$\langle q_1 | v_{1D}^- | q_2 \rangle = -\frac{1}{4\pi} U_{1D}^- q_1 q_2 (e^{iq_2 0} + e^{-iq_2 0}), \tag{2.73}$$

where $|q\rangle = e^{iqz}/\sqrt{2\pi}$. The off-shell two-body T matrix can be determined by the Lippmann–Schwinger equation

$$\langle q_1 | T_{1D}^- | q_2 \rangle = \langle q_1 | v_{1D}^- | q_2 \rangle + \int dq \, \langle q_1 | v_{1D}^- | q \rangle G_0(q) \langle q | T_{1D}^- | q_2 \rangle, \tag{2.74}$$

where

$$G_0(q) = \frac{2\mu}{p_0^2 - q^2 + i0} \tag{2.75}$$

is the 1D free Green function.

Because the pseudopotential (2.73) has a separable form, the solution of Eq. (2.74) can be expressed as

$$\langle q_1 | T_{1D}^- | q_2 \rangle = -\frac{1}{4\pi} \frac{U_{1D}^- q_1 q_2}{1 - i\mu p_0 U_{1D}^-} (e^{iq_2 0} + e^{-iq_2 0}). \tag{2.76}$$

It should be noted that the factor in the brackets in Eq. (2.73) ensures the convergence of the integral in Eq. (2.74). The T matrix (2.76) depends on both the initial and final momenta, unlike the T matrix for δ interactions (2.43). The zero-range nature of δ' interactions is reflected here by the separable form of $\langle q_1 | T_{1D}^- | q_2 \rangle$. The outgoing-wave wavefunction can be expressed in terms of the 1D odd-wave scattering amplitude,

$$f_{1D}^- = -\frac{2\pi i \mu}{p_0} \langle p_0 | T_{1D}^-(p_0) | p_0 \rangle, \tag{2.77}$$

as

$$\varphi(z) = \exp(ip_0 z) + f_{1D}^-(p_0) \frac{z}{|z|} \exp(ip_0 |z|). \tag{2.78}$$

The T matrix can also be extracted from the Green function $\tilde{G}_{1D}^-(z_1, z_2)$ in the coordinate representation for the 1D δ'-potential presented in Albeverio et al. (1988),

$$\tilde{G}_{1D}^-(z_1, z_2) = -i \frac{\mu}{p_0} \exp(ip_0 |z_1 - z_2|)$$

$$-\frac{\mu^2 U_{1D}^-}{1-i\mu p_0 U_{1D}^-}\frac{z_1 z_2}{|z_1||z_2|}\exp(ip_0|z_1|+ip_0|z_2|). \qquad (2.79)$$

A Fourier transform gives the Green function $G_{1D}^-(q_1, q_2)$ in the momentum representation

$$G_{1D}^-(q_1, q_2) = G_0(q_1)\delta(q_1 - q_2) + G_0(q_1)\langle q_1|T_{1D}^-|q_2\rangle G_0(q_2), \qquad (2.80)$$

where $\langle q_1|T_{1D}^-|q_2\rangle$ is expressed by Eq. (2.76). Recognition in Eq. (2.80) of a well-known relation between the Green function and the T matrix provides independent proof of Eq. (2.76), without relying on manipulations of divergent integrals.

Both the 1D even-parity δ-function potential of Eq. (2.41) and the odd-parity δ'-pseudopotential (2.71) are special cases of four-parameter self-adjoint extensions of the 1D Schrödinger operator (Seba, 1986). In general, these have indefinite parity and can be generalized to include multiple components (Albeverio et al., 2001). However, because ultracold atom collisions are either dominated by s-wave scattering, or in the case of spin-polarized fermions by p-wave scattering, interactions with indefinite parity have not found physical applications.

The permutation symmetry of indistinguishable particles requires the coordinate wavefunction to have definite parity. Thus spin-polarized fermions (and multicomponent bosons with odd spin wavefunctions) are described by the odd-parity coordinate wavefunction $\varphi^-(z) = -\varphi^-(-z)$ with $d\varphi^-(z)/dz = d\varphi^-(-z)/dz$. In this case, the boundary condition (2.69) takes the form

$$\varphi^-(0+0) = \mu U_{1D}^- d\varphi^-(0+0)/dz. \qquad (2.81)$$

Spinless bosons (and fermions with odd spin wavefunctions) have an even-parity coordinate wavefunction $\varphi^+(z) = \varphi^-(-z)$ with $d\varphi^+(z)/dz = -d\varphi^+(-z)/dz$. The δ-function potential in Eq. (2.41) is equivalent for this wavefunction to the boundary condition

$$d\varphi^+(0+0)/dz = \mu U_{\text{eff}} \varphi^+(0+0). \qquad (2.82)$$

Thus, the boundary conditions (2.81) and (2.82) are equivalent if the wavefunctions and interaction strengths are related by

$$\varphi^+(z) = \text{sign}\, z \varphi^-(z), \qquad U_{\text{eff}} = 1/(\mu^2 U_{1D}^-). \qquad (2.83)$$

These relations express the so-called Fermi–Bose mapping that was introduced in Girardeau (1960) (see Sections 3.2 and 3.3.5 for more discussion) for free fermions and impenetrable bosons. It was generalized to arbitrary coupling strength in Cheon and Shigehara (1998, 1999), Granger and Blume (2004), Girardeau et al. (2003, 2004), Girardeau and Olshanii (2004). The mapping from the fermionic to the bosonic problem preserves energy eigenvalues and dynamics.

3. Bethe Ansatz and beyond

3.1. Many-Body 1D Problems

Many-body problems can be analyzed using various exact, approximate, and numerical methods. A system of N 1D particles with equal, non-zero masses m can be described in the second-quantized representation by the Hamiltonian

$$\hat{H}_{\mathrm{MB}} = \sum_{\alpha} \int_{-L/2}^{L/2} dz \hat{\Psi}_{\alpha}^{\dagger}(z) \left[-\frac{1}{2m} \frac{\partial^2}{\partial z^2} + V_{\mathrm{ax}}(z, \alpha) \right.$$

$$+ \sum_{\alpha'} \int_{-L/2}^{L/2} dz' \left(1 - \frac{1}{2} \delta_{\alpha\alpha'} \right)$$

$$\times V_{1D}(z - z', \alpha, \alpha') \hat{\Psi}_{\alpha'}^{\dagger}(z') \hat{\Psi}_{\alpha'}(z') \left. \right] \hat{\Psi}_{\alpha}(z), \quad (3.1)$$

where the set of quantum numbers α characterizes the particle type, spin, and internal state, z is the particle coordinate, $\hat{\Psi}_{\alpha}(z)$ are the annihilation operators, and $V_{1D}(z - z', \alpha, \alpha')$ is the potential of interaction between the particles of type α and α'. The axial trapping potential $V_{\mathrm{ax}}(z, \alpha)$ may include an internal particle energy. This Hamiltonian (3.1) can be used to describe multichannel two-body interactions (see Yurovsky, 2005).

A many-body system can be described by the state vector

$$|\Psi^{(N)}\rangle = (N_{\alpha_1}! N_{\alpha_2}! \cdots)^{-1/2} \int d^N z \Psi^{(N)}(\{z\}, \{\alpha\}) \prod_{j=1}^{N} \hat{\Psi}_{\alpha_j}^{\dagger}(z_j) |\mathrm{vac}\rangle, \quad (3.2)$$

where $|\mathrm{vac}\rangle$ is the vacuum state vector and N_{α} is the number of indistinguishable particles of the type α. Substitution of this state vector into the Schrödinger equation $E|\Psi^{(N)}\rangle = \hat{H}_{\mathrm{MB}} |\Psi^{(N)}\rangle$ leads (see Lai and Haus, 1989a) to the stationary Schrödinger equation

$$E \Psi^{(N)}(\{z\}, \{\alpha\}) = \left\{ -\frac{1}{2m} \sum_{j=1}^{N} \left[\frac{\partial^2}{\partial z_j^2} + V_{\mathrm{ax}}(z_j, \alpha_j) \right] \right.$$

$$\left. + \sum_{j<j'} V_{1D}(z_j - z_{j'}, \alpha_j, \alpha_{j'}) \right\} \Psi^{(N)}(\{z\}, \{\alpha\}) \quad (3.3)$$

for the N-body wavefunction $\Psi^{(N)}(\{z\}, \{\alpha\})$, where E is the total energy. If the system contains indistinguishable particles, the wavefunction $\Psi^{(N)}(\{z\})$ has the corresponding Fermi or Bose permutation symmetry. The finite interval

$[-L/2, L/2]$ allows for periodic boundary conditions,

$$\Psi^{(N)}(\{z_1, \ldots, z_j + L, \ldots, z_N\}, \{\alpha\}) = \Psi^{(N)}(\{z_1, \ldots, z_j, \ldots, z_N\}, \{\alpha\}) \quad (3.4)$$

for all $1 \leqslant j \leqslant N$. The infinite interval, $L \to \infty$, can be used as well.

Several classes of many-body 1D problems have exact solutions. The spatially-homogeneous ($V_{ax} = 0$) Luttinger liquid model (Luttinger, 1963; Mattis and Lieb, 1965) of spinless and massless fermions can be solved for arbitrary interaction between the particles [this model is not described by the Hamiltonian (3.1) since $m = 0$]. Hard-core zero-range models [$V_{1D} = U_a \delta(z - z')$, $U_a \to \infty$] have exact solutions for arbitrary axial potentials (see Section 3.2 below). Problems with harmonic and inverse-square interparticle interactions can be solved for spatially homogeneous systems and for harmonic axial potentials. Related higher-dimensional problems have exact solutions as well. Various spatially homogeneous models with zero-range and finite-range interactions have Bethe-ansatz solutions (see Section 3.3 below). The quantum inverse scattering method (see Korepin et al., 1993) is a related and powerful analytical method that is often used to calculate correlation functions.

Several approximate methods have been brought to bear on 1D systems. Mean-field (Gross–Pitaevskii) equations can sometimes be used, and non-mean-field effects can be taken into account using the Bogoliubov–de Gennes and Hartree–Fock–Bogoliubov equations or similar methods (see Section 3.4 below). The bosonization method (Haldane, 1981b, 1981a; Giamarchi, 2004) is based on the Luttinger liquid model. It has been recently extended to atom–molecule models (Sheehy and Radzihovsky, 2005; Citro and Orignac, 2005; Orignac and Citro, 2006). An axial potential can be taken into account using the local density approximation when the solution of the related spatially-homogeneous system is known (see, e.g. Olshanii and Dunjko, 2003).

Direct numerical simulations, such as the quantum Monte-Carlo technique (Blume, 2002; Nho and Blume, 2005; Astrakharchik and Giorgini, 2002, 2003; Astrakharchik et al., 2004a, 2006; Boninsegni et al., 2007), stochastic simulations (Schmidt et al., 2005), and simulations based on the density-matrix renormalization group (Schmidt and Fleischhauer, 2007), allow 1D systems to be analyzed when exact and approximate methods cannot be applied. They can also check the applicability of exact models and approximations. Other methods that have been applied to 1D systems include exact diagonalization (Kanamoto et al., 2003a, 2003b, 2005, 2006) and configuration interactions (Alon et al., 2004).

3.2. TONKS–GIRARDEAU GAS

The first exact solution of a many-body 1D problem was obtained in the hard-core limit (see Girardeau et al., 2004, for a historical review). The solution is based on the Girardeau mapping theorem (Girardeau, 1960, 1965; Girardeau and Wright, 2000a, 2000b; Yukalov and Girardeau, 2005) for an N-body system described by the non-stationary Schrödinger equation

$$i\frac{\partial}{\partial t}\Psi^{(N)}(\{z\},t) = \left[-\frac{1}{2m}\sum_{j=1}^{N}\frac{\partial^2}{\partial z_j^2} + V(\{z\},t)\right]\Psi^{(N)}(\{z\},t), \quad (3.5)$$

where the potential V, including all external and finite interaction potentials, is symmetric under permutations of z_j. It does not include the hard-sphere repulsion, which is treated as a constraint on the wavefunction

$$\Psi^{(N)}(\{z\},t) = 0, \quad |z_j - z_k| < R_0 \quad (3.6)$$

for all j and k, where R_0 is the interaction radius.

The wavefunction for N spin-polarized fermions $\Psi_F^{(N)}(\{z\},t)$ is antisymmetric under all particle permutations. Suppose that $\Psi_F^{(N)}(\{z\},t)$ satisfies the Schrödinger equation (3.5) as well as the constraint (3.6). Define a "unit antisymmetric function"

$$A(\{z\}) = \prod_{1 \leqslant j < k \leqslant N} \text{sign}(z_k - z_j). \quad (3.7)$$

We can then write down a bosonic wavefunction, defined by the Fermi–Bose mapping

$$\Psi_B^{(N)}(\{z\},t) = A(\{z\})\Psi_F^{(N)}(\{z\},t), \quad (3.8)$$

that is symmetric under permutations, and satisfies both the Schrödinger equation (3.5) and the hard-core constraint (3.6). It is therefore manifestly a valid solution of the N-boson problem. The relationship between the Bose and Fermi ground-state stationary wavefunctions can be represented in the even simpler form (Girardeau, 1960)

$$\Psi_{B0}^{(N)}(\{z\}) = |\Psi_{F0}^{(N)}(\{z\})|. \quad (3.9)$$

Now consider the case of zero-range interactions, when

$$V(\{z\},t) = \sum_j V_{\text{ax}}(z_j,t)$$

only includes external potentials and $R_0 \to 0$ in the constraint (3.6). The wavefunctions then only vanish when $z_j = z_k$. The constraint (3.6) is then satisfied by any antisymmetric wavefunction of non-interacting fermions. Note that this mapping is similar to the two-body one (2.83), where $U_{\text{eff}} \to \infty$ implies $U_{1D}^- \to 0$. The solution for non-interacting fermions can be represented as a Slater determinant

$$\Psi_F^{(N)}(\{z\},t) = (N!)^{-1/2}\det\varphi_n(z_j,t), \quad (3.10)$$

where the single-particle orbitals $\varphi_n(z_j, t)$ are solutions of the Schrödinger equation

$$i\frac{\partial}{\partial t}\varphi_n(z,t) = \left[-\frac{1}{2m}\frac{\partial^2}{\partial z^2} + V_{\mathrm{ax}}(z,t)\right]\varphi_n(z,t). \quad (3.11)$$

In certain cases the solutions have simple analytical forms. For instance, in the spatially-homogeneous system ($V_{\mathrm{ax}} = 0$) the orbitals are plane waves $\varphi_n(z) = \exp(ip_n z)$, and the system can be analyzed using the stationary Schrödinger equation. The allowed momenta p_n are determined by the periodic boundary conditions (3.4) on $\Psi_B^{(N)}(\{z\})$, which require periodic (antiperiodic) boundary conditions on $\Psi_F^{(N)}(\{z\})$ for odd (even) N. The ground state is the Fermi sea, filled with momenta $p_n = (-N/2 + n - 1/2)$, $1 \leqslant n \leqslant N$. The bosonic wavefunction can be represented as a pair product of the Bijl–Jastrow form (Girardeau, 1960)

$$\Psi_{B0}^{(N)}(\{z\}) = \mathrm{const} \prod_{1 \leqslant i < j \leqslant N} \left|\sin\left[\frac{\pi}{L}(z_i - z_j)\right]\right|. \quad (3.12)$$

An exact solution can be obtained in this case even for a finite-range hard sphere interaction, $R_0 > 0$ (Girardeau, 1960). The ground-state energy is then

$$E_0 = \frac{\pi^2}{6mL^2}(N^3 - N)\left(1 - N\frac{R_0}{L}\right)^{-2}. \quad (3.13)$$

When there is a harmonic axial potential $V_{\mathrm{ax}}(z) = \frac{m}{2}\omega_\parallel^2 z^2$ with axial frequency $\omega_\parallel$, the ground state is a Fermi sea containing N single-particle oscillator orbitals. The bosonic wavefunction then has the Bijl–Jastrow pair product form (Kolomeisky et al., 2000; Girardeau et al., 2001)

$$\Psi_{B0}^{(N)}(\{z\}) = \mathrm{const}\, \exp\left(\sum_{i=1}^{N}\frac{z_i^2}{2a_\parallel^2}\right) \prod_{1 \leqslant j < k \leqslant N} |z_k - z_j|, \quad (3.14)$$

where $a_\parallel = (m\omega_\parallel)^{-1/2}$ is the axial oscillator length. The ground state energy is

$$E_0 = \frac{N^2}{2}\omega_\parallel. \quad (3.15)$$

The problem of spin-polarized fermions with infinite p-wave interactions (the fermionic Tonks–Girardeau gas, see Girardeau and Olshanii, 2003, and Section 3.3.5 below) can also be solved for an arbitrary axial potential, by mapping onto an ideal Bose gas.

3.3. Bethe Ansatz

3.3.1. Bethe-Ansatz Overview

Consider the spatially-homogeneous case of the Hamiltonian (3.1) ($V_{\text{ax}} = 0$). In two-body 1D scattering, the final momenta are unambiguously determined by momentum and energy conservation. Thus particles with equal masses can only keep or exchange their momenta. The situation is different when there are many bodies involved. In the asymptotic region, where interactions can be neglected, the scattered wavefunction can be represented as

$$\Psi^{(N)}(\{z\}, \{\alpha\}) \sim \sum_{\mathcal{P}} C_N(\mathcal{P}) \exp\left(i \sum_{j=1}^{N} p_{\mathcal{P}j} z_j\right)$$

$$+ \int d^N p' S_{\text{difr}}(\{p'\}, \{p\}) \exp\left(i \sum_{j=1}^{N} p'_j z_j\right), \quad (3.16)$$

where $\mathcal{P} = (\mathcal{P}1, \ldots, \mathcal{P}N)$ are permutations of the numbers $(1, \ldots, N)$. The first term describes non-diffractive scattering, where the particles simply exchange their momenta, and the asymptotic momentum set $\{p_1, \ldots, p_N\}$ is conserved. The second term corresponds to diffractive scattering, where the final momenta $\{p'\}$ are restricted only by energy and momentum conservation. There can be a diffractive term even when all the scattered particles have equal masses.

The Bethe-ansatz solution (see Gaudin, 1983; Korepin et al., 1993; Sutherland, 2004; Giamarchi, 2004) requires that $S_{\text{difr}} = 0$. That is, the Bethe ansatz can only be applied when there is only non-diffractive scattering. For non-diffractive scattering, the coefficients $C_N(\mathcal{P})$ in Eq. (3.16) can each be related to two-body scattering amplitudes. That is, at a large distance from an N-body non-diffractive scattering event, it looks equivalent to a series of two-body collisions. It is useful to first consider a set of distinguishable particles identified by the numbers $1 \leq \mathcal{Q}j \leq N$ ($1 \leq j \leq N$), which are elements of the permutation $\mathcal{Q} = (\mathcal{Q}1, \ldots, \mathcal{Q}N)$. The particle $\mathcal{Q}j$ has the coordinate z_j and a set of quantum numbers $\alpha_{\mathcal{Q}j}$. The coordinates are ordered, such that $z_1 \leq z_2 \leq \cdots \leq z_N$. Thus each permutation $\mathcal{Q}$ determines one of the $N!$ sectors that have a particular arrangement of particle positions, such that the particle $\mathcal{Q}1$ is left of the particle $\mathcal{Q}2$, etc., and the particle $\mathcal{Q}N$ is the rightmost.

We will start with the case of zero-range interactions, keeping in mind that the Bethe ansatz can also be applied to certain finite-range models, as discussed in Section 3.3.4 below. With zero-range interactions, the entirety of sector $\mathcal{Q}$, except for its boundaries, is in the asymptotic region. The Bethe ansatz in this sector can be expressed as

$$\Psi^{(N)}_{\{p\}}(\{z\},\{\alpha\}) = \sum_{\mathcal{P}} C_N(\mathcal{Q},\mathcal{P}) \exp\left(i \sum_{j=1}^{N} p_{\mathcal{P}j} z_j\right). \quad (3.17)$$

Equation (3.17) is the essence of the Bethe ansatz, and it can be generalized to include complex momenta, which then describe bound states.

The coefficients $C_N(\mathcal{Q},\mathcal{P})$ in different sectors can be related through two-body scattering amplitudes. Two-body 1D scattering from zero-range potentials can be described in the center-of-mass system by the wavefunction

$$\begin{aligned}\varphi(z_j, z_{j+1}) &= \exp(ip_0(z_j - z_{j+1})) \\ &\quad + \left[f^+_{1D}(p_0) + f^-_{1D}(p_0)\,\mathrm{sign}(z_j - z_{j+1})\right] \\ &\quad \times \exp(ip_0|z_j - z_{j+1}|),\end{aligned} \quad (3.18)$$

where $p_0 = (p_{\mathcal{P}j} - p_{\mathcal{P}(j+1)})/2$ is the collision momentum. The scattering amplitudes $f^\pm_{1D}(p_0)$ are related by Eqs. (2.45) and (2.77) to the T matrices for even and odd 1D pseudopotentials [see Eqs. (2.43) and (2.76), respectively, with $\mu = m/2$]. Thus if $p_0 > 0$ ($p_{\mathcal{P}j} > p_{\mathcal{P}(j+1)}$) the collision transforms the state $(\mathcal{Q},\mathcal{P})$ to the state $(\mathcal{Q}, \hat{\Pi}^{\mathcal{P}}_{jj+1}\mathcal{P})$ with the amplitude

$$f^{\mathrm{ref}}_{\mathcal{P}j\mathcal{P}(j+1)} = f^+_{1D}(p_0) - f^-_{1D}(p_0) \quad (3.19)$$

and to the state $(\hat{\Pi}^{\mathcal{Q}}_{jj+1}\mathcal{Q}, \hat{\Pi}^{\mathcal{P}}_{jj+1}\mathcal{P})$ with the amplitude

$$f^{\mathrm{tr}}_{\mathcal{P}j\mathcal{P}(j+1)} = 1 + f^+_{1D}(p_0) + f^-_{1D}(p_0). \quad (3.20)$$

Here $\hat{\Pi}^{\mathcal{Q}}_{jj+1}$ and $\hat{\Pi}^{\mathcal{P}}_{jj+1}$ are the permutation operators, acting on $\mathcal{Q}$ and $\mathcal{P}$, respectively. If instead, $p_0 < 0$ ($p_{\mathcal{P}j} > p_{\mathcal{P}(j+1)}$), these equations can be obtained from the complex conjugate of the wavefunction (3.18), leading to

$$f^{\mathrm{ref}}_{\mathcal{P}j\mathcal{P}(j+1)} = f^{+*}_{1D}(|p_0|) - f^{-*}_{1D}(|p_0|), \quad (3.21)$$

$$f^{\mathrm{tr}}_{\mathcal{P}j\mathcal{P}(j+1)} = 1 + f^{+*}_{1D}(|p_0|) + f^{-*}_{1D}(|p_0|). \quad (3.22)$$

As $f^{\pm *}_{1D}(p_0) = f^\pm_{1D}(-p_0)$ [see Eqs. (2.43), (2.45), (2.76), and (2.77)], Eqs. (3.19) and (3.20) are applicable for both positive and negative values of p_0, and the relations between the coefficients $C(\mathcal{Q},\mathcal{P})$ of adjacent sectors can be expressed in a compact form as

$$C_N(\mathcal{Q}, \hat{\Pi}^{\mathcal{P}}_{jj+1}\mathcal{P}) = \hat{Y}^{jj+1}_{\mathcal{P}j\mathcal{P}(j+1)} C_N(\mathcal{Q},\mathcal{P}) \quad (3.23)$$

in terms of the Yang matrices (Yang, 1967)

$$\hat{Y}^{ab}_{jk} = f^{\mathrm{ref}}_{jk} + f^{\mathrm{tr}}_{jk} \hat{\Pi}^{\mathcal{Q}}_{ab}. \quad (3.24)$$

Assuming that the momenta are ordered such that $p_1 > p_2 > \cdots > p_N$, the incident wave, which has never seen scattering, corresponds to $\mathcal{P} = (1, 2, \ldots, N)$. Other waves can be obtained from the incident one by consecutive action of the Yang matrices. It is in this way that the Bethe ansatz relates many-body scattering to a sequence of two-body collisions.

Note that if many-body scattering in a given system is diffractive, then the Bethe ansatz approach cannot be applied to that system. This holds true even when interactions are zero-range, if they are diffractive. The condition (3.27) below illustrates this point; it must be satisfied for the Bethe ansatz to be applied, but it is violated when the interaction strength depends on energy. In general, the non-diffractive scattering picture is consistent only if the different sequences that relate the same states give the same amplitudes (Lieb and Liniger, 1963). This requirement can be expressed as the Yang–Baxter relation (Yang 1967, 1968; Gu and Yang, 1989),

$$\hat{Y}_{jk}^{jj+1} \hat{Y}_{ik}^{j+1j+2} \hat{Y}_{ij}^{jj+1} = \hat{Y}_{ij}^{j+1j+2} \hat{Y}_{ik}^{jj+1} \hat{Y}_{jk}^{j+1j+2}, \tag{3.25}$$

which is the necessary condition for non-diffractive scattering. Since this relation involves three momenta only, we can, without loss of generality, consider a three-body problem and reexpress Eq. (3.25), using Eq. (3.24), as a condition on scattering amplitudes (McGuire, 1964)

$$f_{12}^{\text{ref}} f_{13}^{\text{tr}} f_{23}^{\text{ref}} = f_{23}^{\text{ref}} f_{13}^{\text{ref}} f_{12}^{\text{tr}} + f_{23}^{\text{tr}} f_{13}^{\text{ref}} f_{12}^{\text{ref}}. \tag{3.26}$$

Physically, the left-hand side corresponds to the sequence starting from the collision of the 2nd and 3rd particles, and the right-hand side describes interference of two sequences starting from the collision of the 1st and 2nd particles. If the condition (3.26) is violated, the non-diffractive transition amplitude has a discontinuity when the three particles collide simultaneously, and the diffractive scattering term in Eq. (3.16) is required in order to compensate for this discontinuity. In the case of scattering by δ-function potentials [see Eq. (2.41)], Eqs. (2.43), (2.45), and (3.26) lead to the following condition on the interaction strength:

$$p_1 U_{\text{eff}}\left(\frac{(p_2 - p_3)^2}{8\mu}\right) \left[U_{\text{eff}}\left(\frac{(p_1 - p_3)^2}{8\mu}\right) - U_{\text{eff}}\left(\frac{(p_1 - p_2)^2}{8\mu}\right) \right]$$
$$+ p_2 U_{\text{eff}}\left(\frac{(p_1 - p_3)^2}{8\mu}\right) \left[U_{\text{eff}}\left(\frac{(p_1 - p_2)^2}{8\mu}\right) - U_{\text{eff}}\left(\frac{(p_2 - p_3)^2}{8\mu}\right) \right]$$
$$+ p_3 U_{\text{eff}}\left(\frac{(p_1 - p_2)^2}{8\mu}\right) \left[U_{\text{eff}}\left(\frac{(p_2 - p_3)^2}{8\mu}\right) - U_{\text{eff}}\left(\frac{(p_1 - p_3)^2}{8\mu}\right) \right]$$
$$= 0. \tag{3.27}$$

A constant, energy-independent interaction strength U_{eff} satisfies this condition exactly. The condition is not satisfied, however, for energy-dependent interaction strengths, such as those related to Feshbach resonances and the CIR [see

Eqs. (2.42) and (2.46)]. There will always be diffractive scattering in such many-body problems, so they do not have Bethe-ansatz solutions. Many-body systems with multichannel two-body interactions are also not amenable to Bethe-ansatz solutions, since there is no way to use closed-channel elimination in the related two-body problem to restate these interactions in single-channel form. One must instead eliminate the many-body closed channels, as described in Section 5.1.3 below.

If the asymptotic momentum set contains two equal momenta (e.g. $p_{\mathcal{P}j} = p_{\mathcal{P}(j+1)}$ for any permutation $\mathcal{P}$) then

$$f^{-}_{1D}(p_{\mathcal{P}j} - p_{\mathcal{P}(j+1)}) = f^{-}_{1D}(0) = 0$$

[see Eq. (2.77)] and

$$f^{+}_{1D}(p_{\mathcal{P}j} - p_{\mathcal{P}(j+1)}) = f^{+}_{1D}(0) = -1$$

[see Eq. (2.45)]. Therefore

$$\hat{Y}^{jj+1}_{\mathcal{P}j\mathcal{P}(j+1)} = -1,$$

so the terms proportional to $C_N(\mathcal{Q}, \mathcal{P})$ and $C_N(\mathcal{Q}, \hat{\Pi}^{\mathcal{P}}_{jj+1}\mathcal{P})$ cancel each other in Eq. (3.17), making the wavefunction zero. So the set cannot contain equal asymptotic momenta. This is reminiscent of the behavior of indistinguishable fermions, which cannot occupy the same state. In this situation, however, it even applies to distinguishable particles and bosons, as long as the scattering is non-diffractive. It is this property that prevents the formation of a true BEC in 1D systems.

Periodic boundary conditions

$$\Psi^{(N)}_{\{p\}}(\{z_1, z_2, \ldots, z_N\}, \{\alpha\}) = \Psi^{(N)}_{\{p\}}(\{z_2, \ldots, z_N, z_1 + L\}, \{\alpha\}),$$
$$-L/2 \leqslant z_1 \leqslant z_2 \leqslant \cdots \leqslant z_N \leqslant L/2, \qquad (3.28)$$

can be used to analyze many-body properties (see Section 4 below). The application of a sequence of transformations (Yang, 1967; Gaudin, 1983; Sutherland, 2004), moving the particle from the point z_j across all other particles, and then back to the point z_j, leads to N matrix eigenvalue equations

$$\hat{X}_{j+1j}\hat{X}_{j+2j} \cdots \hat{X}_{Nj}\hat{X}_{1j} \cdots \hat{X}_{j-1j}\xi_0 = \exp(ip_j L)\xi_0,$$
$$j = 1, \ldots, N, \qquad (3.29)$$

where $\hat{X}_{ij} = \hat{\Pi}^{\mathcal{Q}}_{ij}\hat{Y}^{ij}_{ij}$ and the $N!$-component vector ξ_0 contains the coefficients $C_N(\mathcal{Q}, (1, \ldots, N))$ for all permutations $\mathcal{Q}$. Equations (3.29) determine a set of asymptotic momenta p_j that are needed to satisfy the periodic boundary conditions. They depend on the symmetry of the system. Obviously, it is impractical to solve these equations for a large number of distinguishable particles. But when there are certain symmetries, solutions can be obtained.

Both the Bethe-ansatz solution (3.17) and the solution (3.10) for the TG gas with *infinite* interaction strengths in an axial potential have the form of a sum of products of single-particle orbitals (plane waves in the case of Bethe ansatz). However, a solution of this form cannot be obtained for a gas with *finite* interaction strength in an arbitrary axial potential. The reason is that even the two-body problem does not have a solution for anharmonic axial potentials. Even in the special case of a harmonic potential, where the center-of-mass motion can be separated and the two-body problem can be solved, an exact solution of the many-body problem has not been obtained.

3.3.2. Lieb–Liniger–McGuire Model

The first and simplest application of the Bethe ansatz to atomic collisions was the Lieb–Liniger model of N spinless indistinguishable bosons with repulsive interactions (Lieb and Liniger, 1963). The case of attractive interactions was considered by McGuire (1964). Since both repulsive and attractive interactions can be described by the same expressions, we consider the Lieb–Liniger and McGuire models together. This model was also independently solved by Berezin et al. (1964).

Bose symmetry prohibits odd interactions ($f_{1D}^- = 0$), while f_{1D}^+ is determined by Eq. (2.45), where $U_{\text{eff}} = U_a$ is constant for the non-diffractive system and $\mu = m/2$. That is, the system is described by the Schrödinger equation (3.3) with $V_{1D}(z_j - z_{j'}, \alpha_j, \alpha_{j'}) = U_a \delta(z_j - z_{j'})$ and $V_{\text{ax}} = 0$. The wavefunction here is symmetric over all particle permutations, so the coefficients $C_N(\mathcal{Q}, \mathcal{P}) \equiv C_N(\mathcal{P})$ are independent of the permutation $\mathcal{Q}$ that determines the sector. Thus $\hat{\Pi}_{ab}^{\mathcal{Q}}$ in Eq. (3.24) acts as the unit operator, leading to the following relation between the coefficients in Eq. (3.17):

$$C_N(\hat{\Pi}_{jj+1}^{\mathcal{P}}\mathcal{P}) = \hat{Y}_{\mathcal{P}j\mathcal{P}(j+1)}^{jj+1} C_N(\mathcal{P})$$
$$= -\frac{mU_a - i(p_{\mathcal{P}(j+1)} - p_{\mathcal{P}j})}{mU_a - i(p_{\mathcal{P}j} - p_{\mathcal{P}(j+1)})} C_N(\mathcal{P}), \qquad (3.30)$$

which can be explicitly expressed as

$$C_N(\mathcal{P}) = (2\pi)^{-N/2} \exp\left(i \sum_{j<j'} \arctan \frac{mU_a}{p_{\mathcal{P}j} - p_{\mathcal{P}j'}}\right). \qquad (3.31)$$

The functions (3.17) with coefficients (3.31) and $p_1 \geqslant p_2 \geqslant \cdots \geqslant p_N$ form a complete orthonormal set on $-\infty < z_1 \leqslant z_2 \leqslant \cdots \leqslant z_N < \infty$ (see Gaudin, 1983)

$$\int_{z_1 \leqslant z_2 \leqslant \cdots \leqslant z_N} d^N z \, \Psi_{\{p'\}}^{(N)*}(\{z\}) \Psi_{\{p\}}^{(N)}(\{z\}) = \prod_{j=1}^{N} \delta(p'_j - p_j), \qquad (3.32)$$

$$\int_{p_1 \geqslant p_2 \geqslant \cdots \geqslant p_N} d^N p \Psi_{\{p\}}^{(N)*}(\{z'\}) \Psi_{\{p\}}^{(N)}(\{z\}) = \prod_{j=1}^{N} \delta(z'_j - z_j). \tag{3.33}$$

The sum in Eq. (3.17) contains the incident wave with momenta arranged in descending order $p_{\mathcal{P}1} \geqslant p_{\mathcal{P}2} \geqslant \cdots \geqslant p_{\mathcal{P}N}$, which correspond to particles that have never scattered in the past, and the outgoing wave with momenta arranged in ascending order $p_{\mathcal{P}1} \leqslant p_{\mathcal{P}2} \leqslant \cdots \leqslant p_{\mathcal{P}N}$, which correspond to particles that will not collide in the future. The S matrix is the ratio of the coefficients of these waves. It can be expressed as

$$S = \prod_{j<j'} \frac{(p_j - p_{j'}) - imU_a}{(p_j - p_{j'}) + imU_a}. \tag{3.34}$$

For the periodic boundary conditions (3.28), Eq. (3.31) leads to the Bethe equations in the asymptotic momenta p_j

$$Lp_j = 2\pi l_j - 2 \sum_{i=1}^{N} \arctan \frac{p_j - p_i}{mU_a}, \tag{3.35}$$

where l_j are integer or half-integer numbers for odd or even N, respectively. (These equations can also be obtained from Eqs. (3.29) and (3.30) when the elements of the vector ξ_0 are equal.). The system (3.35) has a single real solution for each ordered sequence of the quantum numbers $l_1 < l_2 < \cdots < l_N$ (Yang and Yang, 1969).

The functions

$$\Psi_{\{l\}}^{(N)}(\{z\}) = \left(\frac{D(2\pi\{l\})}{D(\{p\})} \right)^{-1/2} \sum_{\mathcal{P}} C_N(\mathcal{P}) \exp\left(i \sum_{j=1}^{N} p_{\mathcal{P}j} z_j \right) \tag{3.36}$$

form an orthonormal basis on $-L/2 \leqslant z_1 \leqslant \cdots \leqslant z_N \leqslant L/2$, where p_j are solutions of Eq. (3.35), the coefficients $C_N(\mathcal{P})$ are determined by Eq. (3.31), and the Jacobian

$$\frac{D(2\pi\{l\})}{D(\{p\})} = \det\left[\left(L + \sum_{k=1}^{N} \frac{2mU_a}{(p_j - p_k)^2 + (mU_a)^2} \right) \delta_{ij} \right. $$
$$\left. - \frac{2mU_a}{(p_j - p_i)^2 + (mU_a)^2} \right] \tag{3.37}$$

(see Korepin et al., 1993). A system in a box $[-L/2, L/2]$ with infinite walls is exactly related to the corresponding $2L$-periodic system (Gaudin, 1971).

When the interactions are attractive ($U_a < 0$), the particles can form N-bosonic bound states (clusters), where N is arbitrary (McGuire, 1964; Berezin et al.,

1964). These states have complex momenta

$$p_j = p + \frac{i}{2}mU_a(N - 2j + 1), \quad 1 \leqslant j \leqslant N. \tag{3.38}$$

Their wavefunctions (3.17) contain only one term, which is proportional to $C_N((1, 2, \ldots, N))$, because Eq. (3.30) makes all the other coefficients zero. The cluster binding energy is

$$E_B(N) = N\frac{N^2 - 1}{24}mU_a^2. \tag{3.39}$$

Because the asymptotic momenta (3.38) have equal real parts, all particles in this state have the same phase, as in a BEC.

The asymptotic momentum set (3.38) describes an N-particle cluster with total momentum Np. A many-body system can contain several such clusters. Cluster collisions have been analyzed by McGuire (1964) and Yang (1968). Interactions do not change the asymptotic momentum set, as is the case for real momenta, where this property relates to non-diffractive scattering. For clusters, the imaginary parts of p_j depend on the cluster size. Therefore collisions cannot change cluster size, and all association and dissociation processes are forbidden, even if they do not violate energy and momentum conservation. Reflection is forbidden when clusters of different sizes collide, since it would change the real parts of the asymptotic momenta. These various restrictions all relate to the internal symmetry of non-diffractive models. When the symmetry is lifted, the restrictions are violated (see Section 5.1.3 below).

3.3.3. Multicomponent Models

Relations (3.23) allow the coefficients C_N to be determined. They yield the Bethe-ansatz wavefunction (3.17) for energy-independent δ-function interactions even in the case of Boltzmann statistics (of distinguishable particles). The equivalence of masses and the equivalence and energy-independence of interaction strengths are the only conditions needed to satisfy the Yang–Baxter relations. Scattering and bound states of Boltzmann particles were analyzed by Yang (1968). As in the Lieb–Liniger–McGuire model, particles with attractive interactions can form clusters, and inelastic collisions of clusters with different sizes are forbidden.

If some of the particles are indistinguishable fermions, the Pauli principle requires that the wavefunction be zero at their points of contact, so they cannot feel δ-function interactions. For example, the wavefunction of a gas of indistinguishable fermions with such interactions is the same as for free fermions (cf. Section 3.2 above). Thus in a multicomponent system containing fermions we can formally suppose the same interactions between all particles, even though even-parity interactions are forbidden when the fermions are indistinguishable.

The odd-parity interactions, which are related to p-wave scattering, are considered in Section 3.3.5 below.

Periodic boundary conditions (3.28) are satisfied by the asymptotic momenta p_j that are the solutions of Eq. (3.29). Since ξ_0 has the symmetry of an irreducible representation of the permutation group, these equations can be solved for appropriately symmetric systems (see Gaudin, 1983; Sutherland, 2004). Such solutions were obtained by Yang (1967) for two-component Fermi or Bose gases with repulsive interactions. They were generalized to multi-component cases by Sutherland (1968).

When the interactions are attractive, the Pauli principle does not allow coupling of indistinguishable fermions in the same cluster. Thus in Fermi gases the cluster size is restricted by the number of components. Such a two-component gas has been considered by Gaudin (1967). The related problem of a two-component trapped Fermi gas was analyzed in Astrakharchik et al. (2004b).

3.3.4. Finite-Range Interactions

As was demonstrated above, a zero range interaction is not sufficient for non-diffractive scattering. It is also not a necessary condition. As an example, the asymptotic Bethe ansatz can be written out for certain finite-range potentials (Sutherland, 2004). This solution only has the form of Eq. (3.17) asymptotically, while the coefficients $C_N(\mathcal{Q}, \mathcal{P})$ are still expressed by Eq. (3.23) in terms of the Yang matrices (3.24). Therefore a many-body collision can be represented as a sequence of two-body ones. To give an unambiguous result, the Yang matrices must satisfy the Yang–Baxter relations (3.25).

Scattering is non-diffractive for the hyperbolic model (Calogero et al., 1975; Calogero, 1975) with the potential

$$V_{\text{hyp}}(z) = \frac{\lambda(\lambda - 1)}{\sinh^2 z} \qquad (3.40)$$

(a solution of this model was also presented in Bhanu Murti, 1960; Berezin et al., 1964) and for its limiting form, the Calogero–Sutherland model (Calogero, 1969b, 1969a, 1971; Marchioro and Presutti, 1970; Perelomov, 1971; Sutherland, 1971c, 1971b, 1971a, 1972) with an inverse-square potential

$$V_{\text{is}}(z) = \lim_{c \to 0} c^2 V_{\text{hyp}}(cz) = \frac{\lambda(\lambda - 1)}{z^2}. \qquad (3.41)$$

Both these potentials are non-penetrable, i.e., $f_{jk}^{\text{tr}} = 0$. Therefore both sides of Eq. (3.26) are zero, and the Yang–Baxter relations are satisfied. However, when the scattering potentials have finite range, several particles can interact simultaneously over a finite length, which means that the Yang–Baxter relations are not sufficient to imply non-diffractive scattering. (If they were, every non-penetrable potential would be non-diffractive.) A proof of integrability and non-diffractive scattering for many-body problems with hyperbolic potentials is presented in Sutherland (2004). The reflection amplitude for the hyperbolic potential has the form (Sutherland, 2004)

$$f_{jk}^{\text{ref}} = f_{jk}^{\text{hyp}} = -\frac{\Gamma(1 + i(p_j - p_k)/2)\Gamma(\lambda - i(p_j - p_k)/2)}{\Gamma(1 - i(p_j - p_k)/2)\Gamma(\lambda + i(p_j - p_k)/2)}. \quad (3.42)$$

In multicomponent systems, non-penetrable potentials prevent different types of particles from mixing. This limitation is lifted in the case of hyperbolic and inverse square exchange potentials (see Sutherland, 2004)

$$V_{\text{he}}(z) = \frac{\lambda(l - \hat{\Pi}^{\mathcal{Q}})}{\sinh^2 z}, \qquad V_{\text{ie}}(z) = \frac{\lambda(l - \hat{\Pi}^{\mathcal{Q}})}{z^2}, \quad (3.43)$$

where the operator $\hat{\Pi}^{\mathcal{Q}}$ exchanges the identities of colliding particles.

The hyperbolic exchange potential leads to the Yang matrix in the form of Eq. (3.24) with reflection and transmission amplitudes

$$f_{jk}^{\text{ref}} = \frac{2\lambda}{2\lambda + i(p_j - p_k)} f_{jk}^{\text{hyp}}, \qquad f_{jk}^{\text{tr}} = i\frac{p_j - p_k}{2\lambda + i(p_j - p_k)} f_{jk}^{\text{hyp}}. \quad (3.44)$$

These amplitudes are proportional to the amplitudes for the δ-function potential with $\mu U_{\text{eff}} = \lambda$ and therefore satisfy the relation (3.26). The non-diffractive behavior of the hyperbolic exchange potential is proven in Sutherland (2004). The inverse square potential is a limiting case of the hyperbolic one and it is therefore non-diffractive as well.

3.3.5. Fermi–Bose Mapping

The relation between the solutions of fermionic and bosonic problems for hard-core interactions, discussed in Section 3.2 above, can be extended to multicomponent many-body problems with zero-range interactions of arbitrary strength (Cheon and Shigehara, 1999; Granger and Blume, 2004; Girardeau and Olshanii, 2003, 2004; Girardeau et al., 2004; Yukalov and Girardeau, 2005). Consider an N-boson system described by the wavefunction $\Psi_B^{(N)}(\{z\}, \{\alpha\})$ with proper permutation symmetry. It consists of the sum of spatially odd and even components with corresponding spin symmetry and satisfies the Schrödinger equation (3.3) with the zero-range interaction

$$\hat{V}_{1D}^{(B)}(z_j - z_k, \alpha_j, \alpha_k) = U_B^-(\alpha_j, \alpha_k)\delta'(z_j - z_k)\hat{\partial}_{jk} \\
+ U_B^+(\alpha_j, \alpha_k)\hat{\delta}_{jk}, \quad (3.45)$$

which includes odd- and even-parity parts. Here, the operator $\hat{\partial}_{jk}$, which acts on the many-body wavefunction in the following way,

$$\hat{\partial}_{jk}\Psi^{(N)} = \frac{1}{2}\left[\frac{\partial}{\partial z_j}\Psi^{(N)}\bigg|_{z_j=z_k+0} - \frac{\partial}{\partial z_k}\Psi^{(N)}\bigg|_{z_j=z_k-0}\right], \quad (3.46)$$

is the many-body generalization of the two-body operator $\hat{\partial}_{\pm}$ in Eq. (2.71). The symmetric δ-function operator

$$\hat{\delta}_{jk}\Psi^{(N)} = \frac{1}{2}\left[\Psi^{(N)}\big|_{z_j=z_k+0} + \Psi^{(N)}\big|_{z_j=z_k-0}\right]\delta(z_j - z_l) \qquad (3.47)$$

is required because the odd-parity interaction leads to a wavefunction discontinuity.

The transformation

$$\Psi_F^{(N)}(\{z\},\{\alpha\}) = A(\{z\})\Psi_B^{(N)}(\{z\},\{\alpha\}), \qquad (3.48)$$

where the unit antisymmetric function is defined by Eq. (3.7), does not affect the spin part of the wavefunction. It transforms spatially odd components of $\Psi_B^{(N)}(\{z\},\{\alpha\})$ to spatially even components of $\Psi_F^{(N)}(\{z\},\{\alpha\})$ and vice versa. Therefore, $\Psi_F^{(N)}(\{z\},\{\alpha\})$ has the proper fermionic permutation symmetry.

The two terms of the interaction (3.45) act only on the parts of $\Psi^{(N)}$ that are, respectively, odd and even functions of $z_j - z_k$ in the vicinity of $z_j = z_k$. In this vicinity, the transformation (3.48) is the same as the two-body one (2.83). Therefore, $\Psi_F^{(N)}(\{z\},\{\alpha\})$ satisfies the Schrödinger equation (3.3) with the interaction

$$\hat{V}_{1D}^{(F)}(z_j - z_k, \alpha_j, \alpha_k) = U_F^-(\alpha_j, \alpha_k)\delta'(z_j - z_k)\hat{\partial}_{jk} \\ + U_F^+(\alpha_j, \alpha_k)\hat{\delta}_{jk}, \qquad (3.49)$$

where the fermionic interaction parameters are related to the bosonic ones by

$$U_F^-(\alpha_j, \alpha_k) = \frac{4}{m^2 U_B^+(\alpha_j, \alpha_k)}, \quad \text{and}$$

$$U_F^+(\alpha_j, \alpha_k) = \frac{4}{m^2 U_B^-(\alpha_j, \alpha_k)}. \qquad (3.50)$$

This approach maps a spin-polarized Fermi gas with odd-parity interactions onto a gas of δ-function interacting bosons. This mapping avoids the consideration of odd-parity pseudopotentials, which become ambiguous in higher-order perturbation theory. For example, the Fermi–Bose mapping solves the problem of the spin-polarized fermionic Tonks–Girardeau gas when $U_F^- \to \infty$ by mapping it onto the ideal Bose gas (Girardeau and Olshanii, 2003; Girardeau et al., 2004; Girardeau and Minguzzi, 2006).

3.4. Gross–Pitaevskii and Bogoliubov Methods

This section describes approximate methods based on the separation of mean-field effects and quantum fluctuations (see Fetter and Walecka, 1971; Akhiezer and Peletminskii, 1981; Bogolubov and Bogolubov Jr, 1994; Pitaevskii and Stringari,

2003; Dalfovo et al., 1999; Leggett, 2001; Andersen, 2004; Kocharovsky et al., 2006). The many-body Hamiltonian (3.1) leads to the following equation of motion for annihilation operators (in the Heisenberg representation)

$$i\frac{\partial}{\partial t}\hat{\Psi}_\alpha(z,t) = \left[-\frac{1}{2m}\frac{\partial^2}{\partial z^2} + V_{ax}(z,\alpha) \right.$$
$$\left. + \sum_{\alpha'}\int dz' V_{1D}(z-z',\alpha,\alpha')\hat{\Psi}^\dagger_{\alpha'}(z',t)\hat{\Psi}_{\alpha'}(z',t)\right]$$
$$\times \hat{\Psi}_\alpha(z,t). \quad (3.51)$$

It applies to both bosons and fermions. However, it is a non-linear operator equation and cannot be solved directly.

A very general method for analyzing many-body systems uses the quantum Bogoliubov–Born–Green–Kirkwood–Yvon (BBGKY) equations (see Akhiezer and Peletminskii, 1981) for many-body distribution functions

$$f_{kl}(\{z\},\{z'\},\{\alpha\},\{\alpha'\},t)$$
$$= \langle \hat{\Psi}^\dagger_{\alpha_1}(z_1,t)\cdots\hat{\Psi}^\dagger_{\alpha_k}(z_k,t)\hat{\Psi}_{\alpha'_1}(z'_1,t)\cdots\hat{\Psi}_{\alpha'_l}(z'_l,t)\rangle,$$

where $\langle \hat{A}\rangle = \text{Tr}(\hat{\rho}\hat{A})$ and $\hat{\rho}$ is the system density operator. The quantum distribution functions f_{kl} satisfy an infinite set of coupled equations of motion.

For bosons, the lowest-order distribution function is the mean field (Fetter and Walecka, 1971)

$$\Phi_\alpha(z,t) = f_{01}(z,t) = \langle \hat{\Psi}_\alpha(z,t)\rangle. \quad (3.52)$$

In the case of a single component, replacing the annihilation operator in Eq. (3.51) by the corresponding mean (classical) field and using the δ-function interaction $V_{1D}(z-z',\alpha,\alpha') = U_a\delta(z-z')$ lead to the 1D variant

$$i\frac{\partial}{\partial t}\Phi(z,t) = \left[-\frac{1}{2m}\frac{\partial^2}{\partial z^2} + V_{ax}(z,\alpha) + U_a|\Phi(z,t)|^2\right]\Phi(z,t) \quad (3.53)$$

of the famous Gross–Pitaevskii equation (Gross, 1961, 1963; Pitaevskii, 1961) (see also Fetter and Walecka, 1971; Dalfovo et al., 1999; Pitaevskii and Stringari, 2003). The classical field approach, which is a mean-field method in the finite-temperature case, was developed in Scalapino et al. (1972), Castin et al. (2000), Castin (2004). Coupled Gross–Pitaevskii equations have been applied to atom–molecule systems (see Timmermans et al., 1999; Cusack et al., 2001; Alexander et al., 2002; Köhler et al., 2006, and references therein), including systems in 1D (Driben et al., 2007; Oles and Sacha, 2007).

The non-mean-field part of the annihilation operator,

$$\hat{\xi}_\alpha(z,t) = \hat{\Psi}_\alpha(z,t) - \Phi_\alpha(z,t), \quad (3.54)$$

describes quantum fluctuations. Whenever the fluctuations are weak, a linear equation of motion for $\hat{\xi}_\alpha(z,t)$ can be derived from Eq. (3.51), since higher powers of $\hat{\xi}_\alpha(z,t)$ can be neglected. For δ-function interactions and a single component this equation has the form

$$i\frac{\partial}{\partial t}\hat{\xi}(z,t) = \hat{H}_\xi\hat{\xi}(z,t) + U_a\Phi^2(z,t)\hat{\xi}^\dagger(z,t), \tag{3.55}$$

where

$$\hat{H}_\xi = -\frac{1}{2m}\frac{\partial^2}{\partial z^2} + V_{ax}(z,\alpha) + 2U_a|\Phi(z,t)|^2. \tag{3.56}$$

A solution of this operator equation can be expressed as a linear combination of its c-number solutions with operator coefficients. Bogoliubov–de Gennes equations (de Gennes, 1966; see Fetter, 1972, for bosons),

$$\epsilon_k u_k(z) = \hat{H}_\xi u_k(z) - U_a\Phi^2(z,t)v_k(z), \tag{3.57}$$

$$\epsilon_k v_k(z) = U_a[\Phi^*(z,t)]^2 u_k(z) - \hat{H}_\xi v_k(z), \tag{3.58}$$

can be derived if the fluctuation operator can written in the form

$$\hat{\xi}(z,t) = \sum_k\left[u_k(z)\hat{\zeta}_k e^{-i\epsilon_k t} - v_k^*(z)\hat{\zeta}_k^\dagger e^{i\epsilon_k t}\right], \tag{3.59}$$

in terms of field operators $\hat{\zeta}_k$ for Bogoliubov quasiparticles, which obey the conventional Bose commutation relations. A different representation of $\hat{\xi}(z,t)$ was used in Haus and Yu (2000). Bogoliubov–de Gennes equations were applied to 1D systems in Poulsen and Molmer (2003), Sinha et al. (2006), Sacha and Timmermans (2006). In the spatially-homogeneous case, $u_k(z)$ and $v_k(z)$ are plane waves, and Eq. (3.59) is the standard Bogoliubov transformation.

However, a Bogoliubov transformation only exists if the eigenvalues ϵ_k, obtained from Eq. (3.57), are real (Bogolubov and Bogolubov Jr, 1994). Otherwise, e.g., in the case of attractive interactions ($U_a < 0$), other representations of $\hat{\xi}(z,t)$ must be used, such as quantum integrals of motion (see Dodonov and Man'ko, 1989; Man'ko, 1997, and references therein), and the parametric approximation (Yurovsky, 2002). In the spatially homogeneous parametric approximation the fluctuation operator in the momentum representation is expressed as

$$\hat{\tilde{\xi}}(q,t) = (2\pi)^{-1/2}\int dz e^{-iqz}\hat{\xi}(z,t)$$
$$= \xi_c(q,t)\hat{\tilde{\xi}}(q,0) + \xi_s(q,t)\hat{\tilde{\xi}}^\dagger(q,0), \tag{3.60}$$

where the c-number functions $\zeta_{c,s}(q,t)$ are solutions of the equations

$$i\frac{\partial}{\partial t}\xi_{c,s}(q,t) = \left[\frac{p^2}{2m} + 2U_a|\Phi(t)|^2\right]\xi_{c,s}(q,t) + U_a\Phi^2(t)\xi_{s,c}^*(q,t). \tag{3.61}$$

Unlike Eq. (3.59), this representation does not require the existence of quasiparticles. This approach was applied to atom–molecule systems in Yurovsky and Ben-Reuven (2003). A similar method for fermions was recently developed by Kheruntsyan (2006). The parametric approximation was applied to 1D systems in Driben et al. (2007).

All these methods are equivalent to using truncated quantum BBGKY equations, taking into account f_{01}, f_{11}, and f_{02}. The Hartree–Fock–Bogoliubov (HFB) method (Fetter and Walecka, 1971; Griffin, 1996; Proukakis and Burnett, 1996; Proukakis et al., 1998; Walser et al., 1999) uses the normal and anomalous densities defined, respectively, as

$$n_s(z_1, z_2, t) = \langle \hat{\xi}^\dagger(z_1, t)\hat{\xi}(z_2, t) \rangle, \tag{3.62}$$

$$m_s(z_1, z_2, t) = \langle \hat{\xi}(z_1, t)\hat{\xi}(z_2, t) \rangle. \tag{3.63}$$

These densities are also related to quantum distribution functions. Equations of motion for the densities can be derived from the equation of motion for the fluctuation operator (Griffin, 1996)

$$i\frac{\partial}{\partial t}\hat{\xi}(z,t) = \left[\hat{H}_\xi + 2U_a n_s(z,z,t)\right]\hat{\xi}(z,t)$$
$$+ U_a\left[\Phi^2(z,t) + m_s(z,z,t)\right]\hat{\xi}^\dagger(z,t), \tag{3.64}$$

which differs from Eq. (3.55) by non-linear terms proportional to n_s and m_s. Such non-linear terms can also be incorporated into equations that have the form of the parametric approximation used for 1D problems in Buljan et al. (2005), Merhasin et al. (2006). However, the HFB method gives a quasiparticle spectrum with an artificial gap in the limit of long wavelengths, which may be especially important in 1D systems due to infrared divergence, as discussed below (for discussion of conserving and gapless approximations see Griffin, 1996; Andersen, 2004; Yukalov and Kleinert, 2006; Yukalov, 2006, and references therein). The HFB method was applied to atom–molecule systems in Holland et al. (2001), Kokkelmans et al. (2002), Kokkelmans and Holland (2002). The macroscopic quantum dynamics approach (Köhler et al., 2003, 2006; Góral et al., 2004) considers the molecules to be two-atom bound states, and incorporates actual interatomic potentials.

The normal and anomalous densities are second-order quantum cummulants. Quantum cummulants can be defined to arbitrary order (Fricke, 1996; Kocharovsky et al., 2000a, 2000b; Köhler and Burnett, 2002). Corresponding equations have been derived for higher-order cummulants (Fricke, 1996; Proukakis and Burnett, 1996; Proukakis et al., 1998; Köhler and Burnett, 2002) and for 1D systems (Gasenzer et al., 2005).

When the Gross–Pitaevskii equation takes into account the effects of second-order cummulants, it takes the form:

$$i\frac{\partial}{\partial t}\Phi(z,t) = \left[-\frac{1}{2m}\frac{\partial^2}{\partial z^2} + V_{\text{ax}}(z,\alpha)\right.$$
$$\left. + U_a\left(|\Phi(z,t)|^2 + 2n_s(z,z,t)\right)\right]\Phi(z,t)$$
$$+ U_a m_s(z,z,t)\Phi^*(z,t). \tag{3.65}$$

In the 3D case, $m_s(\mathbf{r}_1, \mathbf{r}_2, t)$ diverges when $\mathbf{r}_1 \to \mathbf{r}_2$. This ultraviolet divergence (in the momentum representation) is caused by the 3D δ-function interaction. It can be eliminated by a renormalization procedure, which leads to the replacement of U_a by the two-body T matrix (Beliaev, 1958a, 1958b; Proukakis et al., 1998).

In the 1D case, both the normal and anomalous densities can be estimated according to

$$n_s(z,z,t) \sim m_s(z,z,t) \sim \frac{1}{l_c}\ln\frac{L}{l_c}, \tag{3.66}$$

where the coherence (or healing) length is

$$l_c = (mU_a n_{1D})^{-1/2} \tag{3.67}$$

and $n_{1D} = N/L \approx |\Phi|^2$ is the 1D density. Thus in infinite systems, $L \to \infty$, the quantum fluctuations diverge. This infrared divergence, which is related to long-wavelength fluctuations, prevents BEC formation in infinite 1D systems. Nevertheless, one can divide the system into segments of finite length L, each of which can be described as a BEC with weak fluctuations, as long as the density is large enough that

$$n_{1D}l_c \gg \ln L/l_c. \tag{3.68}$$

In general, the condensates in different segments have different phases. In spite of this fluctuating phase, the quasicondensate has weak density fluctuations. Unlike a true condensate, the one-body correlation function of a quasicondensate decays at long distance (see Section 4 below). The Bogoliubov approach to 1D quasicondensates was developed in Mora and Castin (2003), Castin (2004). This approach takes into account the effect of the Goldstone mode (see also Lewenstein and You, 1996; Castin and Dum, 1998). Number and phase fluctuations in solitons have been analyzed by the linearization method (Haus and Lai, 1990; Haus and Yu, 2000).

The mean-field approach can be applied to high-density 1D systems, unlike in 3D, where high density leads to strong fluctuations. The low-density (or strong-interaction) regime can be analyzed using the Girardeau hard-core model (see Section 3.2 above). A version of the Gross–Pitaevskii equation has also been derived for this case (Kolomeisky et al., 2000). In this version the cubic term in Eq. (3.53) is replaced by the term $\frac{\pi^2}{2m}|\Phi|^4\Phi$. This is similar to the quintic term $24\ln(4/3)a_{el}^2\omega_\perp|\Phi|^4\Phi$, which can be added to the Gross–Pitaevskii equation [Eq. (3.53)] to account for virtual excitations in a transverse waveguide (Muryshev et al., 2002; Cherny and Brand, 2004).

4. Ground State Properties of Short-Range-Interacting 1D Bosons: Known Results and Their Experimental Verification

4.1. THERMODYNAMIC LIMIT

The coefficients of the Bethe-ansatz solution (3.17) of the Lieb–Liniger–McGuire model have the explicit form (3.31). The Bethe equations (3.35) for asymptotic momenta, which satisfy periodic boundary conditions, are also explicit. The ground state wavefunction $\Psi_0(\{z\})$ that corresponds to $l_{j+1} - l_j = 1$ is the Fermi sea, in which all asymptotic momentum states between $p_1 = -p_F$ and $p_N = p_F$ are filled, where p_F is the "Fermi" momentum. Despite the fact that these solutions are explicit, it is practically impossible to solve the set of transcendental equations (3.35) for macroscopic systems with large numbers of particles (although these equations were solved for N up to 50 in Sakmann et al., 2005). It is, however, possible to obtain an approximate solution in the thermodynamic limit, where $N \to \infty$, $L \to \infty$, and the 1D density $n_{1D} = N/L$ is fixed. In this limit, Eq. (3.35) for the ground state can be reduced to the Lieb–Liniger equation (Lieb and Liniger, 1963, see also Gaudin, 1983; Korepin et al., 1993; Sutherland, 2004), for the density of asymptotic momenta

$$g\left(\frac{p_j}{p_F}\bigg|\gamma\right) = \frac{1}{(p_{j+1} - p_j)L}. \tag{4.1}$$

Here the dimensionless parameter

$$\gamma = mU_a/n_{1D} \tag{4.2}$$

is proportional to the interaction strength and inversely proportional to the 1D density n_{1D}. The Lieb–Liniger integral equation has the form

$$g(x|\gamma) - \frac{1}{2\pi}\int_{-1}^{1}\frac{2\lambda(\gamma)}{\lambda^2(\gamma) + (y-x)^2}g(y|\gamma)\,dy = \frac{1}{2\pi}, \tag{4.3}$$

where $\lambda(\gamma) = mU_a/p_F$ and the Fermi momentum p_F is determined by the normalization condition

$$\gamma\int_{-1}^{1}g(x|\gamma)\,dx = \lambda(\gamma). \tag{4.4}$$

The Lieb–Liniger equation does not have an exact analytical solution. The numerically-calculated ground state energy of the system is given by

$$E/N = \frac{1}{2m} n_{1D}^2 e_2(\gamma). \tag{4.5}$$

where the function $e_2(\gamma)$ is the normalized second moment [tabulated in Olshanii, 2002, under $e(\gamma)$] of $g(x|\gamma)$. The normalized mth moment is

$$e_m(\gamma) = \left(\frac{\gamma}{\lambda(\gamma)}\right)^{m+1} \int_{-1}^{1} x^m\, g(x|\gamma)\, dx. \tag{4.6}$$

Approximate solutions of the Lieb–Liniger equation can be obtained in two limiting cases (Lieb and Liniger, 1963). In the Tonks–Girardeau regime, which corresponds to strong interactions and low density, where $\gamma \gg 1$, the system is similar to a gas of hard-core bosons (see Section 3.2). In this case

$$g(x|\gamma) \approx \frac{\lambda}{2\pi\lambda - 4}, \quad \lambda \approx \frac{\gamma + 2}{\pi}, \quad p_F \approx \frac{\pi \gamma n_{1D}}{\gamma + 2},$$
$$e_2(\gamma) \approx \frac{\pi^2}{3}\left(\frac{\gamma}{\gamma+2}\right)^2. \tag{4.7}$$

The opposite, Gross–Pitaevskii, regime of weak interactions and high densities, where $\gamma \ll 1$, can be analyzed using the Bogoliubov method (see Section 3.4). In this case

$$g(x|\gamma) \approx \frac{1}{2\pi\lambda}\sqrt{1 - x^2}, \quad \lambda \approx \frac{1}{2}\sqrt{\gamma},$$
$$p_F \approx 2n_{1D}\sqrt{\gamma}, \quad e_2(\gamma) \approx \gamma. \tag{4.8}$$

Excited states of the Lieb–Liniger McGuire model can be formed from two types of elementary excitations (Lieb, 1963; Yang and Yang, 1969), see also (Korepin et al., 1993; Sutherland, 2004). The first, particle type consists of adding an extra particle with momentum p_{N+1} outside the Fermi sea, so that $|p_{N+1}| > p_F$. The second, hole type consists of removing a particle with momentum p_j from the Fermi sea. Elementary excitations in a two-component Bose gas were analyzed in Fuchs et al. (2005).

4.2. Correlation Functions

We will be interested in the ground-state m-body correlation functions:

$$\rho_m(z_1, \ldots, z_m; z_1', \ldots, z_m') \equiv \langle \hat{\Psi}^\dagger(z_1') \cdots \hat{\Psi}^\dagger(z_m') \hat{\Psi}(z_1) \cdots \hat{\Psi}(z_m) \rangle$$
$$\equiv \frac{N!}{(N-m)!} \underbrace{\int \cdots \int}_{N-m} dz_{m+1} \ldots dz_N$$

$$\times \Psi_0(\{z_1, \ldots, z_m, z_{m+1}, \ldots, z_N\})$$
$$\times \Psi_0^*(\{z_1', \ldots, z_m', z_{m+1}, \ldots, z_N\}). \quad (4.9)$$

A review of exact and approximate results for correlation functions is presented below.

4.2.1. Short-Range Correlation Properties

An expansion of the one-body correlation function in the Tonks–Girardeau limit is published in Lenard (1964), Vaidya and Tracy (1979), Jimbo et al. (1980) up to 10th order,

$$\rho_1(0; z)/n_{1D} \stackrel{\gamma \to \infty}{=} 1 + c_2(n_{1D}z)^2 + c_3(n_{1D}|z|)^3 + c_4(n_{1D}z)^4 + \cdots, \quad (4.10)$$

where

$$c_2 = -\frac{\pi^2}{3!}, \quad c_3 = \frac{\pi^2}{3^2}, \quad c_4 = \frac{\pi^4}{5!}, \quad \ldots. \quad (4.11)$$

An expansion of the one-body correlation function for arbitrary γ was obtained in Olshanii and Dunjko (2003, 2006)

$$\rho_1(0; z)/n_{1D} = 1 + c_2(n_{1D}z)^2 + c_3(n_{1D}|z|)^3 + c_4(n_{1D}z)^4 + \mathcal{O}((n_{1D}z)^5), \quad (4.12)$$

where

$$c_2 = -\frac{1}{2}\{e_2(\gamma) - \gamma e_2'\}, \quad (4.13)$$

$$c_3 = \frac{1}{12}\gamma^2 e_2', \quad (4.14)$$

$$c_4 = \frac{\gamma^2}{12}\left[\frac{1}{\gamma}e_4' - \frac{9}{2\gamma^2}e_4 + \left(1 + \frac{\gamma}{2}\right)e_2' - \frac{2}{\gamma}e_2 - \frac{3}{\gamma}e_2 e_2' + \frac{9}{\gamma^2}e_2^2\right], \quad (4.15)$$

where $e_m' = de_m/d\gamma$

Two-body correlation functions are expressed in Gangardt and Shlyapnikov (2003b) using the Hellmann–Feynman theorem in terms of the scaled energy [the second moment (4.6)] $e_2(\gamma)$,

$$\rho_2(0, 0; 0, 0)/n^2 = e_2'. \quad (4.16)$$

This gives approximate expressions in the Gross–Pitaevskii regime, $\gamma \ll 1$,

$$\rho_2(0, 0; 0, 0) \approx n_{1D}^2, \quad (4.17)$$

and in the Tonks–Girardeau regime, $\gamma \gg 1$,

$$\rho_2(0,0;0,0) \approx 4\pi^2 n_{1D}^2/(3\gamma^2). \tag{4.18}$$

Three-body correlation functions are expressed in Cheianov et al. (2006) in terms of the normalized moments (4.6) as

$$\frac{\rho_3(0,0,0;0,0,0)}{n^3} = \frac{3}{2\gamma}e_4' - \frac{5}{\gamma^2}e_4 + \left(1 + \frac{\gamma}{2}\right)e_2' - \frac{2}{\gamma}e_2$$
$$- \frac{3}{\gamma}e_2 e_2' + \frac{9}{\gamma^2}e_2^2. \tag{4.19}$$

General expressions for m-body correlation functions have been obtained in the Tonks–Girardeau regime (Gangardt and Shlyapnikov, 2003b, 2003a)

$$\frac{\rho_m(0,\ldots,0;0,\ldots,0)}{n_{1D}^m} \stackrel{\gamma \to \infty}{=} \left\{\prod_{m'=0}^{m-1} \pi \frac{2m'+1}{2^{2m'+2}}\left[\frac{\Gamma(m'+2)}{\Gamma(m'+3/2)}\right]^2\right\}\left(\frac{\pi}{\gamma}\right)^{m(m-1)} \tag{4.20}$$

and in the Gross–Pitaevskii regime (Gangardt and Shlyapnikov, 2003a)

$$\frac{\rho_m(0,\ldots,0;0,\ldots,0)}{n_{1D}^m} \stackrel{\gamma \to 0}{=} 1 - \frac{m(m-1)}{\pi}\sqrt{\gamma}. \tag{4.21}$$

Two-body correlation functions can be determined for the two-body problem. They are

$$\rho_{p_1,p_2}^{(2)}(z) \equiv \rho_2(0,0;z,z) = \left[\Psi_{\{p_1,p_2\}}^{(2)}(\{0,0\})\right]^* \Psi_{\{p_1,p_2\}}^{(2)}(\{z,z\})$$
$$= \frac{2}{L^2} \frac{(p_1-p_2)^2}{(p_1-p_2)^2 + m^2 U_a^2} \exp[i(p_1+p_2)z], \tag{4.22}$$

where $\Psi_{\{p_1,p_2\}}^{(2)}(\{z\})$ is the Bethe-ansatz wavefunction (3.17) with coefficients (3.31) for $N = 2$. In the case of $N > 2$, the two-body correlation function can be estimated (Yurovsky and Band, 2007) as the sum of $\rho_{p_j,p_{j'}}^{(2)}(z)$ over all pairs of colliding particles with asymptotic momenta p_j and $p_{j'}$, determined by the periodic boundary conditions

$$\rho_2(0,0;z,z) \approx \sum_{j<j'} \rho_{p_j,p_{j'}}^{(2)}(z)$$
$$\approx \frac{L^2}{2}p_F^2 \int_{-1}^{1} dx_1 \int_{-1}^{1} dx_2 g(x_1|\gamma)g(x_2|\gamma)\rho_{p_Fx_1,p_Fx_2}^{(2)}(z), \tag{4.23}$$

where the summation over pairs is replaced by integration with the asymptotic-momentum densities.

In the Gross–Pitaevskii regime, substitution of Eq. (4.8) into Eq. (4.23) leads to a value for $\rho_2(0, 0; 0, 0)$ that is in full agreement with the exact result (4.17). In the Tonks–Girardeau regime, use of Eq. (4.7) leads to a value for $\rho_2(0, 0; 0, 0)$ that is half of the exact value (4.18). The difference results from the highly-correlated behavior of the Tonks–Girardeau gas. Equation (4.23) does not capture all the correlations, since it includes an average over the independent quasimomentum distributions of the two particles. Still, this estimate demonstrates that short-range correlations in the many-body system are mostly determined by the two-body effect of particle repulsion.

4.2.2. Long-Range Correlation Properties

One- and two-body correlation functions are evaluated in Haldane (1981b, 1981a), Berkovich and Murthy (1989), Korepin et al. (1993), Korepin and Slavnov (1997)

$$\rho_1(0; z)/n_{1D} = \bar{\rho}(n_{1D}z)^{-1/\eta} \times \{1 + \mathcal{O}((n_{1D}z)^{-\eta})\}, \tag{4.24}$$

$$\rho_2(0, z; 0, z)/n_{1D}^2 = 1 + \eta(2\pi n_{1D}z)^{-2} + a_1(n_{1D}z)^{-\eta}\cos(2\pi n_{1D}z)$$
$$+ \mathcal{O}((n_{1D}z)^{-4\eta}), \tag{4.25}$$

where

$$\eta = 2\sqrt{\frac{2\pi^2}{6e_2 - 4\gamma e'_2 + \gamma^2 e''}}. \tag{4.26}$$

In the Tonks–Girardeau regime

$$\eta \stackrel{\gamma \to \infty}{=} 2, \tag{4.27}$$

$$\bar{\rho} \stackrel{\gamma \to \infty}{=} \sqrt{\pi} e^{1/2} 2^{-1/3} A^{-6} \left(1 + (2\mathcal{C} + 4\ln(2) - 5 + 2\ln(\pi))\gamma^{-1}\right.$$
$$\left. + \mathcal{O}(\gamma^{-2})\right)$$
$$= 0.521413(1 + 1.21648\gamma^{-1} + \mathcal{O}(\gamma^{-2})) \tag{4.28}$$

(Creamer et al., 1981; Lenard, 1964; Vaidya and Tracy, 1979; Jimbo et al., 1980), where $A \approx 1.2824$ is Glaisher's constant. In the Gross–Pitaevskii regime (Mora and Castin, 2003)

$$\eta \stackrel{\gamma \to 0}{=} 2\pi/\gamma, \tag{4.29}$$

$$\bar{\rho} \stackrel{\gamma \to 0}{=} \left(e^{2-\mathcal{C}}/(4\gamma)\right)^{1/(2\eta)} = (1.076/\gamma)^{1/(2\eta)}, \tag{4.30}$$

where $\mathcal{C} \approx 0.577216$ is the Euler constant.

The above expansion (4.28) agrees with the all-distance finite-size result (Girardeau, 1960; Girardeau and Wright, 2002)

$$\rho_2(0, 0; z, z)/\left[n_{1D}^2\left(1 - N^{-1}\right)\right] \stackrel{\gamma \to \infty}{=} 1 - \left(\frac{\sin(\pi n_{1D} z)}{\pi n_{1D} z}\right)^2. \qquad (4.31)$$

4.3. FURTHER EXTENSIONS

In the limit of weak interactions, an exhaustive theory of Bose gases for systems without true condensation was developed in Mora and Castin (2003). This theory applies particularly well to the case of 1D gases, where explicit expressions for the correlation functions at all distances for $\gamma \ll 1$ have been obtained.

An impressive body of exact results for finite size corrections to harmonic trap effects on the correlation functions is presented in Forrester et al. (2003, 2006). For long range correlations in a harmonic trap see also Girardeau and Wright (2002), Gangardt and Shlyapnikov (2003b). Atomic correlations at finite temperatures were extensively studied in Gangardt and Shlyapnikov (2003a), Mora and Castin (2003), Kheruntsyan et al. (2003, 2005), Schmidt et al. (2005), Schmidt and Fleischhauer (2007). Dynamic structure factors, which are related to correlation functions, were calculated in Brand and Cherny (2005), Cherny and Brand (2006). Correlation functions have also been analyzed for the Calogero–Sutherland model (see Section 3.3.4 above) (Astrakharchik et al., 2006), for a spin-polarized fermionic gas (Bender et al., 2005), for multicomponent models (Imambekov and Demler, 2006b, 2006a), and for an atom–molecule gas (Sheehy and Radzihovsky, 2005; Citro and Orignac, 2005; Orignac and Citro, 2006).

4.4. EXPERIMENTAL STUDIES OF 1D QUASICONDENSATES

When the parameter γ [see Eq. (4.2)] is small, mean field theory can be used to describe 1D Bose gases (see Section 3.4). This regime is reached with relatively weak transverse trapping and high 1D density. These gases behave like quasicondensates. Their local coherence properties are like 3D BEC's, except that the phase coherence decays with distance. Decaying coherence has been experimentally observed in several ways, all using single magnetic traps. These traps, as opposed to the arrays of traps obtained in an optical lattice, have been the natural choice for these observations for two reasons. First, the quasicondensate phase variations are not averaged out by summing over tubes. Second, these systems have relatively small γ, both due to weak transverse confinement and higher 1D density. High 1D density is made possible by weak transverse confinement, but in fact, there is typically no choice in the matter; high density is dictated by the need for larger signals.

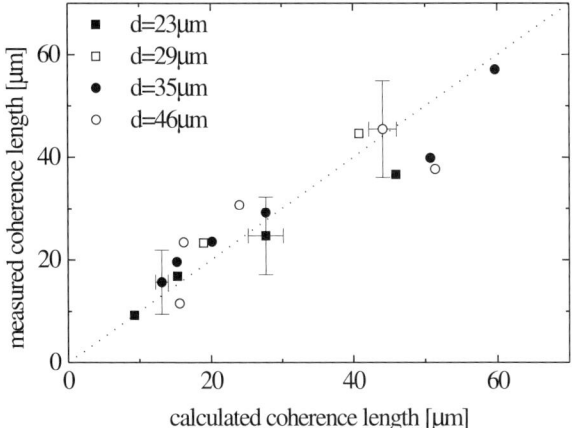

FIG. 4. Phase coherence in a 1D quasicondensate, measured by homodyne spatial interference. The variable d refers to the propagation distance in the interferometer. The quasicondensates in these measurements are larger than 280 µm, much larger than the measured coherence lengths. [Reprinted figure with permission from Hellweg et al. (2003).]

© 2003 American Physical Society

The first evidence for the spatial decay of phase coherence in a quasicondensate was seen by Dettmer et al. (2001). Density fluctuations after release of a 1D BEC were shown to be related to initial phase fluctuations. In Richard et al. (2003), Bragg spectroscopy of a 1D BEC revealed a Lorentzian momentum distribution that is characteristic of power law decaying phase fluctuations. Bragg spectroscopy has also been used to monitor the evolution of the coherence length in a 1D quasicondensate after shock cooling (Hugbart et al., 2007). In Hellweg et al. (2003), progressively distant parts of a 1D condensate were made to spatially interfere with each other. Progressively decreasing fringe contrast was observed. The limited phase correlation length in a quasicondensate is illustrated in Fig. 4. In Schumm et al. (2005), two adjacent 1D BEC's were released from their traps and observed to mutually interfere; decaying phase correlations might ultimately also be observable in this way. Esteve et al. (2006) imaged a 1D Bose gas formed above an atom chip. They observed the reduced spatial fluctuations relative to an ideal gas that are expected for a quasicondensate.

4.5. EXPERIMENTAL TESTS OF THE LIEB–LINIGER EQUATION OF STATE AND CORRELATION FUNCTIONS

Coupling strengths beyond the mean field regime have been experimentally achieved using optical lattices. 1D densities can be made relatively low in each 1D Bose gas, while the signal size is maintained by adding up contributions from many of them. The first experiments were performed with red-detuned optical lattices in the intermediate coupling regime below $\gamma = 1$. In Moritz et al. (2003), the low frequency excitations of 1D Bose gases were studied. Interactions affect the ratio, R, of the frequency of the breathing oscillation in a trap to the dipole oscillation frequency. A TG gas acts the same as a thermal gas, with $R = 4$, while $R = 3$ for a quasicondensate. In the intermediate coupling regime, $R = 3.15 \pm 0.22$ was measured.

In Tolra et al. (2004), the three-body correlation function $g_3 = \rho_3(0, 0, 0; 0, 0, 0)/n_{1D}^3$ [see Eq. (4.9)] was measured by observing loss, Γ_{1D}, in a lattice of 1D Bose gases. The loss was ascribed to contributions from a $K_3^{1D} n_{1D}^3$ three-body loss term and a $K_1^{1D} n_{1D}$ one-body loss term, where the K's are loss rate coefficients. The loss in a tightly confined 3D BEC, Γ_{3D}, was also measured, and the results compared using the identity $K_3^{1D} = K_3^{3D} g_3$. With $\gamma \approx 0.5$, g_3 was measured to be 0.14(9), in reasonable accord with the theoretical estimate

$$g_3 \approx g_2^3(0) \approx 0.25,$$

based on the two-body correlation function $g_2(0) = \rho_2(0, 0; 0, 0)/n_{1D}^2$ calculated in Kheruntsyan et al. (2003).

Some experiments have achieved stronger coupling by using a large waist, blue-detuned optical lattice, with an independent dipole trap to provide axial confinement. With stronger coupling has come the realization of the Tonks–Girardeau gas, and testing of the exact 1D Bose gas theory across coupling regimes. Blue-detuned optical lattices have helped achieve these goals in three ways. First, as illustrated by comparing Eqs. (2.1) and (2.2), they allow for higher trap anisotropy. The transverse confinement can be made tighter without paying the price of higher 1D density as a result of axial compression. Second, since the atoms are trapped nearly in the dark, spontaneous emission is dramatically reduced compared to red-detuned lattices with the same trap depth. Third, and also because of Eq. (2.2), it is possible to turn off only the axial trapping, allowing the atoms to expand while still trapped in 1D. This expands the range of measurements that can be made with 1D gases in blue-detuned lattices.

In Kinoshita et al. (2004), the total energy of a 1D Bose gas was measured. To do this, the axial trapping dipole trap was suddenly turned off and the atoms were detected after a ballistic expansion in 1D. All the 1D energy converts to ballistic energy as the gas expands. It shows up in the final momentum distribution whether it is initially dominated by the mean field energy associated with wavefunction overlap, or by the kinetic energy associated with correlating the wavefunctions to keep them from overlapping. Measurements were taken at both low and high atom densities, scanning the transverse confinement by scanning the lattice depth. The results are summarized in Fig. 5, where the energies have been rescaled according to the TG gas prediction at that point. Only data for lattice depths where there is negligible tunneling between tubes is plotted. With no free parameters, there is

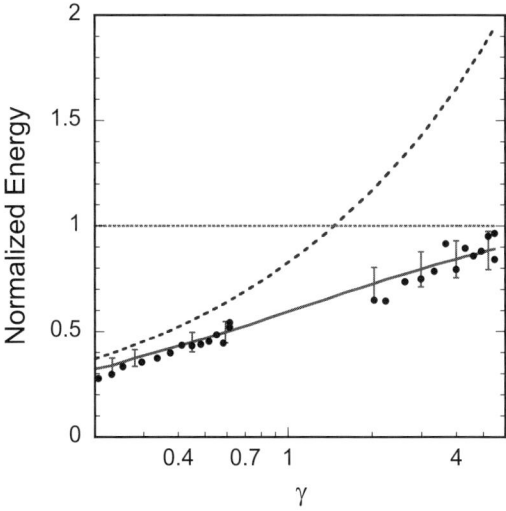

FIG. 5. Total energy of a trapped 1D Bose gas as a function of the coupling strength, averaged over an array of waveguides. The dotted line is the total energy calculated for a TG gas, appropriate for $\gamma \gg 1$; the dashed line is the total energy calculated using mean field theory, as would be appropriate for $\gamma \ll 1$, and the solid line is the exact Lieb–Liniger theory for a 1D Bose gas with delta function interactions. All energies are rescaled to the TG theory result. The points represent the results of ballistic energy measurements of bosons trapped in optical lattice atom waveguides. The lower (upper) points were taken at a relatively high (low) atom density, and within each group the optical lattice trap depth increases from left to right. There are no free parameters in the comparison between theory and experiment. The data is adapted from Kinoshita et al. (2004).

good agreement between the data and the exact Lieb–Liniger theory, modified using the local density approximation to apply to a harmonic trap (Dunjko et al., 2001).

The most direct way to measure the changes to the many-body wavefunction that occur across coupling regimes is to look at the pair correlations, $g_2(z) = \rho_2(0, 0; z, z)/n_{1D}^2$ [see Eq. (4.9)]. The local pair correlation, $g_2(0)$, is particularly easy to interpret. It is one in the weak coupling limit [see Eq. (4.17)], just like in a BEC, and zero in the strong coupling limit [see Eq. (4.18)], just like in a non-interacting Fermi gas. In Kinoshita et al. (2005), local pair correlations were measured by photo-associating pairs of atoms into excited state molecules. Atoms are only excited to the molecular state when their relative distance is ∼2 nm. Since this is much less than the interparticle spacing, which exceeds 200 nm, the production rate of these molecules is nearly proportional to $g_2(0)$. The molecules decay with high probability into untrapped, ground state molecules, so their formation is observed as a loss of atoms.

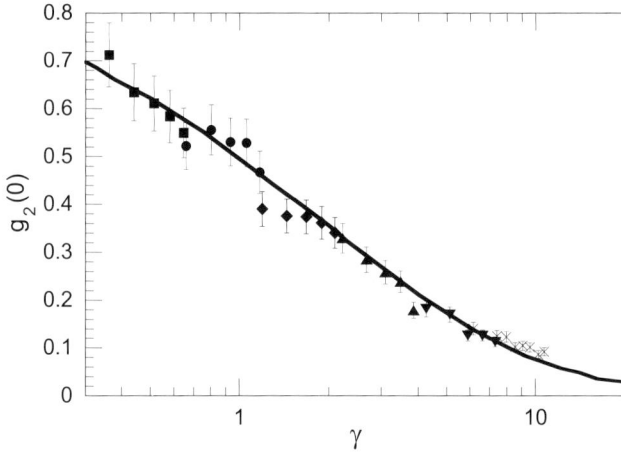

FIG. 6. Local pair correlations (g_2 at zero distance) in 1D Bose gases as a function of coupling strength, measured by photoassociation. The quantities on both axes represent an average over the array of waveguides created by a 2D optical lattice. The different symbols from left to right correspond to progressively lower atom densities, and for each symbol the points to the right have a higher lattice depth. The single fit parameter, the molecular formation matrix element, agrees with the independently measured value to within 1.3σ. For $\gamma = 11$ there is a ten-fold reduction of $g_2(0)$ relative to a BEC. This constitutes a direct observation of the fermionization of bosons in 1D. [Reprinted figure with permission from Kinoshita et al. (2005).]

© 2005 American Physical Society

At the lowest densities (highest γ) obtained in Kinoshita et al. (2005), turning on the 2D lattice actually decreased the molecular formation rate, even though the 3D densities are much higher when the atoms are confined in the tubes. The complete results are shown in Fig. 6, where the averaged $g_2(0)$ across the array of 1D gases is plotted against the average value of γ. Each symbol represents a scan of lattice depths at a fixed number of atoms per tube. There is very good agreement between the measurements and the exact 1D Bose gas theory (Gangardt and Shlyapnikov, 2003b). By $\gamma = 11$, the atoms are well on their way to fermionization, with $g_2(0) = 0.1$.

An alternative way to create strongly-coupled 1D Bose gases is to add an additional 1D lattice in the axial direction (Stoferle et al., 2004; Paredes et al., 2004). When the interatomic distance is comparable to or smaller than the axial lattice spacing, increasing the axial lattice depth makes the atoms undergo a superfluid to Mott-insulator transition at reduced depth compared to the 3D case (Stoferle et al., 2004). For $\gamma \gtrsim 3$ this transition occurs at zero lattice depth, and is best understood as a commensurate-incommensurate transition, which has not yet been observed (Büchler et al., 2003). When the interatomic distance is sufficiently larger than the lattice spacing, one can describe the atoms as quasiparticles with an effective mass, $m_{\text{eff}} > m$, in the lattice. These particles can have a significantly larger effective coupling strength, $\gamma_{\text{eff}} = (m_{\text{eff}}/m)\gamma$. Local correlations are not accessible in these experiments, as the quasiparticle picture fails for distances below a few lattice spacings. The non-local behavior of the wavefunctions is accessible. The tails of the 1D momentum distributions have been measured, and they show the expected TG behavior at large γ_{eff} (Paredes et al., 2004).

Detailed studies of the momentum distributions of 1D gases in the absence of an additional 1D lattice have not been made. Some observations have been made by turning off the 2D lattice upon release from the axial trap, adiabatically on the timescale of the transverse oscillation but suddenly on 1D timescales (Weiss, 2006). The lattice must be turned off to reduce interatomic interactions, so that the momentum distribution does not evolve during the expansion. Still, significant 3D mean field energy remains, and affects the distributions. The cleanest measurement along these lines would be to use a Feshbach resonance to turn off the interactions at release. Then the time-of-flight distribution after expansion in the 1D tubes would be exactly the 1D momentum distribution.

5. What Is Special about Physics in 1D

5.1. Effects of Integrability and Non-Integrability

5.1.1. Physical Manifestations of Integrability (the Quantum Newton's Cradle)

A 1D Bose gas with energy-independent delta-function interactions is a rare example of an integrable many-body model. Integrability means that there are exact solutions, as explicated in Section 3.3. It can be alternatively stated as meaning that there are as many constants of motion as there are degrees of freedom. At first glance, integrability seems a rather abstract concept, having to do with how the theory of these models is implemented. However, the physical implications of a many-body system's integrability are concrete and dramatic. Classically, the phase-space trajectory of integrable system lies on an invariant torus—a surface, where the constants of motion have certain values. The dimension of the invariant torus is equal to the number of degrees of freedom of the system. Although in the case of incommensurable frequencies the trajectory can ergodically sample the invariant torus, it will not ergodically sample the entire phase space. In quantum mechanics, integrability leads to the conservation of the mean values of the integrals of motion, which are as numerous as the degrees of freedom. The latter sets a severe upper limit on the difference between observables in the initial and final states of the system. The result is that the system retains a strong memory of the initial conditions, in contrast to nonintegrable, mixing, systems, which always relax to the state of highest entropy.

In the next subsection we will discuss how to describe the final state to which an integrable many-body system approaches. In this subsection, we will describe the quantum Newton's cradle demonstrated in Kinoshita et al. (2006), an experimental study of thermalization in 1D Bose gases out of equilibrium, where the above-mentioned memory of initial conditions is explicitly evident.

The quantum Newton's cradle experiment (Kinoshita et al., 2006) is performed in a blue-detuned 2D optical lattice. Trapped 1D Bose gases are initially prepared at the center of a nearly harmonic trap in a superposition state of half the particles each with $\pm 2p_0$ of momentum, where $p_0^2/(2m)$ greatly exceeds the interaction energies in the gas. To a first approximation, all the atom start to oscillate with the same amplitude in the trap. The oscillations are observed by releasing the atoms from the trap and detecting the 1D distributions after a ballistic expansion.

These oscillations are complicated by two effects. First, the two initially separated groups of atoms undergo a breathing evolution due to their internal interactions, since each has the length of the initial cloud, but half the density. Second, and ultimately more importantly, atoms with slightly higher energy oscillate slightly slower as a result of trap anharmonicity (the trap is actually Gaussian, not harmonic). As the more energetic atoms dephase with respect to the less energetic ones, the lengths of the two groups of atoms grows. (There is also dephasing among waveguides in the 2D array, but this is a smaller effect.) By the end of 15 cycles, the atoms have dephased so much that the length of the two groups of atoms exceeds the amplitude of the oscillation. By this time, the measured distributions cease to change over the course of an oscillation cycle. The observed dephased distributions are far from what they would have been if these gases had internally thermalized, that is, they are far from the isoenergetic state with the highest entropy.

The non-thermal distribution is then observed over 25 more cycles to see whether it approaches the conventional thermal distribution. No thermalization is observed (see Fig. 7). The experiment sets lower limits on the thermalization rate at three values of the parameter γ [see Eq. (4.2)] spanning the intermediate to strong coupling regimes. The difference in the shapes of the three curves in Fig. 7 do not indicate that any of them experience partial thermalization. These shape differences simply reflect how non-thermal a distribution can be initially prepared. The distributions are more rounded at lower γ because the interaction energies are a larger fraction of $p_0^2/(2m)$ when the density is higher. Still, they are not observed to change with time as a result of thermalization. Where the measurement is most sensitive (at $\gamma = 3.2$), it is inferred that the thermalization time scale exceeds 240,000 1D collisions per atom. Since most collisions entail particles transmitting through each other, this corresponds to 9600 reflections per atom, obviously far greater than the three collisions that are needed to thermalize a 3D gas.

In the TG gas limit (see Section 3.2), as in Fig. 7(a), the quantum Newton's cradle is a strictly integrable system. But in the strong and intermediate coupling

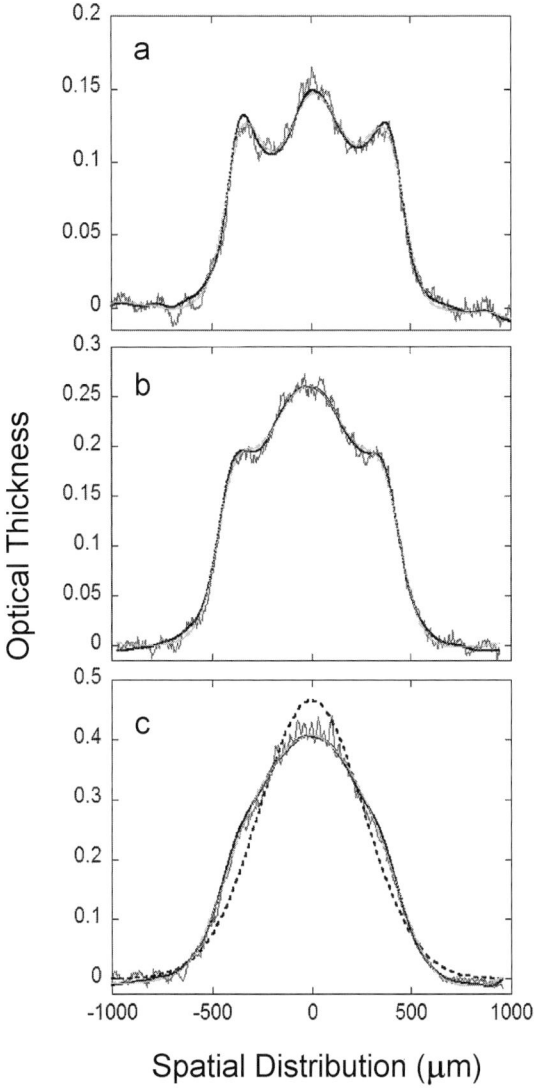

FIG. 7. Evolution of out-of-equilibrium 1D Bose gases. An array of dephased quantum Newton's cradles is observed after 15–25 oscillations, for three different average coupling strengths: (a) $\gamma = 18$, (b) $\gamma = 3.2$, and (c) $\gamma = 1.4$. The curves are spatial distributions after ballistic expansion in 1D; with some small corrections, they correspond to the momentum distributions before release. The smooth lines represent the projected shapes of the initially non-equilibrium atom distributions after ballistic expansion in 1D, taking into account instrumental effects. The noisy lines are the actual measured distributions. [The dotted line in (c) is a comparable energy Gaussian distribution.] In each case, the far greater similarity between the projected and actual curves, compared to the actual curves and Gaussians, allows stringent experimental lower limits to be set on the thermalization rate. [Reprinted figure with permission from Kinoshita et al. (2006).]

© 2006 Nature

regimes, the weak axial harmonic potential lifts integrability (see the discussion at the end of Section 3.3.1). Empirically, however, Fig. 7(b) and (c) set large lower limits on the thermalization rate even in the intermediate coupling regime. Evidently, the weak axial potential that is applied is too small to have an observable effect on thermalization at the current level of experimental sensitivity. This is not surprising, because the size of the gas cloud in this experiment is much greater than the atomic deBroglie wavelengths. The trap potential can be neglected in this limit, so that the system can be approximated by the integrable Lieb–Liniger model (see Section 3.3.2).

The array of 1D Bose gases provides several other ways to lift integrability, including allowing finite range and resonant 1D interactions, allowing tunneling among the tubes, and adding axial potentials. Starting from a demonstrably integrable many-body system, it may be possible to add a new experimental element to the widespread theoretical study of irreversibility in weakly non-integrable systems.

5.1.2. Constrained Thermodynamics

The extreme longevity of the three-peaked momentum distributions of 1D Bose gases that was demonstrated experimentally in Kinoshita et al. (2006) shows unambiguously that the existence of nontrivial conserved quantities affects long-time dynamics. This suggests the question of whether the momentum distribution is exactly preserved, as in the classical case, or the shape of the peaks can be modified due to the quantum effects.

In Rigol et al. (2007), the authors show that the momentum distribution does evolve with time, and after a relaxation period converges to an equilibrium. To deduce the nature of the equilibrium consider a general quantum-mechanical system that possesses several integrals of motion. Maximization of the entropy $S = k_B \, \text{Tr}[\rho \ln(1/\rho)]$ results in the following many-body density matrix:

$$\hat{\rho} = Z^{-1} \exp\left[-\sum_m \lambda_m \hat{\mathcal{I}}_m\right], \tag{5.1}$$

where $\{\hat{\mathcal{I}}_m\}$ is the *full* set of the integrals of motion, $Z = \text{Tr}[\exp[-\sum_m \lambda_m \hat{\mathcal{I}}_m]]$ is the partition function, and $\{\lambda_m\}$ are the Lagrange multipliers, fixed by the initial conditions via

$$\text{Tr}[\hat{\mathcal{I}}_m \hat{\rho}] = \langle \hat{\mathcal{I}}_m \rangle (t=0). \tag{5.2}$$

The above ensemble (5.1) reduces to the usual grand-canonical ensemble in the case of a generic system, where the only integrals of motion are the total energy, the number of particles, and, for periodic systems, the total momentum.

Figure 8 presents results of a numerical experiment that is conceptually close to the Penn State experiment (Kinoshita et al., 2006). A gas of quantum hard-core

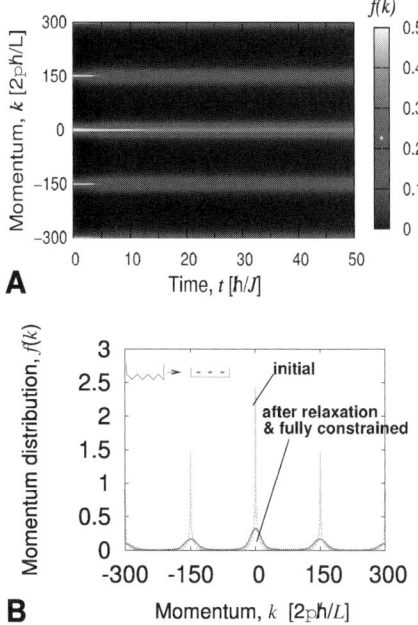

FIG. 8. Relaxation of a gas of 1D hard-core bosons from an initial state with a multi-peak momentum distribution. Results of time propagation are compared to the predictions of the constrained thermodynamical ensemble. [Reprinted figure with permission from Rigol et al. (2007).]

© 2007 American Physical Society

bosons that is initially at rest in a flat-bottomed potential is briefly exposed to a sinusoidally varying potential. The momentum distribution is thereby changed into a sequence of narrow peaks. Over time, the distribution relaxes to a stationary momentum distribution. Although the initial multi-peaked structure is preserved, the final peaks are broader. The shape of the peaks corresponds, to very high accuracy, to the prediction of the generalized equilibrium (5.1).

The central difference between the distributions shown in Figs. 8 and 7 is that the physical experiment is done in a nearly harmonic trap, so that an atom with initial momentum p_0 periodically takes on all momentum values between p_0 and $-p_0$. In the physical experiment, when the oscillating atoms dephase, the associated $2p_0$ spread in momentum dominates the broadening of the peaks due to quantum effects, which are clearly visible in the numerical experiment.

5.1.3. 1D Chemistry

Although three-body problems are rather simple, they can illustrate most of the prominent features of many-body problems. Compared to two-body collisions, which can only be elastic, three-body collisions include chemical reactions, such as three particle association

$$A + B + C \rightarrow AB + C, \tag{5.3}$$

rearrangement,

$$AB + C \rightarrow AC + B, \tag{5.4}$$

dissociation,

$$AB + C \rightarrow A + B + C, \tag{5.5}$$

and particle–dimer elastic collisions

$$AB + C \rightarrow AB + C. \tag{5.6}$$

If a system is non-diffractive and has a Bethe-ansatz solution (see Section 3.3 above), then the dimer states are described by complex momenta. Thus reflection in elastic collisions (5.6) and all chemical reactions are forbidden, since these processes would change the asymptotic momentum set (see Section 3.3.2 above). The existence of a Bethe-ansatz solution can be determined using the Yang–Baxter relation (3.25), which is formulated in terms of three asymptotic momenta. So three-body systems are already large enough to demonstrate the effect of the Yang–Baxter relation being violated.

These effects were first studied by Dodd (1972) for zero-range interactions with energy-independent interaction strengths when either the interaction strengths or the masses are not all the same. The results demonstrated that symmetry violation leads to reflection in elastic atom–dimer collisions.

The Yang–Baxter relations can be violated even in the case of equal masses and interactions, e.g., if the interaction strength is energy-dependent due to a CIR [see discussion following Eq. (3.27)]. Reflection in elastic atom–dimer collisions due to this effect was demonstrated in Mora et al. (2004) for two-component fermions and in Mora et al. (2005) for indistinguishable bosons. The same effect, described by quintic terms in the Bogoliubov–de Gennes equations, leads to atom–soliton reflection (Sinha et al., 2006).

Richer physical phenomena result from the violation of the Yang–Baxter relations due to a Feshbach resonance (Yurovsky et al., 2006). This effect, however, cannot a priori be described by an energy-dependent interaction strength, since the elimination of many-body closed channels is required. The three-body 1D problem with zero-range Feshbach-resonance two-body interactions can be described by coupled equations for the three-atom $\tilde{\varphi}_0^{(3)}(q_1, q_2, q_3)$ and atom–molecule $\tilde{\varphi}_1^{(3)}(q_1, q_m)$ channel wavefunctions in the momentum representation (Yurovsky et al., 2006; Yurovsky, 2006),

$$E\tilde{\varphi}_0^{(3)}(q_1, q_2, q_3) = \frac{1}{2m} \sum_{j=1}^{3} q_j^2 \tilde{\varphi}_0^{(3)}(q_1, q_2, q_3)$$

$$+ \frac{1}{2\pi} U_a \sum_{j=1}^{3} \int d^3 q' \delta(q'_j - q_j) \delta(Q - Q') \tilde{\varphi}_0^{(3)}(q'_1, q'_2, q'_3)$$

$$+ (3\pi)^{-1/2} U^*_{am} \sum_{j=1}^{3} \tilde{\varphi}_1^{(3)}(q_j, Q - q_j), \tag{5.7}$$

$$E \tilde{\varphi}_1^{(3)}(q_1, q_m) = \left(\frac{q_1^2}{2m} + \frac{q_m^2}{4m} + D_{1D} \right) \tilde{\varphi}_1^{(3)}(q_1, q_m)$$

$$+ \left(\frac{3}{\pi} \right)^{1/2} U_{am} \int dq_3 \tilde{\varphi}_0^{(3)}(q_1, q_m - q_3, q_3). \tag{5.8}$$

Here $Q = q_1 + q_2 + q_3$ is the center-of-mass momentum, q_j are the momenta of the atoms, and q_m is the momentum of the 1D closed-channel molecule, which effectively incorporates contributions from excited waveguide modes and the 3D closed channel (see Section 2.3.3 above). The parameters U_a, U_{am}, and D_{1D} are expressed by Eq. (2.47) with $\mu = m/2$. The wavefunction $\tilde{\varphi}_0^{(3)}(q_1, q_2, q_3)$ is symmetric over permutations of the atomic momenta due to the indistinguishability of the bosonic atoms.

Equation (5.8) allows a simple elimination of the atom–molecule channel, leading to the single-channel Schrödinger equation for the three-atom channel wavefunction,

$$E \tilde{\varphi}_0^{(3)}(q_1, q_2, q_3) = \frac{1}{2m} \sum_{j=1}^{3} q_j^2 \tilde{\varphi}_0^{(3)}(q_1, q_2, q_3)$$

$$+ \frac{1}{2\pi} \sum_{j=1}^{3} U_{\text{eff}} \left(E - \frac{1}{2m} q_j^2 - \frac{1}{4m} (Q - q_j)^2 \right)$$

$$\times \int d^3 q' \delta(q'_j - q_j) \delta(Q - Q') \tilde{\varphi}_0^{(3)}(q'_1, q'_2, q'_3), \tag{5.9}$$

where the interaction strength $U_{\text{eff}}(E)$ happens to be the same function (2.42) as in the two-body problem. Many-body problems with $N > 3$ do not allow such a simple elimination, which is based on the neglect of collisions between atoms and 1D closed-channel molecules in Eq. (5.8). Elimination of channels and energy-dependent potentials in 3D three-body problems were considered in Abdurakhmanov and Zubarev (1985), Abdurakhmanov et al. (1987), Vinitskii et al. (1992).

Three-body problems can be described by the Faddeev–Lovelace equations (Lovelace, 1964) for the symmetric transition amplitude $X(p, p_0)$ (see also Schmid and Ziegelmann, 1974; Glöckle, 1983). In the 1D case, this equation has the following form:

$$X(p, p_0) = 2Z(p, p_0) + \frac{\pi m^2}{\kappa^3} \int dq\, Z(p, q) \langle |T_{1D}^+(q)| \rangle X(q, p_0). \quad (5.10)$$

Here

$$Z(p, q) = \frac{2\kappa^3}{\pi m} \frac{1}{mE + i0 - p^2 - pq - q^2}, \quad (5.11)$$

where the reciprocal dimer size κ is a positive solution of Eq. (2.51), and $\langle |T_{1D}^+(q)| \rangle$ is determined by Eq. (2.43) with the initial and final momenta omitted, since the 1D T matrix is independent of them.

Exact analytical (Dodd, 1970; Majumdar, 1972) and numerical (Mehta and Shepard, 2005) solutions of Eq. (5.10) for the Lieb–Liniger–McGuire model agree with the Bethe-ansatz solution. Numerical solution of the Faddeev–Lovelace equations were used in Dodd (1972) for asymmetric systems. The method of Mora et al. (2004, 2005) is based on the approach by Skorniakov and Ter-Martirosian (1957), which is equivalent to the Faddeev equations (see Baz' et al., 1971). Three-body 1D problems were also analyzed by numerical solution of the Schrödinger equation in hyperspherical coordinates (Mehta et al., 2006).

The asymptotic state of a three-body wavefunction can include three well-separated atoms or an atom separated from a two-body bound state. In the latter case, the bound state includes contributions from both two-body open and closed channels. The open-channel component of the properly normalized atom–dimer scattering state is proportional to $\sqrt{W_o^{1D}}$, where the open-channel contribution W_o^{1D} is expressed by Eq. (2.57). The probabilities of reflection, transmission, and dissociation in the collision of an atom with momentum p_0 and a dimer with momentum $-p_0$ are

$$P_{\text{ref}}(p_0) = \left| \frac{4\pi m}{3 p_0} W_o^{1D} X(-p_0, p_0) \right|^2,$$

$$P_{\text{tran}}(p_0) = \left| 1 - i \frac{4\pi m}{3 p_0} W_o^{1D} X(p_0, p_0) \right|^2,$$

$$P_{\text{diss}} = 1 - P_{\text{ref}} - P_{\text{tran}}. \quad (5.12)$$

They include the factor W_o^{1D} since the amplitude $X(p, p_0)$ is determined by Eq. (5.10) in the single-channel picture.

In the case of the association of three atoms with momenta p_1, p_2, and p_3 (in the center-of-mass system $p_1 + p_2 + p_3 = 0$), the momenta of the resulting atom and diatom $\pm p_*$ are determined by energy conservation,

$$E = \frac{1}{2m}(p_1^2 + p_2^2 + p_3^2) = \frac{3}{4m} p_*^2 - E_b. \quad (5.13)$$

The association rate coefficient is given by

$$K_3^{1D}(p_1, p_2, p_3) = \frac{8\pi^4 m^3}{27\kappa^3 p_*} W_o^{1D} \left[\left| \sum_{j=1}^{3} \langle |T_{1D}^+(p_j)| \rangle X(p_j, p_*) \right|^2 \right.$$
$$\left. + \left| \sum_{j=1}^{3} \langle |T_{1D}^+(p_j)| \rangle X(p_j, -p_*) \right|^2 \right]. \quad (5.14)$$

Calculations (Yurovsky et al., 2006) demonstrate that reflection is the dominant output channel of atom–dimer low-energy collisions, as is the case for other mechanisms of Lieb–Liniger–McGuire model symmetry violation, as discussed above. In addition to reflection, Feshbach resonances lead to dissociation (above the threshold) in atom–dimer collisions and association in three-atom collisions. All these processes vanish at large detunings, when the energy-dependence of the interaction strength becomes negligible and where the non-diffractive nature of the system is restored. Chemical processes vanish both at high and low collision energies, persisting at $E \sim m^{1/3}|U_{am}|^{4/3}$. This energy scale is less than the transverse waveguide frequency $\omega_\perp$ for rather weak Feshbach resonances with $\mu_{oc}\Delta < \omega_\perp a_\perp/a_{el}$ [see Eq. (2.47)], which allows for the observation of chemical processes in the quasi-1D regime ($E < \omega_\perp$). The violation of non-diffraction due to a CIR can also be analyzed with this approach using effective Feshbach resonance parameters [see Eq. (2.49)]. In this case, however, the energy scale is $m^{1/3}|U_{am}|^{4/3} > \omega_\perp$, and chemical processes are negligible in the quasi-1D regime.

Three-body methods can also be used to analyze the deactivation of quasi-1D molecules $A_2(\text{in})$,

$$A_2(\text{in}) + A \rightarrow A_2(\text{fin}) + A, \quad (5.15)$$

where the state $A_2(\text{fin})$ lies below $A_2(\text{in})$. If $A_2(\text{in})$ is a tightly bound molecule, the deactivation (5.15) can be considered to be a two-body inelastic collision, which is suppressed at low collision energies [see Eq. (2.62)]. However, broad molecules have a different nature. They can be described by the integrable Lieb–Liniger–McGuire model (see Section 3.3.2) with the negative interaction strength $U_a < 0$. If $A_2(\text{fin})$ is a tightly bound state, the deactivation rate is proportional to the three-atom correlation function $\rho_3(0, 0, 0; 0, 0, 0)$ [see Eq. (4.9)], much like in the free-space case (Petrov et al., 2004, 2005). In the present case, the wavefunction $\Psi_0(z_1, z_2, z_3)$ corresponds to the state where two atoms are bound and one is free (and it is not the ground state). The Lieb–Liniger–McGuire model leads to an exact expression for ρ_3 that has a non-zero limit at low collision energy. [This behavior differs from the free-atom case (Gangardt and Shlyapnikov, 2003b), since bound atoms have non-vanishing imaginary momenta $\pm\frac{i}{2}mU_a$.] This absence of deactivation suppression can be related to non-diffraction of the Lieb–Liniger–McGuire model, which forbids atom–dimer reflection. Therefore the atom and dimer can approach each other, the transmission probability remains finite at low collision energies, and the ratio (2.61) does not vanish.

Calculations (Yurovsky, 2007) demonstrate that when the symmetry of the Lieb–Liniger–McGuire model is lifted due to a Feshbach resonance, the deactivation of broad molecules becomes suppressed. It is surprising that the deactivation rate is proportional to the collision energy, as in collisions of structureless particles (Gangardt and Shlyapnikov, 2003b; Yurovsky and Band, 2007) Thus, both the presence and the suppression of certain processes are among the observable effects of non-integrability.

5.2. Solitons

The 1D Gross–Pitaevskii equation (3.53) is a special case of the non-linear Schrödinger equation, which also describes Langmuir waves in plasmas, thermal waves in solids, and non-linear optical media (see Kivshar and Agrawal, 2003, and references therein). In the spatially-homogeneous ($V_{\text{ax}} = 0$) case of attractive interactions ($U_a < 0$) Eq. (3.53) has an exact solution (see Novikov et al., 1984; Pitaevskii and Stringari, 2003; Kivshar and Agrawal, 2003)

$$\Phi(z,t) = A_0 \exp\left[i\left(mv_0 z - \frac{U_a A_0^2 + mv_0^2}{2}t + \vartheta_0\right)\right]$$
$$\times \text{sech}\left[A_0\sqrt{m|U_a|}(z - z_0 - v_0 t)\right], \tag{5.16}$$

which has the form of a localized wave, or bright soliton. It depends on four parameters: the amplitude A_0, the velocity v_0, the initial phase ϑ_0 and the initial position z_0. General, multisoliton solutions of the non-linear Schrödinger equation can be obtained by the inverse scattering method (see Novikov et al., 1984, and references therein). They show how solitons collide elastically, conserving their shapes, amplitudes, and velocities. There are also spatial solitons in non-linear optics; in that case, t corresponds to the second spatial coordinate.

The quantum non-linear Schrödinger equation is the spatially-homogeneous operator equation of motion (3.51) with zero-range interactions. It describes the Lieb–Liniger–McGuire model, which has an exact Bethe-ansatz solution (see Section 3.3.2 above). In the case of attractive interactions the solution can have the form of a many-body bound state (cluster) or of several clusters, which can only collide elastically, like solitons. Solitonic solutions can be constructed as superpositions of clusters with different numbers of particles (Lai and Haus, 1989b). The Bethe-ansatz method was also used in Shnirman et al. (1994), Cheng and Kurizki (1995, 1996), Mazets and Kurizki (2006). The inverse scattering method has a quantum generalization as well (see Korepin et al., 1993), which is used to analyze correlation properties.

As the peak soliton density is $n_{1D} = |A_0|^2$, its length, $1/(A_0\sqrt{mU_a})$, is simply the coherence length (3.67), and the condition of weak fluctuations (3.68) is satisfied along the soliton whenever it contains large number of atoms. Thus in the mean-field approach, one can treat a soliton as a BEC. In some sense, this is a peculiarity of attractive interactions, since the real parts of all asymptotic momenta in the many-body bound state are equal [see Eq. (3.38)] and, therefore, all particles have the same phase, as in a BEC. However, non-mean field effects are important in certain situations (Haus and Lai, 1990; Haus and Yu, 2000; Poulsen and Molmer, 2003; Buljan et al., 2005; Sinha et al., 2006; Sacha and Timmermans, 2006; Merhasin et al., 2006; Driben et al., 2007).

Solutions of the Gross–Pitaevskii equation with repulsive interactions ($U_a > 0$) can demonstrate dark solitons (Tsuzuki, 1971; Fedichev et al., 1999; Busch and Anglin, 2000; Muryshev et al., 2002; Konotop and Pitaevskii, 2004; Brazhnyi et al., 2006; Buljan et al., 2006; Pitaevskii and Stringari, 2003), which are dips in the density that propagate through the condensate. Two branches of excitations, corresponding to particle and hole excitations in the Lieb–Liniger–McGuire model, have been found by mean-field analysis (Kulish et al., 1976) and identified with dark solitons (Ishikawa and Takayama, 1980). The dark solitons are known in non-linear optics (see Kivshar and Agrawal, 2003), and they were the first solitons that were observed in a BEC (Burger et al., 1999; Denschlag et al., 2000; Anderson et al., 2001; Dutton et al., 2001). Bright solitons in a BEC were observed (see Fig. 9) in Khaykovich et al. (2002), Strecker et al. (2002) using Feshbach resonances to control the interactions (see also reviews Strecker et al., 2003; Abdullaev et al., 2005). Solutions of the non-linear Schrödinger equations under box and periodic boundary conditions (Carr et al., 2000a, 2000b) can have the forms of various solitons. Although the related quantum problem with periodic boundary conditions has an exact Bethe-ansatz solution, an analysis of its properties for macroscopic systems requires solution of a large number of transcendental Bethe equations (3.35). This problem was analyzed using exact diagonalization (Kanamoto et al., 2003a, 2003b, 2005, 2006) and configuration-interaction (Alon et al., 2004) methods.

The integrability of the non-linear Schrödinger equations can be lifted by incorporating additional time-dependent, coordinate-dependent, or non-linear terms. The localized stable solutions of such equations are called solitons as well, although they are not exact solutions. In certain situations a periodic modulation can even improve soliton stability (see Malomed, 2006). Gap solitons can be formed in spatially-periodic potentials (Cheng and Kurizki, 1995, 1996; Merhasin et al., 2006). Quantum effects on soliton scattering by external potentials were analyzed in Mazets and Kurizki (2006), Lee and Brand (2006). Additional quintic terms that can appear in the Gross–Pitaevskii equation due to virtual transverse mode

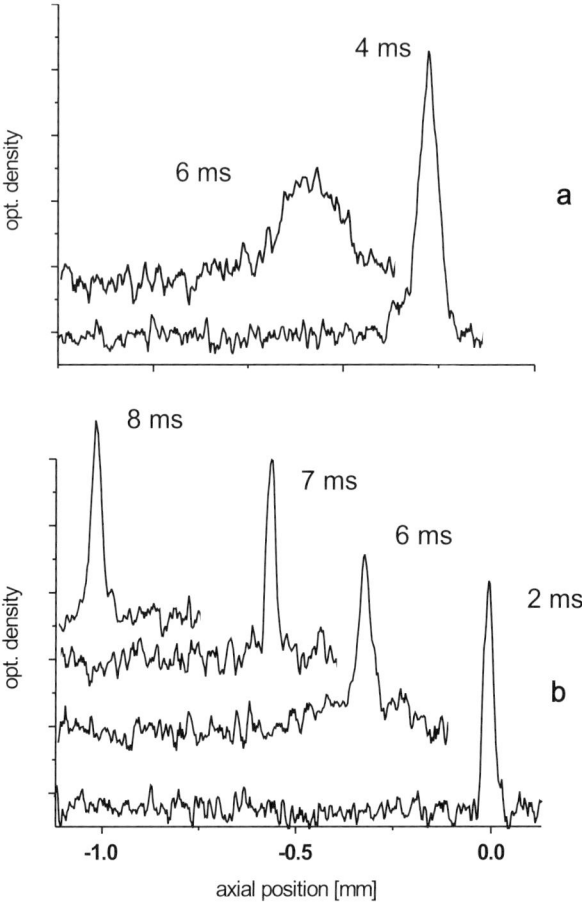

FIG. 9. A bright soliton. The upper frame shows the propagation of a BEC, and the lower frame shows the propagation of soliton in an atom waveguide. The soliton propagates without dispersion for more than 1 mm. [Reprinted figure with permission from Khaykovich et al. (2002).]

© 2002 Science

excitation (see Section 3.4 above) can lead to reflection in atom-soliton collisions (Sinha et al., 2006).

An atom–molecule gas is an analog (Drummond et al., 1998) of a second-harmonic generating medium, or a medium in quantum optics with a quadratic nonlinearity (see Drummond et al., 1999; Etrich et al., 2000; Buryak et al., 2002, and references therein). This system is diffractive and does not have a Bethe-ansatz solution [see the discussion following Eq. (3.27)], although under certain conditions, when the closed channel can be eliminated and the energy-dependence of the interaction strength can be neglected, it can be reduced to a pure atomic system with $U_a = 2\omega_\perp a_{\text{eff}}(0)$ (see Section 2.2). However, this approximation cannot be applied close to the resonance, as was demonstrated by binding-energy calculations in (Steblina et al., 1995). Moreover, atom–molecule bright solitons can exist even for $a_{\text{eff}}(0) < 0$, which corresponds to repulsive effective interatomic interactions (cf. two-body bound states, Section 2.3.3). Analysis of the three-body problem in Yurovsky (2006) demonstrates that even for large negative detunings the binding energy corresponds to an atom coupled to a closed-channel molecule, rather then to three coupled atoms. Other effects of non-integrability are discussed in Section 5.1.3. The effect of detuning modulation on atom–molecule solitons was analyzed in Driben et al. (2007).

6. Summary and Outlook

We have reviewed theoretical and experimental work on atoms in 1D waveguides. We have discussed the experimental and theoretical logistics of trapping atoms in waveguides and of controlling the mutual interactions therein. We have also provided an overview of the physics of 1D many-body systems, and showed how theoretical models can and have been experimentally implemented.

Although some widely studied 1D theoretical models have been brought to life in the lab, most notably the Lieb–Liniger–McGuire model and the Tonks–Girardeau gas, there remain aspects of these systems in equilibrium that are as yet unstudied, including their full momentum distributions. Other features of integrability have been studied, like solitons, but there remains a panoply of effects that are as yet unobserved. There are also other known integrable models, like Yang–Gaudin and Calogero–Sutherland, that have not as yet been realized experimentally. Integrability is a rather sensitive property and we have reviewed a range of mechanisms by which it is lifted, as well as the observable consequences of non-integrability. The set of as yet unobserved 1D phenomena related to either integrability or the lifting of integrability remains rather large.

Among the defining features of fully integrable systems are that they do not thermalize or undergo irreversible reactions. Still, the fact that in a many-body system such behavior can be experimentally observed has been one of the more surprising outcomes of advances in cold atom technology. Since atoms in 1D waveguides provide multiple means for controlling effective 1D interactions, these systems are positioned to provide the first experimental handle on a many-body systems' integrability. There are fundamental issues in statistical mechanics and non-linear dynamics that can now be addressed for the first time in parallel by theory and experiment. This is certainly the way science proceeds most effectively. So while this review illustrates how far theory and experiment have come

7. Acknowledgements

The authors are very grateful to Doerte Blume, Brian Granger, Yvan Castin, and Lev Pitaevskii for helpful discussion. DSW acknowledges support from the National Science Foundation and the US Army Research Office. MO acknowledges support from the Office of Naval Research (grant No. N00014-03-1-0427), the National Science Foundation (grants No. NSF-PHY-0301052, No. NSF-DMR-0240918, No. NSF-DMR-0312261, and the Institute for Theoretical Atomic and Molecular Physics at Harvard University and Smithsonian Astrophysical Observatory).

8. Appendix: Some Useful Properties of the Hurwitz Zeta Function

The Hurwitz zeta function is defined as (see Apostol, 1998; Bateman and Erdely, 1953; Whittaker and Watson, 1996),

$$\zeta(\nu, \alpha) = \lim_{n_c \to \infty} \left[\sum_{n=0}^{n_c} (n+\alpha)^{-\nu} - \frac{1}{1-\nu}(n_c+\alpha)^{1-\nu} \right], \quad (8.1)$$

with $-2\pi < \arg(n+\alpha) \leqslant 0$.

$$\zeta(\nu, \alpha) = \frac{2\Gamma(1-\nu)}{(2\pi)^{1-\nu}} \left[\sin\frac{\nu\pi}{2} \sum_{n=1}^{\infty} \frac{\cos(2\pi\alpha n)}{n^{1-\nu}} \right.$$
$$\left. + \cos\frac{\nu\pi}{2} \sum_{n=1}^{\infty} \frac{\sin(2\pi\alpha n)}{n^{1-\nu}} \right]. \quad (8.2)$$

This expression is applicable to $\nu < 0$, e.g.

$$\zeta\left(-\frac{1}{2}, 1\right) = -\frac{1}{4\pi}\zeta\left(\frac{3}{2}\right), \quad (8.3)$$

where $\zeta(\nu)$ is the Riemann ζ-function (see Bateman and Erdely, 1953).

The Taylor series in term of the second argument is (Moore et al., 2004)

$$\zeta\left(\frac{1}{2}, \alpha\right) = \frac{1}{\sqrt{\alpha}} + \sum_{n=0}^{\infty} \frac{(2n-1)!!}{2^n n!} \zeta\left(n+\frac{1}{2}\right)(-\alpha)^n. \quad (8.4)$$

The first terms of this expansion can be written as

$$\zeta\left(\frac{1}{2},\alpha\right) \underset{\alpha \to 0}{\sim} \frac{1}{\sqrt{\alpha}} - C - C'\alpha, \quad \sqrt{-|\alpha|} = -i\sqrt{|\alpha|}, \tag{8.5}$$

where $C = -\zeta(\frac{1}{2}) \approx 1.4603$, $C' = \frac{1}{2}\zeta(\frac{3}{2}) \approx 1.3062$.

For high values of α, $\zeta(\frac{1}{2},\alpha)$ can be expanded (Yurovsky, 2005)

$$\zeta\left(\frac{1}{2},\alpha\right) \underset{\alpha \to \infty}{\sim} -2\alpha^{1/2} + \frac{1}{2}\alpha^{-1/2} + \frac{1}{24}\alpha^{-3/2}, \tag{8.6}$$

$$\zeta\left(\frac{1}{2},\alpha\right) \underset{\alpha \to -\infty}{\sim} \zeta\left(\frac{1}{2},[|\alpha|]+1-|\alpha|\right)$$
$$+ i\left\{2\sqrt{|\alpha|} + \zeta\left(\frac{1}{2},|\alpha|-[|\alpha|]\right)\right\}. \tag{8.7}$$

9. References

Abdullaev, F.K., Gammal, A., Kamchatnov, A.M., Tomio, L. (2005). Dynamics of bright matter wave solitons in a Bose–Einstein condensate. *Int. J. Mod. Phys. B* **19**, 3415–3473.

Abdurakhmanov, A., Zubarev, A.L. (1985). Energy-dependent potentials in the nuclear three body problem. *Z. Phys. A* **322**, 523–525.

Abdurakhmanov, A., Zubarev, A.L., Latipov, A.S., Nasyrov, M. (1987). Dependence of the 3-particle observables on the description of the 2-particle observables. *Sov. J. Nucl. Phys.* **46**, 217–219.

Abramowitz, M., Stegun, I.A. (1964). "Handbook of Mathematical Functions with Formulas, Graphs, and Mathematical Tables". Dover, New York.

Akhiezer, A.I., Peletminskii, S.V. (1981). "Methods of Statistical Physics". Pergamon, Oxford.

Albeverio, S., Gesztesy, F., Krohn, R.H., Holden, H. (1988). "Solvable Models in Quantum Mechanics". Springer, New York.

Albeverio, S., Fei, S.-M., Kurasov, P. (2001). Many body problems with "spin" related contact interactions. *Rep. Math. Phys.* **47**, 157–199.

Alexander, T.J., Ostrovskaya, E.A., Kivshar, Y.S., Julienne, P.S. (2002). Vortices in atomic–molecular Bose–Einstein condensates. *J. Opt. B* **4**, S33–S38.

Alon, O.E., Streltsov, A.I., Sakmann, K., Cederbaum, L.S. (2004). Continuous configuration-interaction for condensates in a ring. *Europhys. Lett.* **67**, 8–13.

Andersen, J.O. (2004). Theory of the weakly interacting Bose gas. *Rev. Mod. Phys.* **76**, 599.

Anderson, B.P., Haljan, P.C., Regal, C.A., Feder, D.L., Collins, L.A., Clark, C.W., Cornell, E.A. (2001). Watching dark solitons decay into vortex rings in a Bose–Einstein condensate. *Phys. Rev. Lett.* **86**, 2926–2929.

Apostol, T.M. (1998). "Introduction to Analytic Number Theory". Springer, New York.

Arnold, A.S., Garvie, C.S., Riis, E. (2006). Large magnetic storage ring for Bose–Einstein condensates. *Phys. Rev. A* **73**. 041606.

Astrakharchik, G.E., Giorgini, S. (2002). Quantum Monte Carlo study of the three- to one-dimensional crossover for a trapped Bose gas. *Phys. Rev. A* **66**. 053614.

Astrakharchik, G.E., Giorgini, S. (2003). Correlation functions and momentum distribution of one-dimensional Bose systems. *Phys. Rev. A* **68**. 031602.

Astrakharchik, G.E., Blume, D., Giorgini, S., Granger, B.E. (2004a). Quasi-one-dimensional Bose gases with a large scattering length. *Phys. Rev. Lett.* **92**. 030402.

Astrakharchik, G.E., Blume, D., Giorgini, S., Pitaevskii, L.P. (2004b). Interacting fermions in highly elongated harmonic traps. *Phys. Rev. Lett.* **93**. 050402.

Astrakharchik, G.E., Gangardt, D.M., Lozovik, Y.E., Sorokin, I.A. (2006). Off-diagonal correlations of the Calogero–Sutherland model. *Phys. Rev. E* **74**. 021105.

Bateman, H., Erdely, A. (1953). "Higher Transcendental Functions, vol. 2". McGraw-Hill, New York.

Baz', A.I., Zeldovich, Y.B., Perelomov, A.M. (1971). "Scattering, Reactions and Decays in the Nonrelativistic Quantum Mechanics". Nauka, Moscow (in Russian).

Beliaev, S.T. (1958a). Application of the methods of quantum field theory to a system of bosons. *Sov. Phys. JETP* **34**, 289–299.

Beliaev, S.T. (1958b). Energy-spectrum of a non-ideal Bose gas. *Sov. Phys. JETP* **34**, 299–307.

Bender, S.A., Erker, K.D., Granger, B.E. (2005). Exponentially decaying correlations in a gas of strongly interacting spin-polarized 1D fermions with zero-range interactions. *Phys. Rev. Lett.* **95**. 230404.

Berezin, F.A., Pohil, G.P., Finkelberg, V.M. (1964). The Schrödinger equation for a system of one-dimensional particles with point interactions. *Vestnik Moskovskogo Universiteta* **1**, 21–28 (in Russian).

Bergeman, T., Moore, M.G., Olshanii, M. (2003). Atom–atom scattering under cylindrical harmonic confinement: Numerical and analytic studies of the confinement induced resonance. *Phys. Rev. Lett.* **91**. 163201.

Berkovich, A., Murthy, G. (1989). Time-dependent multipoint correlation functions of the nonlinear Schrödinger model. *Phys. Lett. A* **142**, 121–127.

Bethe, H. (1931). Zur theorie der metalle I. Eigenwerte und eigenfunktionen der linearen atomkette. *Z. Phys.* **71**, 205–226.

Bethe, H., Peierls, R. (1935). Quantum theory of the diplon. *Proc. R. Soc. London, Ser. A* **148**, 146–156.

Bhanu Murti, T. (1960). Plancherel's measure for the factor-space $SL(n; R)/SO(n; R)$. *Sov. Math. Dokl.* **1**, 860–862.

Bloch, I., Dalibard, J., Zwerger, W. (2007). Many-body physics with ultracold gases. arXiv:0704.3011.

Blume, D. (2002). Fermionization of a bosonic gas under highly elongated confinement: A diffusion quantum Monte Carlo study. *Phys. Rev. A* **66**. 053613.

Blume, D., Greene, C.H. (2002). Fermi pseudopotential approximation: Two particles under external confinement. *Phys. Rev. A* **65**. 043613.

Bogolubov, N.N., Bogolubov Jr, N.N. (1994). "An Introduction to Quantum Statistical Mechanics". Gordon and Breach, Lausanne.

Bohn, J.L., Julienne, P.S. (1997). Prospects for influencing scattering lengths with far-off-resonant light. *Phys. Rev. A* **56**, 1486–1491.

Bolda, E.L., Tiesinga, E., Julienne, P.S. (2002). Effective-scattering-length model of ultracold atomic collisions and Feshbach resonances in tight harmonic traps. *Phys. Rev. A* **66**. 013403.

Bolda, E.L., Tiesinga, E., Julienne, P.S. (2003). Pseudopotential model of ultracold atomic collisions in quasi-one- and two-dimensional traps. *Phys. Rev. A* **68**. 032702.

Boninsegni, M., Kuklov, A.B., Pollet, L., Prokof'ev, N.V., Svistunov, B.V., Troyer, M. (2007). Luttinger liquid in the core of a screw dislocation in helium-4. *Phys. Rev. Lett.* **99**. 035301.

Bouchoule, I., Kheruntsyan, K.V., Shlyapnikov, G. (2007). Crossover to a quasi-condensate in a weakly interacting trapped 1D Bose gas. *Phys. Rev. A* **75**. 031606.

Boyd, M., Streed, E.W., Medley, P., Campbell, G.K., Mun, J., Ketterle, W., Pritchard, D.E. (2006). Atom trapping with a thin magnetic film. cond-mat/0608370.

Brand, J., Cherny, A.Y. (2005). Dynamic structure factor of the one-dimensional Bose gas near the Tonks–Girardeau limit. *Phys. Rev. A* **72**. 033619.

Brazhnyi, V.A., Konotop, V.V., Pitaevskii, L.P. (2006). Dark solitons as quasiparticles in trapped condensates. *Phys. Rev. A* **73**. 053601.

Büchler, H.P., Blatter, G., Zwerger, W. (2003). Commensurate–incommensurate transition of cold atoms in an optical lattice. *Phys. Rev. Lett.* **90**. 130401.

Buljan, H., Segev, M., Vardi, A. (2005). Incoherent matter-wave solitons and pairing instability in an attractively interacting Bose–Einstein condensate. *Phys. Rev. Lett.* **95**. 180401.

Buljan, H., Manela, O., Pezer, R., Vardi, A., Segev, M. (2006). Dark stationary matter waves via parity-selective filtering in a Tonks–Girardeau gas. *Phys. Rev. A* **74**. 043610.

Burger, S., Bongs, K., Dettmer, S., Ertmer, W., Sengstock, K., Sanpera, A., Shlyapnikov, G.V., Lewenstein, M. (1999). Dark solitons in Bose–Einstein condensates. *Phys. Rev. Lett.* **83**, 5198–5201.

Buryak, A.V., Trapani, P.D., Trillo, D.V.S.S. (2002). Optical solitons due to quadratic nonlinearities: From basic physics to futuristic applications. *Phys. Rep.* **370**, 63–235.

Busch, T., Anglin, J.R. (2000). Motion of dark solitons in trapped Bose–Einstein condensates. *Phys. Rev. Lett.* **84**, 2298–2301.

Calogero, F. (1969a). Ground state of a one-dimensional n-body system. *J. Math. Phys.* **10**, 2197–2200.

Calogero, F. (1969b). Solution of a three-body problem in one dimension. *J. Math. Phys.* **10**, 2191–2196.

Calogero, F. (1971). Solution of the one-dimensional n-body problems with quadratic and/or inversely quadratic pair potentials. *J. Math. Phys.* **12**, 419–436.

Calogero, F. (1975). Exactly solvable one-dimensional many-body problems. *Lett. Nuovo Cimento* **13**, 411–416.

Calogero, P., Ragnisco, O., Marchioro, C. (1975). Exact solution of the classical and quantal one-dimensional many-body-problem with the two-body potential $V_a(x) = g^2 a^2 \sinh^2(ax)$. *Lett. Nuovo Cimento* **13**, 383–387.

Carr, L.D., Clark, C.W., Reinhardt, W.P. (2000a). Stationary solutions of the one-dimensional nonlinear Schrödinger equation. I. Case of repulsive nonlinearity. *Phys. Rev. A* **62**. 063610.

Carr, L.D., Clark, C.W., Reinhardt, W.P. (2000b). Stationary solutions of the one-dimensional nonlinear Schrödinger equation. II. Case of attractive nonlinearity. *Phys. Rev. A* **62**. 063611.

Castin, Y. (2004). Simple theoretical tools for low dimension Bose gases. *J. Phys. IV (France)* **116**, 89–132.

Castin, Y., Dum, R. (1998). Low-temperature Bose–Einstein condensates in time-dependent traps: Beyond the $U(1)$ symmetry-breaking approach. *Phys. Rev. A* **57**, 3008–3021.

Castin, Y., Dum, R., Mandonnet, E., Minguzzi, A., Carusotto, I. (2000). Coherence properties of a continuous atom laser. *J. Mod. Opt.* **47**, 2671–2695.

Cheianov, V.V., Smith, H., Zvonarev, M.B. (2006). Exact results for three-body correlations in a degenerate one-dimensional Bose gas. *Phys. Rev. A* **73**. 051604.

Cheng, Z., Kurizki, G. (1995). Optical "multiexcitons": Quantum gap solitons in nonlinear Bragg reflectors. *Phys. Rev. Lett.* **75**, 3430–3433.

Cheng, Z., Kurizki, G. (1996). Theory of one-dimensional quantum gap solitons. *Phys. Rev. A* **54**, 3576–3591.

Cheon, T., Shigehara, T. (1998). Realizing discontinuous wave functions with renormalized short-range potentials. *Phys. Lett. A* **243**, 111–116.

Cheon, T., Shigehara, T. (1999). Fermion–boson duality of one-dimensional quantum particles with generalized contact interactions. *Phys. Rev. Lett.* **82**, 2536–2539.

Cherny, A.Y., Brand, J. (2004). Self-consistent calculation of the coupling constant in the Gross–Pitaevskii equation. *Phys. Rev. A* **70**. 043622.

Cherny, A.Y., Brand, J. (2006). Polarizability and dynamic structure factor of the one-dimensional Bose gas near the Tonks–Girardeau limit at finite temperatures. *Phys. Rev. A* **73**. 023612.

Chin, C. (2005). A simple model of Feshbach molecules. cond-mat/0506313.

Citro, R., Orignac, E. (2005). Atom–molecule coherence in a one-dimensional system. *Phys. Rev. Lett.* **95**. 130402.

Clement, D., Varon, A.F., Hugbart, M., Retter, J.A., Bouyer, P., Sanchez-Palencia, L., Gangardt, D.M., Shlyapnikov, G.V., Aspect, A. (2005). Suppression of transport of an interacting elongated Bose–Einstein condensate in a random potential. *Phys. Rev. Lett.* **95**. 170409.

Courteille, P.W., Bagnato, V.S., Yukalov, V.I. (2001). Bose–Einstein condensation of trapped atomic gases. *Laser Phys.* **11**, 659–800.

Creamer, D.B., Thacker, H.B., Wilkinson, D. (1981). Some exact results for the two-point function of an integrable quantum field theory. *Phys. Rev. D* **23**, 3081–3084.

Cusack, B.J., Alexander, T.J., Ostrovskaya, E.A., Kivshar, Y.S. (2001). Existence and stability of coupled atomic–molecular Bose–Einstein condensates. *Phys. Rev. A* **65**. 013609.

Dalfovo, F., Giorgini, S., Pitaevskii, L.P., Stringari, S. (1999). Theory of Bose–Einstein condensation in trapped gases. *Rev. Mod. Phys.* **71**, 463–512.

de Gennes, P.G. (1966). "Superconductivity of Metals and Alloys". Benjamin, New York.

Demkov, Y.N., Ostrovskii, V. (1988). "Zero-Range Potentials and Their Applications in Atomic Physics". Plenum, New York.

Denschlag, J., Simsarian, J.E., Feder, D.L., Clark, C.W., Collins, L.A., Cubizolles, J., Deng, L., Hagley, E.W., Helmerson, K., Reinhardt, W.P., Rolston, S.L., Schneider, B.I., Phillips, W.D. (2000). Generating solitons by phase engineering of a Bose–Einstein condensate. *Science* **287**, 97–101.

Derevianko, A. (2005). Revised Huang–Yang multipolar pseudopotential. *Phys. Rev. A* **72**. 044701.

Dettmer, S., Hellweg, D., Ryytty, P., Arlt, J.J., Ertmer, W., Sengstock, K., Petrov, D.S., Shlyapnikov, G.V., Kreutzmann, H., Santos, L., Lewenstein, M. (2001). Observation of phase fluctuations in elongated Bose–Einstein condensates. *Phys. Rev. Lett.* **87**. 160406.

Dodd, L.R. (1970). Exact solution of the Faddeev equations for a one-dimensional system. *J. Math. Phys.* **11**, 207–213.

Dodd, L.R. (1972). Numerical study of a model three-body system. *Australian Journal of Physics* **25**, 507–521.

Dodonov, V.V., Man'ko, V.I. (1989). Generalization of the uncertainty relations in quantum mechanics. In: Markov, M.A. (Ed.), *Invariants and Evolution of Nonstationary Quantum Systems*. In: *Proceedings of Lebedev Physics Institute*, vol. 183. Nova Science, Commack, NY, pp. 3–101.

Driben, R., Oz, Y., Malomed, B.A., Gubeskys, A., Yurovsky, V.A. (2007). Mismatch management for optical and matter-wave quadratic solitons. *Phys. Rev. E* **75**. 026612.

Drummond, P.D., Kheruntsyan, K.V., He, H. (1998). Coherent molecular solitons in Bose–Einstein condensates. *Phys. Rev. Lett.* **81**, 3055–3058.

Drummond, P.D., Kheruntsyan, K.V., He, H. (1999). Novel solitons in parametric amplifiers and atom lasers. *J. Opt. B* **1**, 387–395.

Dunjko, V., Lorent, V., Olshanii, M. (2001). Bosons in cigar-shaped traps: Thomas–Fermi regime, Tonks–Girardeau regime, and in between. *Phys. Rev. Lett.* **86**, 5413–5416.

Dutton, Z., Budde, M., Slowe, C., Hau, L.V. (2001). Observation of quantum shock waves created with ultra-compressed slow light pulses in a Bose–Einstein condensate. *Science* **293**, 663–668.

Esteve, J., Trebbia, J.-B., Schumm, T., Aspect, A., Westbrook, C.I., Bouchoule, I. (2006). Observations of density fluctuations in an elongated Bose gas: Ideal gas and quasicondensate regimes. *Phys. Rev. Lett.* **96**. 130403.

Etrich, C., Lederer, F., Malomed, B.A., Peschel, T., Peschel, U. (2000). Optical solitons in media with a quadratic nonlinearity. In: Wolf, E. (Ed.). *Progress in Optics*, vol. 41. North Holland, Amsterdam, pp. 483–568.

Fedichev, P.O., Kagan, Y., Shlyapnikov, G.V., Walraven, J.T.M. (1996). Influence of nearly resonant light on the scattering length in low-temperature atomic gases. *Phys. Rev. Lett.* **77**, 2913–2916.

Fedichev, P.O., Muryshev, A.E., Shlyapnikov, G.V. (1999). Dissipative dynamics of a kink state in a Bose-condensed gas. *Phys. Rev. A* **60**, 3220–3224.

Fermi, E. (1934). Displacement by pressure of the high lines of the spectral series. *Nuovo Cimento* **11**, 157–166.

Fermi, E. (1936). Motion of neutrons in hydrogenous substances. *Ricerca Sci.* **7**, 13–52.
Fermi, E., Pasta, J., Ulam, S. (1955). Studies of nonlinear problems. Los-Alamos Scientific Report, LA-1940.
Fertig, C.D., O'Hara, K.M., Huckans, J.H., Rolston, S.L., Phillips, W.D., Porto, J.V. (2005). Strongly inhibited transport of a degenerate 1D Bose gas in a lattice. *Phys. Rev. Lett.* **94**. 120403.
Fetter, A.L. (1972). Nonuniform states of an imperfect Bose gas. *Annals of Physics* **70**, 67–101.
Fetter, A.L., Walecka, J.D. (1971). "Quantum Theory of Many-Particle Systems". McGraw-Hill, New York.
Folman, R., Krueger, P., Schmiedmayer, J., Denschlag, J., Henkel, C. (2002). Microscopic atom optics: From wires to an atom chip. In: *Adv. At. Mol. Opt. Phys.*, vol. 48. Academic Press, New York, pp. 263–356.
Forrester, P.J., Frankel, N.E., Garoni, T.M., Witte, N.S. (2003). Finite one-dimensional impenetrable Bose systems: Occupation numbers. *Phys. Rev. A* **67**. 043607.
Forrester, P.J., Frankel, N.E., Makin, M.I. (2006). Analytic solutions of the one-dimensional finite-coupling delta-function Bose gas. *Phys. Rev. A* **74**. 043614.
Fricke, J. (1996). Transport equations including many-particle correlations for an arbitrary quantum system: A general formalism. *Annals of Physics* **252**, 479–498.
Fuchs, J.N., Gangardt, D.M., Keilmann, T., Shlyapnikov, G.V. (2005). Spin waves in a one-dimensional spinor Bose gas. *Phys. Rev. Lett.* **95**. 150402.
Gangardt, D.M., Shlyapnikov, G.V. (2003a). Local correlations in a strongly interacting one-dimensional Bose gas. *New J. Phys.* **5**, 79.
Gangardt, D.M., Shlyapnikov, G.V. (2003b). Stability and phase coherence of trapped 1D Bose gases. *Phys. Rev. Lett.* **90**. 010401.
Gasenzer, T., Berges, J., Schmidt, M.G., Seco, M. (2005). Nonperturbative dynamical many-body theory of a Bose–Einstein condensate. *Phys. Rev. A* **72**. 063604.
Gaudin, M. (1967). Un systeme a une dimension de fermions en interaction. *Phys. Lett. A* **24**, 55–56.
Gaudin, M. (1971). Boundary energy of a Bose gas in one dimension. *Phys. Rev. A* **4**, 386–394.
Gaudin, M. (1983). "La fonction d'onde de Bethe". Masson, Paris.
Gea-Banacloche, J., Rey, A.M., Pupillo, G., Williams, C.J., Clark, C.W. (2006). Mean-field treatment of the damping of the oscillations of a one-dimensional Bose gas in an optical lattice. *Phys. Rev. A* **73**. 013605.
Giamarchi, T. (2004). "Quantum Physics in One Dimension". University Press, Oxford.
Girardeau, M. (1960). Relationship between systems of impenetrable bosons and fermions in one dimension. *J. Math. Phys.* **1**, 516–523.
Girardeau, M.D. (1965). Permutation symmetry of many-particle wave functions. *Phys. Rev.* **139**, B500–B508.
Girardeau, M.D., Minguzzi, A. (2006). Bosonization, pairing, and superconductivity of the fermionic Tonks–Girardeau gas. *Phys. Rev. Lett.* **96**. 080404.
Girardeau, M.D., Olshanii, M. (2003). Fermi–Bose mapping and N-particle ground state of spin-polarized fermions in tight atom waveguides. cond-mat/0309396.
Girardeau, M.D., Olshanii, M. (2004). Theory of spinor Fermi and Bose gases in tight atom waveguides. *Phys. Rev. A* **70**. 023608.
Girardeau, M.D., Wright, E.M. (2000a). Breakdown of time-dependent mean-field theory for a one-dimensional condensate of impenetrable bosons. *Phys. Rev. Lett.* **84**, 5239–5242.
Girardeau, M.D., Wright, E.M. (2000b). Dark solitons in a one-dimensional condensate of hard core bosons. *Phys. Rev. Lett.* **84**, 5691–5694.
Girardeau, M.D., Wright, E.M. (2002). Quantum mechanics of one-dimensional trapped Tonks gases. *Laser. Phys.* **12**, 8–20.
Girardeau, M.D., Wright, E.M., Triscari, J.M. (2001). Ground-state properties of a one-dimensional system of hard-core bosons in a harmonic trap. *Phys. Rev. A* **63**. 033601.

Girardeau, M., Nguyen, H., Olshanii, M. (2004). Effective interactions, Fermi–Bose duality, and ground states of ultracold atomic vapors in tight de Broglie waveguides. *Opt. Comm.* **243**, 3–22.

Glöckle, W. (1983). "Quantim Mechanical Few-Body Problem". Springer, Berlin.

Góral, K., Köhler, T., Gardiner, S.A., Tiesinga, E., Julienne, P.S. (2004). Adiabatic association of ultracold molecules via magnetic-field tunable interactions. *J. Phys. B* **37**, 3457–3500.

Görlitz, A., Vogels, J.M., Leanhardt, A.E., Raman, C., Gustavson, T.L., Abo-Shaeer, J.R., Chikkatur, A.P., Gupta, S., Inouye, S., Rosenband, T., Ketterle, W. (2001). Realization of Bose–Einstein condensates in lower dimensions. *Phys. Rev. Lett.* **87**. 130402.

Granger, B.E., Blume, D. (2004). Tuning the interactions of spin-polarized fermions using quasi-one-dimensional confinement. *Phys. Rev. Lett.* **92**. 133202.

Greiner, M., Bloch, I., Mandel, O., Hänsch, T.W., Esslinger, T. (2001). Exploring phase coherence in a 2D lattice of Bose–Einstein condensates. *Phys. Rev. Lett.* **87**. 160405.

Griffin, A. (1996). Conserving and gapless approximations for an inhomogeneous Bose gas at finite temperatures. *Phys. Rev. B* **53**, 9341–9347.

Gross, E.P. (1961). Structure of quantized vortex in boson systems. *Nuovo Cimento* **20**, 454–477.

Gross, E.P. (1963). Hydrodynamics of a superfluid condensate. *J. Math. Phys.* **4**, 195–207.

Gu, C.H., Yang, C.N. (1989). A one-dimensional N fermion problem with factorized S matrix. *Commun. Math. Phys.* **122**, 105–116.

Gunter, K., Stoferle, T., Moritz, H., Kohl, M., Esslinger, T. (2005). p-wave interactions in low-dimensional fermionic gases. *Phys. Rev. Lett.* **95**. 230401.

Gupta, S., Murch, K.W., Moore, K.L., Purdy, T.P., Stamper-Kurn, D.M. (2005). Bose–Einstein condensation in a circular waveguide. *Phys. Rev. Lett.* **95**. 143201.

Haldane, F.D.M. (1981a). Demonstration of the "Luttinger liquid" character of Bethe-ansatz-soluble models of 1-D quantum fluids. *Phys. Lett. A* **81**, 153–155.

Haldane, F.D.M. (1981b). Effective harmonic-fluid approach to low-energy properties of one-dimensional quantum fluids. *Phys. Rev. Lett.* **47**, 1840–1843.

Hänsel, W., Reichel, J., Hommelhoff, P., Hänsch, T.W. (2001). Trapped-atom interferometer in a magnetic microtrap. *Phys. Rev. A* **64**. 063607.

Haus, H.A., Lai, Y. (1990). Quantum theory of soliton squeezing: A linearized approach. *J. Opt. Soc. Am. B* **7**, 386–392.

Haus, H.A., Yu, C.X. (2000). Soliton squeezing and the continuum. *J. Opt. Soc. Am. B* **17**, 618–628.

Hellweg, D., Cacciapuoti, L., Kottke, M., Schulte, T., Sengstock, K., Ertmer, W., Arlt, J.J. (2003). Measurement of the spatial correlation function of phase fluctuating Bose–Einstein condensates. *Phys. Rev. Lett.* **91**. 010406.

Holland, M., Park, J., Walser, R. (2001). Formation of pairing fields in resonantly coupled atomic and molecular Bose–Einstein condensates. *Phys. Rev. Lett.* **86**, 1915–1918.

Huang, K. (1987). "Statistical Mechanics". Wiley, New York.

Huang, K., Yang, C.N. (1957). Quantum-mechanical many-body problem with hard-sphere interaction. *Phys. Rev.* **105**, 767–775.

Hugbart, M., Retter, J.A., Gerbier, F., Varon, A.F., Richard, S., Thywissen, J.H., Clement, D., Bouyer, P., Aspect, A. (2005). Coherence length of an elongated condensate. *Eur. Phys. J. D* **35**, 155–163.

Hugbart, M., Retter, J.A., Varon, A.F., Bouyer, P., Aspect, A., Davis, M.J. (2007). Population and phase coherence during the growth of an elongated Bose–Einstein condensate. *Phys. Rev. A* **75**. 011602.

Idziaszek, Z., Calarco, T. (2006). Pseudopotential method for higher partial wave scattering. *Phys. Rev. Lett.* **96**. 013201.

Imambekov, A., Demler, E. (2006a). Applications of exact solution for strongly interacting one-dimensional Bose–Fermi mixture: Low-temperature correlation functions, density profiles, and collective modes. *Annals of Physics* **321**, 2390–2437.

Imambekov, A., Demler, E. (2006b). Exactly solvable case of a one-dimensional Bose–Fermi mixture. *Phys. Rev. A* **73**. 021602.

Ishikawa, M., Takayama, H. (1980). Solitons in a one-dimensional Bose system with the repulsive delta-function interaction. *J. Phys. Soc. Jap.* **49**, 1242–1246.

Jessen, P.S., Deutsch, I.H. (1996). Optical lattices. In: *Adv. At. Mol. Opt. Phys.*, vol. 37. Academic Press, New York, pp. 95–138.

Jimbo, M., Miwa, T., Mori, Y., Sato, M. (1980). Density matrix of an impenetrable Bose gas and the fifth Painleve transcendent. *Physica D* **1**, 80–158.

Kanamoto, R., Saito, H., Ueda, M. (2003a). Quantum phase transition in one-dimensional Bose–Feshbach condensates with attractive interactions. *Phys. Rev. A* **67**. 013608.

Kanamoto, R., Saito, H., Ueda, M. (2003b). Stability of the quantized circulation of an attractive Bose–Feshbach condensate in a rotating torus. *Phys. Rev. A* **68**. 043619.

Kanamoto, R., Saito, H., Ueda, M. (2005). Symmetry breaking and enhanced condensate fraction in a matter-wave bright soliton. *Phys. Rev. Lett.* **94**. 090404.

Kanamoto, R., Saito, H., Ueda, M. (2006). Critical fluctuations in a soliton formation of attractive Bose–Feshbach condensates. *Phys. Rev. A* **73**. 033611.

Kanjilal, K., Blume, D. (2004). Nondivergent pseudopotential treatment of spin-polarized fermions under one- and three-dimensional harmonic confinement. *Phys. Rev. A* **70**. 042709.

Kartavtsev, O.I., Macek, J.H. (2002). Low-energy three-body recombination near a Feshbach resonance. *Few-Body Systems* **31**, 249–254.

Ketterle, W., Durfee, D.S., Stamper-Kurn, D.M. (1999). Making, probing and understanding Bose–Einstein condensates. In: Inguscio, M., Stringari, S., Wieman, C.E. (Eds.), *Proceedings of the International School of Physics – Enrico Fermi*. IOS Press, pp. 67–176.

Khaykovich, L., Schreck, F., Ferrari, G., Bourdel, T., Cubizolles, J., Carr, L.D., Castin, Y., Salomon, C. (2002). Formation of a matter-wave bright soliton. *Science* **296**, 1290–1293.

Kheruntsyan, K.V. (2006). Quantum atom optics with fermions from molecular dissociation. *Phys. Rev. Lett.* **96**. 110401.

Kheruntsyan, K.V., Gangardt, D.M., Drummond, P.D., Shlyapnikov, G.V. (2003). Pair correlations in a finite-temperature 1D Bose gas. *Phys. Rev. Lett.* **91**. 040403.

Kheruntsyan, K.V., Gangardt, D.M., Drummond, P.D., Shlyapnikov, G.V. (2005). Finite-temperature correlations and density profiles of an inhomogeneous interacting one-dimensional Bose gas. *Phys. Rev. A* **71**. 053615.

Kim, Y.E., Zubarev, A.L. (2003). Cold Bose gases near Feshbach resonances. *Phys. Lett. A* **312**, 277–286.

Kim, J.I., Schmiedmayer, J., Schmelcher, P. (2005). Quantum scattering in quasi-one-dimensional cylindrical confinement. *Phys. Rev. A* **72**. 042711.

Kinoshita, T., Wenger, T., Weiss, D.S. (2004). Observation of a one-dimensional Tonks–Girardeau gas. *Science* **305**, 1125–1128.

Kinoshita, T., Wenger, T., Weiss, D.S. (2005). Local pair correlations in one-dimensional Bose gases. *Phys. Rev. Lett.* **95**. 190406.

Kinoshita, T., Wenger, T., Weiss, D.S. (2006). A quantum Newton's cradle. *Nature* **440**, 900–903.

Kivshar, Y.S., Agrawal, G. (2003). "Optical Solitons: From Fibers to Photonic Crystals". Academic Press, New York.

Kocharovsky, V.V., Kocharovsky, V.V., Scully, M.O. (2000a). Condensate statistics in interacting and ideal dilute Bose gases. *Phys. Rev. Lett.* **84**, 2306–2309.

Kocharovsky, V.V., Kocharovsky, V.V., Scully, M.O. (2000b). Condensation of N bosons. III. Analytical results for all higher moments of condensate fluctuations in interacting and ideal dilute Bose gases via the canonical ensemble quasiparticle formulation. *Phys. Rev. A* **61**. 053606.

Kocharovsky, V.V., Kocharovsky, V.V., Holthaus, M., Ooi, C.R., Svidzinsky, A.A., Ketterle, W., Scully, M.O. (2006). Fluctuations in ideal and interacting Bose–Einstein condensates: From the laser phase transition analogy to squeezed states and Bogoliubov quasiparticles. In: *Adv. At. Mol. Opt. Phys.*, vol. 53. Academic Press, New York, pp. 291–411.

Köhler, T., Burnett, K. (2002). Microscopic quantum dynamics approach to the dilute condensed Bose gas. *Phys. Rev. A* **65**. 033601.

Köhler, T., Gasenzer, T., Burnett, K. (2003). Microscopic theory of atom–molecule oscillations in a Bose–Einstein condensate. *Phys. Rev. A* **67**. 013601.

Köhler, T., Goral, K., Julienne, P.S. (2006). Production of cold molecules via magnetically tunable Feshbach resonances. *Rev. Mod. Phys.* **78**, 1311.

Kokkelmans, S.J.J.M.F., Holland, M.J. (2002). Ramsey fringes in a Bose–Einstein condensate between atoms and molecules. *Phys. Rev. Lett.* **89**. 180401.

Kokkelmans, S.J.J.M.F., Milstein, J.N., Chiofalo, M.L., Walser, R., Holland, M.J. (2002). Resonance superfluidity: Renormalization of resonance scattering theory. *Phys. Rev. A* **65**. 053617.

Kolomeisky, E.B., Newman, T.J., Straley, J.P., Qi, X. (2000). Low-dimensional Bose liquids: Beyond the Gross–Pitaevskii approximation. *Phys. Rev. Lett.* **85**, 1146–1149.

Konotop, V.V., Pitaevskii, L. (2004). Landau dynamics of a grey soliton in a trapped condensate. *Phys. Rev. Lett.* **93**. 240403.

Korepin, V., Slavnov, N. (1997). Time and temperature dependent correlation functions of 1D models of quantum statistical mechanics. *Phys. Lett. A* **236**, 201–205.

Korepin, V.E., Bogoliubov, N.M., Izergin, A.G. (1993). "Quantum Inverse Scattering Method and Correlation Functions". University Press, Cambridge.

Krämer, M., Menotti, C., Pitaevskii, L., Stringari, S. (2003). Bose–Einstein condensates in 1D optical lattices. *Eur. Phys. J. D* **27**, 247–261.

Kulish, P.P., Manakov, S.V., Faddeev, L.D. (1976). Comparison of the exact quantum and quasiclassical results for a nonlinear Schrödinger equation. *Theor. Math. Phys.* **28**, 615–620.

Lai, Y., Haus, H.A. (1989a). Quantum theory of solitons in optical fibers. I. Time-dependent Hartree approximation. *Phys. Rev. A* **40**, 844–853.

Lai, Y., Haus, H.A. (1989b). Quantum theory of solitons in optical fibers. II. Exact solution. *Phys. Rev. A* **40**, 854–866.

Lee, C., Brand, J. (2006). Enhanced quantum reflection of matter-wave solitons. *Europhys. Lett.* **73**, 321–327.

Leggett, A.J. (2001). Bose–Einstein condensation in the alkali gases: Some fundamental concepts. *Rev. Mod. Phys.* **73**, 307.

Lenard, A. (1964). Momentum distribution in the ground state of the one-dimensional system of impenetrable bosons. *J. Math. Phys.* **5**, 930–943.

Lewenstein, M., You, L. (1996). Quantum phase diffusion of a Bose–Einstein condensate. *Phys. Rev. Lett.* **77**, 3489–3493.

Lieb, E.H. (1963). Exact analysis of an interacting Bose gas. II. The excitation spectrum. *Phys. Rev.* **130**, 1616–1624.

Lieb, E.H., Liniger, W. (1963). Exact analysis of an interacting Bose gas. I. The general solution and the ground state. *Phys. Rev.* **130**, 1605–1616.

Lipkin, H.J. (1973). "Quantum Mechanics: New Approaches to Selected Topics". North-Holland, Amsterdam.

Lovelace, C. (1964). Practical theory of three-particle states. I. Nonrelativistic. *Phys. Rev.* **135**, B1225–B1249.

Luttinger, J.M. (1963). An exactly soluble model of a many-fermion system. *J. Math. Phys.* **4**, 1154–1162.

Macek, J.H., Sternberg, J. (2006). Properties of pseudopotentials for higher partial waves. *Phys. Rev. Lett.* **97**. 023201.

Majumdar, C.K. (1972). Solution of Faddeev equations for a one-dimensional system. *J. Math. Phys.* **13**, 705–708.

Malomed, B.A. (2006). "Soliton Management in Periodic Systems". Springer, New York.

Man'ko, O.V. (1997). Symplectic tomography of nonlinear coherent states of a trapped ion. *Phys. Lett. A* **228**, 29–35.

Marchioro, C., Presutti, E. (1970). Exact statistical model in one dimension. *Lett. Nuovo Cimento* **4**, 488.

Mattis, D.C., Lieb, E.H. (1965). Exact solution of a many-fermion system and its associated boson field. *J. Math. Phys.* **6**, 304–312.

Mazets, I.E., Kurizki, G. (2006). How different are multiatom quantum solitons from mean-field solitons? *Europhys. Lett.* **76**, 196–202.

McGuire, J.B. (1964). Study of exactly soluble one-dimensional N-body problems. *J. Math. Phys.* **5**, 622–636.

Mehta, N.P., Shepard, J.R. (2005). Three bosons in one dimension with short-range interactions: Zero-range potentials. *Phys. Rev. A* **72**. 032728.

Mehta, N., Greene, C., Esry, B. (2006). Three-body recombination in one-dimensional systems. *Bull. Am. Phys. Soc.* **51**. O1.00006.

Menotti, C., Stringari, S. (2002). Collective oscillations of a one-dimensional trapped Bose–Einstein gas. *Phys. Rev. A* **66**. 043610.

Merhasin, I.M., Malomed, B.A., Band, Y.B. (2006). Partially incoherent gap solitons in Bose–Einstein condensates. *Phys. Rev. A* **74**. 033614.

Meyrath, T.P., Schreck, F., Hanssen, J.L., Chuu, C.-S., Raizen, M.G. (2005). Bose–Einstein condensate in a box. *Phys. Rev. A* **71**. 041604.

Moore, M.G., Bergeman, T., Olshanii, M. (2004). Scattering in tight atom waveguides. *J. Phys. IV (France)* **116**, 69–86.

Mora, C., Castin, Y. (2003). Extension of Bogoliubov theory to quasicondensates. *Phys. Rev. A* **67**. 053615.

Mora, C., Egger, R., Gogolin, A.O., Komnik, A. (2004). Atom–dimer scattering for confined ultracold fermion gases. *Phys. Rev. Lett.* **93**. 170403.

Mora, C., Egger, R., Gogolin, A.O. (2005). Three-body problem for ultracold atoms in quasi-one-dimensional traps. *Phys. Rev. A* **71**. 052705.

Moritz, H., Stöferle, T., Köhl, M., Esslinger, T. (2003). Exciting collective oscillations in a trapped 1D gas. *Phys. Rev. Lett.* **91**. 250402.

Moritz, H., Stöferle, T., Günter, K., Köhl, M., Esslinger, T. (2005). Confinement induced molecules in a 1D Fermi gas. *Phys. Rev. Lett.* **94**. 210401.

Morsch, O., Oberthaler, M. (2006). Dynamics of Bose–Einstein condensates in optical lattices. *Rev. Mod. Phys.* **78**, 179.

Muryshev, A., Shlyapnikov, G.V., Ertmer, W., Sengstock, K., Lewenstein, M. (2002). Dynamics of dark solitons in elongated Bose–Einstein condensates. *Phys. Rev. Lett.* **89**. 110401.

Naidon, P., Tiesinga, E., Mitchell, W.F., Julienne, P.S. (2007). Effective-range description of a Bose gas under strong one- or two-dimensional confinement. *New J. Phys.* **9**, 19.

Nho, K., Blume, D. (2005). Superfluidity of mesoscopic Bose gases under varying confinements. *Phys. Rev. Lett.* **95**. 193601.

Novikov, S., Manakov, S.V., Pitaevskii, L.P., Zakharov, V.E. (1984). "Theory of Solitons: The Inverse Scattering Method". Consultants Bureau, New York.

Oles, B., Sacha, K. (2007). Solitons in coupled atomic–molecular Bose–Einstein condensates. *J. Phys. B* **40**, 1103–1116.

Olshanii, M. (1998). Atomic scattering in the presence of an external confinement and a gas of impenetrable bosons. *Phys. Rev. Lett.* **81**, 938–941.

Olshanii, M. (2002). http://physics.usc.edu/~olshanii/DIST/.

Olshanii, M., Dunjko, V. (2003). Short-distance correlation properties of the Lieb–Liniger system and momentum distributions of trapped one-dimensional atomic gases. *Phys. Rev. Lett.* **91**. 090401.

Olshanii, M., Dunjko, V. (2006). Unpublished.

Orignac, E., Citro, R. (2006). Phase transitions in the boson–fermion resonance model in one dimension. *Phys. Rev. A* **73**. 063611.

Orso, G., Pitaevskii, L.P., Stringari, S., Wouters, M. (2005). Formation of molecules near a Feshbach resonance in a 1D optical lattice. *Phys. Rev. Lett.* **95**. 060402.

Paredes, B., Garcia-Ripoll, J.J., Zoller, P., Cirac, I.J. (2004). Strong correlation effects and quantum information theory of low dimensional atomic gases. *J. Phys. IV (France)* **116**, 135–168.

Parkins, A.S., Walls, D.F. (1998). The physics of trapped dilute-gas Bose–Einstein condensates. *Phys. Rep.* **303**, 1–80.

Peano, V., Thorwart, M., Kasper, A., Egger, R. (2005a). Nanoscale atomic waveguides with suspended carbon nanotubes. *Appl. Phys. B* **81**, 1075–1080.

Peano, V., Thorwart, M., Mora, C., Egger, R. (2005b). Confinement-induced resonances for a two-component ultracold atom gas in arbitrary quasi-one-dimensional traps. *New J. Phys.* **7**, 192.

Perelomov, A.M. (1971). Algebraic approach to the solution of a one-dimensional model of N interacting particles. *Theor. Math. Phys.* **6**, 263–282.

Petrov, D.S., Shlyapnikov, G.V., Walraven, J.T.M. (2000). Regimes of quantum degeneracy in trapped 1D gases. *Phys. Rev. Lett.* **85**, 3745–3749.

Petrov, D.S., Shlyapnikov, G.V., Walraven, J.T.M. (2001). Phase-fluctuating 3D Bose–Einstein condensates in elongated traps. *Phys. Rev. Lett.* **87**. 050404.

Petrov, D.S., Salomon, C., Shlyapnikov, G.V. (2004). Weakly bound dimers of fermionic atoms. *Phys. Rev. Lett.* **93**. 090404.

Petrov, D.S., Salomon, C., Shlyapnikov, G.V. (2005). Scattering properties of weakly bound dimers of fermionic atoms. *Phys. Rev. A* **71**. 012708.

Pitaevskii, L., Stringari, S. (2003). "Bose–Einstein Condensation". University Press, Oxford.

Pitaevskii, L.P. (1961). Vortex lines in an imperfect Bose gas. *Sov. Phys. JETP* **13**, 451–454.

Poulsen, U.V., Molmer, K. (2003). Scattering of atoms on a Bose–Einstein condensate. *Phys. Rev. A* **67**. 013610.

Pricoupenko, L. (2006). Modeling interactions for resonant p-wave scattering. *Phys. Rev. Lett.* **96**. 050401.

Pritchard, D.E. (1983). Cooling neutral atoms in a magnetic trap for precision spectroscopy. *Phys. Rev. Lett.* **51**, 1336–1339.

Proukakis, N.P., Burnett, K. (1996). Generalized mean fields for trapped atomic Bose–Einstein condensates. *J. Res. Natl. Inst. Stand. Technol.* **101**, 457–469.

Proukakis, N.P., Burnett, K., Stoof, H.T.C. (1998). Microscopic treatment of binary interactions in the nonequilibrium dynamics of partially Bose-condensed trapped gases. *Phys. Rev. A* **57**, 1230–1247.

Regal, C.A., Jin, D.S. (2006). Experimental realization of the BCS-BEC crossover with a Fermi gas of atoms. In: *Adv. At. Mol. Opt. Phys.*, vol. 54. Academic Press, New York, pp. 1–77.

Rey, A.M., Hu, B.L., Calzetta, E., Roura, A., Clark, C.W. (2004). Nonequilibrium dynamics of optical-lattice-loaded Bose–Einstein-condensate atoms: Beyond the Hartree–Fock–Bogoliubov approximation. *Phys. Rev. A* **69**. 033610.

Rey, A.M., Hu, B.L., Calzetta, E., Clark, C.W. (2005a). Quantum kinetic theory of a Bose–Einstein gas confined in a lattice. *Phys. Rev. A* **72**. 023604.

Rey, A.M., Pupillo, G., Clark, C.W., Williams, C.J. (2005b). Ultracold atoms confined in an optical lattice plus parabolic potential: A closed-form approach. *Phys. Rev. A* **72**. 033616.

Rey, A.M., Satija, I.I., Clark, C.W. (2006). Quantum coherence of hard-core bosons: Extended, glassy, and Mott phases. *Phys. Rev. A* **73**. 063610.

Rey, A.M., Burnett, K., Satija, I.I., Clark, C.W. (2007). Lifshitz-like transition and enhancement of correlations in a rotating bosonic ring lattice. *Phys. Rev. A* **75**. 063616.

Richard, S., Gerbier, F., Thywissen, J.H., Hugbart, M., Bouyer, P., Aspect, A. (2003). Momentum spectroscopy of 1D phase fluctuations in Bose–Einstein condensates. *Phys. Rev. Lett.* **91**. 010405.

Rigol, M., Dunjko, V., Yurovsky, V., Olshanii, M. (2007). Relaxation in a completely integrable many-body quantum system: An ab initio study of the dynamics of the highly excited states of 1D lattice hard-core Bosons. *Phys. Rev. Lett.* **98**. 050405.

Ross, M.H., Shaw, G.L. (1961). Multichannel effective range theory. *Annals of Physics* **13**, 147–186.
Sacha, K., Timmermans, E. (2006). Self-localized impurities embedded in a one-dimensional Bose–Einstein condensate and their quantum fluctuations. *Phys. Rev. A* **73**. 063604.
Sakmann, K., Streltsov, A.I., Alon, O.E., Cederbaum, L.S. (2005). Exact ground state of finite Bose–Einstein condensates on a ring. *Phys. Rev. A* **72**. 033613.
Scalapino, D.J., Sears, M., Ferrell, R.A. (1972). Statistical mechanics of one-dimensional Ginzburg–Landau fields. *Phys. Rev. B* **6**, 3409–3416.
Schmid, E.W., Ziegelmann, H. (1974). "The Quantum Mechanical Three-Body Problem". Pergamon Press, Oxford.
Schmidt, B., Fleischhauer, M. (2007). Exact numerical simulations of a one-dimensional trapped Bose gas. *Phys. Rev. A* **75**. 021601.
Schmidt, B., Plimak, L.I., Fleischhauer, M. (2005). Stochastic simulation of a finite-temperature one-dimensional Bose gas: From the Bogoliubov to the Tonks–Girardeau regime. *Phys. Rev. A* **71**. 041601.
Schreck, F., Khaykovich, L., Corwin, K.L., Ferrari, G., Bourdel, T., Cubizolles, J., Salomon, C. (2001). Quasipure Bose–Einstein condensate immersed in a Fermi sea. *Phys. Rev. Lett.* **87**. 080403.
Schumm, T., Hofferberth, S., Andersson, L.M., Wildermuth, S., Groth, S., Bar-Joseph, I., Schmiedmayer, J., Krüger, P. (2005). Matter-wave interferometry in a double well on an atom chip. *Nature Physics* **1**, 57–62.
Seba, P. (1986). The generalized point interaction in one dimension. *Czech. J. Phys. B* **36**, 667–673.
Shaw, G.L., Ross, M.H. (1962). Analysis of multichannel reactions. *Phys. Rev.* **126**, 806–813.
Sheehy, D.E., Radzihovsky, L. (2005). Quantum decoupling transition in a one-dimensional Feshbach-resonant superfluid. *Phys. Rev. Lett.* **95**. 130401.
Shnirman, A.G., Malomed, B.A., Ben-Jacob, E. (1994). Nonperturbative studies of a quantum higher-order nonlinear Schrödinger model using the Bethe ansatz. *Phys. Rev. A* **50**, 3453–3463.
Sinha, S., Cherny, A.Y., Kovrizhin, D., Brand, J. (2006). Friction and diffusion of matter-wave bright solitons. *Phys. Rev. Lett.* **96**. 030406.
Skorniakov, G.V., Ter-Martirosian, K.A. (1957). Three body problem for short range forces. I. Scattering of low energy neutrons by deutrons. *Sov. Phys. JETP* **4**, 648–661.
Steblina, V.V., Kivshar, Y.S., Lisak, M., Malomed, B.A. (1995). Self-guided beams in a diffractive $\chi^{(2)}$ medium: Variational approach. *Opt. Comm.* **118**, 345–352.
Stock, R., Silberfarb, A., Bolda, E.L., Deutsch, I.H. (2005). Generalized pseudopotentials for higher partial wave scattering. *Phys. Rev. Lett.* **94**. 023202.
Stoferle, T., Moritz, H., Schori, C., Kohl, M., Esslinger, T. (2004). Transition from a strongly interacting 1D superfluid to a Mott insulator. *Phys. Rev. Lett.* **92**. 130403.
Strecker, K.E., Partridge, G.B., Truscott, A.G., Hulet, R.G. (2002). Formation and propagation of matter-wave soliton trains. *Nature* **417**, 150–153.
Strecker, K.E., Partridge, G.B., Truscott, A.G., Hulet, R.G. (2003). Bright matter wave solitons in Bose–Einstein condensates. *New J. Phys.* **5**, 73.
Sutherland, B. (1968). Further results for the many-body problem in one dimension. *Phys. Rev. Lett.* **20**, 98–100.
Sutherland, B. (1971a). Exact results for a quantum many-body problem in one dimension. *Phys. Rev. A* **4**, 2019–2021.
Sutherland, B. (1971b). Quantum many-body problem in one dimension: Ground state. *J. Math. Phys.* **12**, 246–250.
Sutherland, B. (1971c). Quantum many-body problem in one dimension: Thermodynamics. *J. Math. Phys.* **12**, 251–256.
Sutherland, B. (1972). Exact results for a quantum many-body problem in one dimension. II. *Phys. Rev. A* **5**, 1372–1376.
Sutherland, B. (2004). "Beautiful Models: 70 Years of Exactly Solved Quantum Many-Body Problems". World Scientific, New Jersey.

Tiesinga, E., Moerdijk, A., Verhaar, B.J., Stoof, H.T.C. (1992). Conditions for Bose–Einstein condensation in magnetically trapped atomic cesium. *Phys. Rev. A* **46**, R1167–R1170.

Tiesinga, E., Verhaar, B.J., Stoof, H.T.C. (1993). Threshold and resonance phenomena in ultracold ground-state collisions. *Phys. Rev. A* **47**, 4114–4122.

Timmermans, E., Tommasini, P., Hussein, M., Kerman, A. (1999). Feshbach resonances in atomic Bose–Einstein condensates. *Phys. Rep.* **315**, 199–230.

Tolra, B.L., O'Hara, K.M., Huckans, J.H., Phillips, W.D., Rolston, S.L., Porto, J.V. (2004). Observation of reduced three-body recombination in a correlated 1D degenerate Bose gas. *Phys. Rev. Lett.* **92**. 190401.

Tsuzuki, T. (1971). Nonlinear waves in the Pitaevskii–Gross equation. *J. Low Temp. Phys.* **4**, 441–457.

Vaidya, H.G., Tracy, C.A. (1979). One-particle reduced density matrix of impenetrable bosons in one dimension at zero temperature. *Phys. Rev. Lett.* **42**, 3–6.

Vinitskii, S., Kuperin, Y., Motovilov, A., Suz'ko, A. (1992). Three-channel Hamiltonian for the muon catalysis reaction. *Sov. J. Nucl. Phys.* **55**, 245–253.

Voit, J. (1995). One-dimensional Fermi liquids. *Rep. Progr. Phys.* **58**, 977–1116.

Walser, R., Williams, J., Cooper, J., Holland, M. (1999). Quantum kinetic theory for a condensed bosonic gas. *Phys. Rev. A* **59**, 3878–3889.

Weiss D.S. (2006). Unpublished.

Whittaker, E.T., Watson, G.N. (1996). "A Course of Modern Analysis". University Press, Cambridge.

Winoto, S.L., DePue, M.T., Bramall, N.E., Weiss, D.S. (1999). Laser cooling at high density in deep far-detuned optical lattices. *Phys. Rev. A* **59**, R19–R22.

Yang, C.N. (1967). Some exact results for the many-body problem in one dimension with repulsive delta-function interaction. *Phys. Rev. Lett.* **19**, 1312–1315.

Yang, C.N. (1968). S matrix for the one-dimensional N-body problem with repulsive or attractive δ-function interaction. *Phys. Rev.* **168**, 1920–1923.

Yang, C.N., Yang, C.P. (1969). Thermodynamics of a one-dimensional system of bosons with repulsive delta-function interaction. *J. Math. Phys.* **10**, 1115–1122.

Yukalov, V.I. (2004). Principal problems in Bose–Einstein condensation of dilute gases. *Laser Phys.* **1**, 435–462.

Yukalov, V.I. (2006). Self-consistent theory of Bose-condensed systems. *Phys. Lett. A* **359**, 712–717.

Yukalov, V., Girardeau, M. (2005). Fermi–Bose mapping for one-dimensional Bose gases. *Laser Phys. Lett.* **2**, 375–382.

Yukalov, V.I., Kleinert, H. (2006). Gapless Hartree–Fock–Bogoliubov approximation for Bose gases. *Phys. Rev. A* **73**. 063612.

Yurovsky, V.A. (2002). Quantum effects on dynamics of instabilities in Bose–Einstein condensates. *Phys. Rev. A* **65**. 033605.

Yurovsky, V.A. (2005). Feshbach resonance scattering under cylindrical harmonic confinement. *Phys. Rev. A* **71**. 012709.

Yurovsky, V.A. (2006). Properties of quasi-one-dimensional molecules with Feshbach-resonance interaction. *Phys. Rev. A* **73**. 052709.

Yurovsky, V.A. (2007). Effects of nonintegrability on stabilization of Feshbach molecules in atom waveguides. physics/0703168.

Yurovsky, V.A., Band, Y.B. (2007). Control of ultraslow inelastic collisions by Feshbach resonances and quasi-one-dimensional confinement. *Phys. Rev. A* **75**. 012717.

Yurovsky, V.A., Ben-Reuven, A. (2003). Formation of a molecular Bose–Einstein condensate and an entangled atomic gas by Feshbach resonance. *Phys. Rev. A* **67**. 043611.

Yurovsky, V.A., Ben-Reuven, A., Olshanii, M. (2006). One-dimensional Bose chemistry: Effects of nonintegrability. *Phys. Rev. Lett.* **96**. 163201.

MOTRIMS: MAGNETO–OPTICAL TRAP RECOIL ION MOMENTUM SPECTROSCOPY

B.D. DEPAOLA[1,*], R. MORGENSTERN[2] and N. ANDERSEN[3]

[1] Department of Physics, Kansas State University, Manhattan, KS 66506, USA
[2] KVI, Atomic Physics, Rijksuniversiteit Groningen - Zernikelaan 25, NL-9747 AA Groningen, The Netherlands
[3] Niels Bohr Institute, University of Copenhagen, Universitetsparken 5, DK-2100 Copenhagen, Denmark

1. Introduction to MOTRIMS 140
   1.1. What Is MOTRIMS/COLTRIMS 140
   1.2. What Can One Do with MOTRIMS 141
   1.3. General Description of the Apparatus 142
2. Relative Total Electron Transfer Cross Sections 147
3. Case Studies in Total Electron Transfer Collisions 149
   3.1. Removal of Inner- and Outer-Shell Na Electrons by H^+ Impact 150
   3.2. Electron Transfer in O^{6+} + Na Collisions 156
   3.3. Electron Capture from Excited Na Atoms 159
4. Case Studies of Differential Electron Transfer Cross Sections 161
   4.1. Na(3s,3p) Targets 162
   4.2. Rb^+, K^+, Li^+ + Rb(5s,5p) Collisions 170
   4.3. O^{6+} + Na(3s) 171
5. Probing Excitation Dynamics 172
   5.1. Excited Fractions 172
   5.2. Population Dynamics 176
6. Future Applications 180
   6.1. Electron Collisions 181
   6.2. Photo-Ionization and Photo-Associative Ionization 181
   6.3. Anisotropic Targets: Alignment and Orientation 183
   6.4. Stark Effect 184
7. Concluding Comments 184
8. Acknowledgements 185
9. References 185

* Corresponding author. E-mail: depaola@phys.ksu.edu.

1. Introduction to MOTRIMS

1.1. What Is MOTRIMS/COLTRIMS

One of the truly revolutionary techniques to be developed in the past few decades for mapping the dynamics of ion–atom/molecule and photon–atom/molecule interactions is COLTRIMS or cold target recoil ion momentum spectroscopy. The novel idea behind COLTRIMS is to exploit "inverse kinematics" by measuring the momentum vectors of the target fragments that have been charged (either positively or negatively) during the interaction of the target with the projectile. From the momenta of these fragments, measured event by event, one can reconstruct the dynamics of the interaction. For example, the Q-value, or energy defect for single electron transfer is given by

$$Q = -\frac{1}{2}m_e v_p^2 - p_\| v_p, \tag{1}$$

where m_e is the mass of the electron, v_p is the velocity of the projectile, and $p_\|$ is the component of the recoil ion velocity parallel to the initial direction of the projectile ion. The energy defect, Q, can be thought of as the difference between the initial and final potential energies of the transferred electron. The scattering angle of the projectile is given by

$$\theta_p = -\frac{p_\perp}{m_p v_p} \tag{2}$$

where $p_\perp$ is the component of the recoil ion perpendicular to the initial direction of the projectile ion, and m_p is the projectile mass. Equations (1) and (2) are readily derived from classical equations of energy and momentum conservation, and were shown by Mergel (1994) to be relativistically correct as well.

Because the recoil "kick" received by the target in the interaction is typically on the order of one atomic unit of momentum, its initial momentum distribution must be small compared to this amount in order to make a measurement of the recoil momentum meaningful. This typically necessitates cooling the target atoms or molecules by using a combination of supersonic expansion, collimated jets, and cold finger pre-cooling. Excellent review articles have been written on COLTRIMS by Dörner et al. (2000) and Ullrich et al. (2003), and so nothing more directly pertaining to that technique will be mentioned in this work.

In the past decade, several groups have carried out experiments that were heavily influenced by the general concepts of COLTRIMS. In these experiments, however, the target atoms were not cooled using the usual COLTRIMS methods, but rather by laser cooling and trapping in a MOT (magneto–optical trap). In the earliest experiments performed by Wolf and Helm (1997, 2000), a ^{87}Rb MOT was constructed in the constant electric field section of a time-of-flight

(TOF) spectrometer. When the Rb target was made to interact with a picosecond pulse of laser light, the Rb$^+$ produced in the interaction was analyzed by the TOF spectrometer and some photon–atom kinematics could be measured. Related measurements were made in a recent experiment by Coutinho et al. (2004). In a similar arrangement, Huang et al. (2002, 2003) studied the interaction of a beam of electrons with cooled and trapped ^{7}Li.

In none of these experiments was the full COLTRIMS methodology exploited since the simple TOF apparatus integrated over the transverse momentum components of the recoiling ions. However, from the experiments of Wolf and Helm (1997), was born a key component of the topic of this work: MOTRIMS, or magneto–optical trap recoil ion momentum spectroscopy. MOTRIMS is essentially the COLTRIMS technology for the measurement of the momentum vectors of recoiling charged fragments from the target, merged with the technology that has been developed for the cooling and trapping of target atoms using lasers. The technique was independently and, essentially, simultaneously developed by three groups located in Denmark (van der Poel et al., 2001), the Netherlands (Turkstra et al., 2001), and the United States (Fléchard et al., 2001). This article is an attempt to describe the MOTRIMS methodology, to summarize the experiments that have been carried out thus far using this methodology, and to speculate on what may be done with this technique in the near future.

1.2. What Can One Do with MOTRIMS

The questions naturally arise, "Does adding a MOT to the basic COLTRIMS methodology give any advantages?", and "If so, what are they?". The answer to the first question is yes, there are advantages that come about from the use of a MOT. In answer to the second question, these advantages will be briefly summarized in this section. First of all, lower temperature means that one can potentially achieve better recoil momentum resolution, and thereby measure with even finer precision the dynamics of interest. The COLTRIMS technology is presently limited by target temperature; MOTRIMS is not—yet. Although the resolution in recoil momentum has not yet been greatly improved in MOTRIMS, the limitation is no longer due to temperature, but rather to detector spatial and timing resolution. These can be—and no doubt will be—improved as the need arises.

The second advantage of MOTRIMS over COLTRIMS is that an additional set of atomic species can be studied in MOTRIMS. These are elements that have optically active electrons, such as alkali or alkaline-earth atoms. Such elements, upon undergoing COLTRIMS-like cooling such as supersonic expansion, tend to form molecules or clusters. This may be a desirable effect for some studies, but not when experiments with atomic targets are the goal. Furthermore, this class of atoms has the advantage of being readily excited with lasers, thus greatly expanding RIMS (recoil ion momentum spectroscopy) measurements to include targets

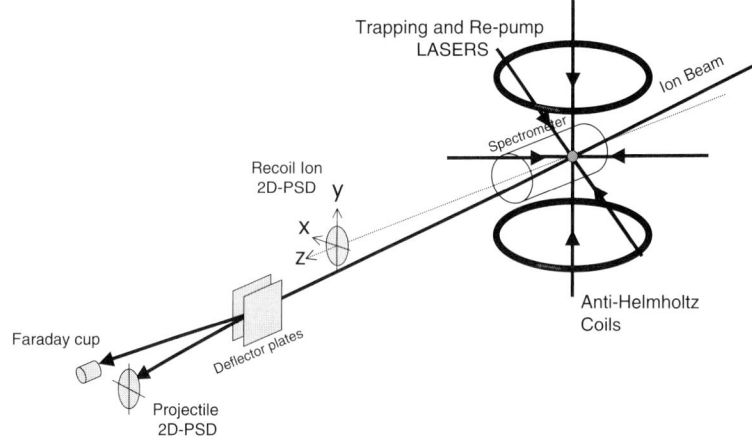

FIG. 1. Schematic of a typical MOTRIMS apparatus. Not shown is the source of projectile ions which enter the apparatus from the right. [Reprinted figure with permission from Camp (2005).]

in which one can vary the initial state of the target, as well as optically preparing anisotropic targets.

Another advantage of MOTRIMS is the study of cold collision dynamics since this regime occurs at far lower temperatures than can be achieved using COLTRIMS cooling methods. In Sections 3, 4, and 5, experiments will be described in which some or all of these advantages have been exploited.

1.3. GENERAL DESCRIPTION OF THE APPARATUS

A schematic of a typical MOTRIMS apparatus is shown in Fig. 1. A beam of projectile ions enters the apparatus from an ion source off figure to the right, and enters the momentum spectrometer portion of the system. Target atoms are trapped and cooled in the interior region of a "stack of plates" spectrometer, using a combination of "anti-Helmholtz" coils and trapping/repumping lasers, as shown in the figure. Laser cooling and trapping of atomic species is very well described in the literature, for example, by Metcalf and van der Straten (1999) (and references cited therein). Therefore, only the special aspects of this technology that one must consider when merging it with RIMS will be discussed here. After passing through the spectrometer and the cloud of trapped and cooled atoms, the projectile beam is charge-state analyzed with the portion of the beam that has captured an electron from the target going to the projectile detector. In the event of an ionizing collision, the recoil ion is extracted from the collision region with essentially 100% efficiency and directed onto the recoil position-sensitive detector (PSD).

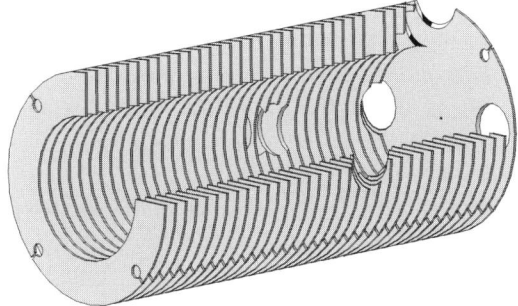

FIG. 2. Cut away view of the spectrometer. [Reprinted figure with permission from Nguyen et al. (2004b).]

A close up cut-away detail of the stack-of-plates portion of the recoil spectrometer is shown in Fig. 2. In this drawing, one can see the access holes through which the trapping lasers pass on their way to the cooling/trapping location in the interior of the spectrometer. The spectrometer is a variant of that developed by Wiley and McLaren (1955) in that there is first order *temporal focussing*. That is, the geometry of the system is such that, to first order, all atoms created in the vicinity of the MOT, and having the same initial velocity, will arrive at the same time at the detector. The system differs from the Wiley–McLauren spectrometer in that there is also *spatial focussing* that comes about from appropriately designed fringing fields in the spectrometer. Spatial focussing means that ions formed in the vicinity of the MOT, and having the same initial velocity, all hit the position-sensitive detector at the same spot. This approach is essential for good momentum resolution; its use is typical in most RIMS systems. In general, spectrometers of this sort are designed using SIMION.[1] The components of the recoil ions' momenta are measured, event by event, by measuring the recoil ion flight time and position on the target PSD. From these momentum components, the capture channel and corresponding scattering angle are determined using Eqs. (1) and (2). Results from this sort of *differential capture cross section* measurement are discussed in Section 4.

In considering how to incorporate laser cooling into a recoil ion momentum spectrometer, three possibilities come to mind. The first is what every MOTRIMS group has at this time done, that is, direct the light from the trapping and re-pump lasers directly into the spectrometer, and then volume-feed the MOT with a source of warm atoms to be trapped and cooled. This portion of the spectrometer

---

[1] The SIMION 3D Version 6.113 software package was developed by D.A. Dahl, MS 2208 Idaho National Engineering and Environmental Laboratory, P.O. Box 1625, Idaho Falls, ID 83415.

is immersed in the magnetic field gradient required for trapping, typically near 10 Gauss/cm. The advantage of this approach is its simplicity. The disadvantage is that one must create access ports into a region of the spectrometer that is very sensitive to small perturbations of the electric field. Thus, the overall resolution of the spectrometer can be degraded because of these access ports. A further disadvantage is that volume feeding raises the overall background pressure in the trapping region. This degrades, to some extent, the MOT performance.

As mentioned by van der Poel (2001), a second approach for introducing laser cooled atoms into the spectrometer is to use a cold atomic beam. That is, one can cool and trap the atoms in a conventional MOT, and then "push" the cold atoms into a secondary MOT located in the interior of the spectrometer using an additional laser beam in a manner similar to that already used in the work of Lu et al. (1996). In a sub-variant of this, one could employ a "pyramid" or an "axicon" MOT, as described, for example, by Williamson et al. (1998), Lee et al. (1996) and Kim et al. (1997). In this configuration, cold atoms would be pushed out of the MOT in a cold, extremely bright beam which would pass through the spectrometer in a manner similar to that of a conventional COLTRIMS setup—with the exception that the atoms are about three orders of magnitude colder.

A third approach, also suggested by van der Poel (2001), is to trap the atoms in a position above the spectrometer and then simply drop them in (gravity feed). The principle advantages of these last two methods is that the magnetic field gradient of the MOT can be distanced from the spectrometer region, eliminating the problem of image distortion that results in the usual MOTRIMS arrangement. Furthermore, the spectrometer region can be differentially pumped, allowing a much lower vacuum in the collision region than is possible with the volume-filled MOT arrangement.

As of this writing, there are 4 MOTRIMS apparatuses in existence: 2 at Kansas State University, one at the University of Groningen, and one at the University of Freiburg (constructed and previously used at the University of Copenhagen). In addition, there are 3 more MOTRIMS setups under construction: one at CNRS in Caen, one at Southern Mississippi University, and one at the University of Wisconsin at Stevens Point. Of the 4 existing setups, the two in Europe employ "transverse" extraction, in which the recoil ions are extracted perpendicular to the projectile beam path, while the setups in Kansas use "longitudinal" extraction, in which the recoil ions are extracted nearly parallel to the projectile beam path. Of the setups under construction, the "continental" preference is maintained with the French apparatus employing transverse extraction and the two American setups using longitudinal extraction. A third option, thus far not employed, is the variation referred to as a "reaction microscope", sometimes used in COLTRIMS experiments. What follows is a brief description of each of these approaches, along with a summary of their advantages and disadvantages.

1.3.1. Longitudinal Extraction

Figure 1 shows the approach used by Kansas State University. The extraction axis of the spectrometer is at a small angle ($\sim 3.5°$) with respect to the projectile beam axis. Thus, the recoil detector can intercept the recoil ions while the projectiles pass by en route to the projectile detector (or Faraday cup). The component of the recoil ion momentum that lies parallel to the projectile axis, $p_\parallel$, is very nearly a linear function of the time difference between the flight times of the once-less-charged projectile and the recoil ion, but for optimal resolution, the small trigonometric corrections of Eqs. (3) must be made.

$$p_\parallel = 0.998 p_z + 0.06 p_x \approx p_z + 0.06 p_x, \quad (3a)$$

$$p_\perp^2 = p_y^2 + (0.998 p_x - 0.06 p_z)^2 \approx p_y^2 + p_x^2 \quad (3b)$$

where the numerical values are just the sine and cosine of $3.5°$, and $p_i, i = x, y, z$ refer to the components of recoil momentum with respect to the spectrometer axis. The principal advantage of longitudinal extraction is that the Q-value of the collision is trivially obtained from the particle flight times alone (except for the fine corrections just mentioned). Furthermore, in most cases the TOF resolution is better than the position resolution of the detectors. Therefore, if the Q-value is considered the most important information in the collision, longitudinal extraction is advantageous. The disadvantage of this geometry is that even with good projectile beam collimation, a significant number of "recoil" counts are actually due to projectile ions that have been scattered onto the recoil detector, thus potentially adding to detector dead time and false coincidences.

1.3.2. Transverse Extraction

Figure 3 shows the approach used by the University of Copenhagen in the construction of their MOTRIMS apparatus. Most of the apparatus is the same as that in Fig. 1, but the recoil ions are extracted perpendicularly to the projectile axis. The principal advantage of this approach is that the aforementioned problem with projectiles striking the recoil detector is eliminated. The disadvantage is that, as already mentioned, it is often easier to obtain better timing resolution than position resolution with PSDs. Therefore, $p_\parallel$ is often measured with lower resolution than in the longitudinal case. On the other hand, with a bit of effort, "delay line" detectors[2] are now capable of achieving position resolution that rivals the corresponding temporal measurements inherent in TOF measurements. Furthermore, with TOF now giving $p_\perp$, scattering angle measurement should be better with the transverse extraction than with longitudinal. Finally, because $p_\parallel$ can be measured

---

[2] RoentDek GmbH: http://www.roentek.com.

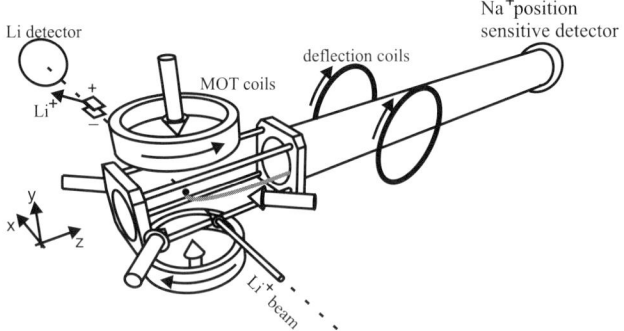

FIG. 3. Schematic of MOTRIMS apparatus with transverse extraction. [Reprinted figure with permission from van der Poel et al. (2002).]

© 2002 Institute of Physics

with both TOF *and* position, the transverse geometry automatically provides a check on both measurements.

1.3.3. Reaction Microscope

The reaction microscope geometry employed, for example, by Moshammer et al. (1996) in COLTRIMS measurements, has yet to be employed for MOTRIMS, however there is no reason it could not. The basic geometry, as shown in Fig. 4, consists of a pair of parallel plates on either side of the collision region. The plates are made of some resistive material such as coated glass, and have potentials

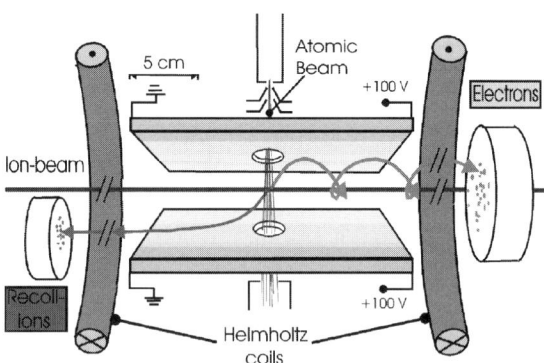

FIG. 4. Schematic of a COLTRIMS apparatus in the "Reaction Microscope" geometry. [Reprinted figure with permission from Moshammer et al. (1996).]

© 1996 Elsevier

applied to each of their four corners. Using these potentials, one can create an electric field that directs recoil ions away from the projectile beam and onto a recoil detector. As the figure shows, electrons are also extracted and, with the aid of a longitudinal B-field, are directed onto a separate PSD. In terms of its application in MOTRIMS, the parallel plate design of the reaction microscope has the advantage of very good optical access. That is, one must drill a single pair of holes for the trapping laser beams that enter perpendicular to the planes of the plates, but the other two pairs of beams are parallel to the planes of the plates and no penetration holes—with the consequent disturbance to the electric field—need be drilled. The major disadvantage of the reaction microscope is that its resolution is not quite as good as that of the "stack of plates" design because it lacks the additional field region required for temporal and spatial focussing as discussed in the beginning of Section 1.3.

2. Relative Total Electron Transfer Cross Sections

On average, some fraction of the atoms in a MOT are in an electronically excited state as a consequence of their interaction with the trapping and repump lasers. Therefore, in an electron transfer cross section measurement some peaks in the Q-value spectrum correspond to capture from the ground state and others correspond to capture from the excited state. For example, the Q-value spectrum of Fig. 5 shows electron transfer from Rb(5s) and Rb(5p) to different final states of Na in the 7 keV Na$^+$ + Rb collision system. The peaks are labelled to show their initial

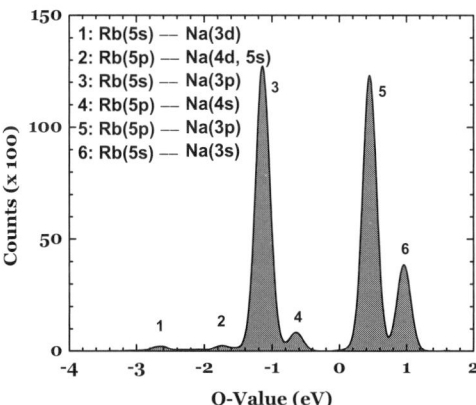

FIG. 5. Typical Q-value spectrum from which relative cross sections can be deduced. The different electron transfer channels are indicated on the figure. [Reprinted figure with permission from Camp (2005).]

and final states. Because the area under each of the peaks in a Q-value spectrum is proportional to the product of the electron transfer cross section and the number of target atoms in the initial state, it would seem that without some independent measurement of the relative populations, a determination relative cross sections is problematic. However, observing that, at least on a short time-scale, the number of atoms in a MOT does not change when the trapping laser is turned off, Fléchard et al. (2003) realized that one could, in fact, separately determine the relative cross section and relative population.

The key notion is that by chopping the trapping and repump lasers and comparing the change in electron transfer rates from the ground and excited states for lasers on and lasers off, one can determine the excited state fraction and, independently, the relative capture cross section from both states. If A_i refers to the area under a Q-value peak corresponding to electron transfer from the target's ith initial state whose relative population is given by n_i, to a particular final state, then

$$A_s \propto \sigma_s n_s, \quad (4a)$$

$$A_p \propto \sigma_p n_p, \quad (4b)$$

where the constant of proportionality contains acquisition time and geometric factors, and σ_i are the electron transfer cross sections. If the trap is not turned off long enough for the atoms to fall out of the path of the ion beam (typically less than a ms),

$$n_s + n_p = \text{constant}, \quad (5)$$

and therefore

$$\Delta A_s \propto \sigma_s \Delta n_s, \quad (6a)$$

$$\Delta A_p \propto \sigma_p \Delta n_p, \quad (6b)$$

$$\Delta n_s + \Delta n_p = 0, \quad (6c)$$

where Δn_i refers to changes in the ith population as the trapping (and/or repumping) laser goes from the on condition to the off condition. Note that it is the low temperature of the target which allows Eqs. (5) and (6c) to be satisfied at relatively low chopping frequencies. Taking the ratio of Eq. (6b) to Eq. (6a), and using Eq. (6c) we obtain

$$\frac{\Delta A_p}{\Delta A_s} = \frac{\sigma_p}{\sigma_s} \frac{\Delta n_p}{\Delta n_s} = -\frac{\sigma_p}{\sigma_s}. \quad (7)$$

Thus, the cross section for electron transfer from Rb(5p) can be measured relative to capture from Rb(5s), where by "5p" and "5s" we really mean specific capture channels for which the initial states were 5p and 5s.

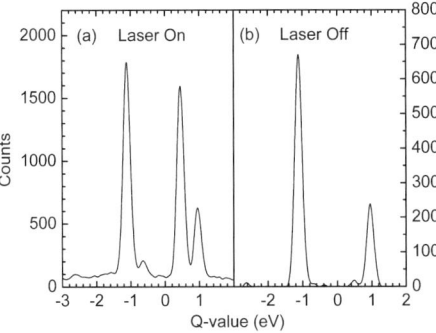

FIG. 6. Typical Q-value spectra for which the trapping lasers are on (a), and off (b). [Reprinted figure with permission from Nguyen (2003).]

In the experiments of Fléchard et al. (2003) an acousto–optical modulator (AOM) was used to chop both the trapping and repump laser beams at a frequency of 50 kHz, with an on-time duty cycle of 75%. A pair of Q-value spectra, one taken with trapping lasers on, the other with the lasers off, is shown in Fig. 6. The difference in the vertical scales for the two sides of this figure are due to the fact that the duty cycle of the trapping laser is *not* 50%.

This procedure was expanded by Fléchard et al. (2003) to determine the relative capture cross section from any initial state of the target by generalizing the above equations to:

$$\frac{\sigma_k}{\sigma_1} = -\Delta A_k \left(\sum_{i=1}^{k-1} \Delta A_i \frac{\sigma_1}{\sigma_i} \right)^{-1} \quad (2 \leqslant k \leqslant N). \tag{8}$$

Using this approach, relative electron transfer cross sections from Rb(4d) in collisions with 7 keV Na^+ were measured by Shah et al. (2005) using the previously measured value of σ_p/σ_s. It should be emphasized that Eq. (8) is completely general and is limited only by ability of the MOTRIMS apparatus to resolve individual capture channels.

3. Case Studies in Total Electron Transfer Collisions

In this section we will discuss how the electron dynamics during collisions of singly and multiply charged ions on Na can be studied by MOTRIMS in a rather detailed way.

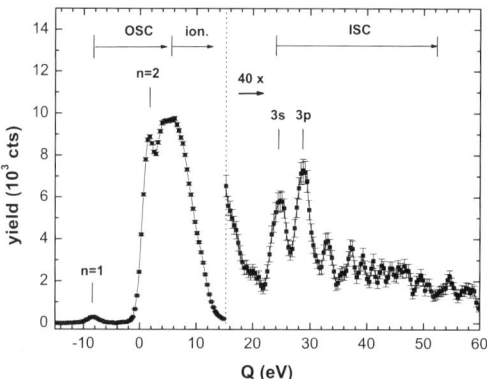

FIG. 7. Q-value spectrum of Na$^+$ ions resulting from 14 keV/amu H$^+$ collisions on Na. Features due to outer-shell capture (OSC), ionization (Ion), and inner-shell capture (ISC) can clearly be distinguished. [Reprinted figure with permission from Knoop et al. (2004).]

© 2004 American Physical Society

3.1. REMOVAL OF INNER- AND OUTER-SHELL NA ELECTRONS BY H$^+$ IMPACT

As a first example we will discuss collisions of protons on Na, whereby electrons from outer or inner shells are captured or transferred into the ionization continuum. In a first approximation this can be regarded as a one-electron system and therefore is especially suited to test quasi-one-electron models. This is in fact one of the reasons why proton–sodium collisions have intensively been studied in the past both experimentally (Aumayr et al., 1987; Aumayr and Winter, 1987; Dowek et al., 1990; Dubois and Toburen, 1985; Ebel and Salzborn, 1987) and theoretically (Fritsch and Lin, 1991; Jain and Winter, 1995; Lundy and Olson, 1996) by various methods, including the determination of total electron capture cross sections. At low collision energies it was found that it is nearly exclusively the Na 3s-orbital which participates in the interaction dynamics, whereas with increasing collision energy the inner 2p- and 2s-orbitals are also involved and the quasi-one-electron approach is no longer justified. This was already suspected from the fact that at collision energies above approximately 20 keV the total electron capture cross section no longer decreases as predicted by quasi-one-electron models, but remains at a much higher level. However, experimentally the contributions from inner and outer shells could not be distinguished. Using the MOTRIMS method the situation could be clarified: Knoop et al. (2004) could not only identify contributions from inner and outer shell to the total capture cross section, but were even able to reveal the active role of the outer-shell electron during inner-shell capture processes. The corresponding experimental result is shown in Fig. 7. This figure shows a Q-value spectrum of Na$^+$ recoil ions which was obtained from

14 keV/amu H^+ collisions on Na. The measured Q-values allow a direct assignment of the various peaks to the underlying processes. The spectrum is dominated by recoil ions with Q-values between 0 and 10 eV, resulting from removal of the outer Na(3s) electron, either by capture into the $n = 2$ ($Q = 1.74$ eV) or higher orbitals of the resulting H atom, or by transfer into the ionization continuum ($Q > 5.4$ eV).

Only a small fraction of capture processes lead to ground state H atoms ($Q = -8.46$ eV). Na^+ ions with Q-values above 24.5 eV arise from processes in which Na(2p) inner-shell electrons are removed from the target. Although transfer of the outer Na(3s) electron higher into the ionization continuum can give rise to the same Q-values, inner-shell capture can be identified since it results in discrete peaks on top of the ionization continuum. The most prominent of these peaks are those at $Q = 24.5$ and 28.7 eV, corresponding to inner-shell capture, whereby the outer electron either remains in the 3s or is promoted to the 3p orbital respectively. The fact that these two peaks have nearly the same intensity indicates that the outer electron in this case is not simply a spectator but plays an important role in the collision-induced electron dynamics.

From these Q-value spectra of recoil ions, taken for a series of collision energies, relative cross sections for the various processes have been determined by Knoop et al. (2004) and put on an absolute scale by comparison with absolute total cross sections measured earlier. At the same time Zapukhlyak et al. (2005) applied the newly developed basis generator method (BGM) to calculate these cross sections and to compare them with the experimental results. The BGM method, developed by Lüdde and Kirchner (Lüdde et al., 1996; Kroneisen et al., 1999), is in fact a close coupling approach, in which the set of basis functions is dynamically adapted to the time dependent status of the collision system.

Figure 8 shows the measured cross sections as functions of the collision energy in comparison with theoretical values. At low collision energies outer-shell capture clearly dominates, but above 40 keV/amu contributions of inner-shell capture are higher than those from outer-shell capture. The spectra even allow a more detailed analysis in terms of final states which are populated by electron capture. In the case of the $H^+ + Na$ collisions discussed here the resolution is sufficient to distinguish capture into $n = 1, n = 2$ and $n \geqslant 3$ and to determine the corresponding partial cross sections as functions of the collision energy on an absolute scale. Figure 9 shows the results in comparison with calculated cross sections, whereby good agreement between experimental and theoretical results is found.

Ionization cross sections are shown in Fig. 10. It should be pointed out that the MOTRIMS measurements allow one to distinguish inner- and outer-shell ionization, whereas in the past, only total ionization cross sections have been reported. However, as can be seen from this figure, for the present example these cross

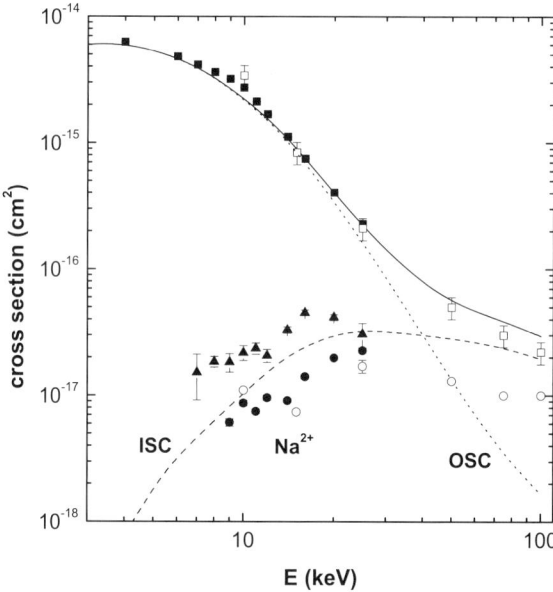

FIG. 8. Experimentally determined absolute cross sections for outer-shell capture (OSC, ■), inner-shell capture (ISC, ▲), and total Na^{2+} production (●), compared with theoretical data: BGM results for single electron capture (SEC, —), contribution from 3s capture (···), and from 2p capture (- - -). Results obtained by Dubois (1986) by measuring charge state-selected target and projectile ions in coincidence: for total SEC (□) and transfer ionization (○). [Reprinted figure with permission from Knoop et al. (2004).]

© 2004 American Physical Society

sections are practically the same. This indicates that here inner-shell ionization is nearly negligible.

With the knowledge that inner-shell ionization is not significantly contributing to the recoil ion yield, one can exploit the fact that the ionization range of the Q-value spectrum reflects the energy distribution of the electrons transferred to the continuum. This allows an investigation of the relative importance of various ionization mechanisms without directly measuring the electrons. In particular, three different ionization processes have been discussed in the literature: (i) electron excitation to the continuum (EEC), implying that the electron is promoted to the continuum, but remains close to the target, i.e. has a near zero energy with respect to the target; (ii) electron capture to the continuum (ECC), implying that the ionized electron stays close to the projectile, and (iii) the so-called saddle point mechanism (SP) (Olson, 1986; Pieksma et al., 1994; Pieksma, 1997), which gives rise to electrons just balancing on the saddle point of

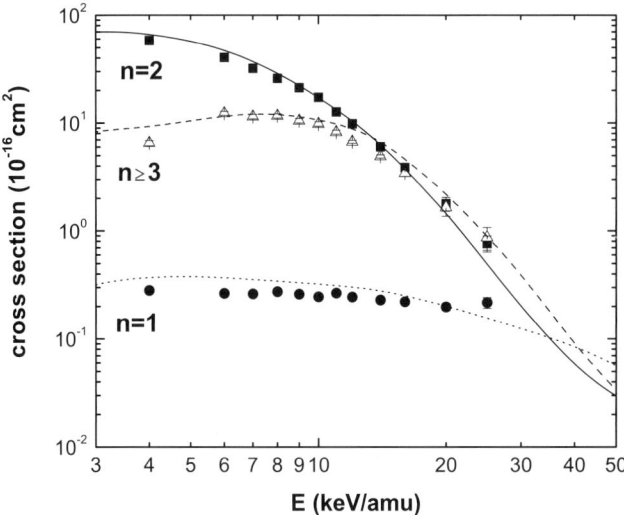

FIG. 9. Partial cross sections for electron capture into excited H with various principal quantum numbers n. The systematic uncertainty of the experimental points is about 20% due to uncertainties in the calibration procedure. The curves represent theoretical results using the Two-Center-BGM method. [Reprinted figure with permission from Knoop (2006).]

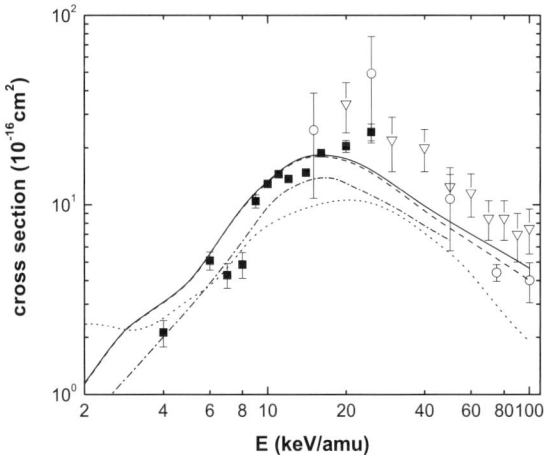

FIG. 10. Cross sections for Na ionization due to H^+ impact as a function of impact energy. Experiments: data for outer-shell single ionization (■) from Knoop (2006); total single ionization (▽) from O'Hare et al. (1975); sum of all pure ionization processes (○) from Dubois (1986). Theory: TC-BGM results for total single ionization (—) and outer-shell ionization (- - -) from Zapukhlyak et al. (2005); single ionization TCSAO70 results (· · ·) from Jain and Winter (1995); and CTMC results (- - -) from Lundy and Olson (1996). [Reprinted figure with permission from Knoop (2006).]

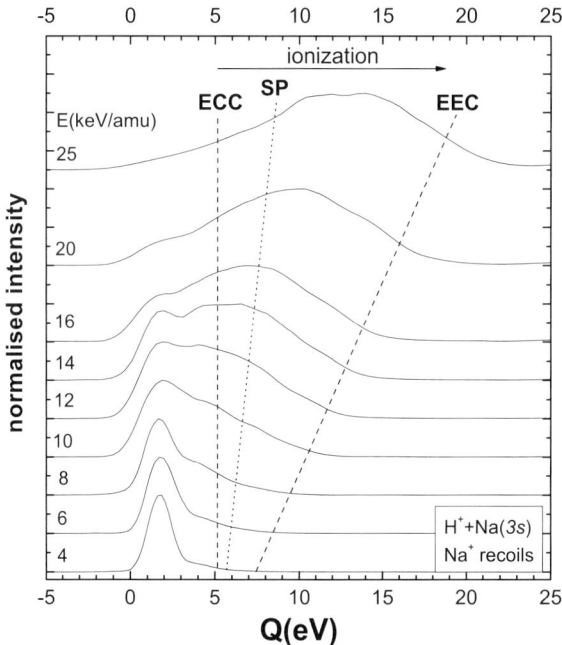

FIG. 11. Q-value spectra of the Na$^+$ recoils resulting from H$^+$ + Na(3s) collisions for different impact energies. The positions of ECC at $Q = IP$ (where IP is the ionization potential) and SP at $Q = IP + \frac{1}{8}v_p^2$ and $Q = IP + \frac{1}{2}v_p^2$, corresponding to ECC, SP, and EEC, respectively, are indicated. The offset of the spectra is chosen such that the variation of the Q-values as a function of the collision energy follows a straight line through these spectra. [Reprinted figure with permission from Knoop (2006)].

the Coulomb potential formed by the combined influence of target and projectile ion.

Figure 11 shows a series of Q-value spectra obtained at various collision energies. The Q-values corresponding to energies of electrons arising from the above-mentioned processes are indicated in the figure. The offsets of the spectra are chosen such that the variation of these Q-values follows a straight line through these spectra. Most remarkable here is the fact that no clear peaks can be identified at the energies corresponding to the three different processes. Obviously at none of the collision energies investigated does one of the above-mentioned mechanisms dominate the ionization process. On the other hand one can clearly see that at low collision energies electron capture into the continuum (ECC) is more important, whereas at high energies the ionized electron tends to remain in the vicinity of the target and the EEC process becomes more important. The

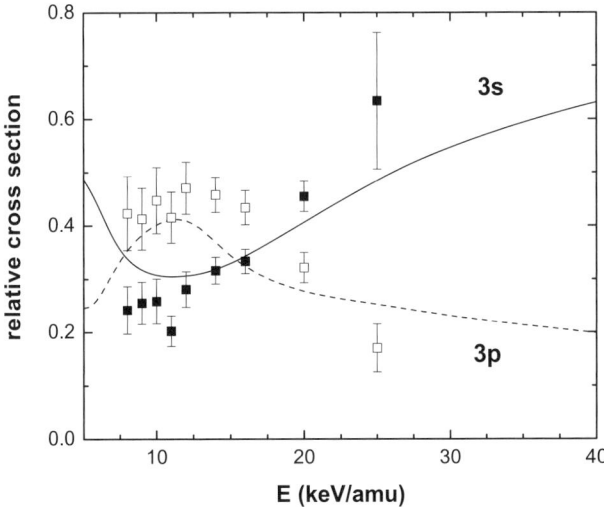

FIG. 12. MOTRIMS and TC-BGM results for the different contributions to ISC into H($n = 1$): Na$^+(2p^53s)$ (■, —) and Na$^+(2p^53p)$ (□, - -). [Reprinted figure with permission from Knoop (2006).]

saddle point mechanism seems to be most relevant at collision energies around 15 keV/amu.

As mentioned before, inner-shell capture becomes more important than outer-shell capture, for collision energies above 40 keV/amu. A more detailed analysis has been performed and separate cross sections for inner-shell capture with or without additional excitation of the outer 3s electron, have been determined. In Fig. 12 these two cross sections are shown as functions of the collision energy.

Whereas at high collision energies the 3s electron behaves like a spectator, at energies below about 16 keV/amu it seems to be actively involved in the collision dynamics and is promoted to the 3p orbital with a relatively high probability. In order to understand this effect Knoop et al. (2004) considered the influence of angular momentum conservation. Since the inner Na(2p) electron is captured into the H(1s) orbital and thus has to change angular momentum, a transition of the outer Na(3s) electron to the 3p orbital could serve to conserve the total angular momentum of the Na atom. The decreasing probability of this process with increasing collision energy could then be understood by the fact that a certain minimum time is necessary for the electron–electron interaction involved in this process. However, as can be seen from Fig. 12, the theory of Zapukhlyak et al. (2005) also predicts a diminishing probability for 3s–3p promotion at collision energies below 5 keV/amu, a feature that can not be accounted for by the above-mentioned interpretation.

3.2. Electron Transfer in O^{6+} + Na Collisions

As a second example we discuss collisions of multi-charged ions on Na, using O^{6+} as a projectile. MOTRIMS is especially well suited for an investigation of the various electron capture processes in such a system. First of all, multiple electron capture plays an important role and the observation of the multiply ionized target is generally more instructive than the observation of the charge exchanged projectile. Since electron capture by multi-charged ions mostly results in multiply excited states, the eventually observed charge state of the projectile is often drastically changed by subsequent Auger or autoionization processes. Therefore, the charge state of the target after the collision generally gives better information on the primary capture and/or ionization processes than that of the projectile. Moreover, the good energy resolution of the MOTRIMS method is essential for the investigation of such collisions. Since electron capture by multi-charged ions proceeds predominantly into closely lying excited states, these processes can only be distinguished from each other with rather good energy resolution.

In the past, collisions between multiply charged ions and alkali atoms have mainly been experimentally investigated by photon emission spectroscopy (PES) (Schlatmann et al., 1993; Schippers et al., 1994, 1995, 1996; Lembo et al., 1985; Hoekstra et al., 1993; Jacquet et al., 2000; Bazin et al., 2002), exploiting the fact that electron capture in these systems proceeds predominantly into excited states. Regarding theory, these systems have mainly been treated by classical trajectory Monte Carlo (CTMC) calculations (Pascale et al., 1990; Schippers et al., 1994, 1996; Hoekstra et al., 1993; Jacquet et al., 2000; Bazin et al., 2002; Olson et al., 1992; Cornelius et al., 2000). As opposed to close coupling calculations, this approach does not have the problem of huge basis sets of wave functions.

First experiments using the MOTRIMS method for the O^{6+} + Na system have been reported by Turkstra et al. (2001), whereas a more detailed study including a comparison with theoretical predictions has been made by Knoop et al. (2005a). The relative importance of single and multiple electron capture can be seen from a simple charge state distribution spectrum of the various Na^{q+} ions, as shown in Fig. 13. Obviously, formation of Na^+, i.e. single electron capture, is by far the dominant process at this collision energy of 7.5 keV/amu. Much less likely is the formation Na^{q+} ions with $q = 2, 3$ and 4, which show up with similar intensities, whereas recoil ions with charges $q > 4$ apparently are not observed at all. A longitudinal momentum spectrum of Na^+ recoil ions is shown in Fig. 14.

As can be expected from the classical over-the-barrier model, single electron capture results in excited $O^{5+}(nl)$ states with principal quantum numbers n around 7. It is worth noting that a more or less identical Q-value spectrum was measured for C^{6+} projectiles. This indicates that the $(1s)^2$ electronic core of O^{6+} has nearly no influence on the electron capture process, implying that theoretical calculations using bare C^{6+} ions can be used to investigate the electron transfer mechanisms induced by O^{6+} ions. The various capture channels leading to

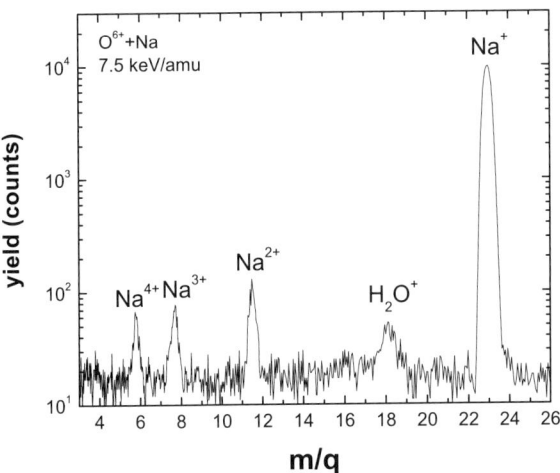

FIG. 13. Relative yield of Na^{q+} recoil ions resulting from 7.5 keV/amu O^{6+} collisions on Na. [Reprinted figure with permission from Knoop (2006).]

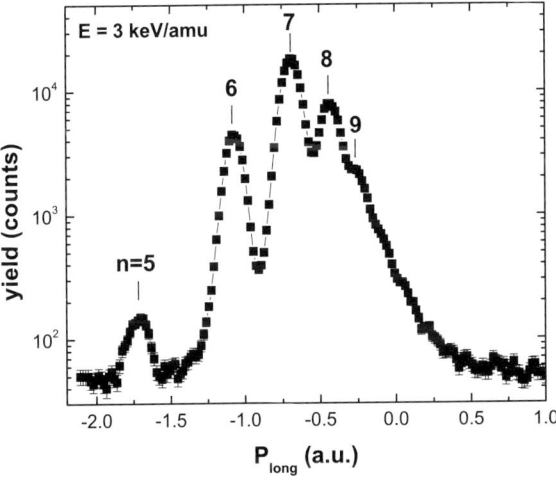

FIG. 14. Longitudinal momentum spectrum of Na$^+$ recoil ions resulting from 3 keV/amu O^{6+} collisions on Na. Note the logarithmic scale. Numbers at the peaks indicate the n-orbital into which electrons are captured. [Reprinted figure with permission from Knoop et al. (2005a).]

© 2005 Institute of Physics

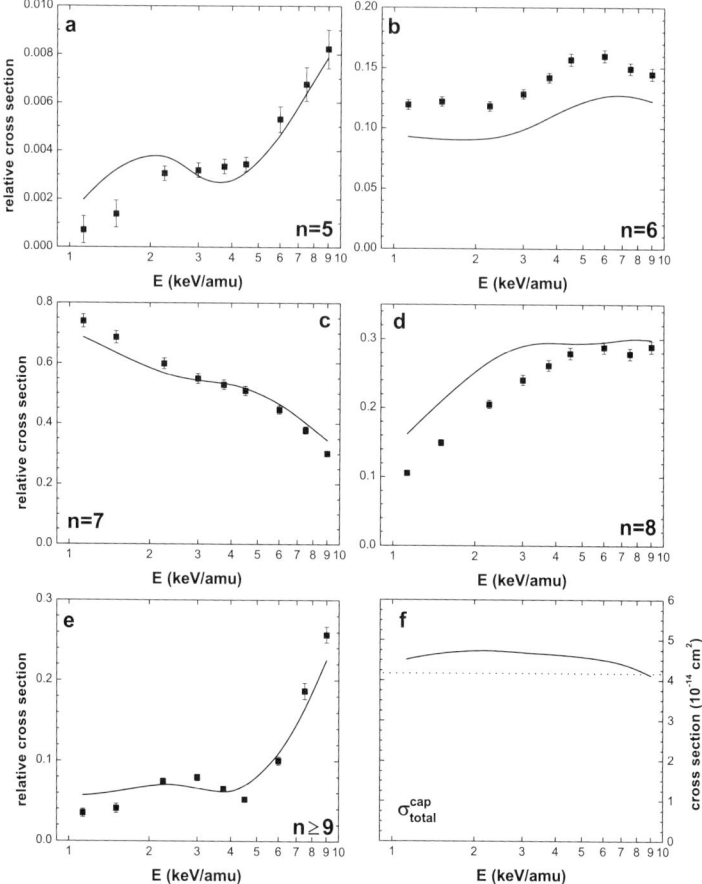

FIG. 15. Energy dependence of relative partial cross sections for electron capture into various n-shells during collisions of O^{6+} on N(3s). Theoretical results obtained with the TC-BGM method are shown as solid lines. Panel (f) shows the absolute TC-BGM cross section for total one-electron capture (solid line) and the cross section estimated from the classical over-the-barrier model (dotted line). [Reprinted figure with permission from Knoop et al. (2005a).]

© 2005 Institute of Physics

population of different n-orbitals are sufficiently resolved (notice the log-scale in Fig. 14) to allow a separate determination of the corresponding partial cross sections. These are shown in Fig. 15.

Whereas the cross section for the dominant $n = 7$ channel decreases with increasing collision energy, those for the other channels increase. Especially orbitals with $n = 5$ and $n \geqslant 9$ are only significantly populated at energies above

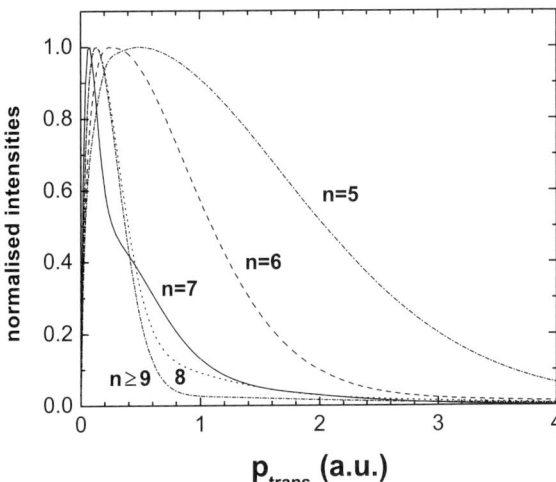

FIG. 16. Distribution of Na$^+$ transverse momenta resulting from electron capture into various O^{5+}(nl) orbitals in 4.5 keV/amu O^{6+} + Na collisions. [Reprinted figure with permission from Knoop (2006).]

6 keV/amu. This is partly reflected in the corresponding distributions of recoil ion momenta perpendicular to the beam direction which are shown in Fig. 16. Especially for the $n = 5$ channel the corresponding perpendicular momenta are very high, indicating that small internuclear distances of the collision partners are required for an effective population of this channel. The same is not the case for the $n \geqslant 9$ channels: although populated only at relatively high collision energies, the corresponding perpendicular recoil momenta are small. This is probably due to the fact that at large internuclear distances these endothermic channels are energetically close to the incoming channel and a kind of Demkov coupling (Demkov, 1964) at large separations is responsible for their population. This is supported by the fact that a significant population of endothermic channels has also been observed for the He^{2+} + Na collision system (Knoop et al., 2005b).

The importance of transverse momentum distributions for the identification of electron transfer mechanisms also becomes very obvious in case of double electron capture, as shown by Knoop et al. (2006a) for the case of O^{4+}(nl, nl') formation in O^{6+} collisions on Na. This will be discussed in Section 4.3.

3.3. ELECTRON CAPTURE FROM EXCITED Na ATOMS

As a third example we will now discuss experiments which exploit the fact that by means of MOTRIMS the target atoms can be prepared in an excited state. During

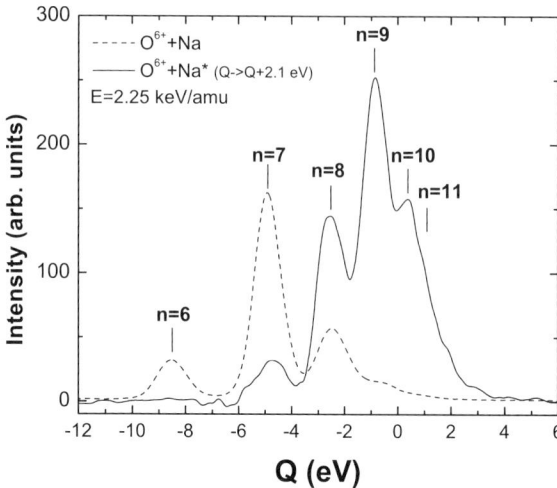

FIG. 17. Q-value spectra of Na$^+$ recoil ions resulting from 2.25 keV/amu O^{6+} collisions on ground state Na(3s) (dotted line) and excited state Na(3p) (full line) respectively. In the figure, the spectra are shifted by 2.1 eV with respect to each other such that peaks corresponding to electron capture from either ground or excited state into a specific n-orbital appear at the same position. [Reprinted figure with permission from Knoop (2006).]

the preparation of the MOT there is always a certain fraction of excited atoms (for a Na MOT estimated to be between 15 and 30%, depending on laser power and detuning). Electrons are therefore captured either from ground or excited state atoms, unless the trapping laser is turned off for a short period of time during the measurements. In order to study capture from excited Na atoms, appropriately weighted recoil ion spectra obtained with "laser off" and "laser on" have to be subtracted from each other. This has in fact been done by Knoop et al. (2006b) to study collisions of He^{2+} and O^{6+} on excited Na(3p) atoms. Figure 17 shows Q-value spectra of Na$^+$ recoil ions obtained for the case of O^{6+} collisions on a pure ground state (dotted line) or a pure excited state target (solid line). In fact the spectra are shifted with respect to each other such that peaks corresponding to electron capture from either ground or excited state into a specific n-orbital appear at the same position. As a consequence of the electron's reduced binding energy in the Na(3p) target, electron capture proceeds into significantly higher levels than for the case of the Na ground state target: capture into the $n = 9$ orbital is the dominant channel for the Na(3p) target as opposed to $n = 7$ for the Na(3s) target. Such Q-value spectra have been use by Knoop (2006) to extract relative partial cross sections for electron capture into specific n-shells. Figure 18 shows these cross sections as a function of collision energy.

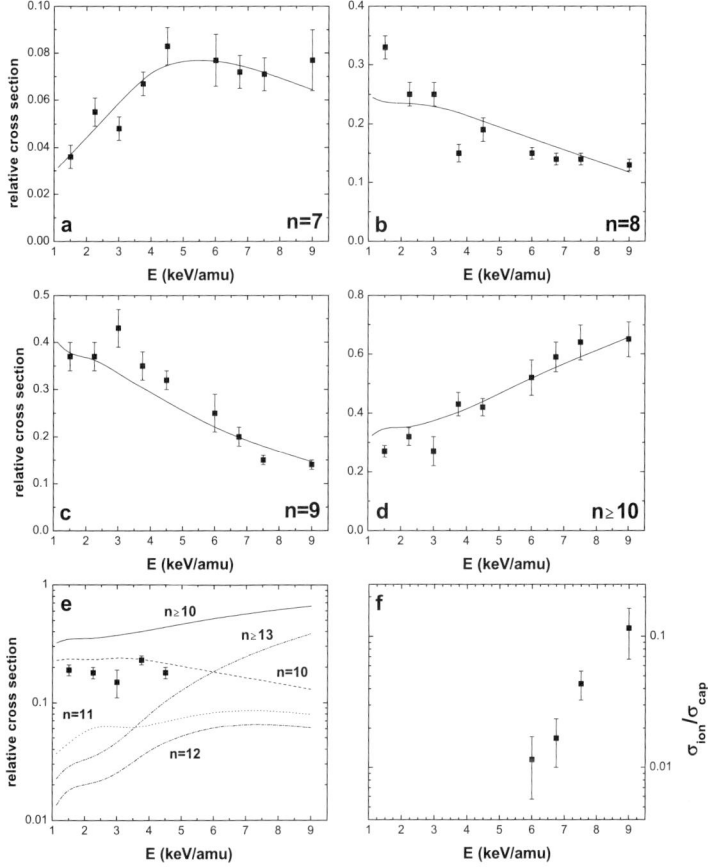

FIG. 18. Energy dependence of relative partial cross sections for electron capture into various n-shells during collisions of O^{6+} on $Na^*(3p)$. Theoretical results obtained with the TC-BGM method are shown as solid lines. Panel (e) shows TC-BGM results for capture into $n = 10$–12 and $n \geqslant 13$ together with experimental data for $n = 10$. Panel (f) shows the experimentally determined ratio of single ionization and one-electron capture. [Reprinted figure with permission from Knoop (2006). Calculations by Keim and Lüdde (2006).]

4. Case Studies of Differential Electron Transfer Cross Sections

The power of the MOTRIMS method compared to more conventional techniques is perhaps nowhere more evident than in the series of angular differential studies of resonant and near-resonant electron transfer processes performed with cold alkali targets and projectile beams of protons, alkali ions, and multiply charged oxygen atoms at keV impact energies. Selected results of these studies will now be presented.

4.1. Na(3s,3p) Targets

Extensive and very detailed studies of electron transfer in near-resonant alkali ion–alkali atom collisions have been performed by a series of theoretical and experimental groups over many years, including so-called quantum-mechanically complete experiments. [See Andersen and Bartschat (2001) for a recent review.] These studies have resulted in a very satisfactory agreement between the very sophisticated experimental investigations and state-of-the-art atomic scattering theories based on the close-coupling approach using large basis sets of atomic and/or molecular orbitals. However, one elusive aspect of the theoretical predictions escaped experimental confrontation, namely some very rapid oscillations in the predicted angular differential cross sections that could not be revealed by conventional experimental techniques due to excessive demands on the angular collimation necessary to resolve the structures. How this obstacle could be overcome by the MOTRIMS technique will now be presented, along with some examples of high-resolution differential cross sections for targets in optically prepared excited states. We start by presenting a series of results using a cold Na target in the Na(3s) ground state or Na(3p) excited state.

4.1.1. $Li^+ + Na(3s)$

Figure 19 shows the longitudinal momentum transfer $p_\parallel$ for 1 keV/amu Li$^+$ projectiles that have captured the 3s electron of Na into the 2s and 2p states, respectively, integrated over all scattering angles (van der Poel et al., 2002).

We note that near-resonant 3s → 2s transfer dominates at this impact energy and that the 2s and 2p exit channels can be nicely separated. Earlier investigations show that the 3s → 2s transfer takes place mainly at large impact parameters, typically at $R = 10$ a.u. The projectiles follow almost straight line trajectories. Typical scattering angles are thus very small and the scattering intensity at a given angle is obtained by integration over a large disc in impact parameter space, in a way analogous to Fraunhofer diffraction, well-known from classical optics. The ^{6}Li$^+$ projectiles used in Fig. 19 have a velocity of 0.2 a.u., corresponding to a de Broglie wavelength of about $\lambda = 150$ fm. With an effective diameter of $2R = 20$ a.u. centered on the Na(3s) target, the usual criterion for the wave aspect of the particles to be prominent, namely that the de Broglie wavelength of the particle be of the same order of magnitude as the object from which it is scattered, is *not* fulfilled. The predicted angular width of the Fraunhofer diffraction rings is the ratio between these two quantities,

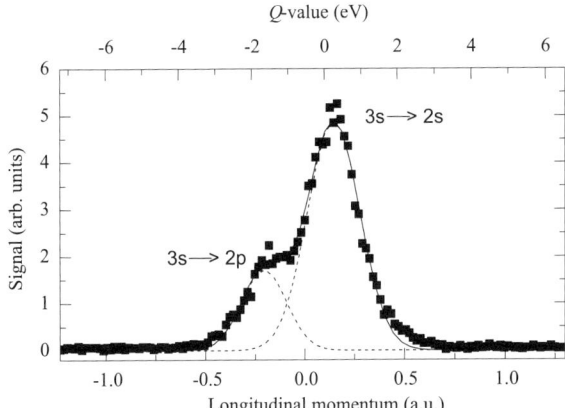

FIG. 19. Longitudinal momentum transferred to the Li$^+$ projectile and Q-value spectrum for Li$^+$ + Na(3s) electron transfer collisions at 1 keV/amu impact energy. [Reprinted figure with permission from van der Poel et al. (2002).]

© 2002 Institute of Physics

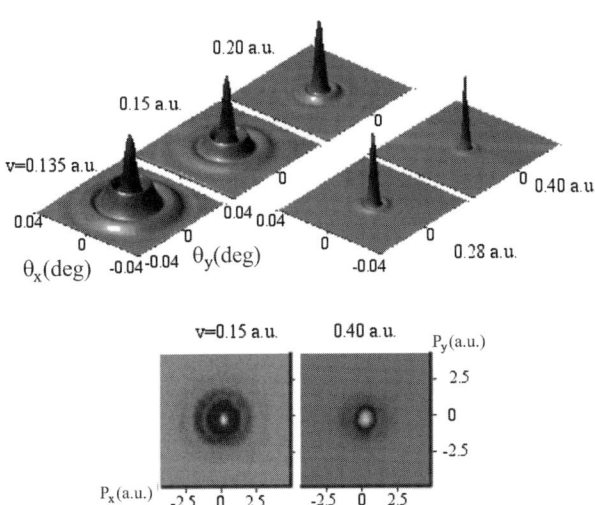

FIG. 20. Upper panel: Li angular scattering pattern for 3s → 2s electron transfer collisions. Fraunhofer rings are clearly seen. Lower panel: The actual measured scattering pattern in momentum space for two selected projectile velocities. [Reprinted figure with permission from van der Poel et al. (2002).]

© 2002 Institute of Physics

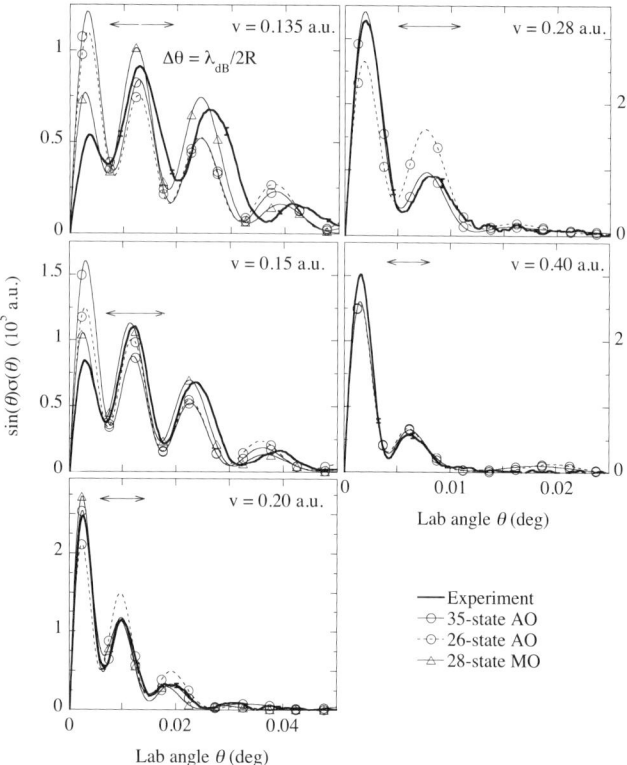

FIG. 21. Reduced differential cross section of the Na(3s) → Li(2s) electron transfer channel. For ease of comparison, all curves are normalized to the 35-state AO total cross section. The theoretical curves have been convoluted with the apparatus response function. The double arrow shows the Fraunhofer ring spacing $\lambda/2R$ for an atomic disc of diameter $2R = 20$ a.u. Note the different scales in the left and right columns. [Reprinted figure with permission from van der Poel et al. (2002).]

© 2002 Institute of Physics

$\lambda/2R = 1.5 \times 10^{-4}$ radians, or about $0.01°$. The MOTRIMS technique is nevertheless able to resolve this, as is illustrated in Fig. 20 which shows measured Li angular scattering patterns at five different impact velocities. Fraunhofer diffraction rings are clearly seen. A quantitative comparison of the experimental reduced differential cross sections with state-of-the-art theoretical predictions is shown in Fig. 21 for the five velocities of Fig. 20. We note that the level of agreement between the atomic orbital (AO) calculation and experiment at the highest collision velocities becomes excellent when the number of basis states is increased from 28 to 35. At the lowest velocities and the smallest angles, however, some disagreement is seen. Under these circumstances, the molecular orbital (MO) cal-

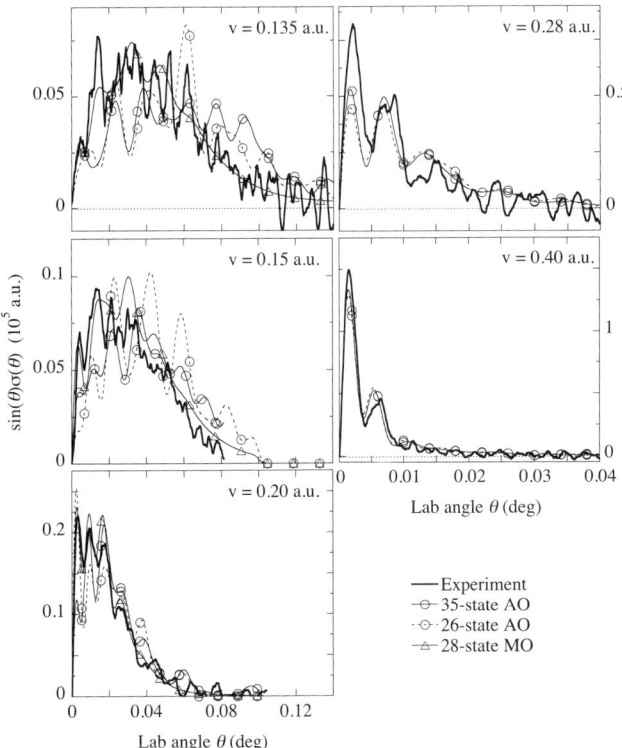

FIG. 22. Reduced differential cross section of the Na(3s) → Li(2p) electron transfer channel. All curves are normalized to the 35-state AO total cross section. Theory has been convoluted with the apparatus response function. [Reprinted figure with permission from van der Poel et al. (2002).]

© 2002 Institute of Physics

culation performs better. For a more detailed discussion see van der Poel et al. (2002).

Figure 22 shows reduced differential cross sections for Na(3s) → Li(2p) electron transfer, summed over the 2p magnetic sublevels. The rapid oscillations in the cross section, most noticeable at the lowest velocities is within the statistical noise of the experiment. The 3s → 2p transition is more difficult to predict with an AO basis set, since this transfer is caused by a rotational coupling in a narrow range of impact parameters around 5 a.u. Such a transition implies a more molecular character of the electronic wave function which is particularly difficult to represent adequately with an AO basis set. It is seen that the measurement tends to follow the MO prediction at the lowest velocities. The importance of smaller impact parameters is reflected in larger scattering angles than seen in Fig. 21.

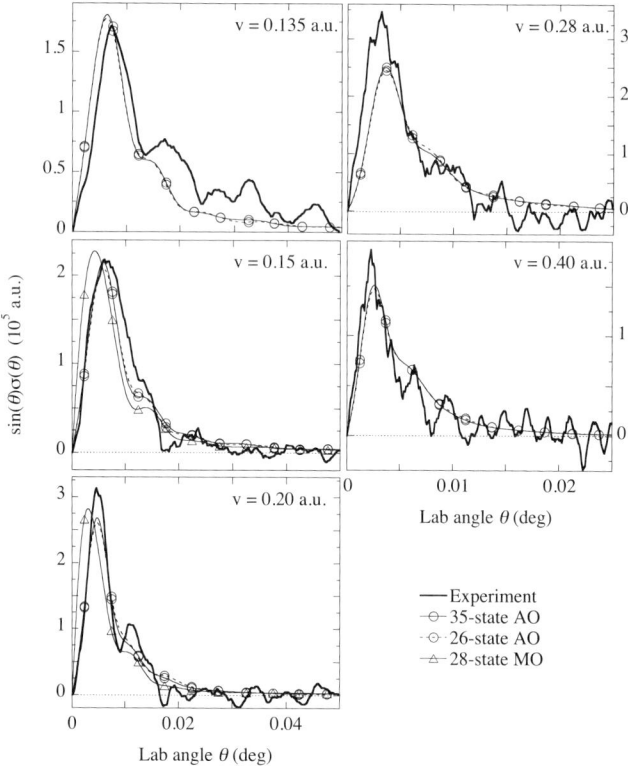

FIG. 23. Reduced differential cross section of the Na(3p) → Li(2p) electron transfer channel. Isotropic excitation of the Na(3p) target is assumed. [Reprinted figure with permission from van der Poel et al. (2002).]

© 2002 Institute of Physics

4.1.2. $Li^+ + Na(3p)$

Electron transfer from the effectively isotropic excited Na(3p) target occurs predominantly into the Li(2p) state. The cross section shown in Fig. 23 is the level-to-level cross section since capture into the magnetic substates of Li(2p) is not resolved. At the largest angles a fast jitter which falls within the statistical noise of the experiment is seen. We note again an overall good agreement with theory even if the MO calculations tend to predict scattering angles that are too small.

4.1.3. $Na^+ + Na(3s,3p)$

For the $Na^+ - Na(3s,3p)$ system only exploratory experimental data exist presently, obtained from a MOT with an estimated 25% of the target atoms in the Na(3p) state. The larger projectile mass compared to the Li case above implies

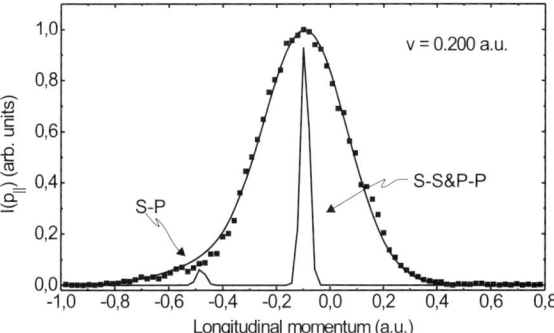

FIG. 24. Q-value spectrum of Na$^+$ + Na at $v = 0.2$ a.u. The experimental results fit well the predictions of a 26-state AO calculation, convoluted with the experimental resolution and assuming a 25% isotropic excitation of the target in the Na(3p) state. [Reprinted figure with permission from van der Poel (2001).]

a correspondingly higher scattering angle resolution. Thus an angular resolution better than 0.001° is achieved for the experimental data. Figure 24 shows the corresponding Q-value spectrum at a collision velocity of 0.20 a.u. summed over all scattering angles. The full line passing through the data points is a fit to a 26-state AO calculation, convoluted with the apparatus response function. For clarity, the same fit is shown again, but with an assumed response function that is much better than that of the actual apparatus. The dominating peak corresponds to 3s → 3s and 3p → 3p electron transfer. Integrated cross section for the 3p → 3p channel is about 1.5 times higher than for the 3s → 3s channel and is the dominant transfer channel from the excited atoms. Figure 25 shows the corresponding differential cross sections at velocities $v = 0.15$ and 0.20 a.u. The overall agreement of the AO calculation and the experiment is reasonable. At $v = 0.15$ a.u. the predicted amplitude of the first oscillations is larger than in the experiment, where the second peak appears as a shoulder on the first one, while the agreement is better at larger angles. At $v = 0.20$ a.u. the situation is almost reversed, with the height of the first peak being underestimated by the calculation and the predicted oscillations at the larger angles almost not visible in the experiment.

4.1.4. $K^+ + Na(3s,3p)$

An exploratory measurement with a K^+ projectile beam on the Na(3s,3p) MOT was also done by van der Poel (2001). Within the experimental limits, no trace of electron transfer from Na(3s) was seen in the Q-value spectrum, as indicated in Fig. 26, taken at a collision velocity of $v = 0.06$ a.u. At this velocity, the dominant peak by far is 3p → 4p transfer, with a few percent 3p → 4s contribution. The differential cross section corresponding to the dominant peak is shown in Fig. 27. For this system, typical scattering angles are less than 0.01°.

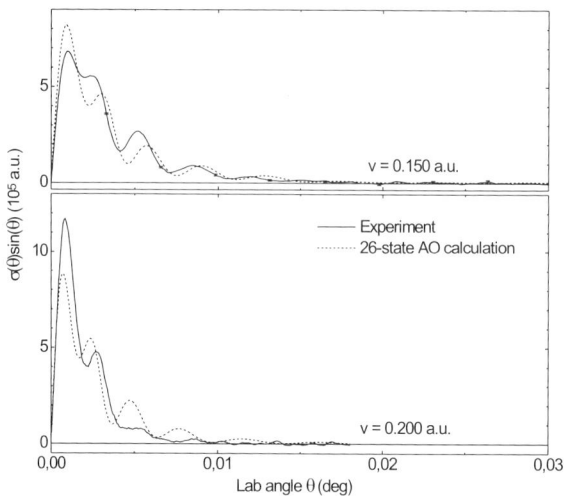

FIG. 25. Reduced differential cross section of the sum of the 3s → 3s and 3p → 3p electron transfer cross section at an impact velocity of 0.15 and 0.20 a.u. The curves are normalized to the AO total cross section. The theoretical curve has been convoluted with the experimental apparatus response function. A 25% excitation of the target to Na(3p) has been assumed. [Reprinted figure with permission from van der Poel (2001).]

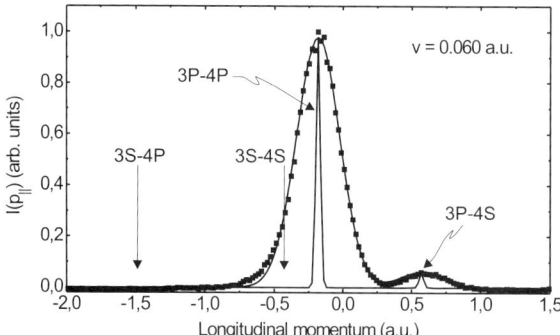

FIG. 26. Q-value spectrum for $K^+ + $Na(3s,3p) at $v = 0.06$ a.u. The fit corresponds to a cross section ratio for 3p → 4s and 3p → 4p electron transfer of 0.55 ± 0.04, convoluted with the experimental resolution. Within the experimental detection limits, no electron transfer from the Na(3s) target was observed. [Reprinted figure with permission from van der Poel (2001).]

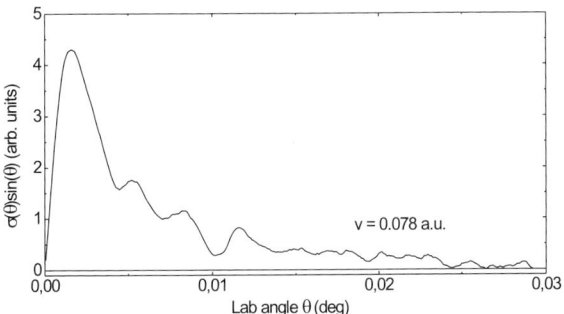

FIG. 27. Reduced differential cross section for 3p → 4p electron transfer at an impact velocity of 0.078 a.u. Isotropic Na(3p) target excitation is assumed. [Reprinted figure with permission from van der Poel (2001).]

4.1.5. $H^+ + Na(3s)$

Differences in electron capture from inner and outer shells for H^+ impact on a Na(3s) target has been studied by Knoop et al. (2006a). Figure 28 compares differential cross sections for the two processes (OSC) $H^+ + Na(3s) \rightarrow H(n=2) + Na^+$ and (ISC) $H^+ + Na(3s) \rightarrow H(1s) + Na(2p^5 3p)$ at a proton impact energy of 14 keV/amu. We note that the OSC electron transfer process corresponds to

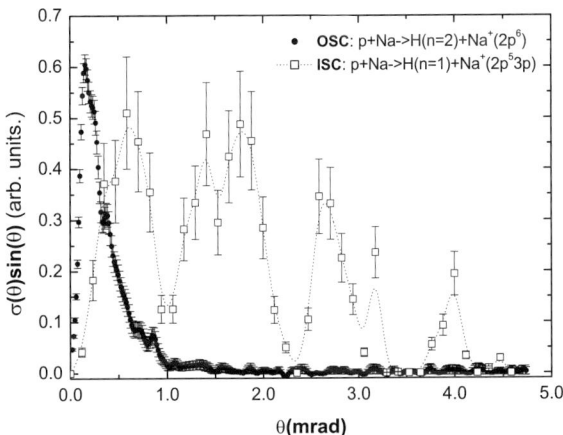

FIG. 28. Reduced differential cross sections for outer-shell Na(3s) electron transfer into $H(n = 2)$ (OSC) and inner-shell Na(2p) electron transfer into H(1s), leaving the target in the Na(2p^{5}3p) state (ISC), respectively, for $H^+ -$ Na(3s) collisions at an impact energy of 14 keV/amu. The line is to guide the eye. [Reprinted figure with permission from Knoop et al. (2004).]

© 2004 American Physical Society

very small scattering angles, while the ISC process leads to larger angles and an oscillatory pattern. Assuming that this ring pattern has an origin similar to the Fraunhofer diffraction studied for the Li$^+$ − Na(3s) system above, and using the de Broglie wavelength $\lambda = 240$ fm for the proton beam, the observed angular spacing $\Delta\theta$ of about 1.2 mrad corresponds to a capture radius R for the inner-shell 2p electron of about 2 a.u., derived from the relation $\Delta\theta = \lambda/2R$.

4.2. Rb$^+$, K$^+$, Li$^+$ + Rb(5s,5p) Collisions

Angular differential studies have also been performed by Nguyen et al. (2004a) for a series of projectiles (Rb$^+$, K$^+$, Li$^+$) colliding with a Rb(5s,5p) target at an impact energy of 7 keV. Results for the most prominent energetically resonant or near-resonant channels are shown in each case for Rb$^+$ + Rb(5s,5p) (Fig. 29), K$^+$ + Rb(5s,5p) (Fig. 30), and Li$^+$ + Rb(5s,5p) (Fig. 31), respectively.

We note in all cases rather smooth variations with scattering angle, in some cases superimposed on undulatory structures, for the latter two systems just be-

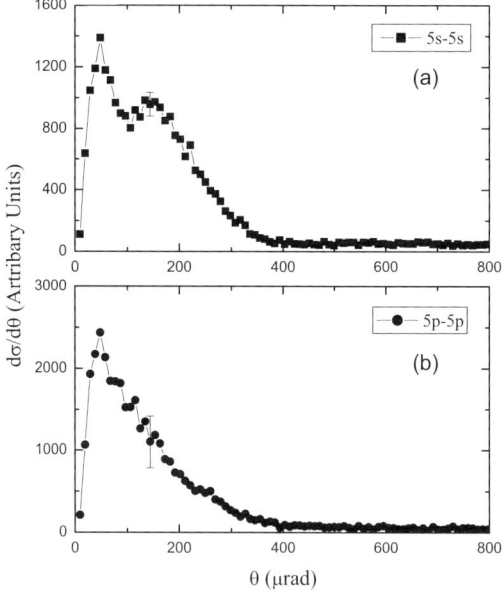

FIG. 29. Reduced differential resonant electron transfer cross sections for 7 keV (a) Rb(5s) → Rb(5s) and (b) Rb(5p) → Rb(5p) in Rb$^+$ + Rb(5s,5p) collisions in arbitrary, but comparable units. [Reprinted figure with permission from Nguyen et al. (2004a).]
© 2004 American Physical Society

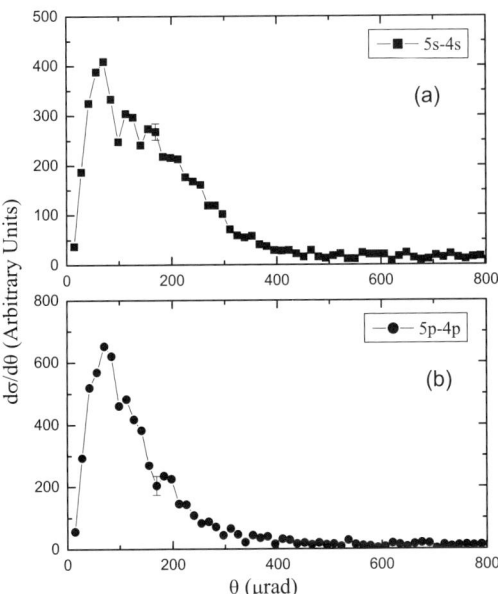

FIG. 30. Reduced differential electron transfer cross sections for 7 keV (a) Rb(5s) → K(4s) and (b) Rb(5p) → K(4p) in K$^+$+Rb(5s,5p) collisions in arbitrary, but comparable units. [Reprinted figure with permission from Nguyen et al. (2004a).]

© 2004 American Physical Society

yond the resolving power of the apparatus. No theoretical predictions are yet available for these systems.

4.3. O^{6+} + NA(3S)

Differential scattering results for the system O^{6+}+Na(3s) can be used to elucidate different population mechanisms leading to two-electron transfer into $O^{4+}(3lnl')$ states as shown by Knoop et al. (2006a). Figure 32 shows the Q-value spectrum of Na^{2+} recoils resulting from 7.5 keV/amu O^{6+}+Na(3s) collisions. The various double-capture states of O^{4+} are indicated. Double capture into states of the type $O^{4+}(3l3l')$ and $O^{4+}(3lnl', n \geqslant 6)$ are clearly separated in two peaks, with very little population of states of the type $O^{4+}(3l4l')$ and $O^{4+}(3l5l')$. The transverse momentum distributions for Na^{2+} recoils for the two peaks are shown in Fig. 33. Capture into $O^{4+}(3l3l')$ corresponds to large scattering angles and thereby relatively small internuclear distances, while the production of $O^{4+}(3lnl', n \geqslant 6)$ is observed at significantly smaller scattering angles, corresponding to larger internuclear distances. Analysis of the potential energy curves for the various channels

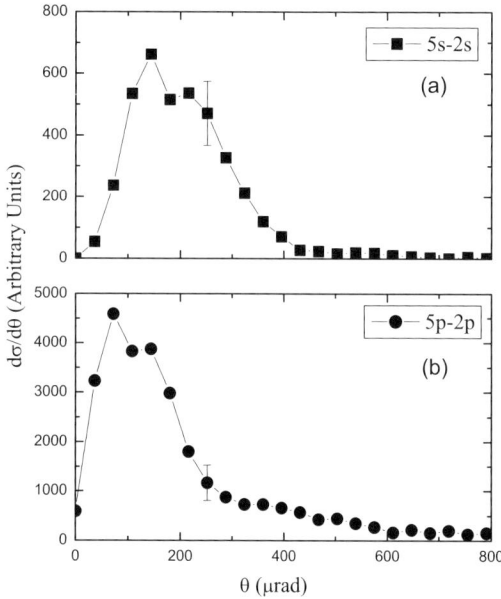

FIG. 31. Reduced differential electron transfer cross sections for 7 keV (a) Rb(5s) → Li(2s) and (b) Rb(5p) → Li(2p) in Li$^+$ + Rb(5s,5p) collisions in arbitrary, but comparable units. [Reprinted figure with permission from Nguyen et al. (2004a).]

© 2004 American Physical Society

reveals that population of the 3l3l' channels occurs at small impact parameters ($R < 5$ a.u.) in a one-step dielectronic process, i.e. by direct coupling with the initial state, while double capture into the asymmetric 3l6l' and 3l7l' configurations takes place at larger distances ($R > 10$ a.u.) via two monoelectronic transitions.

5. Probing Excitation Dynamics

5.1. Excited Fractions

Once the relative cross sections for electron transfer have been measured, it is a simple matter to determine the relative number of target atoms in excited states. For example, in the Q-value spectrum of Fig. 5 the peaks labelled 5 and 6 correspond to capture from Rb(5p) and Rb(5s), respectively, by 7 keV Na$^+$. Using the technique described in Section 2, the relative capture cross sections for these two channels, for this collision system, at this collision energy, has been measured by Lee et al. (2002) to be 11.29 ± 0.66. Then, dividing the areas under peaks 5 and

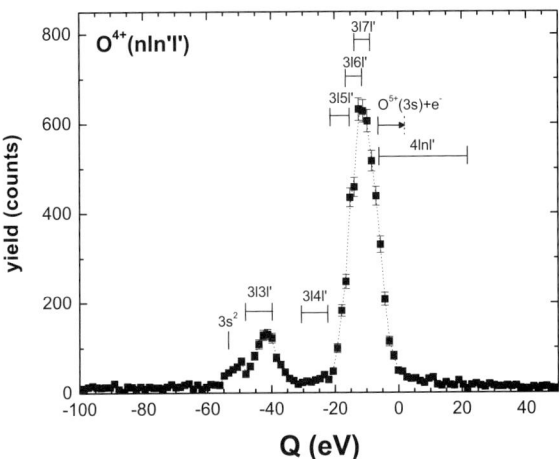

FIG. 32. Q-value spectrum of Na^{2+} recoils resulting from 7.5 keV/amu O^{6+} + Na(3s) collisions. Double-capture states of O^{4+} are indicated as well as the onset for transfer ionization. [Reprinted figure with permission from Knoop et al. (2006a).]

© 2006 Institute of Physics

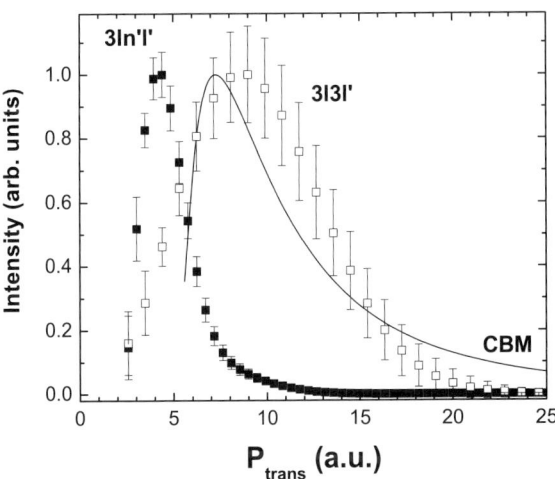

FIG. 33. Experimental transverse momentum distributions for Na^{2+} recoils resulting from 7.5 keV/amu O^{6+} + Na(3s) collisions. Double capture into $O^{4+}(3l3l')$ and $O^{4+}(3lnl', n \geq 6)$ are clearly separated. CBM is the result of a classical over-barrier model calculation using Rutherford scattering. All distributions are normalized to their peak value. [Reprinted figure with permission from Knoop et al. (2006a).]

© 2006 Institute of Physics

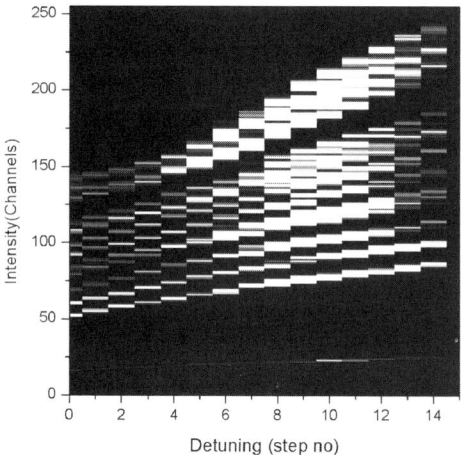

FIG. 34. Measured excited fraction as a function of detuning (horizontal axis, arbitrary units) and trap laser intensity (vertical axis, arbitrary units). The measured fraction is shown in grayscale. The curvature in the plots reflects the dependence of the AOM efficiency on the applied radio frequency. [Reprinted figure with permission from Shah (2006).]

6 by their relative cross sections gives the relative populations of the initial states in Rb. That is, if we designate the area under peak 6 by A_s, and that under peak 5 by A_p, the measured fraction of Rb atoms in the 5p state is given by

$$f_{ex} \equiv \frac{n_p}{n_s + n_p} = \frac{A_p/\sigma_p}{A_s/\sigma_s + A_p/\sigma_p} = \frac{A_p}{RA_s + A_p}, \quad (9)$$

with $R \equiv \sigma_p/\sigma_s$.

Using this approach, Sah et al. (2006, 2007) used MOTRIMS to determine the excited state fraction of atoms in an "active" MOT as a function of trap laser intensity and detuning from resonance over a wide range of other trapping parameters. Because the analysis is so simple, they were able to extract, in real time, the result shown in Fig. 34.

In this experiment, the radio frequency to an acousto–optical modulator, through which the trapping laser passed, was stepped through 16 values; thus the laser detuning was continuously cycled through 16 values. Whenever an electron transfer event took place, the current value of detuning was sent to the data acquisition computer. Because the efficiency of an AOM is frequency-dependent, a known fraction of the trapping light was split off and sent to a power meter, the output of which was also sent to the data acquisition computer whenever an event occurred. This process was repeated for a number of total trap intensities, each measurement resulting in a single curve in Fig. 34. Here, total trap laser intensity is plotted (vertical axis, arbitrary scale) versus detuning (horizontal axis, arbitrary scale) with the excited fraction given in grayscale. The dependence of the AOM efficiency on frequency is seen as a dependency of intensity on frequency in each of the curves. The ultimate goal of this experiment was to determine the veracity of a simple 2-level model for the excited fraction in a system irradiated by coherent light of intensity I, which is detuned from atomic resonance by angular frequency δ. For example, from Demtröder (2002):

$$f_{ex} = \frac{I/I_s}{1 + 2I/I_s + (2\delta/\Gamma)^2}, \tag{10}$$

where I is the total intensity of all six trapping laser beams and Γ is the transition line width. I_s, the so-called saturation intensity, is usually given, for example in Demtröder (2002), by

$$I_s = \frac{2\pi hc\Gamma}{3\lambda^3}, \tag{11}$$

where λ is the transition wavelength. It should be pointed out that there is no reason to expect that Eqs. (10) and (11) should be applicable in a MOT environment since that model was derived for a 2-level system, excited by a single travelling wave of low intensity light. In the case of a MOT, the unknown distribution of m_F levels, split by the B-field gradient, as well as the spatially dependent intensity and phase of the trapping laser, make the applicability of Eq. (10) suspect at best. Nevertheless, in order to test the simple model, the excited fractions of Fig. 34

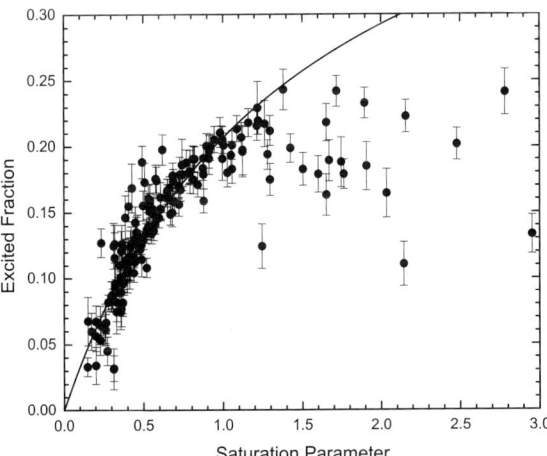

FIG. 35. Measured excited fraction as a function of saturation parameter. The solid circles are data taken with a variety of trapping parameters including different B-field gradients, balance between the trapping laser beams, and different repump laser intensities. The solid curve is a fit of (10) to the data, with I_s as the adjustable parameter. [Reprinted figure with permission from Shah (2006).]

were plotted against the so-called saturation parameter,

$$s \equiv \frac{I/I_s}{1 + (2\delta/\Gamma)^2}, \tag{12}$$

where Eq. (11) was used to define I_s. The result is plotted in Fig. 35. The solid line is a best fit of Eq. (10) to the data using I_s as a free parameter. The simple model is in striking agreement with the data for values of $s < 1.25$, a condition consistent with most MOTs. The value of I_s which allows a best fit of Eq. (10) to the data is $I_s = 9.2 \pm 1.7$ mW/cm^2, compared to the value of $I_s = 3.2$ mW/cm^2 given by Eq. (11). This useful result allows anyone who traps ^{87}Rb to quickly and easily estimate to within a couple of percent the fraction of excited atoms in their MOT, with knowledge only of the total intensity and detuning of their trapping lasers. Thus, the model-independent excitation fraction measurement made using MOTRIMS can be directly used by a large MOT community even though they themselves do not have the MOTRIMS setup. Details of this experiment can be found in Sah et al. (2006, 2007)

5.2. Population Dynamics

The capability for measuring excited fractions in a sample makes MOTRIMS a very powerful technique. The fact that collision times, even in the relatively low energy regime of a few keV, are extremely short means that one can measure population dynamics on time scales that are typically much shorter than the radiative life times of the system. This is done by measuring Q-values as functions of time.

A typical 2-dimensional Q-value spectrum for this sort of measurement is shown in Fig. 36. The collision system is 7 keV Na$^+$ + ^{87}Rb. The horizontal axis is the Q-value (in arbitrary units), the vertical scale is time, measured from some regular stimulus. The electron transfer counts are shown in false grayscale. The vertical stripes correspond to the peaks of the Q-value spectrum of Fig. 5. The stripes near Q-values of 7.1, 8.7, and 9.2 eV correspond, respectively, to the peaks of Fig. 5 labelled 3, 5, and 6. Figure 36 is an example of data from a relative cross section measurement. The gap in the stripe near 8.7 eV is due to the trapping laser being periodically turned off. This results in atoms in the Rb(5p) state decaying to the ground state so that capture from Rb(5p) goes to zero.

As an example of this sort of dynamics experiment, Brédy et al. (2003) measured the population dynamics of a MOT as the repump laser was turned on and off. When the raw data, in a form similar to that of Fig. 36 are converted to populations by dividing the counts contained in the right-most two stripes by the relative cross sections of their respective capture channels, a plot of population versus time, like that of Fig. 37 results. Here, the Rb(5s) and Rb(5p) populations are plotted versus time. At $t = 0$, the repump laser is turned off. The vertical line at approximately 1.25 ms indicates the time at which the repump laser is

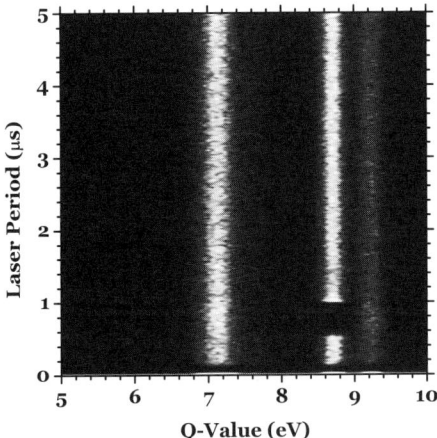

FIG. 36. Two-dimensional Q-value spectrum. Q-value energy is on the horizontal axis, time from some external repetitive event is on the vertical axis and charge transfer counts are shown in false grayscale. [Reprinted figure with permission from Camp (2005).]

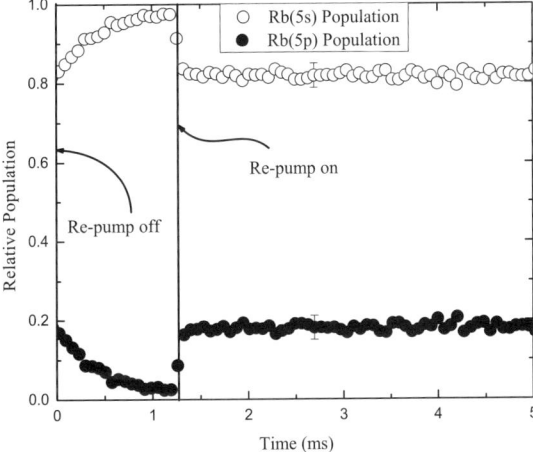

FIG. 37. Plot of Rb(5s) and Rb(5p) populations as a function of time. At $t = 0$, the repump laser is turned off. The vertical line at about 1.25 ms indicates the time at which the repump laser is turned back on. [Reprinted figure with permission from Brédy et al. (2003).]

© 2003 Elsevier

turned back on. Although one could have arrived at the same result by measuring time-dependent fluorescence, the model-independent nature of the MOTRIMS measurement allows the vertical scale to be absolute and without the uncertainty implicit in the interpretation of fluorescence from a sample having an unknown distribution of magnetic sublevels.

The physics of the above example was already understood at the time of the measurement. However, more sophisticated systems can also be probed using the same technique. Because the relative capture cross section from Rb(4d) had already been measured by Shah et al. (2005), population dynamics in the Rb(5s) − Rb(5p) − Rb(4d) ladder system could be studied. Gearba et al. (2007) chose to measure the coherent excitation of this ladder system. The particular excitation scheme they focussed on is known as STIRAP, or stimulated Raman adiabatic passage. This has already been investigated theoretically and experimentally by Bergmann et al. (1998) and Süptitz et al. (1997), among others. However, in all previous measurements, excited fractions were integrated over time. Furthermore, because they were based on measurements of fluorescence from trapped and cooled atoms, the results of Süptitz et al. (1997) were model-dependent. The results of Bergmann et al. (1998) did not suffer from this problem because their "target" was an atomic beam, but because fluorescence was still the principal diagnostic, the temporal resolution was limited by the lifetime of the longest lived state. MOTRIMS does not suffer from this problem.

In the experiments of Gearba et al. (2007), ^{87}Rb was trapped and cooled in the usual way. The trapping and repump lasers were then turned off for 500 ns, allowing the trapped atoms to decay to the electronic ground state. Next, two pulses of light, one from a laser (L1) near 780 nm, the other from a laser (L2) near 1529 nm, were directed onto the target. The timing of these optical pulses was under the control of a computer that sent the appropriate signals to a pair of AOMs such that they occurred during the period of time that the trapping and repump lasers were off; the entire sequence was repeated every 200 microseconds. A beam of 7 keV Na$^+$ ions was continuously directed at the Rb target. Thus, throughout the entire experiment, the population distribution of the Rb atoms was probed by the projectile ions. Two noteworthy aspects of L1 and L2 should be pointed out: First, using the AOMs, the intensities, pulse widths, and relative timing of L1 and L2 could be controlled. For conditions that cause Rb(4d) to be optimally excited, with minimal excitation of the intermediate Rb(5p) state, L1 and L2 were detuned approximately 54 MHz off their individual resonances, L1 to the red and L2 to the blue, maintaining a 2-photon resonance condition. Second, as has been pointed out in previous work, for example, Bergmann et al. (1998) and Süptitz et al. (1997), the optimal conditions for STIRAP require that the pulse from L2 interacts with the target *before* that of L1.

A 2-dimensional plot of electron transfer events is shown in Fig. 38 for a fixed delay between L1 and L2. The Q-value is on the horizontal axis, relative timing

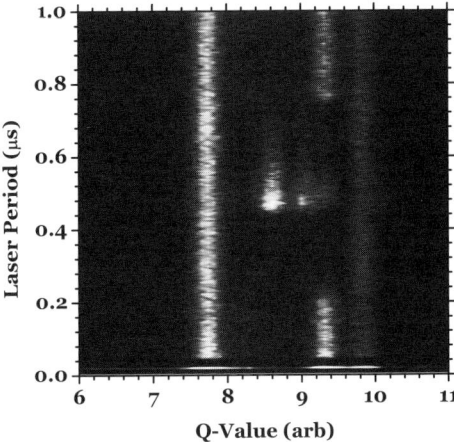

FIG. 38. Raw data for the STIRAP experiment. The horizontal axis is the Q-value; the vertical axis is time with arbitrary offset, with respect to the moment at which the trapping lasers are turned off. [Reprinted figure with permission from Camp (2005).]

from the point at which the trapping lasers were turned off is on the vertical axis, and the number of events is shown in false grayscale. As before, the vertical stripes correspond to peaks in a normal Q-value spectrum. The stripe on the far right corresponds to the Rb(5s) − Na(3s) channel, while the stripe with the gap in it corresponds to the Rb(5p) − Na(3p) channel. The gap corresponds to the period of time during which the trapping lasers were turned off, thus allowing the Rb(5p)

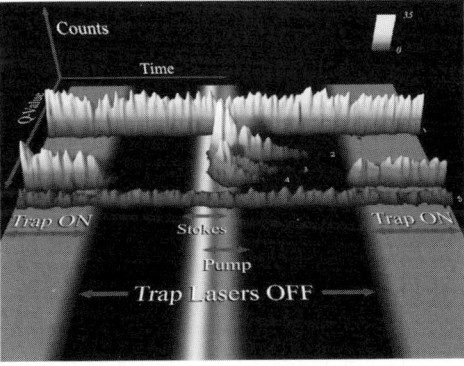

FIG. 39. The same raw data shown in Fig. 38, but in 3-dimensional perspective. The approximate timing of the trapping lasers, as well as pulses coming from L1 and L2, are indicated. Note the axes have been rotated compared to those of Fig. 38.

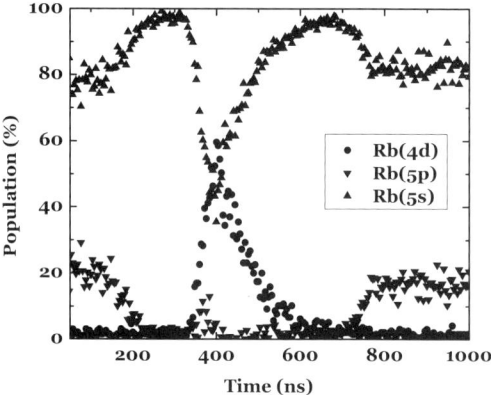

FIG. 40. Population dynamics in a 3-level ladder system in Rb. The upward pointing triangles show the Rb(5s) (ground state) population; the downward pointing triangles show the Rb(5p) (intermediate) population; and the circles show the Rb(4d) (highest level) population. A peak Rb(4d) population is seen near 400 ns. [Reprinted figure with permission from Camp (2005).]

states to rapidly decay away. In the middle of the gap region one can clearly see a set of peaks. The left-most of these is the Rb(4d) − Na(3d) channel, while the center peak is the Rb(4d) − Na(4s) channel. The right-most peak in the gap is due to capture from the small amount of Rb(5p) formed during the excitation process. A better idea of just how large the measured effect is can be had by plotting the data of Fig. 38 as a 3-dimensional perspective plot, shown in Fig. 39.

Dividing the data contained in the different stripes of Figs. 38 and 39 by relative cross sections corresponding to the respective capture channels gives rise to a plot of the population distribution versus time, as shown in Fig. 40. In this figure, the ground state (5s) population is shown by upward pointing triangles; the intermediate state (5p) population is shown by downward pointing triangles; and the upper state (4d) population is shown by circles. The MOTRIMS methodology clearly shows the time evolution of all three levels. Furthermore, unlike other approaches based on fluorescence, this methodology shows this evolution with a temporal resolution of a few nanoseconds.

6. Future Applications

MOTRIMS was originally developed for the measurement of electron transfer cross sections, differential in scattering angle and initial and final states of the projectile. As planned, these measurements have been carried out for singly and multiply charged ions. An unanticipated area of study, also described in this article, is the measurement of population dynamics in the target. In this section we try to anticipate future applications of the basic MOTRIMS apparatus.

6.1. Electron Collisions

In principle, the MOTRIMS technique can also be used in case of electron collisions. A very promising experiment along these lines is presently being prepared by Ott and coworkers in Mainz (Gericke et al., 2006). The aim of this experiment is to reconstruct the spatial correlations of atoms in ultracold quantum gases. To achieve this goal the atoms will be ionized by a focussed electron beam that can be scanned across the trapped atomic cloud. With the beam at a given position the local density and even the temporal correlations of the atoms can be measured. As compared with optical methods, a significantly higher spatial resolution of as good as about 100 nm can be obtained in this way. By extracting the ions and measuring their time-of-flight to a position-sensitive detector, even the position along the electron beam direction at which the ion was created can be reconstructed; thus, a truly three-dimensional image is obtained. Ott and coworkers point out that in case of an optical lattice target single sites can be addressed, which might be an interesting aspect for quantum information processing.

6.2. Photo-Ionization and Photo-Associative Ionization

Thus far we have only discussed MOTRIMS as it pertains to charge particle interactions. However, modern sources of electromagnetic radiation allow other mechanisms for target ionization and, therefore, new opportunities for MOTRIMS. In this section we speculate on future MOTRIMS experiments in which synchrotron facilities could be used to ionize MOT samples through the absorption of a single high energy photon. We will also discuss the possibility of combining MOTRIMS with ultrafast laser technology to study multi-photon interactions with cold trapped atomic targets.

Using apparatus very similar to MOTRIMS along with photons in the energy range 12–40 eV from a synchrotron, Coutinho et al. (2004) studied the angular distribution of photo-ionization fragments from a laser cooled and trapped cesium target. The asymmetry parameter β was observed to deviate significantly from 2, observed for s electrons outside resonances and far from the Cooper minimum. This finding was explained by relativistic effects and interchannel coupling arising from final state configuration mixing.

Modern synchrotron facilities, with their unprecedented intensities of short wavelength radiation offer wonderful opportunities in atomic physics, to which the MOTRIMS methodology could contribute. Similarly, modern lasers having pulse widths in the femtosecond or even attosecond regime have opened up new opportunities in atomic, molecular, and optical physics to which new experiments using MOTRIMS will certainly contribute.

As an example, consider what happens when cold, trapped Rb is illuminated by an intense pulse of light having a central wavelength of about 790 nm, typical for a Ti:saphire ultrafast laser system. For a pulse having a temporal width $\leqslant 30$ fs, the bandwidth of the optical pulse is sufficient to allow *resonant* 3-photon ionization along the $5s - 5p - 5d - \epsilon l$ ladder. An interesting measurement would be that of the Rb ionization rate as the ultrafast pulse is modified in the frequency and phase domains. The MOTRIMS apparatus, with its weak electric field and TOF spectrometer would make this measurement relatively simple. Furthermore, if instead the ultrafast optical pulse were focussed in air, the resulting third harmonic that would be formed could be used to interrogate the MOT population in much the same way as the ion beam was used in Section 5. Thus, one can imagine pump/probe experiments where the fundamental is used to excite and the third harmonic is used to probe using the high resolution in Q-value afforded by the MOTRIMS apparatus.

As described in Salzmann et al. (2006), Weidemüller's group has already started experiments in which shaped femtosecond pulses of light were used to cause photo-association (PA) in a MOT. [For a detailed discussion of PA, for the most part using narrow line width cw lasers, see Weiner (2003).] In Salzmann et al. (2006), an additional laser was used to photo-ionize the molecules formed in the PA process, and the weak electric field of their MOTRIMS apparatus was used to extract the resulting ions. The ions were verified to be molecular in nature by their characteristic flight time in the MOTRIMS spectrometer. The yield of molecular ions was used in a feedback loop to the pulse shaper of their ultrafast laser system in order to maximize the PA process using a genetic algorithm (Assion et al., 1998). Using their MOTRIMS apparatus, Salzmann et al. (2006) were able to demonstrate that by shaping the ultrafast pulses they were able to greatly enhance the production of ultracold molecules in their MOT.

One disadvantage that MOTRIMS suffers from is that molecules can not be cooled and trapped in a MOT. This is because, unlike atomic systems, the almost total lack of selection rules for dipole coupling between ro-vibrational states in molecules means that well-defined cycling transitions do not exist. Nevertheless, as the photo-association process becomes more efficient, either through the use of narrow line width quasi-cw lasers, for example as in Jones et al. (2006), or through the use of ultrafast lasers as discussed above, we may begin to see MOTRIMS experiments with extremely cold molecules as targets. If future photo-association technology advances allow, it may even be possible that the molecular targets could be vibrationally cold, as well as translationally. MOTRIMS could even play a role in developing this technology. One can imagine using collisional dissociation of the cold molecules, and with the exceptional resolution inherent in MOTRIMS one ought to be able to observe vibrational population distributions in the dissociation fragments. The alignment of the cold molecules may also be measured with respect to the polarization axis of the photo-association laser field since the position- and time-sensitive detection of the recoiling molecular fragments can be analyzed to compute alignment (and orientation for heteronuclear systems) parameters.

6.3. ANISOTROPIC TARGETS: ALIGNMENT AND ORIENTATION

A very challenging application of the MOTRIMS technique aims at precision measurements of radioactive beta decay processes in order to observe angular correlations between the emitted electrons or positrons and antineutrinos or neutrinos. Although a direct observation of (anti-)neutrinos in coincidence with the corresponding electrons or positrons is practically impossible, one can exploit energy and momentum conservation in order to deduce these correlations from a coincident measurement between the heavy recoil particle and the electron. First attempts in this direction have been reported by Scielzo et al. (2004) and by Gorelov et al. (2005). Scielzo et al. (2004) studied β decay of laser cooled ^{21}Na in a MOT by measuring the energy of the resulting ^{21}Ne$^+$ recoil nuclei by a time-of-flight method and exploited the fact that the energy distribution is influenced by the angular correlation: emission of the two leptons in the same or in opposite directions leads to a higher or lower recoil energy, respectively. In that investigation no significant deviation of the measured angular correlations from predictions of the standard model was found. Gorelov et al. (2005) studied β-decay of metastable ^{38}K trapped in a MOT by measuring coincidences between the emitted electrons and the recoiling Ar daughter products. Also in these studies, no deviations from the predictions of the standard model were found.

For direct, more detailed, and theoretically more precise measurements of the angular correlations, a nuclear spin polarized target should be used. Experiments like this are, in fact, being prepared at the TRIμP facility in Groningen by Jungmann (2002), Berg et al. (2006). Of course the usual MOT target is generally not polarized since three pairs of mutually perpendicular and circularly polarized laser beams are used for capturing and cooling of the target atoms. Therefore various methods for producing a polarized target are presently being discussed, one of those based on switching off two pairs of laser beams for a certain short period such that the remaining pair produces an electronic circular polarization which eventually results in the desired nuclear spin polarization via hyperfine coupling. In this way the angular distribution of the leptons can not only be measured with respect to each other, but also with respect to the nuclear spin. This gives access to possible time reversal violating processes.

A large number of studies of electron transfer processes, in particular using proton or alkali ion impact on optically prepared anisotropic excited states of alkali atoms such as Na(3p) have revealed dramatic left–right scattering asymmetries at collision velocities comparable to the orbital velocity of the active electron. See Andersen and Bartschat (2004) for an overview. However, none of these studies was yet performed with the superior angular resolution of MOTRIMS which is often crucial to reveal the characteristic structure, as evident from, for example, the data of Fig. 20.

6.4. STARK EFFECT

Collision processes in which impact ionization is accompanied by excitation of target or projectile have recently been a topic of interest since description of the delicate correlated motion of two active electrons often requires detailed attention. Here, the pure Coulomb potential offers some special opportunities due to the special properties of Stark manifolds. For the simplest case of $n = 2$ in hydrogen-like systems, the degeneracy of the 2s and 2p states allows for additional observables in addition to the well-known parameters ($L_\perp$, γ) characterizing coherent excitation of the 2p component. These observables define the size and direction of the electron charge-cloud polarization (Andersen and Bartschat, 2004). For the case of simultaneous ionization-excitation of $He^+(n = 2)$ by electron impact, it is predicted that for ejection angles of the slow electron within 90° of the momentum-transfer direction, the excited electron is preferably located on the side of the nucleus opposite to the ejected electron. For larger electron angles, the situation is reversed. No experimental check of this prediction has yet been performed, but a first, important step in this direction has been taken by Sakhelashvili et al. (2005). For heavy-particle impact, processes such as $H^+ - Na(3s$ or $3p) \rightarrow H(n = 2) + Na^+$ would offer similar opportunities for study.

7. Concluding Comments

In this article we have described the basics of the MOTRIMS methodology and demonstrated its utility using some basic measurements by way of example. We started by describing how MOTRIMS has been used to extract electron transfer cross sections that are differential in projectile charge and energy, scattering angle, initial and final states of the target, and final state of the projectile. We further demonstrated how, once some of these relative cross sections have been measured, the results can be turned around to measure temporally and spatially averaged population distributions in the target species. This sort of measurement was generalized to allow spatially averaged population *dynamics* of the target. Finally, we speculated on future experiments that we may see carried out using

MOTRIMS, including the fertile fields of photo-ionization using radiation from synchrotrons or ultrafast optical sources. We can conclude this article by stating that the MOTRIMS methodology has already demonstrated its utility and predicting that it will continue to be a viable tool in atomic, molecular and optical physics for some time to come.

8. Acknowledgements

The material presented here obviously encompasses the work of a great many individuals. The authors would like to gratefully acknowledge the tremendous effort and ingenuity of the members of their research teams who were involved in designing and performing the experiments as well as in analyzing and interpreting the results as described in this article. These individuals are: X. Fléchard, R. Brédy, M.A. Gearba, H. Nguyen, H.A. Camp, and M.H. Shah of Kansas State University; S. Knoop and R. Hoekstra of the University of Groningen; and M. van der Poel of the University of Copenhagen.

9. References

Andersen, N., Bartschat, K. (2001). "Polarization, Alignment and Orientation in Atomic Collisions". Springer.

Andersen, N., Bartschat, K. (2004). Dipole polarization in coherently excited Stark manifolds. *J. Phys. B: At. Mol. Opt. Phys.* **37**, 3809–3819.

Assion, A., Baumert, T., Bergt, M., Brixner, T., Kiefer, B., Seyfried, V., Strehle, M., Gerber, G. (1998). Control of chemical reactions by feedback-optimized phase-shaped femtosecond laser pulses. *Science* **282**, 919–922.

Aumayr, F., Winter, H. (1987). L_α emission in H^+–Na collisions (1–20 keV). *J. Phys. B: At. Mol. Opt. Phys.* **20**, L803–L807.

Aumayr, F., Lakits, G., Winter, H. (1987). Charge transfer and target excitation in H^+–Na(3s) collisions (2–20 keV). *J. Phys. B: At. Mol. Opt. Phys.* **20**, 2025–2030.

Bazin, V., Boduch, P., Chantepie, M., Jacquet, E., Kucal, H., Lecler, D., Pascale, J. (2002). Excitation and alignment effects in Ar^{8+}–Cs(6s,6p) collisions at low energies. *Phys. Rev. A* **65**. 032712.

Berg, G.P.A., Dermois, O.C., Dammalapati, U., Dendooven, P., Harakeh, M.N., Jungmann, K., Onderwater, C.J.G., Rogachevskiy, A., Sohani, M., Traykov, E., Willmann, L., Wilschut, H.W. (2006). Dual magnetic separator for TRIμP. *Nucl. Instrum. and Meth. in Phys. Res. A* **560**, 169–181.

Bergmann, K., Theuer, H., Shore, B.W. (1998). Coherent population transfer among quantum states of atoms and molecules. *Rev. Mod. Phys.* **70**, 1003–1025.

Brédy, R., Nguyen, H., Camp, H.A., Fléchard, X., DePaola, B.D. (2003). MOTRIMS as a generalised probe of AMO processes. *Nucl. Instrum. and Meth. in Phys. Res. B* **205**, 191–195.

Camp, H.A. (2005). "Measurements of the time evolution of coherent excitation". PhD thesis. Kansas State University.

Cornelius, K.R., Wojtkowski, K., Olson, R.E. (2000). State-selective cross section scalings for electron capture collisions. *J. Phys. B: At. Mol. Opt. Phys.* **33**, 2017–2035.

Coutinho, L.H., Cavasso-Filho, R.L., Rocha, T.C.R., Homem, M.G.P., Figueira, D.S.L., Fonseca, P.T., Cruz, F.C., de Brito, A.N. (2004). Relativistic and interchannel coupling effects in photoionization and angular distributions by synchrotron spectroscopy of laser cooled atoms. *Phys. Rev. Lett.* **93**. 183001.

Demkov, Y.N. (1964). Charge transfer at small resonance defects. *Sov. Phys.-JETP* **18**, 138–142.

Demtröder, W. (2002). "Laser Spectroscopy: Basic Concepts and Instrumentation". Springer.

Dörner, R., Mergel, V., Jagutzki, O., Spielberger, L., Ullrich, J., Moshammer, R., Schmidt-Böcking, H. (2000). Cold target recoil ion momentum spectroscopy: A 'momentum microscope' to view atomic collision dynamics. *Phys. Rep.* **330**, 95–192.

Dowek, D., Houver, J.C., Pommier, J., Richter, C., Royer, T., Andersen, N., Palsdottir, B. (1990). Strong effects of initial orbital alignment observed for electron capture in keV H^+–Na(3p) collisions. *Phys. Rev. Lett.* **64**, 1713–1716.

Dubois, R.D. (1986). Charge transfer leading to multiple ionization of neon, sodium, and magnesium. *Phys. Rev. A* **34**, 2738–2745.

Dubois, R.D., Toburen, L.H. (1985). Electron capture by protons and helium ions from lithium, sodium, and magnesium. *Phys. Rev. A* **31**, 3603–3611.

Ebel, F., Salzborn, E. (1987). Charge transfer of 0.2–5.0 keV protons and hydrogen atoms in sodium-, potassium- and rubidium-vapour targets. *J. Phys. B: At. Mol. Opt. Phys.* **20**, 4531–4542.

Fléchard, X., Nguyen, H., Wells, E., Ben-Itzhak, I., DePaola, B.D. (2001). Kinematically complete charge exchange experiment in the Cs^+ + Rb collision system using a MOT target. *Phys. Rev. Lett.* **87**. 123203.

Fléchard, X., Nguyen, H., Brédy, R., Lundeen, S.R., Stauffer, M., Camp, H.A., Fehrenbach, C.W., DePaola, B.D. (2003). State selective charge transfer cross sections for Na^+ with excited rubidium: A unique diagnostic of the population dynamics of a magneto–optical trap. *Phys. Rev. Lett.* **91**. 243005.

Fritsch, W., Lin, C.D. (1991). The semiclassical close-coupling description of atomic collisions: Recent developments and results. *Phys. Rep.* **202**, 1–97.

Gearba, M.A., Camp, H.A., Trachy, M.L., Veshapidze, G., Shah, M.H., Jang, H.U., DePaola, B.D. (2007). Measurement of population dynamics in stimulated Raman adiabatic passage. *Phys. Rev. A* **76**. 013406.

Gericke, T., Utfeld, C., Hommerstad, N., Ott, H. (2006). A scanning electron microscope for ultracold atoms. *Laser Phys. Lett.* **3**, 415–419.

Gorelov, A., Melconian, D., Alford, W.P., Ashery, D., Ball, G., Behr, J.A., Bricault, P.G., D'Auria, J.M., Deutsch, J., Dilling, J., Dombsky, M., Dubé, P., Fingler, J., Giesen, U., Glück, F., Gu, S., Häusser, O., Jackson, K.P., Jennings, B.K., Pearson, M.R., Stocki, T.J., Swanson, T.B., Trinczek, M. (2005). Scalar interaction limits from the $\beta - \nu$ correlation of trapped radioactive atoms. *Phys. Rev. Lett.* **94**. 142501.

Hoekstra, R., Olson, R.E., Folkerts, H.O., Wolfrum, E., Pascale, J., de Heer, F.J., Morgenstern, R., Winter, H. (1993). Electron capture from Li by B^{5+}, N^{5+} and Be^{4+} ions. *J. Phys. B: At. Mol. Opt. Phys.* **26**, 2029–2040.

Huang, M.-T., Zhang, L., Hasegawaq, S., Southworth, S.H., Young, L. (2002). Measurements of the electron-impact double-to-single ionization ratio using trapped lithium. *Phys. Rev. A* **66**. 012715.

Huang, M.-T., Wong, W.W., Inokuti, M., Southworth, S.H., Young, L. (2003). Triple ionization of lithium by electron impact. *Phys. Rev. Lett.* **90**. 163201.

Jacquet, E., Kucal, H., Bazin, V., Boduch, P., Chantepie, M., Cremer, G., Laulhe, C., Lecler, D., Pascale, J. (2000). Emission-line polarization degrees and nlm_l distributions produced by state-selective electron capture in slow Kr^{8+}–Li(2s) collisions. *Phys. Rev. A* **62**. 022712.

Jain, A., Winter, T.G. (1995). Electron transfer, target excitation, and ionization in H^+ + Na(3s) and H^+ + Na(3p) collisions in the coupled-Sturmian-pseudostate approach. *Phys. Rev. A* **51**, 2963–2973.

Jones, K.M., Tiesinga, E., Lett, P.D., Julienne, P.S. (2006). Ultracold photoassociaton spectroscopy: Long-range molecules and atomic scattering. *Rev. Mod. Phys.* **78**, 483–535.

Jungmann, K. (2002). TRIμP—A new facility to investigate fundamental interactions with optically trapped radioactive atoms. *Acta Phys. Polonica B* **33**, 2049–2057.

Keim, M., Lüdde, H.J. (2006). Private communication.

Kim, J.A., Lee, K.L., Noh, H.R., Jhe, W. (1997). Single-beam atom trap in a pyramidal and conical hollow mirror. *Opt. Lett.* **22**, 117–119.

Knoop, S. (2006). "Electron dynamics in ion–atom interactions". PhD thesis. Rijksuniversiteit Groningen.

Knoop, S., Morgenstern, R., Hoekstra, R. (2004). Direct observation of pure one-electron capture from the target inner shell in low-energy p + Na collisions. *Phys. Rev. A* **70**. 050702(R).

Knoop, S., Keim, M., Lüdde, H.J., Kirchner, T., Morgenstern, R., Hoekstra, R. (2005a). State selective single-electron capture in O^{6+} + Na collisions. *J. Phys. B: At. Mol. Opt. Phys.* **38**, 3163–3172.

Knoop, S., Olson, R.E., Ott, H., Hasan, V.G., Morgenstern, R., Hoekstra, R. (2005b). Single ionization and electron capture in He^{2+} + Na collisions. *J. Phys. B: At. Mol. Opt. Phys.* **38**, 1987–1998.

Knoop, S., Hasan, V.G., Morgenstern, R., Hoekstra, R. (2006a). Identification of distinct two-electron transfer processes in O^{6+} + Na collisions. *Europhys. Lett.* **74**, 992–998.

Knoop, S., Hasan, V.G., Ott, H., Morgenstern, R., Hoekstra, R. (2006b). Single ionization of Na(3s) and Na*(3p) by low energy ion impact. *J. Phys. B: At. Mol. Opt. Phys.* **39**, 2021–2029.

Kroneisen, O.J., Lüdde, H.J., Kirchner, T., Dreizler, R.M. (1999). The basis generator method: Optimized dynamical representation of the solution of time-dependent quantum problems. *J. Phys. A* **32**, 2141–2156.

Lee, K.I., Kim, J.A., Noh, H.R., Jhe, W. (1996). Atom trap in an axicon mirror. *Opt. Lett.* **21**, 1177–1179.

Lee, T.G., Nguyen, H., Fléchard, X., DePaola, B.D., Lin, C.D. (2002). Differential charge-transfer cross sections for Na^+ with Rb collisions at low energies. *Phys. Rev. A* **66**. 042701.

Lembo, L.J., Danzmann, K., Stoller, C., Meyerhof, W.E., Hänsch, T. (1985). Observation of polarized optical radiation following electron capture into slow, highly ionized neon. *Phys. Rev. Lett.* **55**, 1874–1876.

Lu, Z.T., Corwin, K.L., Renn, M.J., Anderson, M.H., Cornell, E.A., Wieman, C.E. (1996). Low-velocity intense source of atoms from a magneto–optical trap. *Phys. Rev. Lett.* **77**, 3331–3334.

Lüdde, H.J., Henne, A., Kirchner, T., Dreizler, R.M. (1996). Optimized dynamical representation of the solution of time-dependent quantum problems. *J. Phys. B: At. Mol. Opt. Phys.* **29**, 4423–4441.

Lundy, C.J., Olson, R.E. (1996). A classical analysis of proton collisions with ground-state and excited, aligned sodium targets. *J. Phys. B: At. Mol. Opt. Phys.* **29**, 1723–1736.

Mergel, V. (1994). PhD thesis. Johann-Wolfgang-Goethe Universität Frankfurt.

Metcalf, H.J., van der Straten, P. (1999). "Laser Cooling and Trapping". Springer.

Moshammer, R., Unverzagt, M., Schmitt, W., Ullrich, J., Schmidt-Böcking, H. (1996). A 4π recoil-ion electron momentum analyzer: A high-resolution 'microscope' for the investigation of the dynamics of atomic, molecular and nuclear reactions. *Nucl. Instrum. and Meth. in Phys. Res. B* **108**, 425–445.

Nguyen, H. (2003). "Magneto optical trap recoil momentum spectroscopy". PhD thesis. Kansas State University.

Nguyen, H., Brédy, R., Camp, H.A., Awata, T., DePaola, B.D. (2004a). Differential charge-transfer cross sections for systems with energetically degenerate or near-degenerate channels. *Phys. Rev. A* **70**, 031401.

Nguyen, H., Fléchard, X., Brédy, R., Camp, H.A., DePaola, B.D. (2004b). Recoil ion momentum spectroscopy using magneto optical trap. *Rev. Sci. Instrum.* **75**, 2638–2647.

O'Hare, B.G., McCullough, R.W., Gilbody, H.B. (1975). Ionization of sodium and potassium vapour by 20–100 keV H^+ and He^+ ions. *J. Phys. B: At. Mol. Opt. Phys.* **8**, 2968–2978.

Olson, R.E. (1986). v/2 electrons in $H^+ + H$ ionizing collisions. *Phys. Rev. A* **33**, 4397–4399.

Olson, R.E., Pascale, J., Hoekstra, R. (1992). Line emission from C^{6+}, O^{8+} + Li electron capture collisions. *J. Phys. B: At. Mol. Opt. Phys.* **25**, 4241–4247.

Pascale, J., Olson, R.E., Reinhold, C.O. (1990). State-selective capture in collisions between ions and ground- and excited-state alkali-metal atoms. *Phys. Rev. A* **42**, 5305–5314.

Pieksma, M. (1997). Saddle-point effects in adiabatic to intermediate-energy ion–atom collisions. *Nucl. Instrum. and Meth. in Phys. Res. B* **124**, 177–185.

Pieksma, M., Ovchinnikov, S.Y., van Eck, J., Westerveld, W.B., Niehaus, A. (1994). Experimental identification of saddle-point electrons. *Phys. Rev. Lett.* **73**, 46–49.

van der Poel, M. (2001). "Construction of a COLTRIMS spectrometer with a laser cooled target for electron transfer studies". PhD thesis. University of Copenhagen.

van der Poel, M., Nielsen, C.V., Gearba, M.A., Andersen, N. (2001). Fraunhofer diffraction of atomic matter waves: Electron transfer studies with a laser cooled target. *Phys. Rev. Lett.* **87**. 123201.

van der Poel, M., Nielsen, C.V., Rybaltover, M., Nielsen, S.E., Machholm, M., Andersen, N. (2002). Atomic scattering in the diffraction limit: Electron transfer in keV $Li^+ - Na(3s,3p)$ collisions. *J. Phys. B: At. Mol. Opt. Phys.* **35**, 4491–4505.

Sakhelashvili, G., Dorn, A., Höhr, C., Ullrich, J., Kheifets, A.S., Lower, J., Bartschat, K. (2005). Triple coincidence $(e, \gamma 2)$ experiment for simultaneous electron impact ionization excitation of helium. *Phys. Rev. Lett.* **95**. 033201.

Salzmann, W., Poschinger, U., Wester, R., Weidemüller, M., Merli, A., Weber, S.M., Sauer, F., Plewicki, M., Weise, F., Esparza, A.M., Wöste, L., Lindinger, A. (2006). Coherent control with shaped femtosecond laser pulses applied to ultracold molecules. *Phys. Rev. A* **73**. 023414.

Schippers, S., Schlatmann, A.R., Wiersema, W.P., Hoekstra, R., Morgenstern, R., Olson, R.E., Pascale, J. (1994). Anisotropy effects in electron capture by O^{6+} from a aligned $Na^*(3p)$. *Phys. Rev. Lett.* **72**, 1628–1631.

Schippers, S., Boduch, P., van Buchem, J., Bliek, F.W., Hoekstra, R., Morgenstern, R., Olson, R.E. (1995). Polarized light emission in keV $He^{2+} + Na(3s)$ collisions. *J. Phys. B: At. Mol. Opt. Phys.* **28**, 3271–3282.

Schippers, S., Hoekstra, R., Morgenstern, R., Olson, R.E. (1996). Anisotropy and polarization in charge changing collisions of C^{4+} with Na(3s) and laser aligned Na(3p). *J. Phys. B: At. Mol. Opt. Phys.* **29**, 2819–2836.

Schlatmann, A.R., Hoekstra, R., Morgenstern, R., Olson, R.E., Pascale, J. (1993). Strong velocity dependence of electron capture in collisions between aligned $Na^*(3p)$ and He^{2+}. *Phys. Rev. Lett.* **71**, 513–516.

Scielzo, N.D., Freedman, S.J., Fujikawa, B.K., Vetter, P.A. (2004). Measurement of the $\beta - \nu$ correlation using magneto–optically trapped ^{21}Na. *Phys. Rev. Lett.* **93**. 102501.

Shah, M.H. (2006). "Model-independent measurement of the excited fraction in a magneto–optical trap (MOT)". PhD thesis. Kansas State University.

Shah, M.H., Camp, H.A., Trachy, M.L., Fléchard, X., Gearba, M.A., Nguyen, H., Brédy, R., DePaola, B.D. (2005). Relative charge transfer cross sections from Rb(4d). *Phys. Rev. A* **72**. 024701.

Shah, M.H., Trachy, M.L., Veshapidze, G., Gearba, M.A., DePaola, B.D. (2007). Model-independent measurement of the excited fraction in a magneto–optical trap. *Phys. Rev. A* **75**. 053418.

Süptitz, W., Duncan, B.C., Gould, P.L. (1997). Efficient 5D excitation of trapped Rb atoms using pulses of diode-laser light in the counterintuitive order. *J. Opt. Soc. Am. B* **14**, 1001–1008.

Turkstra, J.W., Hoekstra, R., Knoop, S., Meyer, D., Morgenstern, R., Olson, R.E. (2001). Recoil momentum spectroscopy of highly charged ion collisions on magneto–optically trapped Na. *Phys. Rev. Lett.* **87**. 123202.

Ullrich, J., Moshammer, R., Dorn, A., Dörner, R., Schmidt, L.P.H., Schmidt-Böcking, H. (2003). Recoil-ion and electron momentum spectroscopy: Reaction microscopes. *Rep. Prog. Phys.* **66**, 1463–1545.

Weiner, J. (2003). "Cold and Ultracold Collisions in Quantum Microscopic and Mesoscopic Systems". Cambridge University Press.

Wiley, W.C., McLaren, I.H. (1955). Time-of-flight mass spectrometer with improved resolution. *Rev. Sci. Instrum.* **26**, 1150–1157.

Williamson, R.S., Voytas, P.A., Newell, R.T., Walker, T. (1998). A magneto–optical trap loaded from a pyramidal funnel. *Opt. Exp.* **3**, 1177–1179.

Wolf, S., Helm, H. (1997). Ion-recoil energy measurement in photoionization of laser-cooled rubidium. *Phys. Rev. A* **56**. 4385.

Wolf, S., Helm, H. (2000). Ion-recoil momentum spectroscopy in a laser-cooled atomic sample. *Phys. Rev. A* **62**. 043408.

Zapukhlyak, M., Kirchner, T., Lüdde, H.J., Knoop, S., Morgenstern, R., Hoekstra, R. (2005). Inner and outer shell electron dynamics in proton collisions with sodium atoms. *J. Phys. B: At. Mol. Opt. Phys.* **38**, 2353–2369.

ALL-ORDER METHODS FOR RELATIVISTIC ATOMIC STRUCTURE CALCULATIONS

M.S. SAFRONOVA[1,*] and W.R. JOHNSON[2,**]

[1] Department of Physics and Astronomy, 223 Sharp Lab, University of Delaware, Newark, Delaware 19716, USA
[2] Department of Physics, University of Notre Dame, Notre Dame, IN 46556, USA

1. Introduction and Overview 192
2. Relativistic Many-Body Perturbation Theory 194
3. Relativistic SD All-Order Method 198
   3.1. Calculation of Transition Matrix Elements 201
   3.2. Calculation of Hyperfine Constants and the SDpT Method 203
4. Motivation for Further Development of the All-Order Method 206
   4.1. Limitations of SD and SDpT Methods 206
   4.2. Estimates of Theoretical Uncertainties 207
   4.3. Availability of New Computational Resources 208
   4.4. Benchmark Comparisons of Theory and Experiments 208
5. Recent Developments in the Calculations of Monovalent Systems: Non-Linear Terms and Triple Excitations 209
6. Many-Particle Systems 213
   6.1. MBPT for Atoms with Two Valence Electrons 213
   6.2. Mixed Configuration Interaction—MBPT Calculations 214
   6.3. All-Order Calculations 217
7. Applications of High-Precision Calculations 219
   7.1. Parity Nonconservation in Heavy Atoms 219
   7.2. Polarizability Calculations and Their Applications. Blackbody Radiation Shift 222
   7.3. C_3 and C_6 Coefficients 225
   7.4. Isotope Shift 227
8. Conclusion 230
9. Acknowledgements 231
10. References 231

Abstract

All-order extensions of relativistic atomic many-body perturbation theory are described and applied to predict properties of heavy atoms. Limitations of relativistic

* E-mail: msafrono@udel.edu.
** E-mail: johnson@nd.edu.

many-body perturbation theory are first discussed and the need for all-order calculations is established. An account is then given of relativistic all-order calculations based on a linearized version of the coupled-cluster expansion. This account is followed by a review of applications to energies, transition matrix elements, and hyperfine constants. The need for extensions of the linearized coupled-cluster method is discussed in light of accuracy limits, the availability of new computational resources, and precise modern experiments. For monovalent atoms, calculations that include non-linear terms and triple excitations in the coupled-cluster expansion are described. For divalent atoms, results from second- and third-order perturbation theory calculations are given, along with results from configuration-interaction calculations and mixed configuration interaction–many-body perturbation theory calculations. Finally, applications of all-order methods to atomic parity nonconservation, polarizabilities, C_3 and C_6 coefficients, and isotope shifts are given.

1. Introduction and Overview

The nonperturbative treatment of relativity in atomic many-body calculations can be traced back to the formulation of relativistic self-consistent field (SCF) equations with exchange by Swirles (1935). The SCF equations, also referred to as Dirac–Hartree–Fock (DHF) equations, are based on a many-electron Hamiltonian in which the electron kinetic and rest energies are from the Dirac equation and the electron–electron interaction is approximated by the Coulomb potential. Numerical solutions of the DHF equations without exchange were obtained during the years 1940–1960 by Williams Jr. (1940), Mayers (1957), and Cohen (1960). The formulation of relativistic SCF theory by Swirles was reexamined by Grant (1961) and the DHF equations were brought into a compact and easily used form. Numerical solutions to the DHF equations with exchange were published by Coulthard (1967), Kim (1967), and Smith and Johnson (1967). The Breit interaction was included in the latter two calculations (Kim, 1967; Smith and Johnson, 1967). Desclaux (1973) published complete DHF studies of atoms with $Z = 1$–120 and Mann and Waber (1973) published DHF studies of the lanthanides, including effects of the Breit interaction. The DHF equations remain as the starting point for relativistic many-body studies of atoms; versatile multi-configuration DHF codes are available publically, notably the codes of Desclaux (1975) and Grant et al. (1980).

Extensions of the DHF approximation have been developed over the past three decades, driven by advances in several areas of experimental atomic physics. Of particular importance in this regard are the precise measurements of energy levels and transition moments for highly-charged ions produced in beam–foil experiments, electron beam ion trap (EBIT) experiments, tokamak plasmas, and astrophysical plasmas (Beiersdorfer, 2003). These measurements have reached such a high level of precision that it has become possible to detect two-loop Lamb-shift corrections to levels in lithium-like U (Beiersdorfer et al., 2005), putting very tight constraints on the accuracy of the underlying atomic structure calculations. An equally important motivating factor in the development of extensions of the DHF approximation are measurements of parity nonconserving (PNC) amplitudes in heavy atoms, especially those designed to test the standard model of the electroweak interaction and to set limits on its possible extensions (Johnson, 2003). For the case of cesium, measurements of PNC amplitudes have reached an accuracy of 0.4% (Wood et al., 1997). To make meaningful tests of the standard model, calculations of the amplitudes must be carried out for heavy neutral atoms to a similar level of accuracy.

One systematic extension of the DHF approximation is relativistic many-body perturbation theory (MBPT). Relativistic MBPT studies of atomic structure start from a lowest-order approximation in which the electron–electron interaction is the "frozen core" DHF potential and include an order-by-order perturbation expansion (in powers of the residual interaction) of energies and wave functions. Relativistic MBPT was used to predict properties of alkali-metal atoms from Li to Cs in Johnson et al. (1987), where energy levels for the ground state and the first few excited states were calculated to second order. In Johnson et al. (1987), electric-dipole matrix elements for the principal transitions and hyperfine constants were calculated through second order and included dominant third-order corrections. Although accurate values for energies, transition matrix elements, and hyperfine constants were obtained for Li, results for heavier alkali-metal atoms were significantly less accurate. The ground-state energy for Cs was accurate to 1.5%, while the Cs transition and hyperfine matrix elements were accurate to about 5% as determined by comparisons with precise experimental data. Later, complete third-order calculations of electric-dipole matrix elements, including all third-order terms were carried out in Johnson et al. (1996) for alkali-metal atoms and for Li-like and Na-like ions. The agreement with available experiments was very good for lighter atoms (within experimental precision for Li and Na), but decreased significantly for Cs and Fr.

To achieve the accuracy required for tests of the standard model in heavy atoms, it is imperative to include contributions beyond third order in MBPT. Although extensions to fourth order represent one possibility, the resulting calculations are formidable; for each first-order matrix element there are four terms in second order, 60 terms in third order, and 3072 terms in fourth order (Cannon and Derevianko, 2004). Owing to this very rapid increase in computational effort with MBPT order, one seeks alternatives to MBPT beyond third order.

One such alternative is the coupled-cluster singles–doubles (CCSD) method in which single and double excitations of the DHF ground state are included to all orders of perturbation theory. A nonrelativistic version of this method was used to calculate precise values of energies and hyperfine constants of $2s$ and $2p$ states of Li by Lindgren (1985). A linearized, but relativistic, version of the coupled-cluster method was later used to obtain energy levels, fine-structure intervals, and dipole matrix elements in Li and Be$^+$ in Blundell et al. (1989). These all-order calculations substantially improved the accuracy of energies and matrix elements compared to older MBPT results (Johnson et al., 1987). A nonrelativistic CCSD calculation for Na was reported in Salomonson and Ynnerman (1991), where energies and hyperfine constants of $3s$ and $3p$ states and the $3s - 3p$ electric-dipole matrix elements were calculated. Partial contributions to the $3s$ energy and hyperfine constant from triple excitations were also included in Salomonson and Ynnerman (1991); the resulting $3s$ energy was accurate to 0.01% and the $3s$ hyperfine constant to 0.2%. A relativistic version of the CCSD method was applied to calculate energy levels of alkali-metal atoms in Eliav et al. (1994) and excellent agreement with experiment was found. A linearized version of the coupled-cluster formalism, including single, double, and partial triple excitations (SDpT) was used to determine atomic properties of Cs in Blundell et al. (1991), where removal energies agreed with experiment to 0.5% and matrix elements agreed with measurements to better than 1%. Properties of Na-like ions ($Z = 11$–16), such as energies, transition matrix elements, and hyperfine constants were studied using the linearized CCSD method in Safronova et al. (1998), and similar studies of alkali-metal atoms including polarizabilities were reported in M.S. Safronova et al. (1999).

Although we concentrate on relativistic all-order coupled-cluster methods in this review, it should be noted that perturbation theory in the screened Coulomb interaction (PTSI) developed by Dzuba et al. (1989a, 1989b), in which important classes of MBPT corrections are summed to all orders, is an alternative method that has been successfully applied to atomic structure calculations for heavy neutral atoms. Moreover, for atoms with more than one valence electron, relativistic configuration-interaction (CI) calculations in an effective Hamiltonian extracted from the linearized SD theory, which has been developed and applied to small systems by Kozlov (2004), is a promising alternative to CCSD methods for large systems.

2. Relativistic Many-Body Perturbation Theory

In the simplest picture of a relativistic many-electron atom, each electron moves independently in a central potential $U(r)$ produced by the remaining electrons. The one-electron orbitals $\phi_a(r)$ describing the motion of an electron with quantum numbers $a = (n_a, \kappa_a, m_a)$ satisfy the one-electron Dirac equation

$$h(r)\phi_a(r) = \epsilon_a \phi_a(r), \tag{1}$$

where

$$h(r) = c\alpha \cdot p + \beta mc^2 - \frac{Z}{r} + U(r). \tag{2}$$

The quantities α and β in Eq. (2) are 4×4 Dirac matrices. The Dirac eigenvalues ϵ_a range through values: $\epsilon_a > mc^2$ for electron scattering states, $mc^2 > \epsilon_a > 0$ for electron bound states, and $-mc^2 > \epsilon_a$ for positron states.

The point of departure for our discussions of many-electron atoms is the *no-pair* Hamiltonian obtained from QED by Brown and Ravenhall (1951) and illuminated in Mittleman (1971, 1972, 1981), Sucher (1980). In this Hamiltonian, the electron kinetic and rest energies are from the Dirac equation and the potential energy is the sum of Coulomb and Breit interactions. Contributions from negative-energy (positron) states are projected out of this Hamiltonian. The no-pair Hamiltonian can be written in second-quantized form as $H = H_0 + V$, where

$$H_0 = \sum_i \epsilon_i [a_i^\dagger a_i], \tag{3}$$

$$V = \frac{1}{2} \sum_{ijkl} (g_{ijkl} + b_{ijkl}) [a_i^\dagger a_j^\dagger a_l a_k]$$
$$+ \sum_{ij} (V_{\text{HF}} + B_{\text{HF}} - U)_{ij} [a_i^\dagger a_j] + \frac{1}{2} \sum_a (V_{\text{HF}} + B_{\text{HF}} - 2U)_{aa}. \tag{4}$$

In Eqs. (3)–(4), $a_i^\dagger$ and a_i are creation and annihilation operators for an electron state i, and the summation indices range over electron bound and scattering states only, since, as mentioned above, contributions from negative energy states are absent in the no-pair Hamiltonian. Products of operators enclosed in brackets, such as $[a_i^\dagger a_j^\dagger a_l a_k]$, designate normal products with respect to a closed core. The summation index a in the last term in (4) ranges over states in the closed core. The quantity ϵ_i in Eq. (3) is the eigenvalue of the Dirac equation (1). The quantities g_{ijkl} and b_{ijkl} in Eq. (4) are two-electron Coulomb and Breit matrix elements, respectively

$$g_{ijkl} = \left\langle ij \left| \frac{1}{r_{12}} \right| kl \right\rangle, \tag{5}$$

$$b_{ijkl} = -\left\langle ij \left| \frac{\alpha_1 \cdot \alpha_2 + (\alpha_1 \cdot \hat{r}_{12})(\alpha_2 \cdot \hat{r}_{12})}{2r_{12}} \right| kl \right\rangle. \tag{6}$$

In Eq. (4), the core DHF potential is designated by V_{HF} and its Breit counterpart is designated by B_{HF}; thus,

$$(V_{\text{HF}})_{ij} = \sum_b [g_{ibjb} - g_{ibbj}], \tag{7}$$

$$(B_{\text{HF}})_{ij} = \sum_b [b_{ibjb} - b_{ibbj}], \tag{8}$$

where b ranges over core states. For neutral atoms, the Breit interaction is often a small perturbation that can be ignored compared to the Coulomb interaction. In such cases, it is particularly convenient to choose the starting potential $U(r)$ to be the core DHF potential $U = V_{\text{HF}}$, since with this choice, the second term in Eq. (4) vanishes. The third term in (4) is, of course, a c-number and provides an additive constant to the energy of the atom.

It should be noted that, although the no-pair Hamiltonian is a useful starting point for relativistic many-body calculations, certain small contributions to wave functions and energies, including frequency-dependent corrections to the Breit interaction, self-energy and vacuum-polarization corrections, and corrections from crossed-ladder diagrams, are omitted in this approach. Perturbation theory based directly on the Furry representation of QED includes all such omitted effects (Sapirstein, 1998). In calculations based on the no-pair Hamiltonian, contributions from these omitted terms are usually estimated and added as an afterthought. Recently, however, an energy-dependent formulation of MBPT that includes QED corrections completely has been developed by Lindgren et al. (2006) and applied to helium-like ions.

Let us return to MBPT and concentrate on the simplest atoms, those with a single valence electron. For monovalent atoms, we write the lowest-order state vector as

$$\left|\Psi_v^{(0)}\right\rangle = a_v^\dagger |0_c\rangle, \tag{9}$$

where $|0_c\rangle = a_a^\dagger a_b^\dagger \cdots a_n^\dagger |0\rangle$ is the state vector for the closed core, $|0\rangle$ being the vacuum state vector and $a_v^\dagger$ being a valence-state creation operator. If we ignore the Breit interaction and start our calculation using DHF wave functions for one-electron states ($U = V_{\text{HF}}$), then the lowest-order energy of the atom, obtained from $H_0 \Psi_v^{(0)} = E^{(0)} \Psi_v^{(0)}$, is

$$E^{(0)} = \epsilon_v + \sum_a \epsilon_a, \tag{10}$$

and the first-order energy is

$$E^{(1)} = \left\langle \Psi_v^{(0)} \middle| V \middle| \Psi_v^{(0)} \right\rangle = -\frac{1}{2} \sum_a (V_{\text{HF}})_{aa}. \tag{11}$$

We see that through first order, the energy separates into a core contribution and a valence contribution, with

$$E^{(0+1)}_{\text{core}} = \sum_a \epsilon_a - \frac{1}{2}\sum_a (V_{\text{HF}})_{aa}$$
$$= \sum_a (h_0)_{aa} + \frac{1}{2}\sum_{ab}(g_{abab} - g_{abba}), \quad (12)$$

$$E^{(0+1)}_v = \epsilon_v. \quad (13)$$

The summation indices a and b in Eqs. (11) and (12) range over core states. The quantity $(h_0)_{aa}$ is the matrix element in state a of the sum of the kinetic energy and nuclear potential terms in the Dirac Hamiltonian (2). The sum of zeroth- plus first-order energies in (12) is precisely the DHF energy of the core.

The energy of a one-electron atom splits order-by-order into core and valence contributions

$$E^{(k)} = E^{(k)}_{\text{core}} + E^{(k)}_v.$$

Since the core contribution is the same for each valence state, it is sufficient to consider valence contributions only when studying excitation or ionization energies of one-electron atoms using MBPT. The second-order contribution to the valence energy is found to be (Blundell et al., 1987)

$$E^{(2)}_v = \sum_{nab} \frac{\tilde{g}_{abvn} g_{vnab}}{\epsilon_v + \epsilon_n - \epsilon_a - \epsilon_b} - \sum_{mnb} \frac{\tilde{g}_{vbmn} g_{mnvb}}{\epsilon_m + \epsilon_n - \epsilon_v - \epsilon_b}. \quad (14)$$

Here and in the following sections, we adopt the convention that letters near the start of the alphabet $(a, b, c, \ldots)$ designate core states, letters in the middle of the alphabet $(m, n, o, \ldots)$ designate virtual states, and letters near the end of the alphabet $(v, w, x, \ldots)$ designate valence states. We let the letters $(i, j, k, \ldots)$ designate either core or virtual (general) states. In Eq. (14), we have also used the notation $\tilde{g}_{ijkl} = g_{ijkl} - g_{ijlk}$ to designate anti-symmetrized two-particle matrix elements. The much longer expression for the third-order contribution to the valence energy for a monovalent atom $E^{(3)}_v$ is given in Blundell et al. (1987) and is not repeated here.

To evaluate the expressions for second- and third-order energies, we first sum over magnetic quantum numbers analytically to obtain expressions involving radial Dirac wave functions and angular momentum coupling coefficients, then we sum over the remaining principal and angular quantum numbers numerically. To aid in the numerical work, we replace the spectrum of the radial Dirac equation, which consists of bound states, a positive-energy continuum of scattering states, and a negative-energy continuum of positron states, by a finite pseudospectrum. For the calculations discussed in this review, the pseudospectrum was constructed from B-splines confined to a large but finite cavity, as described in Johnson et al. (1988b).

In Table I, we give a breakdown of the zeroth- or lowest-order (recall that there is no first-order valence contribution to the energy), second-order, and third-order MBPT contributions to ionization energies of alkali-metal atoms and compare the sum with various all-order calculations and with experiment. The row labeled CC refers to a coupled-cluster calculation (Salomonson and Ynnerman, 1991) that includes some contributions beyond the single and double states. Differences between third-order MBPT calculations and experiment range from fractions of 1% for Li and Na to about 3% for Cs. Moreover, for Cs, including third-order corrections actually worsens the agreement with measured energies found in second order, emphasizing the need for all-order methods.

3. Relativistic SD All-Order Method

As an introduction to relativistic all-order calculations, we briefly describe the relativistic singles–doubles (SD) method, a linearized version of coupled-cluster theory; a more detailed description can be found in Blundell et al. (1989), Safronova et al. (1998). In the coupled-cluster theory, the exact many-body wave function is represented in the form (Coester and Kümmel, 1960)

$$|\Psi\rangle = \exp(S)|\Psi^{(0)}\rangle, \tag{15}$$

where $|\Psi^{(0)}\rangle$ is the lowest-order atomic state vector. The operator S for an N-electron atom consists of "cluster" contributions from one-electron, two-electron, ..., N-electron excitations of the lowest-order state vector $|\Psi^{(0)}\rangle$:

$$S = S_1 + S_2 + \cdots + S_N. \tag{16}$$

The exponential in Eq. (15), when expanded in terms of the n-body excitations S_n, becomes

$$|\Psi\rangle = \left\{1 + S_1 + S_2 + S_3 + \frac{1}{2}S_1^2 + S_1 S_2 + \frac{1}{2}S_2^2 + \cdots\right\}|\Psi^{(0)}\rangle. \tag{17}$$

In the linearized coupled-cluster method, all non-linear terms are omitted and the wave function takes the form

$$|\Psi\rangle = \{1 + S_1 + S_2 + S_3 + \cdots + S_N\}|\Psi^{(0)}\rangle. \tag{18}$$

The SD method is the linearized coupled-cluster method restricted to single and double excitations only. The all-order singles–doubles–partial triples (SDpT) method is an extension of the SD method in which the dominant part of S_3 is treated perturbatively. A detailed description of the SDpT method is given in

Blundell et al. (1991), M.S. Safronova et al. (1999). Inclusion of the non-linear terms in the relativistic SD formalism and a more complete treatment of the triple excitations is given in Porsev and Derevianko (2006), Pal et al. (2007) and will be considered later.

Restricting the sum in Eq. (18) to single and double excitations yields the following expansion for the SD state vector of a monovalent atom in state v:

$$|\Psi_v\rangle = \left[1 + \sum_{ma} \rho_{ma} a_m^\dagger a_a + \frac{1}{2} \sum_{mnab} \rho_{mnab} a_m^\dagger a_n^\dagger a_b a_a \right.$$

$$\left. + \sum_{m \neq v} \rho_{mv} a_m^\dagger a_v + \sum_{mna} \rho_{mnva} a_m^\dagger a_n^\dagger a_a a_v \right] |\Psi_v^{(0)}\rangle, \quad (19)$$

where $|\Psi_v^{(0)}\rangle$ is the lowest-order atomic state vector given in Eq. (9). In Eq. (19), the indices m and n range over all possible virtual states while indices a and b range over all occupied core states. The quantities ρ_{ma}, ρ_{mv} are single-excitation coefficients for core and valence electrons and ρ_{mnab} and ρ_{mnva} are double-excitation coefficients for core and valence electrons, respectively. It should be noted that the operator products in Eq. (19) are normally ordered as they stand.

To derive equations for the excitation coefficients, the state vector $|\Psi_v\rangle$ is substituted into the many-body Schrödinger equation $H|\Psi_v\rangle = E|\Psi_v\rangle$, and terms on the left- and right-hand sides are matched, based on the number and type of operators they contain, leading to the following equations for the single and double valence excitation coefficients:

$$(\epsilon_v - \epsilon_m + \delta E_v)\rho_{mv} = \sum_{bn} \tilde{g}_{mbvn} \rho_{nb} + \sum_{bnr} g_{mbnr} \tilde{\rho}_{nrvb}$$
$$- \sum_{bcn} g_{bcvn} \tilde{\rho}_{mnbc}, \quad (20)$$

$$(\epsilon_{vb} - \epsilon_{mn} + \delta E_v)\rho_{mnvb} = g_{mnvb} + \sum_{cd} g_{cdvb} \rho_{mncd} + \sum_{rs} g_{mnrs} \rho_{rsvb}$$
$$+ \left[\sum_r g_{mnrb} \rho_{rv} - \sum_c g_{cnvb} \rho_{mc} + \sum_{rc} \tilde{g}_{cnrb} \tilde{\rho}_{mrvc} \right] + \left[\begin{array}{c} v \leftrightarrow b \\ m \leftrightarrow n \end{array} \right], \quad (21)$$

where $\delta E_v = E_v - \epsilon_v$, the correlation correction to the energy of the state v, is given in terms of the excitation coefficients by

$$\delta E_v = \sum_{ma} \tilde{g}_{vavm} \rho_{ma} + \sum_{mab} g_{abvm} \tilde{\rho}_{mvab} + \sum_{mna} g_{vbmn} \tilde{\rho}_{mnvb}. \quad (22)$$

In Eq. (21), we use the abbreviation $\epsilon_{ij} = \epsilon_i + \epsilon_j$, and in Eq. (22), we use the notation $\tilde{\rho}_{mnvb} = \rho_{mnvb} - \rho_{nmvb}$. Equations for core excitation coefficients ρ_{ma}

and ρ_{mnab} are obtained from the above equations by removing δE_v from the left-hand side of the equations and replacing the valence index v by a core index a. The core correlation energy is given by

$$\delta E_c = \frac{1}{2} \sum_{mnab} g_{abmn} \tilde{\rho}_{mnab}. \tag{23}$$

After removing the dependence on magnetic quantum numbers, Eqs. (20) and (21) are solved iteratively. To this end, states a, b, m, and n are represented in a finite B-spline basis, identical to that used in the MBPT calculations discussed in Section 2. As a first step, equations for the core single- and double-excitation coefficients ρ_{ma} and ρ_{mnab} are solved iteratively; the core excitation coefficients are stored after the core correlation energy has converged to a specified accuracy. Thus, the calculation of the core excitation coefficients, which is the most demanding in terms of the computational time, is done only once for each atom. As a next step, equations for the valence single- and double-excitation coefficients ρ_{mv} and ρ_{mnva} are iterated using the previously stored core excitation coefficients. The resulting values are stored for each valence electron, ready to be used for the calculation of the matrix elements.

While the valence correlation energy δE_v includes important higher-order terms in the MBPT expansion, including all second-order corrections, those contributions to the third-order energy associated with triple excitations of the DHF ground state are missing. This difficulty was remedied by identifying the missing terms, calculating them separately, and adding them to the SD values (Safronova

Table I
Comparison of the Removal Energies of Li, Na, and Cs Calculated in Different Approximations with Experiment in cm^{-1}

	Li (2s)		Na (3s)		Cs (6s)	
	E_{th}	Δ	E_{th}	Δ	E_{th}	Δ
Lowest-order	43087	400	39952	1497	27954	3453
Second-order	43449	38	41245	204	31865	−458
Third-order	43476	11	41325	124	30529	878
All-order[a]	43492	−5	41447	2	31262	145
CCSD[b]	43483	4	41352	97	31443	36
CC[c]			41452	3		
Expt.	43487		41449		31407	

$\Delta = E_{\text{expt}} - E_{\text{th}}$. The experimental values are from Moore (1971).
[a]Blundell et al. (1989), M.S. Safronova et al. (1999).
[b]Eliav et al. (1994).
[c]Salomonson and Ynnerman (1991).

et al., 1998). Such extension of the SD method led to accurate energies for the s and p states of the alkali-metal atoms and ions with one valence electron (M.S. Safronova et al., 1998, 1999). As an illustration, in Table I we compare zeroth-, second-, and third-order MBPT calculations, and all-order energies for Li, Na, and Cs with other all-order calculations and with experiment. The all-order results are from SD calculations and include the missing third-order terms perturbatively. For Cs, the difference between theory and measurement is reduced from 3% in third order to 0.5% in all-order calculations.

3.1. Calculation of Transition Matrix Elements

The perturbation expansion for state vectors leads immediately to a perturbation expansion for matrix elements. Thus, for the one-particle operator,

$$Z = \sum_{ij} z_{ij} a_i^\dagger a_j, \tag{24}$$

perturbation theory leads to an expansion

$$\langle \Psi_w | Z | \Psi_v \rangle = Z_{wv}^{(1)} + Z_{wv}^{(2)} + \cdots, \tag{25}$$

for the matrix element of Z between states v and w of an atom with one valence electron. One finds

$$Z_{wv}^{(1)} = z_{wv}, \tag{26}$$

$$Z_{wv}^{(2)} = \sum_{am} \frac{z_{am} \tilde{g}_{wmva}}{\epsilon_a - \epsilon_m - \omega} + \sum_{am} \frac{\tilde{g}_{wavm} z_{ma}}{\epsilon_a - \epsilon_m + \omega}, \tag{27}$$

where $\omega = \epsilon_w - \epsilon_v$.

The expression for the third-order matrix element $Z_{wv}^{(3)}$, which contains 60 terms as mentioned earlier, is given in Blundell et al. (1987), Johnson et al. (1996), Savukov and Johnson (2002b). These third-order terms can be grouped into random-phase approximation (RPA), Brueckner orbital (BO), structural radiation (SR), and normalization contributions. The first- and second-order matrix elements together with the third-order RPA contribution form the first three terms in the iterative expansion of the RPA matrix element Z_{vw}^{RPA}, discussed, for example, in Johnson et al. (1996). The RPA matrix element for dipole transitions has the property of being gauge independent; RPA matrix elements evaluated using the "length-form" z of the transition operator are identical to those evaluated using the "velocity-form" v_z/ω. The third-order BO contributions are those associated with the second-order energy. It follows from Eq. (14) that the second-order energy is the diagonal matrix element of the second-order self-energy operator $\Sigma^{(2)}(\epsilon)$:

$$E_v^{(2)} = \left[\Sigma^{(2)}(\epsilon_v) \right]_{vv}, \tag{28}$$

where

$$\left[\Sigma^{(2)}(\epsilon)\right]_{ij} = \sum_{nab} \frac{\tilde{g}_{abjn}g_{inab}}{\epsilon + \epsilon_n - \epsilon_a - \epsilon_b} - \sum_{mnb} \frac{\tilde{g}_{ibmn}g_{mnjb}}{\epsilon_m + \epsilon_n - \epsilon - \epsilon_b}. \quad (29)$$

One can include the self-energy operator along with the DHF potential in the one electron Dirac equation,

$$\left[h_0 + V_{\text{HF}} + \Sigma(\epsilon)\right]\psi(r) = \epsilon\psi(r). \quad (30)$$

Solutions to this equation are referred to as Brueckner orbitals. The BO contributions to third-order matrix elements are obtained by replacing each valence orbital in the first-order matrix element by the corresponding second-order Brueckner orbital; the resulting BO terms often dominate the correlation corrections to matrix elements. The third-order SR contributions are those in which the dipole operator Z connects to internal lines in the self-energy operator Σ (Johnson et al., 1996). Finally, third-order normalization contributions consist of terms arising from normalization of the second-order state vector and to contributions from "backward diagrams" (Blundell et al., 1987).

In the all-order method, matrix element of the operator Z is given by

$$Z_{wv} = \frac{\langle \Psi_w | Z | \Psi_v \rangle}{\sqrt{\langle \Psi_v | \Psi_v \rangle \langle \Psi_w | \Psi_w \rangle}}, \quad (31)$$

where $|\Psi_v\rangle$ and $|\Psi_w\rangle$ are given by the expansion (19). The resulting expression for the numerator of Eq. (31) consists of the sum of the DHF matrix element z_{wv} and twenty other terms that are linear or quadratic functions of the excitation coefficients ρ_{mv}, ρ_{ma}, ρ_{mnva}, and ρ_{mnab} given by Eqs. (20) and (21) and their core counterparts. The complete expression for the matrix elements can be found in Blundell et al. (1989). The expression in Eq. (31) does not depend on the nature of the operator Z, only on its rank and parity. Therefore, electric and magnetic multipole transition matrix elements, magnetic-dipole, electric-quadrupole, and magnetic-octupole hyperfine matrix elements, and nuclear spin-dependent and spin-independent PNC matrix elements, are all calculated using the same general code. We note that the SD expression for the matrix elements contains third-order MBPT terms completely. The SD implementation of the all-order method described above yielded very accurate results for the principal electric-dipole transitions in alkali-metal atoms (M.S. Safronova et al., 1999), where precise experimental results are available for comparison.

In Table II, we present results of first-, second- and third-order MBPT calculations, and all-order SD calculations of reduced dipole matrix elements for $ns_{1/2} - np_{1/2}$ transitions in the alkali-metal atoms Na, Rb, Cs, and Fr. Contributions labeled "RPA 4+" are corrections of fourth and higher order in the RPA sequence. Those labeled "BO 4+" are semi-empirical corrections obtained by scaling the $\Sigma^{(2)}(\epsilon_v)$ to give the experimental energy for the state v. The final scaled

ALL-ORDER METHODS

Table II

MBPT and All-Order Calculations of the $np_{1/2} - ns_{1/2}$ Reduced Electric-Dipole Matrix Elements, Where $n = 3, 5, 6, 7$ for Na, Rb, Cs, and Fr, Respectively

Approximation	Reference	Na	Rb	Cs	Fr
First-order		3.6906	4.8189	5.2777	5.1437
Second-order	Johnson et al. (1996)	3.6521	4.5952	4.9433	4.7301
Third-order	Johnson et al. (1996)	3.5446	4.1986	4.4314	4.1969
Third-order RPA 4+	Johnson et al. (1996)	3.5433	4.1813	4.3868	4.1317
Third-order BO 4+	Johnson et al. (1996)	3.5271	4.2047	4.4550	4.2277
All-order SD	M.S. Safronova et al. (1999)	3.531	4.221	4.478	4.256
CC	Salomonson and Ynnerman (1991)	3.538			
MBPT (PTSI)	Dzuba et al. (1989b)			4.494	
Expt.		3.5246(23)	4.231(3)	4.4890(65)	4.277(8)

The experimental values for Na and Rb are from Volz and Schmoranzer (1996), for Cs from Rafac et al. (1999), and those for Fr are from Simsarian et al. (1998).

MBPT matrix elements differ from measured values by about 1%. Dipole matrix elements evaluated in the SD approximation are significantly more accurate, where differences with measurement range from 0.2 to 0.5%. For comparison, results from all-order PTSI calculations (Dzuba et al., 1989b) and all-order CCSD calculations (Salomonson and Ynnerman, 1991) are also shown in the table.

3.2. Calculation of Hyperfine Constants and the SDpT Method

Despite the success of the SD method in predicting accurate ns and np energies and transition matrix elements, results for the ground state hyperfine constants, especially for heavier systems, were found to be poor. This poor agreement with experiment can be traced to the omission of triple excitations in the SD all-order method. The correlation contribution to the ground state hyperfine constant is dominated by the single term in the numerator of Eq. (31):

$$Z^{(c)} = \sum_m z_{wm} \rho_{mv} + \sum_m z_{mv} \rho_{mw}^*, \qquad (32)$$

where z_{vm} are DHF matrix elements (26) and ρ_{mv} are single valence excitation coefficients (20).

Therefore, the equation for the valence single-excitation coefficients ρ_{mv} was modified to include the dominant part of the valence triple excitations perturbatively (Blundell et al., 1991; M.S. Safronova et al., 1999). The valence excitation term,

$$\frac{1}{6} \sum_{mnrab} \rho_{mnrvab} a_m^\dagger a_n^\dagger a_r^\dagger a_b a_a a_v |\Psi_v^{(0)}\rangle, \qquad (33)$$

was added to the wave function expansion (19). The resulting triple equation is (M.S. Safronova et al., 1999)

$$(\varepsilon_{abv} - \varepsilon_{mnr} + \delta E_v)\rho_{mnrvab}$$
$$= \frac{1}{2} \sum_{\substack{123=\{mnr\} \\ 1'2'3'=\{vab\}}} \left(-\sum_c g_{1c1'2'}\rho_{23c3'} + \sum_s g_{23s3'}\rho_{1s1'2'}\right) + [\text{triples}]. \quad (34)$$

Terms containing ρ_{mnrvab} or ρ_{mnrabc} are grouped together as [triples] on the right-hand side of this equation. In Eq. (34), the notation $123 = \{mnr\}$ designates symbolically that the indices 123 range over all six permutations of the indices mnr; even permutations contribute with a positive sign while odd permutations contribute with a negative sign. This equation was solved in the approximation that all [triples] terms on the right-hand side were omitted. The results were used to substitute triple contributions from ρ_{mnrvab} into the expression for the valence correlation energy δE_v and into ρ_{mv}; no modifications were made to the core equations. This procedure is described in detail in Blundell et al. (1991), M.S. Safronova et al. (1999). We refer to this extension of the SD method as the SDpT method. We note that even such a minimal inclusion of triple excitations significantly increased the number of terms in the equations for ρ_{mv}. The ground state hyperfine constant for Cs calculated using this method is accurate to 1%. The SDpT extension also automatically includes those third-order terms in the correlation energy omitted in the SD expansion. The SDpT method is also very successful for predicting the matrix elements in other cases where the term given by Eq. (32) is dominant, such as the $3d - 4s$ electric quadrupole matrix elements in Ca^+ (Kreuter et al., 2005) discussed below.

In Table III, we present first-, second-, and third-order corrections to the magnetic hyperfine constants of the $3s_{1/2}$ and $3p_{1/2,3/2}$ states of ^{23}Na and the $6s_{1/2}$ and $6p_{1/2,3/2}$ states of ^{133}Cs. The lowest-order hyperfine constant is just the DHF matrix element of the hyperfine operator and differs from experiment by 40–50%. The second-order terms listed in the table actually include all third- and higher-order RPA corrections and bring the theoretical values into slightly better agreement with experiments. The residual third-order terms are dominated by BO contributions for all states considered. Upon inclusion of third-order contributions, differences with experiment are found to be about 5% for the Cs $6s_{1/2}$ state. This relatively large difference provides evidence for the need to go beyond third order. The all-order SD calculation is in relatively good agreement with experiment for the $n = 3$ states of Na, but deviates from experiment by about 6% for the $6s_{1/2}$ state of Cs. As seen in the table, the difference with experiment is reduced to about 1% in the SDpT approximation. We also include comparisons with the CCSD calculations of Salomonson and Ynnerman (1991) and the SDpT calculations of Blundell et al. (1991), which is labeled "SDpT (dr)" in the table.

Table III
Comparison of the ^{23}Na and ^{133}Cs Hyperfine Constants A (MHz) Calculated in Different Approximations with Experiment

Term	Reference	Na			Cs		
		$3s$	$3p_{1/2}$	$3p_{3/2}$	$6s$	$6p_{1/2}$	$6p_{3/2}$
First-order	M.S. Safronova et al. (1999)	624	63.4	12.6	1424	161	23.9
Second-order		767	82.3	18.0	1716	202	42.8
Third-order		867	92.5	18.8	2404	291	51.2
All-order SD		888	95.0	18.8	2439	311	51.9
All-order SDpT	M.S. Safronova et al. (1999)	888	95.1	18.8	2279	290	48.5
All-order SDpT (dr)	Blundell et al. (1991)				2291	293	49.8
CC	Salomonson and Ynnerman (1991)	884	93.0	18.3			
Expt.		886	94.4(1)	18.53(2)	2298	291.89(8)	50.275(3)

The experimental values are from Happer (1974), Tanner and Wieman (1988a), Wijngaarden and Li (1994), Rafac and Tanner (1997).

In these later calculations, RPA corrections were included to all orders and the difference with experiment for the $6s_{1/2}$ state of Cs is reduced to about 0.3%.

As a second example, we show in Table IV a comparison of various calculations of lifetimes of the $3d_{3/2}$ and $3d_{5/2}$ states of Ca$^+$ (Kreuter et al., 2005), which decays by single-photon E2 emission to the $4s_{1/2}$ ground state. The DHF and third-order MBPT values differ significantly from experiment. As in the case of the $6s_{1/2}$ hyperfine constant discussed above, the results of an SD calculation are also in significant disagreement with experiment and, as for hyperfine constants, the term in Eq. (32) dominates the correlation correction. The SDpT approximation, as expected, is found to be in much better agreement with experiment. We also present scaled results SD$_{sc}$ obtained by multiplying the valence single-excitation coefficients ρ_{mv} by the ratio of the experimental to SD correla-

Table IV
Comparison of the $3d$ Ca$^+$ Lifetimes τ (µs) Calculated in Different Approximations with Experiment

	DHF	III	All-order				Theory	Expt.
			SD	SDpT	SD$_{sc}$	SDpT$_{sc}$	Final	
$\tau(3d_{3/2})$	790	1390	1257	1199	1196	1207	1196(11)	1176(11)
$\tau(3d_{5/2})$	771	1351	1224	1168	1165	1177	1165(11)	1168(9)

The lowest order and third-order values are listed in columns labeled "DHF" and "III", respectively. All theoretical and experimental values are from Kreuter et al. (2005).

tion energy. Also shown in the table, are scaled results with triples SDpT$_{sc}$, which are not much different from the scaled SD results, or from *ab initio* SDpT values. The theoretical predictions in this case are taken to be the scaled SD values and differences between the scaled results with and without triples are used to estimate the errors in theoretical values. The final all-order results agree with the experimental values within the corresponding uncertainties.

4. Motivation for Further Development of the All-Order Method

4.1. Limitations of SD and SDpT Methods

While the SDpT extension was successful in producing accurate results for Cs hyperfine constants, it has serious flaws. Whereas the SDpT method leads to satisfactory results for matrix elements in cases where the term in Eq. (32) is dominant, it leads to poor results when other terms, especially the term

$$Z^{(a)} = \sum_{ma} z_{am} \tilde{\rho}_{wmva} + \sum_{ma} z_{ma} \tilde{\rho}^*_{vmwa}, \tag{35}$$

are large or have an opposite sign to the term given in (32). Many atomic properties, including matrix elements for principle transitions in alkali-metal atoms, fall into this category. In fact, the SDpT method often produces poorer results than the original SD method in such cases, owing to cancellations between higher-order terms. Therefore, the usefulness of the SDpT extension is limited to the specific case where the term given by Eq. (32) is dominant.

Both SD and SDpT methods fail to produce accurate results when the correlation corrections are particularly large. For example, we list a comparison of the SD all-order results with the experimental values for the $nd_{3/2}$ and $nd_{5/2}$ hyperfine constants in Cs in Table V (Auzinsh et al., 2007). The lowest-order values are also presented to demonstrate the size of correlation corrections. Correlation contributes on the order of 50% to the hyperfine values for $nd_{3/2}$ states, and the difference between the theory and experiment is on the order of 7–10%. Disagreement for $nd_{5/2}$ states is even more severe, since correlation corrections are larger than the lowest-order values and contribute with an opposite sign. As a result, the difference between all-order results for the $5d_{5/2}$ hyperfine constant in Cs and experiment is 23%. No precise experimental values are available for other $nd_{5/2}$ hyperfine constants, but we still observe significant systematic discrepancies between all-order results and experimental values. The problem described above is not limited to hyperfine constants. The electric-dipole matrix elements for $np - nd$ transitions were also reported (Safronova et al., 2004) to have extremely large correlation corrections. In these cases, no experimental data are

Table V

Hyperfine Constants A (MHz) for $nd_{3/2}$ and $nd_{5/2}$ States in Cesium

State	DHF	All-order	Expt.	State	DHF	All-order	Expt.
$5d_{3/2}$	18.2	52.3	48.78(7)	$5d_{5/2}$	7.47	−16.4	−21.24(8)
$6d_{3/2}$	9.27	17.8	16.30(15)	$6d_{5/2}$	3.73	−3.89	−3.6(10)
$8d_{3/2}$	2.65	4.20	3.94(8)	$8d_{5/2}$	1.06	−0.684	−0.85(20)
$10d_{3/2}$	1.07	1.62	1.51(2)	$10d_{5/2}$	0.428	−0.238	−0.35(10)

The lowest-order and all-order values are compared with experiment. The experimental data are taken from Arimondo et al. (1977).

available for comparison. Neither is it possible to provide very accurate data with the SD all-order method in these cases, nor is it possible to assign recommended values for these transitions since the accuracy of the all-order method cannot be determined when correlation corrections are very large.

4.2. Estimates of Theoretical Uncertainties

Another important motivation for further development of the all-order method is a need for a mechanism to estimate uncertainties of theoretical data. Firstly, the evaluation of the theoretical uncertainty is necessary for the analysis of the PNC experiments. Secondly, recommended values that are produced with the all-order methods have more value when they contain uncertainty estimates. Uncertainty bounds of recommended values are important for use by both experimentalists and other theorists in their research and for making benchmark comparisons. Finally, accurate evaluation of the uncertainty of theoretical values may allow one to cross-check experimental values obtained by different methods and may aid in the evaluation of new experimental data. Accuracy estimates of theoretical data are particular important when significant discrepancies exist between results from different experiments.

Evaluation of theoretical uncertainties is a very difficult problem since it essentially involves evaluation of the quantity that it is not known beforehand. Several strategies can be used in evaluating the uncertainties of the all-order results, such as approximate evaluation of the size of the correlation correction, evaluation of the size of the higher-order corrections, study of the order-by-order convergence of perturbation theory, study of the breakdown of the various all-order contributions, identification of the most important terms, and semi-empirical determination of missing important contributions. In certain cases, a comparison with a large body of reliable and confirmed experiments may be used to aid the purely theoretical procedures listed above. A detailed knowledge of the more important correlation corrections is crucial in developing procedures for estimating uncertainties in theoretical data, especially in cases where one wishes to provide recommended values. Further development of the all-order method will provide additional information on contributions which presently cannot be estimated.

4.3. AVAILABILITY OF NEW COMPUTATIONAL RESOURCES

The development of the all-order method was limited, until recently, by insufficient readily available computational resources, both in speed and memory. For example, the triple valence excitations ρ_{mnrvab} involve three excited state indices and two core indices for each valence electron, and must be calculated iteratively for all possible combinations of m, n, r, a, and b. The B-spline basis sets that are employed in the all-order calculations described to this point utilize approximately 35 orbitals for each partial wave. Since the present method is intrinsically relativistic, each index m contains $35 \times (2 \times l_{max} + 1)$ orbitals, where l_{max} is the maximum number of the partial waves taken into account. Therefore, a number of the valence triple excitation coefficients for Cs, which has 17 core shells, with $l_{max} = 4$ can be roughly estimated to be $(35 \times 9)^3 \times (17)^2 = 9 \times 10^9$ coefficients for each angular momentum channel. (Three extra indices appear after angular reduction. We refer to a single combination of such indices as a channel.) As a result, the memory requirement for the triple excitations easily can exceed 100 GB using the above estimate. This number can be reduced significantly if the symmetry of the excitation coefficients is used, data types are defined as single precision, core shells are partially omitted, and so forth. [See Porsev and Derevianko (2006), Derevianko and Porsev (2006) for more detailed description of this issue.] Nevertheless, the memory requirements are still significant even after all simplifications are made. Note further that each simplification has to be carefully investigated for possible numerical errors. Another major problem is that the triple equations themselves are rather complicated and contain an additional sum over the excited states. The case of the core excitations ρ_{mnrabc} is even more complicated owing to the addition of the extra core index c beyond the already large number of valence excitations. Significant improvement of computational resources in recent years finally permits iteration of the triple equations (Porsev and Derevianko, 2006; Derevianko and Porsev, 2006).

4.4. BENCHMARK COMPARISONS OF THEORY AND EXPERIMENTS

Recent developments in experimental methodologies led to new high-precision measurements of various atomic properties (for example, Kreuter et al., 2005; Letchumanan et al., 2005; Sherman et al., 2005; Gomez et al., 2005). Experimental developments of atomic clocks and quantum information research with atoms and ions both stimulated the need for precise atomic data such as ac polarizabilities, quadrupole shifts, and blackbody radiation shifts, and have produced many new experimental results. Further development of the high-precision theoretical methodologies allows one to provide data needed for the analysis of specific experiments as well as provide data for benchmark comparisons and critical evaluation of the experimental data. One of the main problems of the all-order method is that it is currently developed only for monovalent systems while many interesting experiments are conducted for divalent systems to which the all-order method can be extended.

Accurate theoretical calculations accompanied by corresponding uncertainty estimates may be used to cross-check different types of experiments. For example, the experimental values of the Stark shift of the $6s-6p_{1/2}$ and $6s-6p_{3/2}$ lines and lifetimes of the $5d_j$ states in Cs were checked for consistency in Safronova andClark (2004). From the standpoint of experiment, these are unrelated data produced by entirely different techniques. From the standpoint of theory, both Stark shifts and lifetimes under consideration depend on the values of the $5d_j - 6p_{j'}$ matrix elements. In the case of the lifetimes, there are no other significant contributions since the $5d_j - 6s$ transitions are too weak. The dc Stark shifts, i.e. static atomic polarizabilities, can be calculated from the relevant values of matrix elements and energies using a sum-over-states approach (see Section 7.2). The $5d - 6p$ transition matrix elements dominate the $6p_j$ polarizabilities and the remaining small terms can be calculated accurately using the all-order matrix elements and experimental energies (Moore, 1971). Therefore, it is possible to cross-check these two experiments with minimal (and well-understood) theoretical input by deriving $5d - 6p$ matrix elements from lifetime data, substituting them into the polarizability calculations, and comparing the results with experimental Stark shift values. This procedure was used in Safronova andClark (2004) to demonstrate that the Stark shift and lifetime data differed by $3-4\sigma$.

It is also possible to use experimental dc and ac polarizability values to accurately determine specific matrix elements that dominant contributions to these polarizabilities, provided an accurate calculation of the remaining contributions can be carried out and its uncertainty is known. For example, the Stark-shift value for the $6s - 7s$ transition in Cs was used in M.S. Safronova et al. (1999) to derive the values of the $7p_{1/2} - 7s$ and $7p_{3/2} - 7s$ electric-dipole reduced matrix elements with 0.1% accuracy.

5. Recent Developments in the Calculations of Monovalent Systems: Non-Linear Terms and Triple Excitations

In the linearized SD coupled-cluster approach, the wave function given by Eq. (17) reduces to

$$|\Psi\rangle = \{1 + S_1 + S_2\}|\Psi^{(0)}\rangle, \tag{36}$$

where single and double excitations are separated into the core and valence parts:

$$S_1 = S_{1c} + S_{1v} = \sum_{ma} \rho_{ma} a_m^\dagger a_a + \sum_{m \neq v} \rho_{mv} a_m^\dagger a_v, \tag{37}$$

$$S_2 = S_{2c} + S_{2v} = \frac{1}{2} \sum_{mnab} \rho_{mnab} a_m^\dagger a_n^\dagger a_b a_a + \sum_{mnb} \rho_{mnvb} a_m^\dagger a_n^\dagger a_b a_v. \tag{38}$$

The complete SD coupled-cluster wave function contains six more non-linear terms:

$$|\Psi\rangle = \left\{ 1 + S_1 + S_2 + \frac{1}{2} S_1^2 + S_1 S_2 + \frac{1}{6} S_1^3 + \frac{1}{2} S_2^2 \right.$$
$$\left. + \frac{1}{2} S_1^2 S_2 + \frac{1}{24} S_1^4 \right\} |\Psi^{(0)}\rangle. \tag{39}$$

No other non-linear terms contribute to the SD equations given by Eqs. (20)–(22). Separating core and valence parts of the operators S_1 and S_2, we find that the non-linear terms contributing to the core single and double equations are $\frac{1}{2} S_{1c}^2$, $S_{1c} S_{2c}$, $\frac{1}{6} S_{1c}^3$, $\frac{1}{2} S_{2c}^2$, $\frac{1}{2} S_{1c}^2 S_{2c}$, and $\frac{1}{24} S_{1c}^4$, and the non-linear terms contributing to the valence single and double equations are $S_{1c} S_{1v}$, $\{S_{1v} S_{2c}, S_{1c} S_{2v}\}$, $\frac{1}{2} S_{1c}^2 S_{1v}$, $S_{2c} S_{2v}$, $\{S_{1v} S_{1c} S_{2c}, \frac{1}{2} S_{1c}^2 S_{2v}\}$, and $\frac{1}{6} S_{1c}^3 S_{1v}$. Quadratic valence non-linear (NL) terms were included in the calculation of atomic properties of Na and Cs in Porsev and Derevianko (2006), Derevianko and Porsev (2006), respectively. A complete treatment of all six non-linear terms is given in Pal et al. (2007), where the properties of alkali-metal atoms from Li to Cs are calculated using the complete coupled-cluster SD wave function given in Eq. (39), and where the relative importance of various non-linear terms is investigated.

The addition of the non-linear terms in the all-order wave function significantly complicates the all-order equations for the single and double valence excitation coefficients (20)–(21) and their core counterparts. For example, the single valence excitation coefficient equation has six additional terms:

$$(\epsilon_v - \epsilon_m + \delta E_v) \rho_{mv} = (\text{SD}) + \sum_{drs} \tilde{g}_{mdrs} \rho_{rv} \rho_{sd} - \sum_{cds} \tilde{g}_{cdvs} \rho_{mc} \rho_{sd}$$
$$- \sum_{cdrs} (\tilde{g}_{cdsr} \rho_{srvd} \rho_{mc} + \tilde{g}_{cdsr} \rho_{smcd} \rho_{rv}$$
$$- \tilde{g}_{cdrs} \tilde{\rho}_{mrvc} \rho_{sd} + \tilde{g}_{cdsr} \rho_{mc} \rho_{rd} \rho_{sv}), \tag{40}$$

where (SD) contains linearized SD terms given by Eq. (20). The equation for the core excitation coefficients ρ_{ma} is obtained from the valence equation above by replacing the index v by the index a. We note that only three NL terms in the wave function, $\frac{1}{2} S_1^2$, $S_1 S_2$, and $\frac{1}{6} S_1^3$ contribute to the single excitation coefficient equations. All six NL terms contribute to the equations for the double valence excitation coefficients ρ_{mnab} and ρ_{mnva}. The resulting double core equations must be symmetrized in order to preserve the symmetry relation $\rho_{mnab} = \rho_{nmba}$. Again, the core and valence double equations are identical with the replacement of the valence index v by a core index a. While there are six distinct NL terms contributing to the single equations, together with three linear single terms, there are 35 distinct NL terms contributing to the double equation together with the nine linear terms given by Eq. (21). We count direct + exchange terms as a single term. Therefore, the complete coupled-cluster calculation is far more complicated than the original linearized one. Complete SD coupled-cluster equations containing all terms are given in Pal et al. (2007). The complete SD core energy equation contains only one NL term:

$$\delta E_c = (\text{SD}) + \sum_{abmn} \frac{1}{2} \tilde{g}_{abmn} \rho_{ma} \rho_{nb}, \tag{41}$$

whereas the complete valence coupled-cluster SD energy equation contains six NL terms:

$$\delta E_v = (\text{SD}) - \sum_{cdt} \tilde{g}_{cdvt} \rho_{td} \rho_{vc} + \sum_{dtu} \tilde{g}_{vdtu} \rho_{tv} \rho_{ud}$$
$$- \sum_{cdtu} (\tilde{g}_{cdut} \rho_{vc} \rho_{utvd} + \tilde{g}_{cdut} \rho_{tv} \rho_{uvcd}$$
$$- \tilde{g}_{cdtu} \tilde{\rho}_{vtvc} \rho_{ud} + \tilde{g}_{cdut} \rho_{td} \rho_{uv} \rho_{vc}). \tag{42}$$

The non-linear terms give no additional contributions to the matrix element formula (31). However, the values of the matrix elements change when NL terms are added owing to modified values of the excitation coefficients. Contributions of the various non-linear terms to the removal energies and electric-dipole matrix elements are summarized in Table VI. We note that these are data from Pal et al. (2007). The result of the linearized coupled-cluster calculation is listed in Table VI in row labeled "SD". All calculations are carried out with $l_{\max} = 6$. The contribution of the core terms, listed in the next row, is obtained as the difference of the results obtained with NL terms added only into the core equations and SD values. In this case, the valence data change only because of the change in the core ρ_{ma} and ρ_{mnab} excitation coefficients. Each successive contribution is obtained by adding NL terms specified at the corresponding row of Table VI into the valence equations, redoing the valence iterations and subtracting the previous result from the new one. Firstly, only the S_2^2 term is added to the valence equation and the difference of this new calculation with the previous one is determined. Secondly, other quadratic terms (of S_1^2 and $S_1 S_2$ type) are added to the equations. Finally, all of the NL terms are added into the valence equations. While the contribution from S_2^2 was dominant, as expected, contributions from

Table VI
Contribution of the Non-Linear Terms to the ns Ground State Removal Energies and $np_{1/2} - ns$ Electric-Dipole Reduced Matrix Elements for Na and Cs

Contributions	Na δE_{3s}	Cs δE_{6s}	Na $3s - 3p_{1/2}$	Cs $6s - 6p_{1/2}$
SD	1483	3882	3.53099	4.48157
Core NL terms	1	44	0.00005	−0.01057
$S_{2c}S_{2v}$	−44	−224	0.00487	0.04585
$S_{1c}S_{1v}$, $\{S_{1v}S_{2c}, S_{1c}S_{2v}\}$	−24	−162	0.00211	0.02762
$\frac{1}{2}S_{1c}^2 S_{1v}$, $\{S_{1v}S_{1c}S_{2c}, \frac{1}{2}S_{1c}^2 S_{2v}\}$, $\frac{1}{6}S_{1c}^3 S_{1v}$	0	1	0.00000	−0.00015
Total	1416	3540	3.53802	4.54432

The SD values are calculated using the linearized SD all-order method. The energies are in cm^{-1} and the E1 matrix elements are in atomic units ($a_0 e$). Preliminary data from Pal et al. (2007).

other quadratic terms were also found to be significant and approximately half the size of the S_2^2 contribution. Contributions from cubic and fourth-order terms are negligible (Pal et al., 2007). The contribution of the NL core terms is small for Na but rather significant for Cs and cannot be neglected in accurate calculations. The total contribution from the NL terms is remarkably large, especially for the energies, where the SD result for the ground state correlation energy is modified by 10%. This finding resolves long-standing differences between linearized SD coupled-cluster results from M.S. Safronova et al. (1999) and full coupled-cluster calculation in Eliav et al. (1994). The results of Pal et al. (2007) are in reasonably good agreement with Eliav et al. (1994). The remaining small differences can be explained by the different number of partial waves used in the calculations (Pal et al., 2007).

As demonstrated in Porsev and Derevianko (2006), Derevianko and Porsev (2006), the non-linear terms are canceled to a large extent by certain triple contributions. Therefore, it is necessary to add both NL terms and complete valence triples to increase the accuracy of the all-order method. The addition of the triple contribution ρ_{mnrvab} from Eq. (33) is described in Porsev and Derevianko (2006), where iteration of the valence triple excitation coefficient equation is carried out and Na results are presented. The inclusion of the complete valence triples still omits certain core fourth-order contributions to the matrix elements. The complete fourth-order matrix element calculation is presented in Cannon and Derevianko (2004). The most complete Cs all-order calculation, that includes valence quadratic NL terms, valence triple contributions, and fourth-order core triples, is carried out in Derevianko and Porsev (2006). The resulting energies, electric-dipole matrix elements, and hyperfine constants agree with experimental values at the 0.1% level, with exception of the $6p_{1/2}$ hyperfine constant, which differs from the experimental value by 1%.

6. Many-Particle Systems

6.1. MBPT FOR ATOMS WITH TWO VALENCE ELECTRONS

For atoms and ions with two valence electrons, the basic single-particle orbitals are chosen to be solutions to the Dirac equation in the $N-2$ electron Hartree–Fock potential $V_{\text{HF}}^{(N-2)}$ of the closed core. Lowest-order wave functions are constructed from linear combinations of degenerate or nearly degenerate eigenstates of H_0. This collection of eigenstates is referred to as a model space. As an example, a model space appropriate to the description of the ground-state of Be is the collection of three nearly degenerate states $\Psi_1 = |2s_{1/2}2s_{1/2}\rangle$, $\Psi_2 = |2p_{1/2}2p_{1/2}\rangle$, and $\Psi_3 = |2p_{3/2}2p_{3/2}\rangle$, all coupled to give angular momentum 0.

Expanding the lowest-order wave function in terms of model-space wave functions leads to an eigenvalue problem for the first-order energy:

$$\sum_k \langle \Phi_l | H_0 + V | \Phi_k \rangle c_k = E_l^{(1)} c_l. \tag{43}$$

The first-order energy here corresponds to the sum of the zeroth- and first-order energy for systems with one valence electron; there is no well-defined counterpart of the zeroth-order energy in cases where the model space consists of non-degenerate states. The expansion coefficients are used to construct the lowest wave function associated with a given energy.

Second-order corrections to energies of two particle states have been worked out within the framework of relativistic MBPT by Safronova et al. (1996) and applied to various atomic systems in Johnson et al. (1997). More recently, third-order relativistic MBPT calculations for Be- and Mg-like ions have been carried out by Ho et al. (2006). These third-order calculations are complex, requiring the evaluation of 302 Bethe–Goldstone diagrams. Third-order calculations lead to energies accurate to better than 1% for neutral and near neutral Be-like and Mg-like ions, but (based on experience with monovalent atoms) are expected to be much less precise for heavier systems. As a specific example, we compare third-order calculations of energies of low-lying odd-parity states of Be-like C ($Z = 6$) from Ho et al. (2006) with experiment in Table VII. The row labeled $B^{(2)}$ is the contribution from the second-order Breit interaction.

For atoms with two electrons beyond a closed core, it is of considerable interest to examine correlation corrections to transition matrix elements. Two distinct effects can be distinguished: correlation corrections arising from interaction between the valence electrons and corrections arising from interactions between

Table VII
Comparison of Third-Order MBPT Calculations of Low-Lying ^3P States for Be-Like C ($Z=6$) with Experiment

Term	$(2s2p)\,^3P_0$	$(2s2p)\,^3P_1$	$(2s2p)\,^3P_2$
$E^{(0+1)}$	54204.5	54223.4	54272.9
$E^{(2)}$	−2344.0	−2342.2	−2338.4
$B^{(2)}$	−6.7	−3.3	0.4
$E^{(3)}$	412.9	412.7	412.3
E_{Lamb}	−8.4	−8.3	−8.2
E_{tot}	52258.4	52282.3	52339.1
E_{expt}	52367.1	52390.8	52447.1
$\Delta E\%$	−0.2	−0.2	−0.2

Units: cm^{-1}.

valence electrons and the atomic core. As discussed above, the lowest-order states are chosen as linear combinations of degenerate or nearly degenerate two electron states in a model space. These states are coupled to a specific angular momentum.

The first-order matrix element between an *uncoupled* final state $|xy\rangle$ and an *uncoupled* initial state $|vw\rangle$ is

$$Z^{(1)} = \langle xy|Z|vw\rangle = z_{xv}\delta_{yw} - z_{xw}\delta_{yv} + z_{yw}\delta_{xv} - z_{yv}\delta_{xw}, \quad (44)$$

and the corresponding second-order correction consists of RPA valence–core corrections and second-order valence–valence corrections. Derivative terms also occur in second-order accounting for the correlation correction to the transition energy. Gauge-independent second-order relativistic MBPT calculations of matrix elements for Be-like and Mg-like ions were carried out in U.I. Safronova et al. (1999, 2000), respectively.

6.2. Mixed Configuration Interaction—MBPT Calculations

Relativistic configuration-interaction (CI) calculations, based on the no-pair Hamiltonian provide an important alternative to relativistic coupled-cluster calculations, especially for atoms with two or more valence electrons. Such calculations have been carried out for low-lying states in He-like ions in Chen et al. (1993), Cheng and Chen (2000), for states in Be-like ions in Chen and Cheng (1997a), and for states in Mg-like ions in Chen and Cheng (1997b). Formulation of CI calculations is relatively simple. For example, in the case of a helium-like ion, one expands the two-electron wave function as a linear combination of all distinct

Table VIII
The First and Second Eigenvalues of the Hamiltonian Matrix for the $J = 0$ Even-Parity State of Helium are Tabulated with Respect to the Maximum Value l_{max} of l Included in the Basis

l_{max}	First	Second	$2\,^3S_0$
0	−0.879005	−0.144198	−0.734807
1	−0.900506	−0.145768	−0.754738
2	−0.902758	−0.145916	−0.756842
3	−0.903313	−0.145951	−0.757362
4	−0.903511	−0.145963	−0.757548
5	−0.903598	−0.145968	−0.757630
∞			−0.757702
RM			0.000104
CI			−0.757598
NIST			−0.757616

The difference, which is the energy of the $2\,^3S_0$ state, is extrapolated to $l_{max} = \infty$. Reduced-mass (RM) corrections are added and the resulting theoretical energy is compared with the NIST tabulation. The residual difference is accounted for primarily by the ground-state Lamb shift.

two-electron states of a given angular symmetry

$$\Psi_J = \sum_{k \leq l} c_{kl} \Phi_J(kl), \qquad (45)$$

where

$$\Phi_J(kl) = \eta_{kl} \sum_{m_k, m_l} \langle j_k m_k, j_l m_l | JM \rangle a_k^\dagger a_l^\dagger |0\rangle. \qquad (46)$$

In this equation, $\langle j_1 m_1, j_2 m_2 | JM \rangle$ is a Clebsch–Gordan coefficient and η_{kl} is a normalization factor. The expansion coefficients c_{kl} are obtained from the matrix eigenvalue equation:

$$\sum_{i \leq j} \left[(\epsilon_i + \epsilon_j) \delta_{ik} \delta_{jl} + \langle \Phi_J(kl) | V | \Phi_J(ij) \rangle \right] c_{ij} = E c_{kl}. \qquad (47)$$

Configuration-interaction calculations require substantial computing facilities even in the simplest cases. For example, in calculating odd-parity $J = 1$ states of helium-like ions, including orbitals with $l \leq 5$ using a basis having 40 basis functions for each single-particle state, one is led to matrices having dimension greater than 10,000. One seeks the lowest few eigenvalues of these large matrices. In Table VIII, the first and second eigenvalues for the $J = 0$ even-parity state of helium are tabulated with respect to the maximum value of l included in the basis. The difference, which is the energy of the $2\,^3S_0$ state, is extrapolated to ∞. Reduced-mass (RM) corrections are added and the resulting theoretical energy is compared with the NIST tabulation. The residual difference between experiment and theory seen in the table is accounted for primarily by the ground-state Lamb shift.

A variant of the CI method that has been applied successfully to study energies and transition amplitudes in two-electron atoms is the CI-MBPT method (Dzuba et al., 1996; Kozlov and Porsev, 1997; Dzuba and Johnson, 1998; Porsev et al., 2000; Kozlov et al., 2001a; Savukov and Johnson, 2002a). There are several versions of the CI-MBPT method that utilize an effective Hamiltonian derived from second-order MBPT. Here we concentrate on the version developed in Savukov and Johnson (2002a). For atoms with two valence electrons, second-order correlation corrections divide into one-particle and two-particle parts. The second-order one-particle contributions to the energy can be obtained, as explained earlier, from one-particle Brueckner orbitals. The second-order two-particle part consists of valence–valence (VV), valence–core (VC), and (numerically much smaller) core–core (CC) terms,

$$\mathrm{VV}_{xy,vw} = \sum_{mn} \frac{g_{xymn}\tilde{g}_{mnvw}}{\epsilon_v + \epsilon_w - \epsilon_m - \epsilon_n}, \tag{48}$$

$$\mathrm{VC}_{xy,vw} = \sum_{ma} \frac{\tilde{g}_{xmva}\tilde{g}_{yawm}}{\epsilon_v + \epsilon_a - \epsilon_x - \epsilon_m} \times \bigl[1 - P(vw)\bigr]\bigl[1 - P(xy)\bigr], \tag{49}$$

$$\mathrm{CC}_{xy,vw} = \sum_{ab} \frac{g_{xyab}\tilde{g}_{abvw}}{\epsilon_a + \epsilon_b - \epsilon_x - \epsilon_y}. \tag{50}$$

The operator $P(vw)$ in Eq. (49) indicates that the indices v and w in the sum to the left are to be interchanged.

In the CI-MBPT method, second-order one-particle corrections are included by replacing the HF orbitals and energies in the basis set by second-order Brueckner orbitals and energies. Two-particle VV correlation corrections are accounted for automatically in the CI calculation. The two-particle VC contributions are accounted for by modifying the Hamiltonian matrix to include the VC matrix given in Eq. (49). Core–core contributions are ignored. Thus, in CI-MBPT calculations for atoms with two valence electrons, we use second-order Brueckner orbitals instead of HF orbitals as basis functions and modify the interaction Hamiltonian to include corrections from the second-order VC interaction.

As an example, energies, hyperfine constants, and dipole matrix elements for the isotope ^{25}Mg, obtained from CI-MBPT calculations, are shown in Table IX. We see that the energies agree with measured values at the level of 0.1%. Theoretical hyperfine constants for the accurately measured $(3s3p)\,^3P$ states agree

Table IX
CI-MBPT Energies E_{CI} (cm^{-1}), Hyperfine Constants A_{CI} (MHz), and (Length, Velocity) Form Reduced Dipole Matrix Elements $D(L, V)$ (a.u.) for ^{25}Mg Obtained From CI-MBPT Calculations

State	E_{CI}	E_{exp}	ΔE	A_{CI}	A_{exp}	$D(L)$	$D(V)$
$(3s4s)\,^1S_0$	43451	43503	52				
$(3s5s)\,^1S_0$	52526	52556	23				
$(3s4s)\,^1S_1$	41148	41197	49	−324.4			
$(3s5s)\,^1S_1$	51837	51872	35	−309.29			
$(3s3p)\,^3P_0$	21812	21850	38			1.530	1.531
$(3s3p)\,^3P_1$	21830	21870	40	−143.83	−144.945(5)	2.652	2.654
$(3s3p)\,^3P_2$	21886	21911	25	−128.05	−128.440(5)	3.431	3.434
$(3s3p)\,^1P_1$	35056	35051	−5	−8.86	−7.7(5)	4.027	4.017
$(3s4p)\,^3P_0$	47800	47848	48			4.661	4.650
$(3s4p)\,^3P_1$	47804	47848	44	−152.58		8.072	8.054
$(3s4p)\,^3P_2$	47812	47851	39	−148.29		10.417	10.393
$(3s4p)\,^1P_1$	49314	49347	33	−3.04		0.845	0.841

The nuclear magnetic moment is $\mu = -0.85545\mu_N$ and the nuclear spin is $I = 5/2$. The initial states for the dipole matrix elements are listed on the left; the final state for transitions from triplet states is $(3s4s)\,^3S_1$ and from singlet states is $(3s)^2\,^1S_0$.

with experiment at the level of 0.3%. Length- and velocity-form reduced matrix elements agree with each other at a level of 0.5% or better. It should be mentioned that RPA corrections were included in the evaluation of matrix elements.

From the point of view of future development, a more interesting version of the CI-MBPT method has been discussed recently by Kozlov (2004) and applied to a "toy" atom. In this variant, core–core interactions and valence–core interactions are accounted for using the all-order SD approximation while the strong valence–valence interactions are accounted for by a CI calculation.

6.3. ALL-ORDER CALCULATIONS

All-order SD calculation of excitation energies, hyperfine constants, E1, E2, M1 transitions rates, and lifetimes of $6s^2nl$ states in Tl and Pb II were carried out in Safronova et al. (2005). While Tl and Pb II are systems with three valence electrons, the $6s^2nl$ states in those systems can be described within the framework of the all-order method for monovalent systems described in Section 3, [Nd]$5s^25p^65d^{10}6s^2$ being taken as the closed core. The resulting values were found to be in remarkably good agreement with experiment for most of the atomic properties considered. The values were also compared with theoretical results calculated using other methods, where data were available. Neutral thallium is of particular interest owing to studies of parity nonconservation. Third-order MBPT

Table X

Comparison of the Ground State Removal Energies of Ga, Ga-Like Ge ($Z = 32$), Ga-Like Se ($Z = 34$), and Tl Calculated Using Third-Order MBPT and All-Order SD Methods with Experiment in cm^{-1}

	Ga		Ga-like Ge		Ga-like Se		Tl	
	E_{th}	Δ	E_{th}	Δ	E_{th}	Δ	E_{th}	Δ
Lowest-order	43033	5355	121728	6793	338575	8000	43824	5440
Third-order	48687	−299	128738	−217	346928	−353	49205	59
All-order	48575	−187	128545	−34	346747	−172	49266	−2

$\Delta = E_{expt} - E_{th}$. The theoretical data are from U.I. Safronova et al. (2006b, 2005). The experimental values are from Moore (1971).

Table XI

Comparison of the SD All-Order Values of Lifetimes (ns) and Hyperfine Constants (MHz) in Tl with Experimental Values

	$\tau(7s)$	$\tau(7p_{1/2})$	$\tau(7p_{3/2})$	$A(6p_{1/2})$	$A(6p_{3/2})$	$A(7s)$
Theory	7.43	61.8	47.3	21390	353	12596
Experiment	7.45(2)	63.1(1.7)	48.6(1.3)	21311	265	12297(2)

The theoretical values are from Safronova et al. (2005). The experimental data are taken from the compilation carried out in the same work.

calculations were also carried out in Safronova et al. (2005), in an attempt to analyze the importance of the higher-order contributions. In M.S. Safronova et al. (2006b), the Stark-induced amplitude for the $6P_{1/2} - 7P_{1/2}$ transition and the Stark shifts in the $6P_{1/2} - 7P_{1/2}$ and $6P_{1/2} - 7S_{1/2}$ transitions in Tl I were calculated using the relativistic SD all-order method and again good agreement with experimental values was found. This calculation is discussed in Section 7.2. A comprehensive all-order study of atomic properties of the $4s^2nl$ states of neutral gallium and $4s^24p$ states of Ga-like ions was conducted in U.I. Safronova et al. (2006a, 2006b). A comparison of the third-order and all-order SD results for the ground state energies of Ga, Ga-like ions, and Tl with experiment is given in Table X. We note that third-order energies are also in remarkably good agreement with experiment, unlike the third-order values for heavy alkali-metal atoms (see Cs data in Table I). The examples of the SD all-order calculations of the Tl lifetimes and hyperfine constants (Safronova et al., 2005) are given in Table XI.

Energies of many of two-particle systems were calculated using the coupled-cluster method in Landau et al. (2000) and references therein. We note that the implementation of the coupled-cluster method in Landau et al. (2000) is very different from the all-order method described here and is limited to the calculations of the energies. A very promising method for the calculation of properties of many-particle systems, involving a combination of the CI and all-order techniques (Kozlov, 2004) was mentioned in the previous subsection.

7. Applications of High-Precision Calculations

7.1. Parity Nonconservation in Heavy Atoms

As mentioned earlier, one important motivation for developing relativistic many-body methods has been the desire to calculate parity nonconserving (PNC) amplitudes in heavy atoms with high accuracy. The dominant part of the PNC interaction between the bound electrons in the atom and the nucleus, which is mediated by exchange of a neutral Z boson, is described by the interaction Hamiltonian (Johnson, 2003)

$$H^{(1)} = \frac{G}{2\sqrt{2}} Q_W \gamma_5 \rho(r). \tag{51}$$

In this equation, G is the universal weak coupling constant and γ_5 is the 4×4 Dirac matrix element associated with pseudoscalar interactions. The quantity Q_W is the weak charge defined by

$$Q_W = -N + Z(1 - 4\sin^2\theta), \tag{52}$$

where N is the neutron number, Z is the proton number, and θ is the Weinberg angle. Since $\sin^2\theta \approx 1/4$, it follows that $Q_W \approx -N$. The factor $\rho(r)$, which describes the interaction density, is a weighted sum of the neutron and proton densities; approximately the neutron density.

Electric-dipole transitions between states of the same nominal parity, such as the $6s_{1/2} \to 7s_{1/2}$ transition in Cs, which are forbidden by parity conservation, become allowed in the presence of the pseudoscalar interaction $H^{(1)}$. By comparing measurements of the amplitude $Z_{6s7s} = \langle 6s_{1/2}|z|7s_{1/2}\rangle$ with calculations based on Eq. (51), one can extract an experimental value of Q_W and compare with predictions of the Standard Model. The PNC amplitude is found to be very sensitive to correlation corrections. In lowest-order MBPT, one first solves the perturbed Dirac equation

$$(h_0 + V_{\text{HF}} - \epsilon_v)\tilde{\phi}_v = -H^{(1)}\phi_v \tag{53}$$

to determine the PNC correction $\tilde{\phi}_v$ to the valence orbital ϕ_v, then evaluates the forbidden dipole matrix element as

$$Z_{vw}^{(1)} = \langle \phi_v|z|\tilde{\phi}_w\rangle + \langle \tilde{\phi}_v|z|\phi_w\rangle. \tag{54}$$

For the case of ^{133}Cs, this leads to $Z^{(1)}_{6s7s} = -0.739$ in units $ieQ_W/(-N)$. Inasmuch as the weak interaction modifies core orbitals as well as the valence orbital, a proper lowest-order treatment requires one to modify the perturbed equation above to include $\tilde{V}_{\text{HF}}$, the perturbation of the DHF potential induced by the weak interaction:

$$(h_0 + V_{\text{HF}} - \epsilon_v)\tilde{\phi}_v = -(H^{(1)} + \tilde{V}_{\text{HF}})\phi_v. \tag{55}$$

The perturbed core orbitals $\tilde{\phi}_a$ satisfy similar equations. The system of perturbed core and valence equations is solved self-consistently leading to a modified value of the first order PNC amplitude in Cs:

$$Z^{(1)}_{6s7s} = -0.927 \, ieQ_W/(-N).$$

The correction to the first-order amplitude from weakly perturbed core orbitals is similar to the RPA correction to an allowed transition amplitude; it is referred to as the "weak" RPA correction. Starting from the weak RPA amplitude in lowest order, one evaluates the corrections in second and third orders, using the weakly perturbed orbitals appearing in expressions for the second- and third-order dipole matrix elements. The second-order matrix element in Cs plus all higher-order RPA corrections gives

$$Z^{(2)}_{6s7s} = 0.037 \, ieQ_W/(-N)$$

(Dzuba et al., 1987; Mårtensson-Pendrill, 1985; Johnson et al., 1987). This leads to a value $-0.890 \, ieQ_W/(-N)$ through second order. The third-order Brueckner-orbital correction, which is expected to dominate the residual third-order terms is

$$Z^{(3)}_{6s7s} = -0.061 \, ieQ_W/(-N)$$

(Johnson et al., 1988a). Combining second and third orders, one obtains $-0.951 \, ieQ_W/(-N)$. The fluctuations in the value of the PNC amplitude from order to order are large, leading one to seek all-order methods to evaluate PNC amplitudes.

An alternative expression for the PNC amplitude is

$$Z_{vw} = \sum_n \frac{\langle\Psi_v|H^{(1)}|\Psi_n\rangle\langle\Psi_n|Z|\Psi_w\rangle}{E_n - E_v}$$
$$+ \sum_n \frac{\langle\Psi_v|Z|\Psi_n\rangle\langle\Psi_n|H^{(1)}|\Psi_w\rangle}{E_n - E_w}. \tag{56}$$

For the $6s_{1/2} \to 7s_{1/2}$ transition in Cs, the intermediate states range over all $np_{1/2}$ states. We start from this alternative expression in our all-order SD evaluation of the PNC amplitude. In particular, we evaluate the energies and matrix elements for the terms $n = 6, 7, 8, 9$ in the sum using all-order SD wave functions. The

Table XII
Contributions to E_{PNC} from All-Order SD Calculations of Energies and Matrix Elements

n	$\langle 7s|Z|np \rangle$	$\langle np|H^{(1)}|6s \rangle$	$E_{6s} - E_{np}$	Contrib.
6	1.7291	−0.0562	−0.05093	1.908
7	4.2003	0.0319	−0.09917	−1.352
8	0.3815	0.0215	−0.11714	−0.070
9	0.1532	0.0162	−0.12592	−0.020
n	$\langle 7s|H^{(1)}|np \rangle$	$\langle np|Z|6s \rangle$	$E_{7s} - E_{np}$	Contrib.
6	−1.8411	0.0272	0.03352	−1.493
7	0.1143	−0.0154	−0.01472	0.120
8	0.0319	−0.0104	−0.03269	0.010
9	0.0171	−0.0078	−0.04147	0.003
$n = 6$–9				−0.894(4)
Tail				−0.015(1)
Total				−0.909(4)

We designate the sum of contributions from $n = 2$–5 and $n = 10$–∞, which were calculated in the weak RPA approximation by "tail". Energies and dipole matrix elements are in a.u., and PNC matrix elements and contributions to E_{PNC} are in $-i|e|10^{-11} a_0 Q_W/N$.

theoretical error in these calculations is estimated by replacing *ab initio* theoretical data in the sums by precisely known experimental data or modified theory values, that include semi-empirical estimates of the omitted correlation effects, and noting the changes in the partial sum (Blundell et al., 1992). Contributions from terms with $n = 2$–5 and $n = 10$–∞ are evaluated in the weak RPA approximation. Values of all-order matrix elements and experimental energies used in the evaluation of Eq. (56) (Blundell et al., 1992), together with the residual weak RPA contributions to the sum over states are listed in Table XII. The resulting all-order value for the PNC amplitude in Cs in the table, estimated to be accurate to about 0.5% (Bennett and Wieman, 1999), is in close agreement with theoretical results from PTSI calculations (Dzuba et al., 2002), relativistic CI calculations (Kozlov et al., 2001b), multiconfiguration DHF calculations (Shabaev et al., 2005), and with preliminary values from CCSD calculations (Das et al., 2006). There are small residual corrections to the theoretical PNC amplitude in Table XII from the Breit interaction, vacuum polarization, and the nuclear skin effect. These corrections are enumerated, for example, in Derevianko and Porsev (2006). Combining the corrected theoretical amplitude with the measurements of Wood et al. (1997), one infers the "experimental" value $[Q_W]_{exp} = -72.73(46)$, which differs from the Standard Model prediction $Q_W = -73.09(3)$ by less than 1 standard deviation.

7.2. Polarizability Calculations and Their Applications. Blackbody Radiation Shift

An application of MBPT that is closely related to atomic PNC is the evaluation of atomic polarizabilities. All-order calculations of polarizabilities can be carried out using a sum-over-states approach similar to that used in the PNC calculations [Eq. (56)]. Neglecting hyperfine structure, the energy shift of an atom subjected to an electric field E in z direction is $\Delta E = -\alpha e^2 E^2/2$, where α is the atomic polarizability given by Angel and Sandars (1968):

$$\alpha = \alpha_0 + \alpha_2 \frac{3m_v^2 - j_v(j_v + 1)}{j_v(2j_v - 1)} \tag{57}$$

for the monovalent atom in a state v. The valence state contribution to the scalar frequency-dependent polarizability α_0 is given by

$$\alpha_0^v = \frac{2}{3(2j_v + 1)} \sum_k \frac{\langle v\|D\|k\rangle^2 (E_k - E_v)}{(E_k - E_v)^2 - \omega^2}, \tag{58}$$

where $\langle v\|D\|k\rangle$ is a reduced matrix element of the dipole operator. Intermediate states k in this case are restricted to single valence electron states. The total scalar polarizability is given by the sum of the valence contribution (58), the polarizability of the ionic core α_0^c, and small α_0^{vc} term compensating for violation of the Pauli principle by the core term (M.S. Safronova et al., 1999). The tensor polarizability α_2 is given by

$$\alpha_2 = -4C \sum_k (-1)^{j_v + j_k + 1} \begin{Bmatrix} j_v & 1 & j_k \\ 1 & j_v & 2 \end{Bmatrix} \frac{\langle v\|D\|k\rangle^2 (E_k - E_v)}{(E_k - E_v)^2 - \omega^2},$$

$$C = \left(\frac{5 j_v (2j_v - 1)}{6(j_v + 1)(2j_v + 1)(2j_v + 3)} \right)^{1/2}, \tag{59}$$

where $\{:::\}$ is a $6 - j$ symbol. It is important to note that α_2 is non-zero only for the states with $|m_v| \geq 1$.

Static polarizabilities are given by the above formulas with $\omega = 0$. Calculations of the sums over k in Eqs. (58) and (59) follow the pattern of the PNC sum-over-state calculation described earlier (M.S. Safronova et al., 1999, 2006a). Briefly, the sums are separated into a "main" term and a remainder "tail" term. Such a division is based on the relatively rapid convergence of the sums over k. The speed of the convergence depends on a particular state and frequency ω. The main term contains the dominant contribution, and is generally limited to the sum over the four lowest principal quantum numbers n_k for each value of κ_k allowed by selection rules. All electric-dipole matrix elements in the main term are calculated using the all-order method (SD, SDpT, or their scaled versions) or taken from experiment in cases where high-precision values are available. The tail term is

Table XIII

Polarizabilities of the Noble-Gas Atoms Calculated in the RPA (Kolb et al., 1982) Compared with Recommended Values (Miller and Bederson, 1977)

	He	Ne	Ar	Kr	Xe
RPA	1.32	2.38	10.8	16.5	27.0
Rec	1.38	2.67	11.1	16.7	27.3
Δ%	5	12	3	1	1

generally very small and is calculated in either DHF or RPA approximations. The core contribution to the scalar polarizability α_0^c is small but significant in some cases and is generally calculated in RPA approximation (Kolb et al., 1982). The core term contributes 0.5, 2, 3, and 4% to the scalar static ground state polarizabilities of Na, K, Rb, and Cs, respectively. In precise calculations, it is important to evaluate the uncertainty in the RPA calculation. For this purpose, we compare in Table XIII RPA calculations of the polarizabilities of noble gas atoms (Kolb et al., 1982) with the recommended values compiled in Miller and Bederson (1977).

If we approximate the accuracy of the ionic core polarizability by that of the neighboring closed-shell atom, then we expect that calculating α_0^c in the RPA will induce an error of about 0.06% in the polarizability of Na and K, and an error of 0.03% in the polarizability of Rb and Cs. Given that theoretical results for ions are expected to be even more accurate than those for neutral atoms, one expects that using the RPA values for α_0^c should induce errors smaller than 0.1% in ground state static polarizabilities of alkali-metal atoms. The term α_0^{vc} is very small and is calculated in either DHF or RPA approximation. As a result, the accuracy of the polarizability calculations is generally limited by the accuracy of the calculation of the electric-dipole matrix elements in the main term, as the energy levels are generally experimentally known.

Below, we list some of the all-order polarizability calculations carried out using the method described above. In some of the Ground state static and frequency-dependent polarizabilities of the alkali-metal atoms were calculated in M.S. Safronova et al. (1999, 2004, 2006a), Derevianko et al. (1999). All-order calculations of excited state $np_{1/2}$ and $np_{3/2}$ polarizabilities of alkali-metal atoms are being prepared for publication (Arora et al., in preparation). In Arora et al. (in preparation), polarizabilities between hyperfine levels are calculated. The static scalar polarizabilities of the $6p_j$ states and tensor polarizability of $6p_{3/2}$ state in Cs were calculated in Safronova andClark (2004), and the scalar and tensor polarizabilities of the $7d_{3/2}$, $7d_{5/2}$, $9d_{3/2}$, $9d_{5/2}$, $10d_{3/2}$, and $10d_{5/2}$ states of Cs were reported in Auzinsh et al. (2007). Scalar and tensor polarizabilities of the Fr-like Th IV in it ground state ($5f_{5/2}$) were calculated in U.I. Safronova et al. (2006c).

Table XIV
Comparison of the All-Order Polarizability Results with Experiment

	Cs		Cs		Tl	
	$\alpha_0(6p_{1/2})$ $-\alpha_0(6s)$	Reference	$\alpha_2(10D_{5/2})$	Reference	$\alpha_0(6p_{1/2})$ $-\alpha_0(7s)$	Reference
All-order	1248	Safronova and Clark (2004)	6867(32)	Auzinsh et al. (2007)	−830	M.S. Safronova et al. (2006b)
Expt.	1240.2(24)	Tanner and Wieman (1988b)	6815(20)	Xia et al. (1997)	−829.7(3.1)	Doret et al. (2002)

Units: a_0^3.

While Tl is an atom with three valence electrons, it can be approximated as a system with one valence electron above the closed $[1s^2 \ldots 6s^2]$ core. In this approximation, the Tl polarizabilities for the $6s^2 nl$ states may be determined using the approach described above. Such calculations were carried out in M.S. Safronova et al. (2006b) for the $6s^2 6p_{1/2}$, $6s^2 7s_{1/2}$, and $6s^2 7p_{1/2}$ states. The resulting all-order values were found to be in excellent agreement with experiment. We present a few examples of the all-order polarizability calculations in Table XIV.

Scalar and vector transitions polarizabilities α_S and β_S defined, for example, in M.S. Safronova et al. (1999, 2006b) can be calculated using the sum-over-states approach described above. These atomic parameters were calculated for alkali-metal atoms and Tl in M.S. Safronova et al. (1999, 2006b). The vector transition polarizability β_S is needed for the analysis of some of the PNC experiments, including the PNC experiment in Cs (Wood et al., 1997) discussed earlier in this paper.

Polarizability calculations have useful applications beyond high-precision tests of the atomic methodology, providing benchmark data for comparison with experiment, and studies of parity nonconservation. As we mentioned previously, the Cs $6s$ and $6p_j$ polarizability calculations were used to cross-check the Stark-shift and lifetime experiments, and significant discrepancies were found (Safronova and Clark, 2004). A proposal to minimize heating in a quantum logic gate (Safronova et al., 2003) relied on calculations of the polarizability of Rb in its ground and an excited Rydberg state. In M.S. Safronova et al. (2006a), the calculation of frequency-dependent polarizabilities in alkali-metal atoms was used to predict the oscillation frequencies of optically trapped alkali-metal atoms, and particularly the ratios of frequencies of different species held in the same trap. In the same work, which was motivated by recent experiments involving simultaneous optical trapping of two different alkali-metal species, wavelengths at which two different alkali-metal atoms have the same oscillation frequency were identified.

In Arora et al. (in preparation), the frequency-dependent polarizabilities between the hyperfine states of alkali-metal atoms were used to identify "magic" wavelengths for which ns and $np_{1/2,3/2}$ atomic levels have the same ac-Stark shifts, enabling state-insensitive optical cooling and trapping (McKeever et al., 2003). The magic wavelength is determined as the wavelength at which the Stark shifts of the upper and lower level for a specific transition are the same. In Porsev et al. (2004), polarizability calculation were conducted to assess the possibility of an optical clock using the $6\,^1S_0 \to 6\,^3P_0^o$ transition in Yb atoms held in an optical lattice. In that work, the magic wavelength was identified at which the shift in this proposed clock transition due to the trapping laser light is zero.

Another application of polarizability calculations to the development of the ultra-precise atomic clocks is the evaluation of the blackbody radiation (BBR) shift in the ^{133}Cs primary frequency standard (Beloy et al., 2006; Angstmann et al., 2006). The BBR shift of the energy level is proportional to the static scalar electric-dipole polarizability of the hyperfine level F. Therefore, the calculation of the BBR shift for the Cs microwave clock transition reduces to the calculation of the difference of the scalar polarizabilities of the $6s\ F = 4$ and $6s\ F = 3$ levels. To lowest order in the interaction potential, these polarizabilities are the same. Therefore, the BBR shift calculation involves calculation of the third-order F-dependent scalar polarizability $\alpha_0^{(3)}$. This calculation can also be done using the sum-over-state approach, with the sums containing two electric-dipole matrix elements and one hyperfine matrix element. The complete formulas are given in Beloy et al. (2006). The calculation of Beloy et al. (2006), Angstmann et al. (2006) resolved the discrepancy between the previous theoretical calculations and precise experimental measurements.

7.3. C_3 AND C_6 COEFFICIENTS

Parameters closely related to the polarizabilities are the Lennard-Jones C_3 coefficient that describes the long-range potential between an atom and a wall, $-C_3/R^3$, and the van der Waals C_6 coefficient that describes the long-range potential between two atoms through the interaction between induced dipoles, $-C_6/R^6$. Both of these coefficients can be described in terms of the frequency-dependent polarizability $\alpha(i\omega)$

$$\alpha(i\omega) = \frac{2}{3} \sum_n \frac{(E_n - E_v)|\langle\Psi_v|\mathbf{R}|\Psi_n\rangle|^2}{(E_n - E_v)^2 + \omega^2}, \tag{60}$$

where n ranges over all possible states. The frequency-dependent polarizability can itself be evaluated using the all-order methods described in the previous subsection. One finds:

$$C_3 = \frac{1}{4\pi} \int_0^\infty \alpha(i\omega)\,d\omega, \tag{61}$$

$$C_6 = \frac{3}{\pi} \int_0^\infty [\alpha(i\omega)]^2\,d\omega. \tag{62}$$

Equation (61) reduces to $C_3 = \langle \Psi_v | R^2 | \Psi_v \rangle / 12$ on integration.

An MBPT calculations of C_3 for alkali-metal atoms based directly on the expectation value of R^2 was carried out by Derevianko et al. (1998) and Johnson et al. (2004). The operator R^2 was decomposed into a single-particle operator S and a two-particle operator T: $R^2 = S + 2T$, where

$$S = \sum_j (r^2)_{ij} a_i^\dagger a_j,$$

$$T = \frac{1}{2} \sum_{ijkl} \langle ij | \boldsymbol{r}_1 \cdot \boldsymbol{r}_2 | kl \rangle a_i^\dagger a_j^\dagger a_l a_k.$$

In Derevianko et al. (1998), MBPT contributions to $\langle R^2 \rangle$ were carried out through third order for the dominant term S but were limited to second order for T. The MBPT result for Na, $C_3 = 1.890 a_0^2$, is in good agreement with the recommended (SD all-order) value $C_3 = 1.886 a_0^2$ (Derevianko et al., 1998, 1999). For Cs, the difference between the MBPT value $C_3 = 3.863 a_0^2$ and the recommended value $4.143 a_0^2$ (Derevianko et al., 1999) increases to about 7%. Once again, the need for all-order methods in precise atomic structure calculations for heavy atoms is apparent. In Johnson et al. (2004) a finite-field method, in which the basic electron–electron interaction Hamiltonian is modified to include a scaled contribution from R^2, was used in connection with an SD calculation of the correlation energy to obtain all-order values for C_3 for heavy atoms.

All-order values of both C_3 and C_6 for alkali-metal atoms were also obtained directly from Eqs. (61)–(62) in Derevianko et al. (1999) using all-order values of frequency-dependent polarizabilities such as those discussed in the previous subsection.

In the upper panel of Table XV, we compare values of C_3 obtained in three different ways; evaluating $\langle \Psi_v | R^2 | \Psi_v \rangle / 12$ using SD wave functions, evaluating C_3 using the finite-field method in conjunction with the SD method, and, finally, evaluating the dominant contributions to $\alpha(i\omega)$ with SD wave functions and energies and utilizing Eq. (61). The consistence between these three approaches deteriorates rapidly with increasing nuclear charge indicating the need for further studies.

Table XV

Calculations of the Lennard-Jones Atom–Wall Interaction Constant C_3 and Comparison of SD Calculations of the Atom–Atom Interaction Constant C_6 with Recommended Values

Method	References	Na	K	Rb	Cs	Fr
C_3 coefficient						
$\frac{1}{12}\langle R^2\rangle$	Derevianko et al. (1998, 1999)	1.886	2.860	3.362	4.143	4.281
Finite field	Johnson et al. (2004)	1.887	2.966	3.529	4.499	4.711
Eq. (61)	Derevianko et al. (1999)	1.875	2.877	3.410	4.247	4.427
C_6 coefficient						
Eq. (62)	Derevianko et al. (1999)	1564	3867	4628	6899	5174
Rec. values	Derevianko et al. (1999)	1556(4)	3897(15)	4691(23)	6851(74)	5256(89)

In the lower panel of Table XV, we compare all-order results for the C_6 coefficients of alkali-metal atoms with semi-empirical recommended values (Derevianko et al., 1999). The recommended values are obtained from calculations making use of experimental energies and electric-dipole matrix elements for the principal transitions. The all-order values were determined by evaluating the dominant contributions to $\alpha(i\omega)$ with SD wave functions and energies, then utilizing Eq. (61). In this case, values obtained in two different ways show good consistency.

7.4. ISOTOPE SHIFT

In view of the fact that experimental values of isotope shifts are not available for many atoms of interest in applications, there is a pressing need for accurate calculations. By comparing experimental and theoretical isotope shifts one can extract changes in nuclear charge radii from one isotope to another. These changes are of intrinsic interest in nuclear physics; they are also important in the analysis of possible future PNC experiments involving measurements in chains of isotopes. Accurate values of isotope shifts are also needed for studies of the variation of the fundamental constant α in the spectra of quasistellar objects (Murphy et al., 2001), since changes in isotopic ratios in the early universe lead to systematics that can mask changes in α.

Isotope shifts of energy levels consist of two parts, one associated with changes in nuclear volume from isotope to isotope, and the other associated with nuclear recoil. The nuclear size correction, referred to as the field shift (FS) is obtained from matrix elements of the operator $dV/d\langle r^2\rangle$, where $\langle r^2\rangle$ is the mean-square radius of the nucleus. The recoil contribution is subdivided into the normal mass shift (NMS), which can be accurately determined in terms of the experimental energy, and the specific mass shift (SMS), which is obtained from the expectation value of the two-particle operator $\boldsymbol{p}_1 \cdot \boldsymbol{p}_2$:

$$\delta E_{\text{NMS}} = -\frac{m}{M} E_{\text{expt}}, \tag{63}$$

$$\delta E_{\text{SMS}} = \frac{1}{M} \left\langle \sum_{i<j} \boldsymbol{p}_i \cdot \boldsymbol{p}_j \right\rangle. \tag{64}$$

In the above, m and M are masses of the electron and nucleus, respectively. The FS dominates the isotope shift for heavy atoms such as francium but is relatively unimportant for lighter atoms such as lithium and sodium.

Evaluation of the specific mass shift using MBPT follows a pattern similar to that shown above for the C_3 coefficient. One expresses the SMS operator $P = \sum_{ij} \boldsymbol{p}_i \cdot \boldsymbol{p}_j$ in second quantization as $P = T + S$, where

$$T = \sum_{ijkl} t_{ijkl} [a_i^\dagger a_j^\dagger a_l a_k], \tag{65}$$

$$S = \sum_{ij} t_{ij} [a_i^\dagger a_j], \tag{66}$$

with $t_{ijkl} = \langle ij|\boldsymbol{p}_1 \cdot \boldsymbol{p}_2|kl\rangle$ and $t_{ij} = -\sum_b t_{ibbj}$. As discussed by Bauche and Champeau (1976), expressions for the expectation value of P can be inferred easily from corresponding expressions for the energy. The expansion of the energy is carried out explicitly in the relativistic case through third order in Blundell et al. (1987). With the aid of this expansion, we find:

$$P^{(1)} = S^{(1)} = t_{vv}, \tag{67}$$

$$P^{(2)} = S^{(2)} + T^{(2)} = \sum_{am} \frac{t_{am}\tilde{g}_{vmva} + \tilde{g}_{vavm}t_{ma}}{\epsilon_a - \epsilon_m}$$

$$- 2\sum_{mab} \frac{t_{mvab}\tilde{g}_{mvab}}{\epsilon_a + \epsilon_b - \epsilon_m - \epsilon_v} + 2\sum_{amn} \frac{t_{mnva}\tilde{g}_{mnva}}{\epsilon_v + \epsilon_a - \epsilon_m - \epsilon_n}. \tag{68}$$

Explicit formulas for the third-order corrections $S^{(3)}$ and $T^{(3)}$ can be found in Safronova and Johnson (2001).

As an example of the convergence of the MBPT expansion, we present results from the first three orders of MBPT for the SMS constants for the $3s$, $3p_{1/2}$, and $3d_{3/2}$ states of Na and $4s$, $4p_{1/2}$, and $3d_{3/2}$ states of K in Table XVI. The third-order values $S^{(3)}$ include RPA contributions to all orders. These results are from Safronova and Johnson (2001), but similar results for Na were obtained by Lindroth and Mårtensson-Pendrill (1983) and by Mårtensson and Salomonson (1982) for Li and K. The poor convergence of perturbation expansion seen in the table emphasizes the need for alternative theoretical approaches. A similar slow

Table XVI

Contributions from First-, Second-, and Third-Order Perturbation Theory to the Specific-Mass Shift Constants (GHz Amu) of the $3s$, $3p_{1/2}$ and $3d_{3/2}$ States of Na and $4s$, $4p_{1/2}$ and $3d_{3/2}$ States of K

Term	Na			K		
	$3s_{1/2}$	$p_{1/2}$	$3d_{3/2}$	$4s_{1/2}$	$4p_{1/2}$	$3d_{3/2}$
$S^{(1)}$	−222.0	−115.6	−4.84	−387.9	−118.7	−113.8
$S^{(2)}$	167.9	48.4	0.96	192.3	59.1	43.5
$S^{(3)}$	28.1	1.2	0.95	−2.2	0.1	−46.0
$S^{(\text{SDpT})}$	205.3	51.6	2.79	202.9	66.3	−26.9
$T^{(2)}$	95.0	28.2	−0.42	143.2	35.8	19.0
$T^{(3)}$	−24.4	−7.5	−0.48	−32.9	−8.2	−13.3
$P_{\text{tot}}^{(3)}$	44.7	−45.4	−3.83	−87.5	−31.9	−110.6
$P_{\text{tot}}^{(\text{SDpT})}$	53.9	−43.4	−2.95	−74.4	−24.7	−135.0

All-order $S^{(\text{SDpT})}$ matrix elements of the S are also given. $P_{\text{tot}}^{(3)} = S^{(1)} + S^{(2)} + S^{(3)} + T^{(2)} + T^{(3)}$. $P_{\text{tot}}^{(\text{SDpT})}$ is obtained by replacing $S^{(2)} + S^{(3)}$ by $S^{(\text{SDpT})}$ in the above sum. All data are from Safronova and Johnson (2001).

convergence of the perturbation expansion of the field shift was also found in Safronova and Johnson (2001).

Since S is a one-body operator, its matrix elements can be evaluated using the SD or SDpT all-order approach described earlier. Such SDpT calculations were carried out in Safronova and Johnson (2001) and the corresponding results are shown in Table XVI in row labeled $S^{(\text{SDpT})}$. There are no results for SDpT matrix elements of T; therefore, we include only third-order values in Table XVI. For comparison, we list two totals, $P_{\text{tot}}^{(3)}$ and $P_{\text{tot}}^{(\text{SDpT})}$, in the last two rows of Table XVI. Significant discrepancies between these numbers are attributed to the severe cancellations between the lowest- and second-order values and significant contributions from higher orders, especially for nd states. Comparison of the selected values of the total isotope shifts for Na, K, and Ca$^+$ (Safronova and Johnson, 2001) calculated using the method described above with experimental values is given in Table XVII. Significant discrepancies with experiment exist, especially for the isotope shifts involving nd states. These differences are not unexpected, since third-order corrections are particularly large for the nd states as seen in Table XVI.

A more complete calculation of isotope shifts was reported in Dzuba et al. (2005), where the isotope shifts of K, Rb, Cs, and Fr were calculated using two high-precision methods. The calculation of the FS was reduced to a calculation of the energy using a finite-field approach. The finite-field energy was calculated using an all-order correlation potential method as well as the all-order SD method.

Table XVII
Comparison of the Total Isotope Shifts with Experiment for Na ($\delta\nu^{22,23}$), K ($\delta\nu^{41,39}$), and Ca$^+$ ($\delta\nu^{43,40}$)

	Na	K		Ca$^+$	
	$3p_{1/2} - 3s$	$4p_{1/2} - 4s$	$4d_{3/2} - 4s$	$4p_{1/2} - 4s$	$3d_{3/2} - 4p_{1/2}$
Theory	−733	192	569	592	3843
Expt.	−758.5(7)	235.25(75)	585(9)	672(9)	3464(3)

The theoretical results are from Safronova and Johnson (2001); the experimental data are taken from the compilation carried out in the same work. Units: MHz.

Unfortunately, the same method cannot be used for the calculation of the SMS for reasons explained in Dzuba et al. (2005). The SMS shift was calculated in Dzuba et al. (2005) using both the approach described earlier (including $T^{(3)}$) and by a combination of perturbation theory and the finite-field approach in which only core polarization diagrams are included to all orders. The FS constants calculated by the two different methods were found to be in good agreement with each other. However, the SMS constants calculated by the two different MBPT methods, which included different classes of terms, were found to be in severe disagreement, demonstrating a pressing need to develop all-order techniques for calculating the SMS.

8. Conclusion

The relativistic all-order method has been described and applications to calculate properties of various atomic systems have been given. While the all-order SD method gives accurate results for many properties, further development is clearly needed for other properties. Recent improvements to the SD method, such as inclusion of the non-linear terms and of valence triple excitations, were discussed. Strong cancellation between such contributions makes it necessary to consider both effects simultaneously if one is to increase the accuracy of the all-order calculations. Applications of the high-precision atomic calculations, ranging from studies of parity nonconservation in heavy atoms to the calculation of the blackbody radiation shifts relevant to the development of ultra-precise frequency standards, were described. It was demonstrated that the improvement of all-order techniques for the calculation of the C_3 coefficients and the specific mass shifts is needed for accurate calculation of these properties. All-order methods for calculating properties of divalent and trivalent systems were discussed and examples of the corresponding results were presented.

9. Acknowledgements

Works of MAS and WRJ were supported in part by the U.S. National Science Foundation grants No. PHY-04-57078 and No. PHY-04-56828, respectively.

10. References

Angel, J.R.P., Sandars, P.G.H. (1968). *Proc. Roy. Soc. London Series A* **305**, 125.
Angstmann, E.J., Dzuba, V.A., Flambaum, V.V. (2006). *Phys. Rev. Lett.* **97**. 040802.
Arimondo, E., Inguscio, M., Violino, P. (1977). *Rev. Mod. Phys.* **49**, 31.
Arora, B., Safronova, M.S., Clark, C.W. (in preparation).
Auzinsh, M., Bluss, K., Ferber, R., Gahbauer, F., Jarmola, A., Safronova, M.S., Safronova, U.I., Tamanis, M. (2007). *Rev. Phys. A* **75**. 022502.
Bauche, J., Champeau, R.J. (1976). In: Bates, D., Bederson, B. (Eds.), *Advances in Atomic and Molecular Physics*, vol. 12. Academic Press, Inc., San Diego, pp. 39–86.
Beiersdorfer, P. (2003). *Annu. Rev. Astron. Astrophys.* **41**, 343.
Beiersdorfer, P., Chen, H., Thorn, D.B., Trabert, E. (2005). *Phys. Rev. Lett.* **95**. 233003.
Beloy, K., Safronova, U.I., Derevianko, A. (2006). *Phys. Rev. Lett.* **97**. 040801.
Bennett, S.C., Wieman, C.E. (1999). *Phys. Rev. Lett.* **82**, 2484.
Blundell, S.A., Guo, D.S., Johnson, W.R., Sapirstein, J. (1987). *At. Data Nucl. Data Tables* **37**, 103.
Blundell, S.A., Johnson, W.R., Liu, Z.W., Sapirstein, J. (1989). *Phys. Rev. A* **40**, 2233.
Blundell, S.A., Johnson, W.R., Sapirstein, J. (1991). *Phys. Rev. A* **43**, 3407.
Blundell, S.A., Sapirstein, J., Johnson, W.R. (1992). *Phys. Rev. D* **45**, 1602.
Brown, G.E., Ravenhall, D.G. (1951). *Proc. Roy. Soc. A* **208**, 552.
Cannon, C.C., Derevianko, A. (2004). *Phys. Rev. A* **69**. 030502.
Chen, M.H., Cheng, K.T. (1997a). *Phys. Rev. A* **55**, 166.
Chen, M.H., Cheng, K.T. (1997b). *Phys. Rev. A* **55**, 3440.
Chen, M.H., Cheng, K.T., Johnson, W.R. (1993). *Phys. Rev. A* **47**, 3692.
Cheng, K.T., Chen, M.H. (2000). *Phys. Rev. A* **6104**. 044503.
Coester, F., Kümmel, H. (1960). *Nucl. Phys.* **17**, 477.
Cohen, S. (1960). *Phys. Rev.* **118**, 489.
Coulthard, M.A. (1967). *Proc. Roy. Soc. London* **91**, 44.
Das, B.P., Sahoo, B.K., Gopakumar, G., Chaudhuri, R.K. (2006). *Theochem* **768**, 141.
Derevianko, A., Porsev, S. (2006). hep-ph/0608178.
Derevianko, A., Johnson, W.R., Fritzsche, S. (1998). *Phys. Rev. A* **57**, 2629.
Derevianko, A., Johnson, W.R., Safronova, M.S., Babb, J.F. (1999). *Phys. Rev. Lett.* **82**, 3589.
Desclaux, J.P. (1973). *At. Data Nucl. Data Tables* **12**, 311.
Desclaux, J.P. (1975). *Comput. Phys. Commun.* **9**, 31.
Doret, S.C., Frieberg, P.D., Speck, A.J., Richardson, D.S., Majumder, P.K. (2002). *Phys. Rev. A* **66**. 52504.
Dzuba, A., Flambaum, V., Sushkov, O.P. (1989a). *Phys. Lett. A* **140**, 493.
Dzuba, A., Flambaum, V.V., Sushkov, O.P. (1989b). *Phys. Lett. A* **142**, 373.
Dzuba, V.A., Johnson, W.R. (1998). *Phys. Rev. A* **57**, 2459.
Dzuba, V.A., Flambaum, V.V., Silvestrov, P.G., Sushkov, O.P. (1987). *J. Phys. B* **20**, 3297.
Dzuba, V.A., Flambaum, V.V., Kozlov, M.G. (1996). *Phys. Rev. A* **54**, 3948.
Dzuba, V.A., Flambaum, V.V., Ginges, J.S.M. (2002). *Phys. Rev. D* **66**. 076013.
Dzuba, V.A., Johnson, W.R., Safronova, M.S. (2005). *Phys. Rev. A* **72**. 022503.
Eliav, E., Kaldor, U., Ishikawa, Y. (1994). *Phys. Rev. A* **50**, 1121.

Gomez, E., Orozco, L.A., Galvan, A.P., Sprouse, G.D. (2005). *Phys. Rev. A* **71**. 62504.
Grant, I.P. (1961). *Proc. Roy. Soc. London A* **262**, 555.
Grant, I.P., McKenzie, B.J., Norrington, P.H., Mayers, D.F., Pyper, N.C. (1980). *Comput. Phys. Commun.* **21**, 207.
Happer, W. (1974). "Atomic Physics 4". Plenum Press, New York.
Ho, H.C., Johnson, W.R., Blundell, S.A., Safronova, M.S. (2006). *Phys. Rev. A* **74**. 022510.
Johnson, W.R. (2003). In: Sadeghpour, H.R., Heller, E.J., Pritchard, D.E. (Eds.), *Proceedings of the XVIII International Conference on Atomic Physics*. World Scientific, pp. 327–340.
Johnson, W.R., Idrees, M., Sapirstein, J. (1987). *Phys. Rev. A* **35**, 3218.
Johnson, W.R., Blundell, S.A., Liu, Z.W., Sapirstein, J. (1988a). *Phys. Rev. A* **37**, 1395.
Johnson, W.R., Blundell, S.A., Sapirstein, J. (1988b). *Phys. Rev. A* **37**, 307.
Johnson, W.R., Liu, Z.W., Sapirstein, J. (1996). *At. Data Nucl. Data Tables* **64**, 279.
Johnson, W.R., Safronova, M.S., Safronova, U.I. (1997). *Physica Scripta* **56**, 252.
Johnson, W.R., Dzuba, V.A., Safronova, U.I., Safronova, M.S. (2004). *Phys. Rev. A* **69**. 22508.
Kim, Y.K. (1967). *Phys. Rev.* **54**, 17.
Kolb, D., Johnson, W.R., Shorer, P. (1982). *Phys. Rev. A* **26**, 19.
Kozlov, M.G. (2004). *Int. J. Quantum Chem.* **100**, 336.
Kozlov, M.G., Porsev, S.G. (1997). *Sov. Phys. JETP* **84**, 461.
Kozlov, M.G., Porsev, S.G., Johnson, W.R. (2001a). *Phys. Rev. A* **64**. 052107.
Kozlov, M.G., Porsev, S.G., Tupitsyn, I.I. (2001b). *Phys. Rev. Lett.* **86**, 3260.
Kreuter, A., Becher, C., Lancaster, G.P.T., Mundt, A.B., Russo, C., Häffner, H., Roos, C., Hänsel, W., Schmidt-Kaler, F., Blatt, R., et al. (2005). *Phys. Rev. A* **71**. 032504.
Landau, A., Eliav, E., Ishikawa, Y., Kaldor, U. (2000). *J. Chem. Phys.* **113**. 9905.
Letchumanan, V., Wilson, M.A., Gill, P., Sinclair, A.G. (2005). *Phys. Rev. A* **72**. 12509.
Lindgren, I. (1985). *Phys. Rev. A* **31**, 1273.
Lindgren, I., Salomonson, S., Hedendahl, D. (2006). *Phys. Rev. A* **73**. 062502.
Lindroth, E., Mårtensson-Pendrill, A.-M. (1983). *Z. Phys. A* **309**, 277.
Mann, J.B., Waber, J.T. (1973). *At. Data Nucl. Data Tables* **5**, 201.
Mårtensson, A.-M., Salomonson, S. (1982). *J. Phys. B* **15**, 2115.
Mårtensson-Pendrill, A.-M. (1985). *J. Phys. (Paris)* **46**, 1949.
Mayers, D.F. (1957). *Proc. Roy. Soc. London A* **241**, 93.
McKeever, J., Buck, J.R., Boozer, A.D., Kuzmich, A., Naegerl, H.-C., Stamper-Kurn, D.M., Kimble, H.J. (2003). *Phys. Rev. Lett.* **90**. 133602.
Miller, T.M., Bederson, B. (1977). *Adv. At. Mol. Phys.* **13**, 1.
Mittleman, M.H. (1971). *Phys. Rev. A* **4**, 893.
Mittleman, M.H. (1972). *Phys. Rev. A* **5**, 2395.
Mittleman, M.H. (1981). *Phys. Rev. A* **24**, 1167.
Moore, C.E. (1971). "Atomic Energy Levels, NSRDS-NBS 35". U.S. Government Printing Office, Washington, DC.
Murphy, M.T., Webb, J.K., Flambaum, V.V., Churchill, C.W., Prochaska, J.X. (2001). *Mon. Not. R. Astron. Soc.* **327**, 1223.
Pal, R., Safronova, M.S., Johnson, W.R., Derevianko, A., Porsev, S.G. (2007). *Rev. Phys. A* **75**. 042515.
Porsev, S.G., Derevianko, A. (2006). *Phys. Rev. A* **73**. 012501.
Porsev, S.G., Kozlov, M.G., Rahlina, Y.G. (2000). *JETP Lett.* **72**, 595.
Porsev, S.G., Derevianko, A., Fortson, E.N. (2004). *Phys. Rev. A* **69**. 021403R.
Rafac, R.J., Tanner, C.E. (1997). *Phys. Rev. A* **56**, 1027.
Rafac, R.J., Tanner, C.E., Livingston, A.E., Berry, H.G. (1999). *Phys. Rev. A* **60**, 3648.
Safronova, M.S., Clark, C.W. (2004). *Phys. Rev. A* **69**. 040501R.
Safronova, M.S., Johnson, W.R. (2001). *Phys. Rev. A* **64**. 052501.
Safronova, M.S., Johnson, W.R., Safronova, U.I. (1996). *Phys. Rev. A* **53**. 004036.

Safronova, M.S., Derevianko, A., Johnson, W.R. (1998). *Phys. Rev. A* **58**, 1016.
Safronova, M.S., Johnson, W.R., Derevianko, A. (1999). *Phys. Rev. A* **60**, 4476.
Safronova, M.S., Williams, C.J., Clark, C.W. (2003). *Phys. Rev. A* **67**. 040303.
Safronova, M.S., Williams, C.J., Clark, C.W. (2004). *Phys. Rev. A* **69**. 022509.
Safronova, M.S., Arora, B., Clark, C.W. (2006a). *Phys. Rev. A* **73**. 022505.
Safronova, M.S., Johnson, W.R., Safronova, U.I., Cowan, T.E. (2006b). *Phys. Rev. A* **74**. 022504.
Safronova, U.I., Johnson, W.R., Safronova, M.S., Derevianko, A. (1999). *Physica Scripta* **59**, 286.
Safronova, U.I., Johnson, W.R., Berry, H.G. (2000). *Phys. Rev. A* **61**. 052503.
Safronova, U.I., Safronova, M.S., Johnson, W.R. (2005). *Phys. Rev. A* **71**. 052506.
Safronova, U.I., Cowan, T.E., Safronova, M.S. (2006a). *J. Phys. B* **39**, 749.
Safronova, U.I., Cowan, T.E., Safronova, M.S. (2006b). *Phys. Lett. A* **348**, 293.
Safronova, U.I., Johnson, W.R., Safronova, M.S. (2006c). *Phys. Rev. A* **74**. 042511.
Salomonson, S., Ynnerman, A. (1991). *Phys. Rev. A* **43**, 2233.
Sapirstein, J. (1998). *Rev. Mod. Phys.* **70**, 55.
Savukov, I.M., Johnson, W.R. (2002a). *Phys. Rev. A* **65**. 042503.
Savukov, I.M., Johnson, W.R. (2002b). *Phys. Rev. A* **62**. 052512.
Shabaev, V.M., Tupitsyn, I.I., Pachucki, K., Plunien, G., Yerokin, V.A. (2005). *Phys. Rev. A* **72**. 062105.
Sherman, J.A., Koerber, T.W., Markhotok, A., Nagourney, W., Fortson, E.N. (2005). *Phys. Rev. Lett.* **94**. 243001.
Simsarian, J.E., Orozco, L.A., Sprouse, G.D., Zhao, W.Z. (1998). *Phys. Rev. A* **57**, 2448.
Smith, F.C., Johnson, W.R. (1967). *Phys. Rev.* **160**, 136.
Sucher, J. (1980). *Phys. Rev. A* **22**, 348.
Swirles, B. (1935). *Proc. Roy. Soc. London* **152**, 625.
Tanner, C.E., Wieman, C. (1988a). *Phys. Rev. A* **38**, 1616.
Tanner, C.E., Wieman, C. (1988b). *Phys. Rev. A* **38**, 162.
Volz, U., Schmoranzer, H. (1996). *Phys. Scr. T* **65**, 48.
Wijngaarden, W.A., Li, J. (1994). *Z. Phys. D* **32**, 67.
Williams Jr., A.O. (1940). *Phys. Rev.* **58**, 723.
Wood, C.S., Bennett, S.C., Cho, D., Masterson, B.P., Roberts, J.L., Tanner, C.E., Wieman, C.E. (1997). *Science* **275**, 1759.
Xia, J., Clarke, J., Li, J., van Wijngaarden, W. (1997). *Phys. Rev. A* **56**, 5167.

B-SPLINES IN VARIATIONAL ATOMIC STRUCTURE CALCULATIONS

CHARLOTTE FROESE FISCHER[*]

Department of Electrical Engineering and Computer Science, Vanderbilt University, Box 1679 B, Nashville, Tennessee, 37235, USA

1. Introduction 236
2. The Hartree–Fock Approximation 238
   2.1. A Case with no Orthogonality Constraints 240
   2.2. Cases with Orthogonality Constraints 242
3. Multiconfiguration Hartree–Fock Approximation 249
4. B-Spline Theory 251
   4.1. Spline Methods for the Solution of Differential Equations 253
   4.2. Spline Grid for Radial Functions 254
   4.3. Integration Methods 254
   4.4. Slater Integrals 256
5. B-Spline Methods for the Many-Electron Hartree–Fock Problem 261
   5.1. Notation 261
   5.2. Derivation of B-Spline Equations 263
   5.3. Iterative Solutions of B-Spline Equations 266
6. B-Spline MCHF Equations 275
   6.1. With Orthonormality Constraints 275
   6.2. Without Normalization Constraints 277
   6.3. Without Orthogonality Constraints 282
   6.4. Spline Orbital Basis Methods: CI and Perturbation Theory 283
   6.5. Direct Expansions in Products of B-splines for the Many-Electron Case . 284
7. Conclusion 288
8. Acknowledgements 289
9. References 289

Abstract

Many of the problems associated with the use of finite differences for the solution of variational Hartree–Fock or Dirac–Hartree–Fock equations are related to the orthogonality requirement and the need for node counting to control the solution of the two-point boundary value problem. By expanding radial functions in a B-spline basis, the differential equations are replaced by non-linear systems of equations. Hartree–Fock orbitals become solutions of generalized eigenvalue problems where orthogonality requirements can be dealt with through projection operators. When

---

[*] E-mail: Charlotte.F.Fischer@Vanderbilt.Edu.

expressed as banded systems of equations, all orbitals may be improved simultaneously using singular value decomposition or the Newton–Raphson method for faster convergence. Computational procedures are described for non-relativistic multiconfiguration Hartree–Fock variational methods and extensions to the calculation of Rydberg series. The effective completeness of spline orbitals can be used to combine variational methods with many-body perturbation theory or first-order configuration interaction. Many options are available for improving atomic structure calculations. Although spline methods are discussed only in connection with non-relativistic theory, they apply equally to Dirac–Hartree–Fock theory.

1. Introduction

In Hartree's central field theory the radial wave function $P(nl; r)$ for an nl electron is the solution of the differential equation (in atomic units),

$$\left(-\frac{1}{2}\frac{d^2}{dr^2} - \frac{1}{r}[Z - Y(nl; r)] + \frac{l(l+1)}{2r^2} - \varepsilon_{nl,nl}\right)P(nl; r) = 0, \tag{1}$$

where Z is the nuclear charge and $Y(nl; r)$ describes the screening of the nucleus by the other electrons. This differential equation can be solved numerically by finite difference methods in which the radial wave function is represented by a vector of numerical values $P(nl, r_i), i = 1, \ldots, N$. The boundary conditions, $P(nl; 0) = 0$ and $P(nl; r) \to 0$ as $r \to \infty$, make this a two-point boundary value problem for which a solution exists only for selected values of $\varepsilon_{nl,nl}$. Inward and outward integration often are used and $\varepsilon_{nl,nl}$ adjusted until the two solutions match (Froese Fischer, 1977). Node counting is needed to ensure that a solution has the required number of oscillations. However, when the solution is represented in terms of B-spline basis functions $B_i(r)$, namely

$$P(nl; r) = \sum_{i=1}^{n_s} a_i B_i(r), \tag{2}$$

and the expansion coefficients are determined by the Galerkin condition (Fletcher, 1984) for differential equations, the solution of Eq. (1) becomes a solution of the symmetric, generalized matrix-eigenvalue problem

$$(H - \varepsilon S)a = 0. \tag{3}$$

Here $H = (H_{ij})$ and $S = (S_{ij})$ are matrices with

$$H_{ij} = \langle B_i(r)| -\frac{1}{2}\frac{d^2}{dr^2} - \frac{1}{r}[Z - Y(nl; r)] + \frac{l(l+1)}{2r^2} |B_j(r)\rangle,$$
$$S_{ij} = \langle B_i(r)|B_j(r)\rangle, \tag{4}$$

respectively. This generalized eigenvalue problem with eigenvalues ε_i, $i = 1, \ldots, n_s - 2$ (after boundary conditions have been applied) can be solved using routines from the LAPACK library. Node counting is not necessary. If the eigenvalues are ordered so that $\varepsilon_i < \varepsilon_{i+1}$, then the index i determines the principal quantum number n to be associated with the solution. Furthermore, in a B-spline basis, the matrices are banded and eigenvalues can be determined to high accuracy, even for the equivalent Dirac equation (Froese Fischer and Parpia, 1993).

As soon as exchange is introduced, extra terms appear in the radial equation which may be written as

$$\left(-\frac{1}{2}\frac{d^2}{dr^2} - \frac{1}{r}[Z - Y(nl; r)] + \frac{l(l+1)}{2r^2} - \varepsilon_{nl,nl} \right) P(nl; r)$$
$$= \frac{1}{r} X(nl; r), \tag{5}$$

provided no Lagrange multipliers need to be introduced to ensure orthogonality. In differential equation methods, the functions $Y(nl; r)$ and $X(nl; r)$ are computed from current estimates and a solution of the differential equation determined by outward and inward integration. Because now the equation is non-homogeneous, solutions exist for all values of $\varepsilon_{nl,nl}$ and the latter often is adjusted for a normalized solution. Many problems may occur (Griffin et al., 1971). By expanding orbitals in a B-spline basis, the equation for each orbital again has the form

$$(H - \varepsilon S)a = 0, \tag{6}$$

where now the matrix H is a dense symmetric matrix. Iteration is still needed for a self-consistent solution in which case only one eigenvalue and eigenvector are physically significant. However, in a fixed core calculation for, say the $2p$ radial function of $1s^2 2p$, both the potential and exchange contributions to the H matrix depend only on $1s$ and the eigenvectors, $i = 1, 2 \ldots$ represent fixed core solutions for the Rydberg series $2p, 3p, \ldots$.

The differential equation methods Hartree (1929) developed and used primarily for ground-state configurations were extended and generalized to excited states (Froese Fischer, 1987) and multiconfiguration approximations (Froese Fischer, 1991a). In some instances, approximations were made to simplify computational difficulties, as in the HFX and HFR methods of the Cowan code (Cowan, 1981). With the use of B-spline expansions, differential equations with their many special conditions are replaced by non-linear systems of equations of eigenvalue type that may be solved accurately and reliably without such assumptions.

An extensive review of the applications of B-splines in atomic and molecular physics has been reported by Bachau et al. (2001). Many methods are orbital basis set methods that use the orbitals defined by the eigenvectors of a generalized matrix-eigenvalue problem such as Eq. (3) with an appropriate Hamiltonian. This finite set of orbitals forms a complete *orthonormal* basis for piecewise polynomial spline approximations and an "effectively" complete basis in Hilbert space for a given region. Spline orbitals computed by solving the Dirac equation for an electron in a model potential are the basis of relativistic many-body perturbation theory (RMBPT) methods. The first such study was reported by Johnson and Saperstein (1986). Recently, spline orbitals were used in large-scale relativistic configuration-interaction calculations for C III (Chen et al., 2001) to explore various relativistic and QED effects on some transition probabilities. Expansions in non-orthogonal B-splines have been used directly in the calculation of atomic processes such as photoionization (Xi and Froese Fischer, 1999b) and Rydberg series (Brage and Froese Fischer, 1994a) (see Bachau et al., 2001, for many more examples). In the more recent B-spline R-matrix (BSR) method for collision studies, the one-electron channel functions are expanded in B-splines, a formulation that again leads to a generalized eigenvalue problem (Zatsarinny, 2006). Not fully investigated are B-spline approximations of the Hartree–Fock (HF) and multiconfiguration Hartree–Fock (MCHF) methods though some initial investigations were reported in 1992 (Froese Fischer et al., 1992). Considerable progress has been made since then.

This paper begins with a review of the Hartree–Fock (HF) and multiconfiguration Hartree–Fock (MCHF) theory as well as approximation theory using B-splines, and then outlines the application of B-spline theory to the solution of the HF and MCHF equations. In differential equation theory, equations are written so that the coefficient of the highest derivative is unity. In this review we will not adhere to this practice in order to provide a view better suited for the matrix methods when spline expansions are used. In particular, equations will not be divided by the occupation number of an orbital. This change will affect the definition of the diagonal energy parameter but has the consequence that systems of equations are expressed in terms of symmetric matrices. For the latter a notation is introduced for deriving the equations for variational solutions. Advantages and disadvantages of different computational methods will be discussed. Finally, an extension to the multiconfiguration Hartree–Fock wave function will be presented including methods for Rydberg series, the use of spline orbital methods for capturing the numerous remaining effects to first order, and the use of non-orthogonal tensor products of B-spline basis functions for pair-correlation functions.

2. The Hartree–Fock Approximation

In theoretical quantum chemistry, the Hartree–Fock wave function is defined as being a single Slater determinant. In atomic physics, the general definition is a single configuration state function (CSF) $\Phi(\gamma LS)$ for a configuration γ that is anti-symmetric and an eigenfunction of the total orbital-angular momentum and spin-angular operators $\mathbf{L}^2$, L_z, $\mathbf{S}^2$ and S_z. In some cases the two definitions may be equivalent but, in general, a CSF is a linear combination of determinants for the same configuration (see Froese Fischer et al., 1997; Cowan, 1981, for more details).

In both cases, the wave function for an N-electron system is constructed from one-electron spin-orbitals of the form

$$\phi(nlm_lm_s) = \frac{1}{r} P(nl\ldots;r) Y_{lm_l}(\theta,\varphi) \chi_{m_s}, \tag{7}$$

where $P(nl\ldots;r)$ is a real-valued radial function, $Y_{lm_l}(\theta,\varphi)$ a complex-valued spherical harmonic function, and χ_s a spin function. Theories based on determinants derive equations for complex orbitals and may differ depending on how the radial functions depend on quantum numbers m_l and m_s (Szabo and Ostlund, 1982). The orbitals themselves are determined by applying the variational principle to the energy, $E^{\text{HF}} = \langle \Psi(\gamma) | \mathcal{H} | \Psi(\gamma) \rangle$, subject to orthonormality constraints where $\Psi(\gamma)$ is the Slater determinant for configuration γ and $\mathcal{H}$ is the N-electron Hamiltonian for the system. Equations derived by Roothaan in 1951 (Roothaan, 1951) expressed the energy in terms of one-electron integrals and direct and exchange matrix elements of the Coulomb operator. In contrast, when the HF wave function is a configuration state function, the radial functions $P(nl;r)$ depend only on nl quantum numbers, and angular integrations can be performed using Racah algebra producing an energy expression involving only real radial integrals. Examples of expressions can be found in Froese Fischer et al. (1997) and the associated code is available from Froese Fischer (2000).

Let us define the operator, $\mathcal{I}$, as

$$\mathcal{I} = -\frac{1}{2}\frac{d^2}{dr^2} - \frac{Z}{r} + \frac{l(l+1)}{2r^2}. \tag{8}$$

Then the contribution from the one-electron part of the Hamiltonian (in coordinate space) for a CSF consisting of subshells, a^{q_a}, is

$$\langle \gamma LS | \sum_{i=1}^{N} \left(-\frac{1}{2}\nabla_i^2 - \frac{Z}{r_i} \right) | \gamma LS \rangle = \sum_a q_a I(a,a), \tag{9}$$

where $I(a,a) = \langle a | \mathcal{I} | a \rangle$ and a denotes the nl quantum numbers of an orbital. Contributions from the two-electron Coulomb-repulsion operator $\sum_{i<j} 1/r_{ij}$ lead to Slater integrals $R^k(a,b;c,d)$.

The introduction of these integrals presented many computational difficulties in the 1930s (McDougall, 1932), but Hartree soon realized that a feasible solution was to first compute the function

$$Y^k(ac;r) = r \int_0^\infty \frac{r_<^k}{r_>^{k+1}} P(a;s) P(c;s)\, ds$$

$$= \int_0^r \left(\frac{s}{r}\right)^k P(a;s)P(c;s)\,ds$$

$$+ \int_r^\infty \left(\frac{r}{s}\right)^{k+1} P(a;s)P(c;s)\,ds, \tag{10}$$

where $r_< = \min(r,s)$ and $r_> = \max(r,s)$. Then the general Slater integral $R^k(a,b;c,d)$ could be evaluated as

$$R^k(a,b;c,d) = \int_0^\infty P(b;r)P(d;r)\left(\frac{1}{r}\right)Y^k(ac;r)\,dr. \tag{11}$$

In a direct integral $cd = ab$ whereas in an exchange integral, $cd = ba$ and the integrals are denoted as $F^k(a,b)$ and $G^k(a,b)$ respectively.

Let us consider some cases that illustrate the nature of the Hartree–Fock problem. For brevity, we will refer to orbitals simply as $a, b, \ldots$ and only give the spectroscopic notation when needed for clarity. We shall also use the notation $\mathcal{E}$ for the energy expression, E for the value of the energy in atomic units, ε for Lagrange multipliers also in atomic units, $\mathcal{C}$ for conditions, and $\mathcal{F}$ for the energy functional,

$$\mathcal{F} = \mathcal{E}(\gamma LS) + \sum_a \sum_b \varepsilon_{ab}\delta(l_a, l_b)\mathcal{C}_{ab}, \tag{12}$$

where $\delta(l,l')$ is the Dirac delta function and the conditions are

$$\mathcal{C}_{ab} = \begin{cases} (1 - \langle a|a\rangle), & a = b \text{ (for normalization)}, \\ -\langle a|b\rangle, & a \neq b \text{ (for orthogonality)}. \end{cases} \tag{13}$$

Since $\mathcal{C}_{ab} = \mathcal{C}_{ba}$, it follows that $\varepsilon_{ab} = \varepsilon_{ba}$.

2.1. A Case with no Orthogonality Constraints

For the $1s^2 2p\,^2P$ CSF, the energy expression is

$$\mathcal{E}(1s^2 2p\,^2P) = 2I(a,a) + I(b,b) + F^0(a,a) + 2F^0(a,b)$$
$$- (1/3)G^1(a,b), \tag{14}$$

where a refers to $1s$, and b to $2p$. The orbitals are orthogonal through their angular factors, and no additional constraint is needed. The energy expression was derived under the assumption that all orbitals were normalized. For such a constrained variation, Lagrange multipliers ε_{aa} and ε_{bb} need to be introduced and the variational principle applied to the functional,

$$\mathcal{F}(a,b) = \mathcal{E}(1s^2 2p\,^2P) + \varepsilon_{aa}(1 - \langle a|a\rangle) + \varepsilon_{bb}(1 - \langle b|b\rangle). \tag{15}$$

The stationary condition with respect to variations of $P(a;r)$ and $P(b;r)$ respectively, after dividing by two for Lagrange multipliers in atomic units, lead to the equations (see Froese Fischer et al., 1997, for more details)

$$\left(2\mathcal{I} + \frac{2}{r}\left[Y^0(aa;r) + Y^0(bb;r)\right] - \varepsilon_{aa}\right)P(a;r)$$
$$- \frac{1}{3r}Y^1(ab;r)P(b;r) = 0 \tag{16}$$

and

$$\left(\mathcal{I} + \frac{2}{r}Y^0(aa;r) - \varepsilon_{bb}\right)P(b;r) - \frac{1}{3r}Y^1(ba;r)P(a;r) = 0. \tag{17}$$

Let us define an integral equation operator,

$$Y^k(b\bullet;r) = r\int_0^\infty \frac{r_<^k}{r_>^{k+1}} P(b;s) \bullet ds \tag{18}$$

so that $Y^k(b\bullet;r)P(a;r) = Y^k(ba;r)$. Then these equations become

$$\left(2\mathcal{I} + \frac{2}{r}\left[Y^0(aa;r) + Y^0(bb;r)\right]\right.$$
$$\left. - \frac{1}{3r}P(b;r)Y^1(b\bullet;r) - \varepsilon_{aa}\right)P(a;r) = 0 \tag{19}$$

and, since $Y^k(ba;r) = Y^k(ab;r)$,

$$\left(\mathcal{I} + \frac{2}{r}Y^0(aa;r) - \frac{1}{3r}P(a;r)Y^1(a\bullet;r) - \varepsilon_{b,b}\right)P(b;r) = 0. \tag{20}$$

These can be expressed as the system of equations,

$$\left(\mathcal{H}^a - \varepsilon_{aa}\right)P(a;r) = 0,$$
$$\left(\mathcal{H}^b - \varepsilon_{bb}\right)P(b;r) = 0, \tag{21}$$

where we have introduced the notion of a Hartree–Fock operator that is different for each radial function. Let $\mathcal{E}(\gamma;a)$ be the contribution from the subshell a^{q_a} to $\mathcal{E}(\gamma LS)$ and $\mathcal{E}(\gamma;\underline{a})$ the remaining contribution which, in the single configuration Hartree–Fock approximation, is the energy expression for the configuration that remains after a^{q_a} has been removed. Then,

$$E(\gamma LS) = E(\gamma;a) + E(\gamma;\underline{a}). \tag{22}$$

(Here we distinguish between the expression $\mathcal{E}(\gamma LS)$ for which orbitals need not be known and the value of the expression $E(\gamma LS)$.) The diagonal Lagrange multiplier is derived from $\mathcal{E}(\gamma; a)$ and it can readily be confirmed that

$$\varepsilon_{aa} = E(\gamma; a) + \sum_k c_k F^k(a, a) \tag{23}$$

where the sum represents the Coulomb "self-energy" correction

$$\left\langle a^{q_a} \Big| \sum_{ij} 1/r_{ij} \Big| a^{q_a} \right\rangle$$

due to the fact that the $F^k(a,a)$ terms in the energy expression are multiplied by a factor two larger than integrals $F^k(a, b)$ where $a \neq b$. From Eq. (22), Koopmans' theorem follows, namely

$$\varepsilon_{aa} = E(\gamma LS) - E(\gamma; \underline{a}) + \left\langle a^{q_a} \Big| \sum_{ij} 1/r_{ij} \Big| a^{q_a} \right\rangle. \tag{24}$$

For a singly occupied shell the self-energy correction is zero and ε_{aa} becomes the binding energy of orbital a. In the case of a multiply occupied shell, ε_{aa} is the binding energy of the shell plus the self-energy correction. Because of this relationship to the binding energy the Lagrange multipliers are also referred to as diagonal and off-diagonal energy parameters. In earlier publications (Froese Fischer et al., 1997; Dyall et al., 1989) energy parameters ε_{ab} were related to Lagrange multipliers λ_{ab} in that $\varepsilon_{ab} = \lambda_{ab}/q_a$. Consequently the ε-matrix was not symmetric, in general. In this present paper, their definition is the same. We will refer to Lagrange multipliers when discussing their role in the derivation of the variational equations, and energy parameters when referring to the numerical values of these quantities. In order for the latter to be in atomic units, it is necessary to divide the stationary condition by a factor of two. This step often has not been taken leading to energy parameters in Rydbergs (Froese Fischer et al., 1997; Cowan, 1981).

Notice that the two equations are not explicitly coupled, only implicitly through the fact that $\mathcal{H}^a$, for example, depends on $P(b; r)$. The solutions of these equations satisfy Brillouin's theorem in that the single excitation $1s \to ns$ (without a change in coupling) or $2p \to np$ lead to configuration states for which $\langle 1s^2 2p\,^2P|\mathcal{H}|1sns(^1S)2p\,^3P \rangle = 0$ and $\langle 1s^2 2p\,^3P|\mathcal{H}|1s^2 np\,^3P \rangle = 0$, respectively. In a more complex general case, this is an excitation of a radial function that does not affect the spin-angular factor of the configuration state, only the radial functions. Another way of viewing this result is that the HF approximation has already included these single excitations to unoccupied orbitals to first order.

2.2. Cases with Orthogonality Constraints

Discussions about the Hartree–Fock equations for the $1s2s\,^{1,3}S$ CSFs have a long history. Whereas the triplet state was straight forward the singlet state was problematic. Trefftz et al. (1957) had shown in 1957 that good results could be obtained for $1s2s\,^1S$ using non-orthogonal orbitals. Sharma and Coulson (1962) claimed that a non-orthogonal approach was in fact needed, that orthogonality lead to some serious inconsistencies. Sharma (1967) went on to prove that the requirement of orthogonality was inconsistent with the zero-order orbitals being hydrogenic when $1/Z$ was treated as a perturbation parameter in the expansion of the orbitals. But by 1972, the idea that the orbitals were each a linear superposition of the two hydrogenic orbitals was investigated by Rebelo and Sharma (1967) and the existence of *two* solutions was reported. In the meantime, it was also shown that results improved by requiring the excited state to be orthogonal to the ground state (Froese, 1967). These papers, along with others, were very specific to $1s2s\,^{1,3}S$. Here we will present a discussion that can be generalized to other cases.

The $1s2s\,^3S$ and ^1S cases differ in that the former is the lowest $nsn's\,^3S$ CSF whereas the latter is the second $nsn's\,^1S$ CSF so that its energy is not a minimum but rather a stationary point. Also, the wave function for the former is a single Slater determinant whereas the latter is not. The energy expressions are

$$\mathcal{E}(1s2s\,^{1,3}S) = I(a,a) + I(b,b) + F^0(a,b) \pm G^0(a,b), \tag{25}$$

and the radial parts of the configuration state functions are

$$\frac{1}{\sqrt{2}}\bigl[P(a;r_1)P(b;r_2) \pm P(b;r_1)P(a;r_2)\bigr], \tag{26}$$

where the $+$ sign applies to the ^1S state, and the $-$ to the ^3S. These expressions are based on the assumption that radial functions are orthonormal, and so the energy functional now becomes

$$\begin{aligned}\mathcal{F}(a,b) = {}&\mathcal{E}\bigl(1s2s\,^{3,1}S\bigr) + \varepsilon_{aa}\bigl(1 - \langle a|a\rangle\bigr) \\ &+ \varepsilon_{bb}\bigl(1 - \langle b|b\rangle\bigr) - 2\varepsilon_{ab}\langle a|b\rangle.\end{aligned} \tag{27}$$

The stationary condition with respect to variations of $P(a;r)$ and $P(b;r)$ leads to the equations:

$$\begin{aligned}&\left(\mathcal{I} + \frac{1}{r}\bigl[Y^0(bb;r) \pm P(b;r)Y^1(b\bullet;r)\bigr] - \varepsilon_{aa}\right)P(a;r) \\ &\quad - \varepsilon_{ab}P(b;r) = 0,\end{aligned} \tag{28}$$

and a similar equation with a and b interchanged. These are coupled equations of the form

$$\left(\begin{bmatrix}\mathcal{H}^a & 0 \\ 0 & \mathcal{H}^b\end{bmatrix} - \begin{bmatrix}\varepsilon_{aa} & \varepsilon_{ab} \\ \varepsilon_{ba} & \varepsilon_{bb}\end{bmatrix}\right)\begin{bmatrix}P(a;r) \\ P(b;r)\end{bmatrix} = 0, \tag{29}$$

where the Lagrange multipliers form a symmetric ε-matrix. The solution of these equations is greatly simplified when the ε-matrix is diagonal.

From the orthonormality of the radial functions, it readily can be shown that the elements of ε are matrix elements of the one electron Hartree–Fock Hamiltonians:

$$\varepsilon_{aa} = \langle a|\mathcal{H}^a|a\rangle, \qquad \varepsilon_{ab} = \langle b|\mathcal{H}^a|a\rangle,$$
$$\varepsilon_{bb} = \langle b|\mathcal{H}^b|b\rangle, \qquad \varepsilon_{ba} = \langle a|\mathcal{H}^b|b\rangle. \tag{30}$$

The condition that $\varepsilon_{ab} = \varepsilon_{ba}$ then becomes the condition,

$$\langle b|\mathcal{H}^a|a\rangle = \langle a|\mathcal{H}^b|b\rangle, \tag{31}$$

and affects the ^3S and ^1S states quite differently. When several radial functions with the same angular symmetry are present, the solutions of the HF equations not only are stationary with respect to single excitations to unoccupied radial functions, they also are stationary with respect to orthogonal transformations.

Let us rewrite Eq. (29) in matrix vector form as

$$(\mathcal{H} - \varepsilon)\mathbf{P} = 0, \tag{32}$$

where $\mathcal{H}$ and ε are 2×2 matrices and $\mathbf{P}$ is the column vector $[P(a;r), P(b;r)]^t$. Let $\mathbf{O}$ be an orthogonal matrix. Multiplying Eq. (32) on the left by $\mathbf{O}^t$ and using the property that $\mathbf{OO}^t = \mathbf{I}$ (the identity matrix), we get

$$(\mathbf{O}^t\mathcal{H}\mathbf{O} - \mathbf{O}^t\varepsilon\mathbf{O})\mathbf{O}^t\mathbf{P} = 0 \tag{33}$$

or, in terms of the transformed quantities,

$$(\mathcal{H}^* - \varepsilon^*)\mathbf{P}^* = 0, \tag{34}$$

where $\mathbf{P}^* = \mathbf{O}^t\mathbf{P}$ are the transformed radial functions and, equivalently, since $\mathbf{O}^t = \mathbf{O}^{-1}$, $\mathbf{P} = \mathbf{OP}^*$.

Orthogonal transformations represent the rotation of orbitals in orbital space that in 2-dimensional space can be defined in terms of a single parameter $\eta \in [-1, 1]$. Larger values of η correspond to an interchange of the orbitals. Let

$$\mathbf{O} = \begin{bmatrix} 1 & -\eta \\ \eta & 1 \end{bmatrix} / \sqrt{1 + \eta^2}, \tag{35}$$

where $1/\sqrt{1 + \eta^2} = \cos(\theta)$ and θ represents the angle of rotation $\theta \in [-45, 45]$ degrees. Then the transformation

$$\begin{bmatrix} P^*(a;r) \\ P^*(a;r) \end{bmatrix} = \begin{bmatrix} 1 & \eta \\ -\eta & 1 \end{bmatrix} \begin{bmatrix} P(a;r) \\ P(b;r) \end{bmatrix} / \sqrt{1 + \eta^2} \tag{36}$$

allows us to study the effect of a rotation as a function of η.

2.2.1. The $1s2s\,^3S$ Case

In this case the wave function is a single determinant which is invariant under orthogonal transformations (Szabo and Ostlund, 1982) and consequently the total energy will not change. Using Eq. (36), it is easy to confirm the invariance of the radial factor of the wave function, namely

$$P^*(a; r_1)P^*(b; r_2) - P^*(b; r_1)P^*(a; r_2)$$
$$= P(a; r_1)P(b; r_2) - P(b; r_1)P(a; r_2), \qquad (37)$$

from which it follows that the energy expression

$$\mathcal{E}(1s^*2s^{*\,3}S) = \mathcal{E}(1s2s\,^3S) \qquad (38)$$

and the operator $\mathcal{H}^* = \mathcal{H}$.[1,2] Thus, for a given energy, the radial functions are not unique.

Transformations do not leave the scalar ε-matrix of Eq. (33) invariant. For all orthonormal orbitals (a, b), it can be shown by direct substitution that $\langle b|\mathcal{H}^a|a\rangle = \langle a|\mathcal{H}^b|b\rangle$. Consequently the ε-matrix is symmetric for all transformations. The trace of this matrix (sum of diagonal elements) is invariant under orthogonal transformations and for the particular transformation that diagonalizes the matrix, the diagonal elements assume their extreme values and $\varepsilon_{ab} = \varepsilon_{ba} = 0$. This solution ensures that the $1s$ is the most bound and $2s$ the least bound solution. This choice was suggested by Koopmans (1933) and is part of his theorem that relates diagonal energy parameters to binding energies. What characterizes the ^3S case is the fact that the wave function and energy are unchanged by an orthogonal transformation of the radial functions of a given symmetry and that the ε-matrix is symmetric for all orthonormal orbitals.

2.2.2. The $1s2s\,^1S$ Case

Now the rotation of the orbital basis changes the energy expression as well as the wave function. In the present case, it can be shown by direct substitution into

---

[1] The $\mathcal{H}$-matrix also appears to be transformed but it can be shown that $\mathbf{O}^t\mathcal{H}(P)\mathbf{O} = \mathcal{H}(P^*) = \mathcal{H}(P)$. This is most readily shown by introducing the Fock operator used in determinantal Hartree–Fock theory of quantum chemistry (Szabo and Ostlund, 1982), namely

$$f = \mathcal{L} + \frac{2}{r}\bigl[Y^0(aa; r) + Y^0(bb; r) - P(a; r)Y^0(a\bullet; r) - P(b; r)Y^0(b\bullet; r)\bigr]. \qquad (39)$$

In this form, one direct and one exchange contribution cancel in the application of the operator to an operand depending on whether the operand is a or b. Notice that the operator itself depends on both a and b but a simple substitution shows that $f(\mathbf{P}^*) = f(\mathbf{P})$ and finally, also that $\mathcal{H}^* = \mathcal{H}$.

[2] To better represent the diagonal energy parameters as energies in atomic units, the $\mathcal{L} = -(1/2)\mathcal{I}$ operator has been abandoned and variational equations divided by a factor of two from those appearing in Froese Fischer et al. (1997).

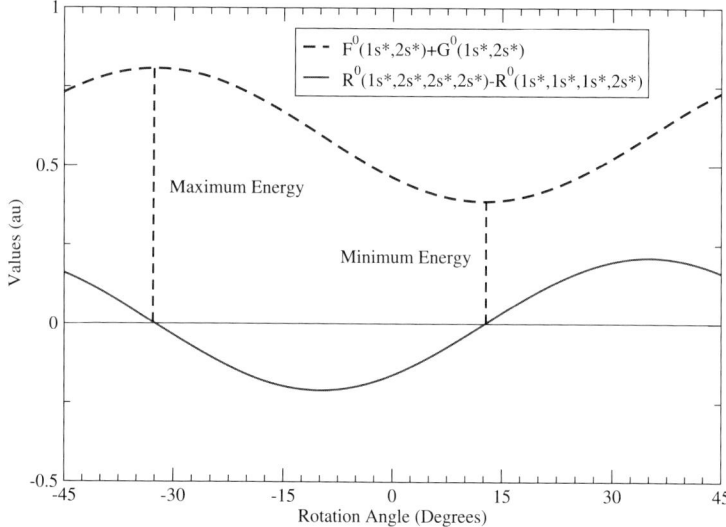

FIG. 1. The values of $F^0(1s^*, 2s^*) + G^0(1s^*, 2s^*)$ and the condition $R^0(1s^*, 2s^*; 2s^*, 2s^*) - R^0(1s^*, 1s^*; 1s^*, 2s^*)$ for rotated orbitals $\{1s^*, 2s^*\}$ as a function of the degree of rotation of hydrogenic orbitals $\{1s, 2s\}$ for $Z = 2$.

Eq. (26) that, in terms of CSFs,

$$\left|1s^*2s^{*\ 1}S\right\rangle = c_1\left|1s2s\ ^1S\right\rangle + c_2\left(\left|2s^{2\ 1}S\right\rangle - \left|1s^{2\ 1}S\right\rangle\right)/\sqrt{2}, \quad (40)$$

where $c_1 = (1 - \eta^2)/(1 + \eta^2)$ and $c_2 = 2\eta/(1 + \eta^2)$. Stationary solutions can be determined by diagonalizing the 2×2 interaction matrix with the result that there will be *two* eigenstates, one where the rotation lowers the energy and another where the energy is raised.

The requirement that the ε-matrix be symmetric reduces to the condition

$$R^0(a, b; b, b) - R^0(a, a; a, b) = 0. \quad (41)$$

Figure 1 shows how $F^0(a^*, b^*) + G^0(a^*, b^*)$ and $R^0(a^*, b^*; b^*, b^*) - R^0(a^*, a^*; a^*, b^*)$ vary as a function of the degree of rotation as $\{1s, 2s\}$ hydrogenic orbitals for helium ($Z = 2$) are rotated. Notice that there are exactly two stationary values of the former and that the orbitals associated with these stationary points satisfy the required condition. The stationary condition of the former is related directly to a stationary point of the total energy since $I(a, a) + I(b, b)$ is invariant under these rotations.

It is important to note that, starting with a CSF, the rotation of the radial functions (which does not affect the spin-angular factor) produced a "perturbation" as shown in Eq. (40) that is a linear combination of two CSFs that may be thought of as a "pseudo-state" (not an approximation to an observed state). A Hartree–Fock solution requires that the energy be stationary with respect to the rotation of orbitals, or equivalently, that the effect of this interaction already be included in the solution, at least to first order. It can readily be verified that in Z-dependent perturbation theory, this pseudo-state is degenerate with the $1s2s\ ^1S$ CSF, and that the zero-order radial functions should then be the rotated hydrogenic orbitals for the smallest rotation.

A numerical multiconfiguration Hartree–Fock (MCHF) program (Froese Fischer, 1986) treats the energy E as a function of η, namely

$$E(\eta) = E(0) + \eta \Delta E + \eta^2 \Delta^2 E + \cdots. \tag{42}$$

Then the stationary condition, neglecting higher order terms, becomes

$$\Delta E + 2\eta \Delta^2 E = 0, \tag{43}$$

from which it follows that $\eta = -\Delta E/(2\Delta^2 E)$ (Froese Fischer, 1986). This process selects the smaller rotation that satisfies the stationary condition.

From expressions for $I(a,a)$ and Slater integrals of rotated orbitals in terms of the original orbitals, rules can be derived correct to order $O(\eta^2)$ such as

$$I(a^*, a^*) = I(a,a) + 2\eta I(a,b) + \eta^2 \big[I(b,b) - I(a,a)\big]$$
$$+ \text{higher order terms},$$
$$I(b^*, b^*) = I(b,b) - 2\eta I(a,b) + \eta^2 \big[I(a,a) - I(b,b)\big]$$
$$+ \text{higher order terms}. \tag{44}$$

Thus $q_a I(a,a) + q_b I(b,b)$ is invariant under rotation for equally occupied, orthogonal subshells ($q_a = q_b$), but not when the subshells are unequally occupied as in $2p^2 3p$.

A direct Slater integral $F^k(a,b)$ has the same transformation properties as $(aa)_1(bb)_2$ where the subscripts indicate the radial coordinates and the symmetry rules are those of Slater integrals: symmetry with regard to interchange of orbitals with a given coordinate, or the interchange of the coordinates themselves. Then

$$(a^*a^*)_1(b^*b^*)_2 = \big[(a+\eta b)_1^2 (b-\eta a)_2^2\big]/(1+\eta^2)^2 \tag{45}$$
$$= \big[(aa + 2\eta ab + \eta^2 bb)_1 (bb - 2\eta ab + \eta^2 aa)_2\big]/(1+\eta^2)^2. \tag{46}$$

Collecting terms by order in η and converting to the notation of Slater integrals we get

$$F^0(a^*, b^*) = F^0(a,b) + 2\eta\big[R^0(a,b;b,b) - R^0(a,a;a,b)\big] \tag{47}$$
$$+ \eta^2 \big[F^0(a,a) + F^0(b,b) - 4G^0(a,b) - 2F^0(a,b)\big] + \cdots. \tag{48}$$

The Slater exchange integrals have the same transformational properties as $(a^*b^*)_1(a^*b^*)_2$ and, expanding in powers of η, the first and second order terms are exactly the same as those for the direct integral. Thus

$$\mathcal{E}(a^*b^{*\,1}S) = \mathcal{E}(ab^{\,1}S) + 4\eta\left[R^0(a,b;b,b) - R^0(a,a;a,b)\right] \quad (49)$$
$$+ 2\eta^2\left[F^0(a,a) + F^0(b,b) - 4G^0(a,b) - 2F^0(a,b)\right] + \cdots \quad (50)$$

and this equation defines the rotation parameter needed for a stationary solution with respect to the current orbital pair (a, b). When this parameter is zero, the ε-matrix will satisfy the symmetry requirement of Eq. (31) and the off-diagonal energy parameter can be eliminated, becoming a projection operator for the orbital Hartree–Fock operator. (We shall prove this for spline-matrix methods.) The solution then is the solution of a pair of eigenvalue problems where the operators are integro–differential operators.

An important consequence of orbital rotations for finite difference methods is that rotations introduce extra nodes into the radial functions, making node-counting an art. See Froese Fischer (1986) for a description of numerical algorithms for solving the HF and MCHF radial equations.

2.2.3. The General Case

When m orbitals of the same angular symmetry are present in the Hartree–Fock configuration state, $m \times m$ orthogonal transformations need to be considered. Vilkas et al. (1998) have reported a quadratically convergent method for finding a unitary transformation for variational solutions of the Dirac–Coulomb Hamiltonian, that transforms an initial set of orthonormal one-electron orbitals to stationary solutions. A similar procedure could be used for finding the needed orthogonal transformations. However in the MCHF program (Froese Fischer, 1986) it was found sufficient to consider each constrained orbital pair separately, and express the m-dimensional orthogonal transformation as $m(m-1)/2$ pairs of rotations.

It should be mentioned that rotation analysis is computationally intensive. In our $1s2s\,^1S$ example, the first-order variation of η required the evaluation of two Slater integrals, whereas the second order variation required four additional integrals. For the multiple shell case, the energy expressions are not as simple. What is essential is that the stationary condition be satisfied, at least approximately at each iteration, and that the process be convergent. Thus there are trade-offs between iterative procedures with a minimum number of self-consistent field (SCF) iterations and extensive computation per iteration and simpler calculations requiring more iterations. In the final analysis, an accurate, stable and robust process is of prime importance. In the early 1950s, Hartree spent a lot of effort on trying to get good initial estimates (Hartree, 1955). A decade or two later, computers were sufficiently powerful that simpler, general procedures were sufficient and preferred.

3. Multiconfiguration Hartree–Fock Approximation

In a multiconfiguration Hartree–Fock approximation, the wave function $\Psi(\gamma LS)$ for a state labelled γLS is a linear combination of orthonormal configuration state functions (CSF) $\Phi(\gamma LS)$ so that

$$\Psi(\gamma LS) = \sum_{i=1}^{M} c_i \Phi(\gamma_i LS), \quad \text{where} \sum_{i=1}^{M} c_i^2 = 1, \tag{51}$$

and the energy is required to be stationary both with respect to the variation of the radial functions and the expansion coefficients. When all the radial functions are known, this becomes a configuration-interaction (CI) problem. where the expansion coefficients are an eigenvector of an interaction matrix $H = (H_{ij})$ where $H_{ij} = \langle \gamma_i LS | \mathcal{H} | \gamma_j LS \rangle$ and $\mathcal{H}$ is an N-electron Hamiltonian.

Many of the difficulties caused by the orthogonality condition in the HF approximation may not be present in the multiconfiguration approximation. Orthogonality conditions are avoided in the more complex non-orthogonal theory but wave functions expressed in terms of non-orthogonal orbitals can be shown to be equivalent to an orthogonal MCHF theory. Consider the $1s2s\,^1S$ case. In a non-orthogonal $1s2s'\,^1S$ approximation the $2s'$ orbital need not be orthogonal to $1s$, so no off-diagonal Lagrange multiplier is needed for the derivation of the orbital equations though the expression for the energy now includes overlap integrals. Solutions can readily be obtained (Froese Fischer, 1966). However, projecting out the $1s$ orbital from a non-orthogonal $2s'$, so that $2s' = a_1 2s + a_2 1s$, and $\langle 1s | 2s \rangle = 0$, we get a multiconfiguration expansion,

$$\left| 1s2s'\,^1S \right\rangle = a_1 \left| 1s2s\,^1S \right\rangle + \sqrt{2} a_2 \left| 1s^2\,^1S \right\rangle. \tag{52}$$

In general, by expressing a non-orthogonal set of orbitals in terms of an orthonormal set, expansions in terms of the former can be shown to be equivalent to larger multiconfiguration expansions in terms of the latter. Thus the MCHF calculation for the second eigenstate of

$$\Psi\left(1s2s\,^1S\right) = c_1 \left| 1s^2\,^1S \right\rangle + c_2 \left| 1s2s\,^1S \right\rangle \tag{53}$$

is computationally simpler than the HF calculation for $1s2s\,^1S$. Furthermore, the energy is now an upperbound to the exact energy. Though non-orthogonal computations are robust, the multiconfiguration calculations generally do not present a problem as long as node-counting can be avoided.

Many different models may be used for generating an expansion. An important concept is that of an active space method, where all configuration states that can be generated from a given set of orbitals are included. In such cases, rotations again will leave the wave function invariant, giving the calculation some degrees of freedom. They could be used to set off-diagonal energy parameters to zero but now there also is the possibility of setting certain coefficients in the expansion to zero or, equivalently, removing a CSF and reducing the size of the problem.

Consider an expansion for the $1s^2$ ground state over the set of 55 CSFs $|nsms\,^1S\rangle$, for orbitals $ns, ms \in \{1s, \ldots, 10s\}$. There are 45 orthogonality conditions and so 45 coefficients can be set to zero, leading to the natural orbital expansion,

$$\Psi(1s^2\,^1S) = \sum_{n=1}^{10} c_n \Phi(ns^2\,^1S), \qquad (54)$$

first proposed by Löwdin (1955) where all CSFs differ by two electrons. In the case of $1s2s\,^1S$, the expansion is similar, but with $2s^2$ replaced by $1s2s$. Such optimizations are very state selective (Froese Fischer, 1977; Froese Fischer et al., 1997).

In the Hartree–Fock approximation, the advantage of a variational method lay in the fact that Brillouin's theorem was satisfied for perturbations in the form of an excitation of an occupied orbital to an unoccupied orbital and sometimes also for rotations. Often, as in $1s2s\,^1S$, the perturbation produced a pseudo-state that is a constrained linear combination of CSFs (Froese Fischer, 1973). To get the correct combination of each CSF, they must be included explicitly in the expansion. Furthermore, in MCHF a given occupied orbital may appear in many CSFs, so any radial perturbation of an occupied orbital to a virtual orbital will also be in terms of combinations of CSFs. So the advantage of Brillouin's theorem is reduced and the CSFs included in the expansion become a more important factor in the accuracy of a calculation.

In the discussion so far, it has always been assumed that optimization is for a single HF or MCHF eigenstate, but an energy expression can also be a weighted linear combination of energy expressions for different LS terms and/or different eigenstates for a given LS term. Such procedures are referred to as "simultaneous optimization" in MCHF (Tachiev and Froese Fischer, 1999), "extended optimal level" in the general relativistic atomic structure package (GRASP) (Parpia et al., 1996) or "state-averaged" in multi-reference Möller–Plesset (MR-MP) methods (Vilkas and Ishikawa, 2004b). In the calculation of energy levels associated with a spectrum (where energy differences are more important than the total energies themselves) independently optimized calculations introduce an imbalance in correlation and prevent an accurate prediction of level separation, an imbalance that would disappear only for "exact" calculations (Vilkas and Ishikawa, 2004b).

Variational methods are adept at determining radial functions for the larger components of a wave function, those that contribute significantly to the energy. Such wave functions may be considered the zero-order approximation and the CSFs in this set are referred to as the "multi-reference set". This set needs to account for near-degeneracy effects and strong interactions within the complex of

CSFs with the same principal quantum numbers. Additional CSFs (and orbitals) may then be introduced to represent the first-order approximation through single- (S) and double- (D) excitations from the reference set. Since the Coulomb operator is a 2-electron operator, this set will include all CSFs obtained through SD excitations that interact with one or more members of the multi-reference set and defines the contributions to the first-order correction to the wave function for the orbital set. Normally these orbitals too are determined variationally, but the effective completeness of spline orbitals may be used to combine variational methods with many-body perturbation theory as done by Vilkas et al. (1998) using an analytic basis.

4. B-Spline Theory

Let us now consider a method in which continuous functions are approximated locally by polynomials to arbitrary accuracy.

Consider an interval $[a, b]$, partitioned into subintervals $[t_i, t_{i+1}]$ defined by the knot sequence, t_i, $i = 1, \ldots$, also called "break-points". A B-spline, say $B_i(r)$, of order k_s is a positive function over k_s adjacent intervals beginning at t_i. In the first interval it increases from zero as $(r - t_i)^{k_s - 1}$ as r increases and in the last, it decreases to zero as $(t_{i+k_s} - r)^{k_s - 1}$. In each intervening subinterval the B-spline is a polynomial of degree $k_s - 1$, and in going from one subinterval to the next the B-spline function and all derivatives up to order $k_s - 2$ are continuous. The only discontinuity is in the derivative of order $k_s - 1$. With sufficiently large values of k_s, a spline has a high degree of smoothness. For each subinterval, there are k_s non-zero B-splines defined in the interval and define the basis for any polynomial approximation in the interval. The leftmost spline requires $k_s - 1$ extra knots at points $t \leqslant a$ and similarly the rightmost spline requires $k_s - 1$ extra knots at points $t \geqslant b$. Thus with n_v intervals, there are $n_t = n_v + 2k_s - 1$ knots and $n_s = n_v + k_s - 1$ basis functions, $B_i(r)$, $i = 1, \ldots, n_s$. When the additional knots are selected to be $t = a$ or $t = b$, the end-points become knots of multiplicity k_s and the B-spline basis acquires some useful properties. At $t = a$, only $B_1(a)$, $B_2'(a)$, $B_3''(a)$, ... are non-zero for derivatives up to $k_s - 1$. Similarly, at $t = b$, only $B_{n_s}(b)$, $B'_{n_s-1}(b)$, $B''_{n_s-2}(b)$, ... are non-zero. Thus applying a zero boundary condition (either to the function or a derivative) can be interpreted as setting the coefficient of a basis function to zero.

Given the set of knots, the B-splines and their derivatives can be generated from a recurrence relation. Let $B_{i,k_s}(r)$ be the spline of order k_s usually denoted simply as $B_i(r)$. Then

$$B_{i,1} = \begin{cases} 1, & t_i \leqslant r \leqslant t_{i+1}, \\ 0, & \text{otherwise} \end{cases} \tag{55}$$

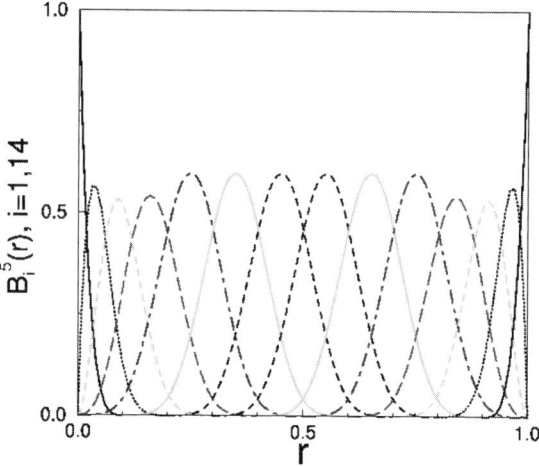

FIG. 2. B-splines of order 5 on the interval [0, 1] with 10 subintervals of equal length and knots of multiplicity 5 at each end point (from Qiu and Froese Fischer, 1999).

with

$$B_{i,k_s}(r) = \frac{r - t_i}{t_{i+k_s-1} - t_i} B_{i,k_s-1}(r) + \frac{t_{i+k_s} - r}{t_{i+k_s} - t_{i+1}} B_{i+1,k_s-1}(r) \quad (56)$$

and

$$B'_{i,k_s}(r) = \frac{k_s - 1}{t_{i+k_s-1} - t_i} B_{i,k_s-1}(r) - \frac{k_s - 1}{t_{i+k_s} - t_{i+1}} B_{i+1,k_s-1}(r) \quad (57)$$

The splines of a given order are normalized so that $\sum_i B_i(r) = 1$. Algorithms for generating splines and their derivatives were developed by de Boor (1985) and published in his book. Figure 2 shows some splines of order 5 on the interval [0, 1] with 10 subintervals of equal length. Note the boundary effect on the shape of the B-splines.

A function that can be expanded exactly in a the B-spline basis, say $F(r) = \sum_j a_j B_j(r)$, is a spline function of order k_s. It is a polynomial of degree $k_s - 1$ in each subinterval with continuous derivatives up to order $k_s - 2$ inside $[a, b]$.

Spline functions may accurately approximate continuous functions such as radial functions. Let $P(r)$ be an arbitrary radial function for which $P(0) = 0$ and $P(R) = 0$, where R is sufficiently large for the boundary condition to apply to the accuracy required. Given a knot sequence and the order of the spline, say k_s, the "best" spline is the one for which the error is a minimum with respect to some norm of the error, $\|P(r) - F(r)\|$. Let us define

$$\|P(r) - F(r)\|^2 = \int_0^R (P(r) - F(r))^2 \, dr. \tag{58}$$

Then the "best" spline in the least squares norm is one for which (in the Dirac notation)

$$\langle B_i(r) | P(r) - F(r) \rangle = 0, \quad i = 1, \ldots, n_s. \tag{59}$$

Expressed another way, the error $P(r) - F(r)$ is orthogonal to each of the spline basis functions. This requirement leads to a system of equations

$$\sum_j a_j \langle B_i(r) | B_j(r) \rangle = \langle B_i(r) | P(r) \rangle, \quad i = 1, \ldots, n_s \tag{60}$$

or, in matrix–vector form, as

$$Sa = b, \tag{61}$$

where the matrix $S = (S_{ij})$, $S_{ij} = \langle B_i(r) | B_j(r) \rangle$ is a symmetric, positive-definite, banded overlap matrix for the B-spline basis of bandwidth $2k_s - 1$, b is the overlap of the radial function with each basis function, and a is the vector of expansion coefficients.

4.1. Spline Methods for the Solution of Differential Equations

B-spline methods can also be used to solve differential equations. Consider the central field equation (1) and assume $P(nl; r) = \sum_j a_j B_j(r)$. The boundary conditions require $a_1 = 0$ and we also assume $a_{n_s} = 0$. Substituting the expansion into Eq. (1) we get

$$\left(-\frac{1}{2} \frac{d^2}{dr^2} - \frac{1}{r} [Z - Y(nl; r)] + \frac{l(l+1)}{2r^2} - \varepsilon_{nl,nl} \right) \sum_j a_j B_j(r)$$
$$= \mathrm{res}(r), \tag{62}$$

where $\mathrm{res}(r)$ is the residual or amount by which the equation is not satisfied. In the Galerkin method (Fletcher, 1984), the solution minimizes $\|\mathrm{res}(r)\|$ and satisfies the equations

$$\langle B_i(r) | \mathrm{res}(r) \rangle = 0, \quad i = 2, \ldots, n_s - 1 \tag{63}$$

which can be written in the matrix–vector form of Eq. (1). Both matrices are symmetric, banded matrices of bandwidth $2k_s - 1$.

4.2. SPLINE GRID FOR RADIAL FUNCTIONS

From the fact that in each interval the linear combination of k_s B-splines non-zero over this interval defines a local polynomial approximation to the radial function, it is apparent from the polynomial behavior of radial functions near the origin that small intervals are not required in spite of the singularity at the origin, and that an exponential grid can be used for large r. In atomic structure applications, it is convenient to define the grid in terms of the variable $t = Zr$ (the hydrogenic case) and then transform to the current value of Z. Let $h = 1/2^m$. This is not strictly necessary, but it ensures that the arithmetic is exact in binary arithmetic. Then

$$\begin{aligned}
t_i &= 0 \quad \text{for } i = 1, \ldots, k_s, \\
t_i &= t_{i-1} + h \quad \text{for } i = k_s + 1, \ldots, k_s + m, \\
t_i &= t_{i-1}(1 + h) \quad \text{for } i = k_s + m + 1, \ldots, n_v + k_s, \\
t_i &= t_{i-1} \quad \text{for } i = n_v + k_s + 1, \ldots, n_t, \\
t_i &= t_i/Z \quad \text{for } i = 1, n_t.
\end{aligned} \quad (64)$$

Thus the grid is linear near the origin and linear in $\log(r)$ thereafter. The knots of multiplicity k_s at $t = 0$ are followed by m intervals of length h. The rightmost knot of this last interval is at $t = 1$ which is where the $1s$ radial function for hydrogen achieves its maximum value and begins an exponential decay. The grid reflects this behavior with intervals that increase in length for a total of n_v intervals. The rightmost knot is then repeated to have multiplicity k_s and a scaling for the nuclear charge is taken into account. This is the grid used in the examples reported here. It is not an optimal grid but, as we will see later, it has some computational advantages. We will refer to it as a logarithmic grid in order to differentiate it from a commonly used exponential grid (see Bachau et al., 2001, for a summary of commonly used grids).

For a given Z, the grid has a parameter h that can be made arbitrarily small. In the absence of round-off, numerical experiments show that

$$\max_{r \in (0,R)} |P(r) - F(r)| = \mathcal{O}(h^{k_s}) \quad (65)$$

for $1s$ and $2s$ radial functions. Thus larger intervals (smaller number of basis functions) can be used with higher-order spline approximations.

4.3. INTEGRATION METHODS

The evaluation of B-spline matrix elements of differential operators or functions in the least squares norm of Eq. (58), requires integration. Since B-splines are polynomials these integrations can be performed symbolically. However numerical integration is faster and can be almost as accurate.

Instead of representing a polynomial of order $k_s - 1$ by k_s coefficients, we represent the polynomial by k_s values at the Gaussian points for Gauss–Legendre integration formula,

$$\int_0^1 f(x)\,dx = \sum_{i=1}^{k_s} g_w(x_i) f(x_i), \tag{66}$$

where $g_w(x_i)$ are the Gaussian weights at the Gaussian points x_i. This formula is exact for integrands $f(x)$ that are polynomials of degree $2k_s - 1$ and, when applied to integration over an interval of length h, the error is $\mathcal{O}(f^{2k_s}(x) h^{2k_s+1})$, where x is some point in the interval. Thus the overlap matrix, for example, is computed exactly, except for round-off and without any cancellation. The matrix elements of $1/r$ and $1/r^2$ are evaluated exactly in the first interval because of the boundary conditions, but integration in the second interval is no longer exact. The polynomial form of the product of the B-splines will now have a constant, say c. Because the derivatives of $f(r) = 1/r$ increase rapidly with order, this region may be the limiting factor for accuracy. Solutions to the Dirac equation are not polynomial near the origin. Errors can only be controlled through the use of small subintervals (Froese Fischer and Parpia, 1993) and possibly lower-order splines that reduce the size of the derivative in the expression for the error.

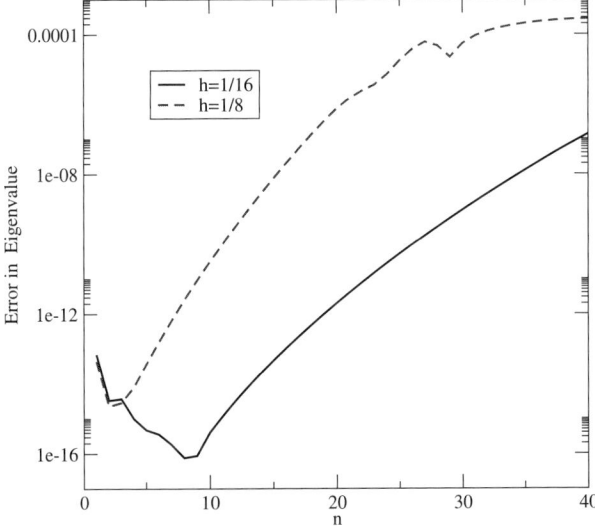

FIG. 3. The error in the eigenvalues as a function of n for two spline calculations for hydrogen with $l = 0$ and $k_s = 8$. (i) dashed curve: $h = 1/8$, $R = 1000 a_0$; (ii) solid curve: $h = 1/16$, $R = 10{,}000 a_0$.

Figure 3 shows the accuracy to which the eigenvalues of the radial equation for hydrogen can be computed as a function of n for s-orbital calculations with $k_s = 8$. Two sets of results are presented, one for $h = 1/8$, $R = 1000a_0$, and another for $h = 1/16$ and $R = 10{,}000a_0$. Some limitations in accuracy are noticed for $n = 1$–3, but for higher n the improvement in accuracy is better by 4 or more orders of magnitude primarily due to the much larger maximum radius. The size of the spline basis was $n_s = 75$ and 175, respectively. Thus, for sufficiently large R, solutions can be obtained for highly excited states. In the smaller calculation with the dashed curve, the eigenvalues started to represent continuum states at $n = 27$ where the irregularity in the curve appears.

4.4. Slater Integrals

The computational process Hartree suggested for determining Slater integrals involved a two-step process: first the Y^k functions of Eq. (10) were determined and then the Slater integral through direct integration. With splines, it is possible to compute Y^k functions at all the Gaussian points from the integral definition, but it is not straight-forward. On the other hand, with the Galerkin method, they can readily be determined from the differential equation,

$$\frac{d^2}{dr^2} Y^k(r) = \frac{k(k+1)}{r^2} Y^k(r) - \frac{2k+1}{r} P_{nl}(r) P_{n'l'}(r), \qquad (67)$$

and the boundary conditions,

$$Y^k(0) = 0,$$
$$\frac{d}{dr} Y^k(r) = \frac{-k}{r} Y^k(r) \quad \text{as } r \to \infty. \qquad (68)$$

Similar equations have been derived for all the Breit–Pauli operators (Brage and Froese Fischer, 1994a; Zatsarinny, 2006). Extensive studies of accuracy have been performed by Zatsarinny (2006) using the B-spline R-matrix (BSR) code.

In variational methods, the derivation of differential equations can be avoided by starting with the assumption that the radial function of an orbital will be expanded in a B-spline basis. Let a, b, c, d denote a set of nl quantum numbers with

$$P(a; r) = \sum_i a_i B_i(r) \qquad (69)$$

and similar expansions for b, c, and d. Then

$$R^k(a, b; c, d) = \sum_i \sum_j \sum_{i'} \sum_{j'} a_i b_j c_{i'} d_{j'} R^k(i, j; i', j') \qquad (70)$$

where

$$R^k(i,j;i',j') = \int_0^R \int_0^R \frac{r_<^k}{r_>^{k+1}} B_i(r_1) B_j(r_2) B_{i'}(r_1) B_{j'+1}(r_2) \, dr_1 \, dr_2 \quad (71)$$

and will be referred to as B-spline Slater integrals or as Slater matrix elements. Many symmetries exist which can reduce the range of summations. From the definition, it follows that the elements are symmetric with respect to the exchange of indices i and i', j and j', and $\{i, i'\}$ and $\{j, j'\}$, i.e.

$$\begin{aligned}
R^k(i,j;i',j') &= R^k(i',j;i,j') \\
&= R^k(i,j';i',j) \\
&= R^k(j,i;j',i') \\
&= R^k(j',i;j,i') \\
&= R^k(j,i';j',i) \\
&= R^k(j',i';j,i) \\
&= R^k(i',j';i,j).
\end{aligned} \quad (72)$$

In addition,

$$R^k(i,j;i',j') = \begin{cases} 0 & \text{if } |i - i'| > k_s, \\ 0 & \text{if } |j - j'| > k_s. \end{cases} \quad (73)$$

This symmetry can be used to reduce the number of elements that need to be computed and stored in memory. Equation (70) can be written as

$$R^k(a,b;c,d) = \sum_i \sum_{i'} a_i c_{i'} \left(\sum_j \sum_{j'} b_j d_{j'} R^k(i,j;i',j') \right). \quad (74)$$

Using symmetry relations these sums can be rearranged to use only the values $i \leq i'$ and $j \leq j'$, namely

$$\sum_j \sum_{j'} b_j d_{j'} R^k(i,j,i',j') = \sum_j b_j d_j R^k(i,j,i',j)$$
$$+ \sum_j \sum_{j'>j} (b_j d_{j'} + b_{j'} d_j) R^k(i,j,i',j') \quad (75)$$

with similar rearrangements for summations over (i, i'). Combining this with the finite support of B-splines of Eq. (73), only the values $i \leq i' \leq i + k_s - 1$ and $j \leq j' \leq j + k_s - 1$ need to be stored. Defining "direct density" matrices,

$$A(i,i') = \begin{cases} a_i c_i & \text{if } i' = i, \\ a_i c_{i'} + a_{i'} c_i & \text{if } i' > i \text{ and } i' \leq n_s, \end{cases}$$

$$B(j, j') = \begin{cases} b_j d_j & \text{if } j' = j, \\ b_j d_{j'} + b_{j'} d_j & \text{if } j' > j \text{ and } j' \leqslant n_s, \end{cases} \quad (76)$$

we get

$$R^k(a, b; c, d) = \sum_{i=1}^{n_s} \sum_{i'=i}^{i+k_s-1} A(i, i') \sum_{j=1}^{n_s} \sum_{j'=j}^{j+k_s-1} B(j, j') R^k(i, j; i', j'). \quad (77)$$

The above analysis has taken into account the (i, i') and (j, j') symmetry but not the symmetry from the interchange of the (i, i') pair with the (j, j') pair.

An accurate and efficient evaluation of Slater integrals in the B-spline basis is essential in a Hartree–Fock calculation, consuming most of the computational time. Altenbeger-Siczek and Gilbert (1976) investigated the use of a spline basis for the two-electron helium problem. Though initially optimistic about the use of splines in earlier publications, they concluded that splines were not suitable for atomic structure calculations when Slater integrals needed to be evaluated. An important factor in this decision was their reliance on splines of order 4 for which a large number of basis functions are needed to achieve high accuracy. With the publication by de Boor (1985) of general purpose codes for B-splines, this limitation has been removed.

At first sight, one might expect the computation of the $R^k(i, j; i', j')$ array to be a task of order $\mathcal{O}(n_s^2 k_s^2)$ but special properties can be exploited. In one dimension, a spline approximation subdivides an interval into subintervals. In two dimensions, the region is divided into patches or "cells". Thus the fundamental process is one of integration over a cell, $r_{i_v} \leqslant r_1 \leqslant r_{i_v+1}$ and $r_{j_v} \leqslant r_2 \leqslant r_{j_v+1}$, namely

$$\int_{r_{i_v}}^{r_{i_v+1}} \int_{r_{j_v}}^{r_{j_v+1}} \frac{r_<^k}{r_>^{k+1}} B_i(r_1) B_j(r_2) B_{i'}(r_1) B_{j'+1}(r_2) \, dr_1 \, dr_2. \quad (78)$$

But for an off-diagonal cell, the above two-dimensional integral is separable and equal to (for $i_v < j_v$)

$$\int_{r_{i_v}}^{r_{i_v+1}} Bi(r_1) r_1^k B_{i'}(r_1) \, dr_1 \int_{r_{j_v}}^{r_{j_v+1}} Bj(r_2) (1/r_2^{k+1}) B_{j'}(r_2) \, dr_2$$

$$\equiv r^k(i, i', i_v) \times r^{-(k+1)}(j, j'; j_v) \quad (79)$$

where $r^k(i, i', i_v)$ and $r^{-(k+1)}(j, j'; j_v)$ are referred to as moments.

The diagonal cells reduce to integrations over upper and lower triangles or, through an interchange of arguments, as

$$R^k(i, j; i', j'; i_v) = R_\Delta^k(i, j; i', j'; i_v) + R_\Delta^k(j, i; j', i'; i_v) \quad (80)$$

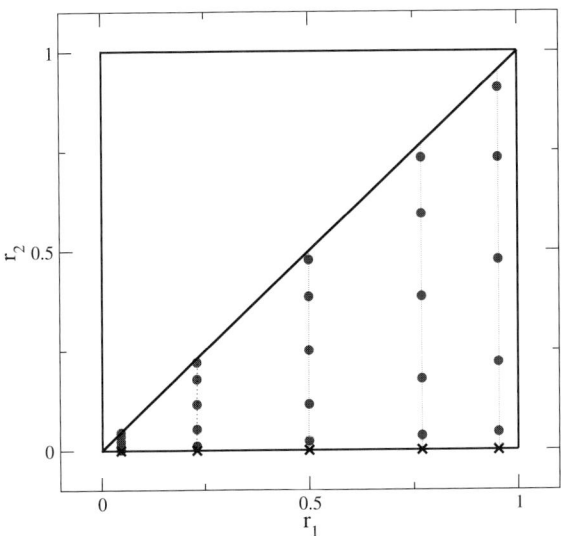

FIG. 4. Gaussian integration points for integration over the lower portion of a diagonal cell of unit dimensions for splines of order $k_S = 5$.

and

$$R_\Delta^k(i, j; i', j'; i_v)
= \int_{r_{i_v}}^{r_{i_v+1}} \frac{1}{r_1^{k+1}} B_i(r_1) B_{i'}(r_1) \, dr_1 \int_{r_{i_v}}^{r_1} r_2^k B_j(r_2) B_{j'}(r_2) \, dr_2. \tag{81}$$

The routines in the BSR spline libraries (Zatsarinny, 2006) are based on algorithms developed by Qiu (1999), and Qiu and Froese Fischer (1999). They assume that integration with respect to r_1 is a summation over the usual Gaussian points (crosses) and that integration with respect to r_2 is from the boundary of the cell to the diagonal as shown in Fig. 4. This requires the introduction of new Gaussian points (circles). Thus the computational complexity of an integration over a diagonal cell is $\mathcal{O}(k_s^2)$. Once the moments and integrals over the diagonal cells have been evaluated, the results are assembled to produce the $R^k(i, j; i', j')$ array. The computational complexity of this process is $\mathcal{O}(n_s k_s^2)$.

Here the logarithmic grid can be used to advantage. In the region where $t_{i+1} = (1 + h)t_i$, scaling laws can be derived. Let the leftmost knot defining $B_i(r)$ be t_i and let $r = t_i + s$, then,

$$B_i(t_i + s) = B_{i+1}\big((1 + h)(t_i + s)\big) = B_{i+1}\big(t_{i+1} + s(1 + h)\big), \tag{82}$$

and it can be shown that

$$\langle B_{i+1}(r)|r^k|B_{j+1}(r)\rangle = (1+h)^{1+k}\langle B_i(r)|r^k|B_j(r)\rangle,$$
$$\langle B_{i+1}(r)|1/r^{k+1}|B_{j+1}(r)\rangle = (1+h)^{-k}\langle B_i(r)|1/r^{k+1}|B_j(r)\rangle,$$
$$R^k(i+1, j+1; i'+1, j'+1) = (1+h)R^k(i, j; i', j'). \tag{83}$$

The BSR library routines take advantage of scaling laws in generating moments and diagonal contributions but do not use scaling to minimize the assembly time which is the more time-consuming task. Note that for $k = 0$, $\langle B_i(r)|1/r^{k+1}|B_j(r)\rangle$ is constant for $i + k_s \leqslant j \leqslant n_s - k_s$. This can reduce the number of multiplications needed for evaluating Slater integrals, but only when $k = 0$.

Assembly can be avoided when the Slater matrix elements are not needed. An example is the calculation of orbital Slater integrals. Zatsarinny (2006) has shown that given the B-spline moments over intervals and diagonal contributions, the orbital Slater integrals are a simple sum of products. Orbital Slater integrals can also be evaluated directly by integrating over cells without the need for introducing the B-spline moments. What is needed are the values of the radial functions at all the Gaussian points (including the triangle ones). An important property can be used to advantage, namely the fact that spline functions can easily be evaluated at any point.

Table I compares the accuracy of differential equation and cell methods. Though the differential equation results are of high accuracy, direct integration is always better. For some Breit–Pauli integrals, the differential equation results suffer a greater loss of accuracy than indicated in Table I for Slater integrals. The integrals in this table are for $k = 0$ or $k = 1$. The factor appearing in Eq. (78), namely

$$\frac{r_<^k}{r_>^{k+1}} = \frac{1}{r_>}\left(\frac{r_<}{r_>}\right)^k \tag{84}$$

is a product of two weighting factors. The first factor reduces the contribution as $r_>$ increases and the second reduces the contribution as the radial coordinates (r_1, r_2) move away from the diagonal where $r_1 = r_2$. This factor decreases rapidly for large k making the region near the diagonal the most important region for the Slater integral. More comparisons with exact values may be found in Froese Fischer et al. (1992), Qiu and Froese Fischer (1999).

The same Slater matrix elements can be used in Dirac–Hartree–Fock calculations. What differs is the grid and the definition of the density matrix which now includes both the large and small components of the orbitals (Froese Fischer and Parpia, 1993).

Table I

Comparison of Some Hydrogenic Slater Integrals from Differential Equation (DE) Methods and Cell Integration (Cell) Methods with Exact Values for Some Slater Integrals Using Splines of Order $k_s = 8$ and $h = 1/8$ (Qiu and Froese Fischer, 1999)

F^k/G^k	Exact value	DE	Cell
$F^0(1s, 1s)$	5/8	−4.2(−15)	4.4(−16)
$F^0(1s, 2s)$	17/81	−1.4(−15)	1.4(−16)
$F^0(1s, 2p)$	59/243	−1.9(−15)	−1.4(−16)
$F^0(2s, 2s)$	77/512	−7.2(−16)	5.6(−17)
$F^0(2s, 2p)$	83/512	−4.7(−16)	1.9(−16)
$F^0(2p, 2p)$	93/512	−8.9(−16)	0.0(+00)
$F^0(4s, 4s)$	19541/524288	−1.6(−14)	−6.8(−16)
$G^0(1s, 2s)$	16/729	−6.6(−17)	2.8(−17)
$G^0(2p, 3p)$	96768/9765625	−1.1(−16)	5.2(−18)
$G^0(2p, 4p)$	560/177147	−7.0(−17)	1.3(−18)
$G^1(1s, 2p)$	112/2187	−5.3(−15)	2.1(−17)
$G^1(2s, 2p)$	45/512	−8.2(−14)	0.0(+00)
$G^1(2p, 3s)$	92016/9765625	−3.0(−15)	2.1(−17)
$G^1(2p, 3d)$	1824768/48828125	−6.9(−14)	−2.1(−17)

5. B-Spline Methods for the Many-Electron Hartree–Fock Problem

5.1. NOTATION

Let us begin with the assumption that radial functions for orbitals $a, b, c, d \ldots$ are spline functions with expansions

$$P(a; r) = \sum_i a_i B_i(r), \quad P(b, r) = \sum_i b_i B_i(r), \quad \text{etc.} \tag{85}$$

Thus a is both a reference to nl quantum numbers and a reference to the column vector of expansion coefficients for the radial function. At this point, it also is convenient to introduce a notation which allows us to relate orbital quantities to basic B-spline matrix elements of operators. Let us define

$$\begin{aligned}
I(a, b) &= \langle a|\mathcal{I}|b\rangle \\
&= \sum_i a_i \sum_j b_j \langle B_i(r)|\mathcal{I}|B_j(r)\rangle \\
&= a^t I(.,.) b \equiv a^t I b,
\end{aligned} \tag{86}$$

$$\begin{aligned} S(a,b) &= \langle a|b\rangle \\ &= \sum_i a_i \sum_j b_j \langle B_i(r)|B_j(r)\rangle \\ &= a^t S(.,.)b \equiv a^t Sb, \end{aligned} \quad (87)$$

where $I(.,.)$ and $S(.,.)$ are matrices in the B-spline basis usually denoted simply as I and S.

In spline expansions, the energy functional $\mathcal{F}(a,b,\ldots)$ needs to be stationary with respect to all variations of the expansion coefficients of orbitals, i.e. $\partial \mathcal{F}/\partial a_i = 0$ and $\partial \mathcal{F}/\partial b_i = 0$, for $i = 1,\ldots,n_s$. This condition leads to a system of equations that we would like to express in matrix–vector form. Let us represent the partial differentiation with respect to each component of a vector a simply as the column vector $\partial \mathcal{F}/\partial a$. Then, from the definition in terms of summations, it follows that

$$\partial I(a,b)/\partial a = \sum_j b_j I(.,.) = Ib. \quad (88)$$

Let us define $I(.,.)b = I(.,b)$, so the effect of variation with respect to a vector of values is to remove the expansion vector as an argument and yield a column vector. When the argument occurs twice in an expression, then it can be removed in more than one way. Thus $\partial I(a,a)/\partial a = 2I(.,a)$. The summation over the column indices can be replaced by matrix–vector multiplication so that $2I(.,a) = 2Ia$. The latter process can be viewed as removing a vector argument to yield a matrix times that vector and does not depend on multiplicity.

This notation is hardly needed for quantities with only two orbital arguments but is particularly convenient when applied in a similar manner to Slater integrals with four arguments. Consider the variation of $R^k(a,b;c,d)$ with respect to the vector a which according to Eq. (70) is

$$\sum_j b_j \sum_{i'} c_{i'} \sum_{j'} d_{j'} R^k(i,j;i',j'), \quad i = 1,\ldots,n_s.$$

Then it follows that the vector of values, according to our notation, is $R^k(.,b;c,d)$ and can be written in matrix–vector form in three different ways,

$$\begin{aligned} R^k(.,b;c,d) &= R^k(.,.;c,d)b \\ &= R^k(.,b;.,d)c \\ &= R^k(.,b;c,.)d, \end{aligned}$$

where $R^k(.,.;c,d)$, $R^k(.,b;.,c)$, and $R^k(.,b;c,.)$ are matrices whose elements are linear combinations of Slater matrix elements. The matrices $R^k(.,.;c,d)$ and $R^k(.,b;c,.)$ are full matrices and symmetric only if c,d or b,c respectively are equivalent, whereas $R^k(.,b;.,d)$ is symmetric and banded with a bandwidth of

$2k_s - 1$. It is important that the row-index from the result of variation not be confused with the column index from the removal of an argument. By convention, the leftmost unfilled position designates the row. Thus the variation of $R^k(a, b; c, d)$ with respect to c would be the column vector $R^k(a, b; ., d) = R^k(., b; a, d)$, by symmetry of the Slater integral. Then $R^k(., b; a, d) = R^k(., .; a, d)b$, for example.

5.2. Derivation of B-Spline Equations

5.2.1. Case without Orthogonality Constraints

Now let us consider the $1s^2 2p\ ^2P$ configuration state. The energy expression for this three-electron system is

$$\mathcal{E}(1s^2 2p\ ^2P) = 2I(a, a) + I(b, b) + F^0(a, a) + 2F^0(a, b) \\ - (1/3)G^1(a, b), \qquad (89)$$

where $1s$ and $2p$ are denoted by a and b respectively. Since there are only normalization constraints, the energy functional can be written as

$$\mathcal{F}(a, b, \varepsilon_{aa}, \varepsilon_{bb}) = \mathcal{E}(1s^2 2p\ ^2P) + \varepsilon_{aa}(1 - S(a, a)) \\ + \varepsilon_{bb}(1 - S(b, b)). \qquad (90)$$

Applying the variational procedure to this functional (and dividing by two), we get two generalized matrix-eigenvalue problems,

$$\left[2I + 2R^0(., a; ., a) + 2R^0(., b; ., b) - (1/3)R^1(., b; b, .) - \varepsilon_{aa}S\right]a = 0, \\ \left[I + 2R^0(., a; ., a) - (1/3)R^1(., a; a, .) - \varepsilon_{bb}S\right]b = 0, \qquad (91)$$

where the exchange contribution makes each matrix a full matrix.

However, the equations could also have been written as one system of equations with banded submatrices, namely

$$\begin{bmatrix} 2I + D^a - \varepsilon_{aa}S & X^a \\ X^b & I + D^b - \varepsilon_{bb}S \end{bmatrix} \begin{bmatrix} a \\ b \end{bmatrix} = 0, \qquad (92)$$

where D^a, D^b represent the contributions from the direct F^k integrals and X^a, X^b those from the G^k integrals. In the present case,

$$D^a = 2R^0(., a; ., a) + 2R^0(., b; ., b), \\ D^b = 2R^0(., a; ., a), \\ X^a = -(1/3)R^1(., b; ., a), \\ X^b = -(1/3)R^1(., a; ., b). \qquad (93)$$

In this form, all submatrices are symmetric and banded, and the coefficient matrix itself for the pair of orbitals is symmetric since $(X^b)^t = X^a$. This symmetry would have been destroyed if, as has been the practice in differential equation methods, the equations for an orbital were divided by the orbital occupation. Solutions of these equations can be obtained through singular-value decomposition (SVD) (Press et al., 1992).

Equation (92) can be rewritten as one matrix-eigenvalue problem by using Eq. (24) to define a deviation d_{aa} for each diagonal energy parameter, namely $d_{aa} = E - \varepsilon_{aa}$. In the case of $1s^2 2p\,^2P$,

$$d_{aa} = I(b, b) - F^0(a, a) \quad \text{and} \quad d_{bb} = 2I(a, a) + F^0(a, a). \tag{94}$$

Substituting into Eq. (92) we get one generalized symmetric eigenvalue problem for the pair of orbitals,

$$\left(\begin{bmatrix} 2I + D^a + d_{aa}S & X^a \\ X^b & I + D^b + d_{bb}S \end{bmatrix} - E \begin{bmatrix} S & 0 \\ 0 & S \end{bmatrix}\right) \begin{bmatrix} a \\ b \end{bmatrix} = 0. \tag{95}$$

Let **a** be the column vector $(a^t, b^t)^t$. Then these equations have the form

$$(\mathbf{H} - E\mathbf{S})\mathbf{a} = 0, \tag{96}$$

where **H** is a matrix of dimension $2n_s \times 2n_s$ that consists of banded blocks. and **S** is a block matrix whose diagonal blocks are the overlap matrix S.

Thus the orbitals are solutions of either implicitly coupled generalized eigenvalue problems of full matrices that can be solved independently or explicitly coupled equations consisting of banded blocks of matrices that are solved simultaneously by either singular value decomposition or a generalized eigenvalue problem. It will be shown later that the two variations of the latter expressed by Eqs. (92) and (95) are closely related.

5.2.2. Case with Orthogonality Constraints

The Hartree–Fock energy expression for $1s2s\,^{1,3}S$ is

$$\mathcal{E}(1s2s\,^3S) = I(a, a) + I(b, b) + F^0(a, b) \pm G^1(a, b). \tag{97}$$

In addition to normalization constraints, there also is an orthogonality constraint so the energy functional becomes

$$\mathcal{F}(a, b, \varepsilon_{aa}, \varepsilon_{bb}, \varepsilon_{ab}) = \mathcal{E}(1s2s\,^{1,3}S) + \varepsilon_{aa}(1 - S(a, a))$$
$$+ \varepsilon_{bb}(1 - S(b, b)) - 2\varepsilon_{ab}S(a, b) \tag{98}$$

and in matrix-eigenvalue form, the equations become

$$[I + R^0(., b; ., b) \pm R^0(., b; b, .) - \varepsilon_{aa}S]a - \varepsilon_{ab}Sb = 0,$$

$$[I + R^0(.,a;.,a) \pm R^0(.,a;a,.) - \varepsilon_{bb}S]b - \varepsilon_{ba}Sa = 0 \tag{99}$$

or

$$\begin{bmatrix} H^a & \\ & H^b \end{bmatrix}\begin{bmatrix} a \\ b \end{bmatrix} - \begin{bmatrix} \varepsilon_{aa}S & \varepsilon_{ab}S \\ \varepsilon_{be}S & \varepsilon_{bb}S \end{bmatrix}\begin{bmatrix} a \\ b \end{bmatrix} = 0. \tag{100}$$

From the orthonormality of solutions, it follows that

$$\begin{aligned}\varepsilon_{aa} &= a^t H^a a, & \varepsilon_{ab} &= b^t H^a a, \\ \varepsilon_{ba} &= a^t H^b b, & \varepsilon_{bb} &= b^t H^b b,\end{aligned} \tag{101}$$

and from the requirement $\varepsilon_{ab} = \varepsilon_{ba}$, the condition that

$$b^t H^a a - a^t H^b b = 0. \tag{102}$$

In the present case, since $I(a,b) = I(b,a)$, this condition becomes

$$\begin{aligned}&[R^0(b,b;a,b) \pm R^0(b,b;b,a)] \\ &- [R^0(a,a;b,a) \pm R^0(a,a;a,b)] = 0.\end{aligned} \tag{103}$$

For the ^3S where the minus sign applies, the condition is satisfied by any pair of orthogonal orbitals. In this case, by Koopmans' theorem, the physically meaningful solutions are the extremum eigenvalues of the ε-matrix characterized by $\varepsilon_{ab} = \varepsilon_{ba} = 0$. This choice uncouples the two equations and makes each a generalized matrix-eigenvalue problem.

However, the situation is different for $1s2s\,^1S$. A stationary solution requires that

$$R^0(a,b;b,b) - R^0(a,a;a,b) = 0, \tag{104}$$

and can be satisfied through rotation (see Section 2.2.2). When the condition is satisfied and $\varepsilon_{ab} = \varepsilon_{ba}$, the off-diagonal Lagrange multipliers can be eliminated (Bentley, 1994). Using first the fact that ε_{ab} is a constant, and then that $\varepsilon_{ab} = b^t H^a a$, we have

$$\varepsilon_{ab}Sb = Sb\varepsilon_{ab} = Sbb^t H^a a. \tag{105}$$

When a and b are orthogonal, $b^t Sa = 0$ so that

$$a = (1 - bb^t S)a, \tag{106}$$

where 1 is the identity operator (not to be confused with I). With similar results when a and b are interchanged, it follows that Eq. (100) can be rewritten as

$$\begin{aligned}&[(1 - Sbb^t)H^a(1 - bb^t S) - \varepsilon_{aa}S]a = 0, \\ &[(1 - Saa^t)H^b(1 - aa^t S) - \varepsilon_{bb}S]b = 0.\end{aligned} \tag{107}$$

This transformation uses both Eqs. (105) and (106), thereby preserving the symmetry of the generalized eigenvalue problems. Earlier methods (Brage and Froese Fischer, 1994a) only eliminated the off-diagonal Lagrange multiplier thereby destroying the symmetry. This matrix method of enforcing orthogonality, does not require that $\varepsilon_{ab} = \varepsilon_{ba}$. It simply defines the new orbital a to be orthogonal to an existing orbital b (or vice versa). If orbital a is fixed, this would be an effective method for obtaining orbital b subject to orthogonality to orbital a, but when both orbitals are varied, the total energy would not be a stationary point unless $\varepsilon_{ab} = \varepsilon_{ba}$.

5.3. Iterative Solutions of B-Spline Equations

5.3.1. Updating Single Orbitals Successively

We have shown that the radial orbital for a Hartree–Fock wave function can be formulated as a solution of a generalized matrix-eigenvalue problem, one for each orbital. This orbital Hartree–Fock Hamiltonian matrix, however, is defined in terms of radial distributions of other orbitals and thus the systems of equations are coupled implicitly and need to be solved iteratively by a self-consistent field (SCF) process. Basically, after orbitals have been rotated (if necessary), for each orbital a in the orbital set, H^a is computed from current estimates of all orbitals, projection operators are applied to ensure orthogonality (as needed), and the generalized eigenvalue problem is solved for the desired eigensolution. The eigenvalue of the solution defines the diagonal energy parameter and the eigenvector the improved expansion coefficients for the orbital. Such a process may converge or diverge.

Table II shows convergence by reporting the change, $\Delta E = E_i - E_{i-1}$, from one iteration to the next as the iteration number i increases for cases where no off-diagonal energy parameters were needed. In all cases, the initial estimates of orbitals were unscreened hydrogenic functions. This facilitates the checking of the calculations since the values of Slater integrals are then known. The calculations for $1s2p\,^1P$ and $1s2s\,^3S$ converge rapidly and quadratically with the results reaching the round-off limit in 4 or 5 iterations. Convergence for $1s^22s^2\,^1S$ is linear but with the accuracy of the total energy increasing by about one digit per iteration. Calculations for the $1s^22s^22p^5$ ground state of fluorine are different in that convergence is much slower.

With five equivalent $2p$ electrons, fluorine has a strong self-interaction. Given estimates of $2p$ that screen the nucleus too effectively the next $2p$ orbital will be too defuse. As a result, the diagonal energy parameter oscillates as shown in Fig. 5. Such oscillations can be reduced through the use of damping factors (also call accelerating factors) and may be needed for convergence. The oscillations can

Table II
Convergence of the Orbital Generalized Eigenvalue Method for Several Cases without Off-Diagonal Lagrange Multipliers

i	$1s2p\,^1P$	$1s2s\,^3S$	$1s^22s^2\,^1S$	$1s^22s^22p^5\,^2P$
2	−3.4(−05)	−1.0(−05)	−2.8(−02)	−2.7
3	−1.6(−11)	−4.6(−09)	−1.6(−03)	−4.5(−01)
4	2.2(−14)	−6.5(−12)	−1.2(−04)	−2.5(−01)
5	−	−2.8(−14)	−9.1(−06)	−8.9(−02)
6	−	−	−6.9(−07)	−5.1(−02)
7	−	−	−5.2(−08)	−2.2(−02)
8	−	−	−3.9(−09)	−1.1(−02)
9	−	−	−3.0(−10)	−5.6(−03)
10	−	−	−2.3(−11)	−2.9(−03)

All calculations are for the neutral atom. Shown is the change in the total energy, $\Delta E_i = E_i - E_{i-1}$, for iterations $i = 2, 3, \ldots$.

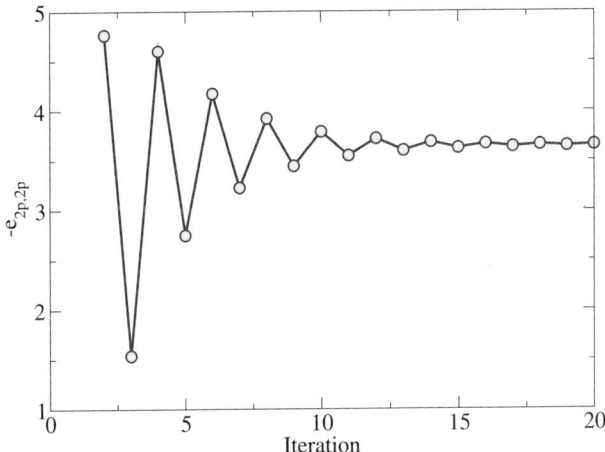

FIG. 5. The oscillating behavior of the value of the diagonal energy parameter, $-\varepsilon_{2p,2p}$, as a function of the iteration number for the $1s^22s^22p^5\,^2P$ ground state of fluorine.

be detected by monitoring the diagonal energy parameter and can be dealt with automatically (Froese Fischer, 1986).

Though it is convenient to pose the orbital problem in terms of an eigenvalue problem where the index of the desired eigenvalue (first, second, etc., with first being the lowest) makes it easy to determine solutions for excited states, explicit rotation is needed to obtain stationary energies when off-diagonal Lagrange multipliers are needed. Such rotations can be included implicitly when solving simultaneously for orbitals that are connected through orthogonality.

5.3.2. Updating Several Orbitals Simultaneously

When all (or several) orbitals are updated simultaneously, there is greater flexibility in how the system of equations are expressed. In particular, it is possible to express the system of equations in terms of only banded $n_s \times n_s$ submatrices that can be generated more efficiently than full matrices.

The stationary condition requires that the gradient vector of the energy functional be zero. For a CSF with m radial functions, this leads to a system of $M \times M$ equations where $M = m \times n_s$ (before the application of boundary conditions). The gradient vector for each orbital can always be expressed as a sum of matrix–vector operations. The gradient vector for orbital a contains contributions from $I(.,a)$, $S(.,a)$, $R^k(.,b;a,b)$, and $R^k(.,b;b,a)$. Then $I(.,a) = Ia$, $S(.,a) = Sa$, $R^k(.,b;a,b) = R^k(.,b;.,b)a$ lead to symmetric, banded matrices, but the contribution from the exchange term can only be expressed in banded form as $R^k(.,b;b,a) = R^k(.,b;.,a)b$. For this contribution, an off-diagonal block needs to be introduced and both a and b determined simultaneously.

Let A_i be the expansion vector for orbital i. Then the system of equations can be written as

$$\begin{bmatrix} F^{11} & F^{12} & \cdots & F^{1m} \\ F^{21} & F^{22} & \cdots & F^{2m} \\ \cdots & \cdots & \cdots & \cdots \\ F^{m1} & F^{m2} & \cdots & F^{mm} \end{bmatrix} \begin{bmatrix} A_1 \\ A_2 \\ \cdots \\ A_m \end{bmatrix} = 0. \qquad (108)$$

where F^{ii} contains the contributions from one-electron integrals, the direct Slater integrals for orbital i, as well as $-\varepsilon_{ii}S$, and F^{ij} ($i \neq j$) contains the contribution from exchange integrals between orbitals i and j and possible orthogonality constraints $-\varepsilon_{ij}S$. This method does not require that orbitals be explicitly rotated but the ε-matrix used should be symmetric. A possible definition is the average of the two computed values.

This formulation leads to a homogeneous system of equations with a symmetric coefficient matrix and could be solved using singular-value decomposition (SVD). But as long as exchange terms are present, the diagonal matrices, F^{ii}, will not be singular so that a banded inverse exists and a block LU-factorization scheme may be used. Consider the 2×2 case. Multiplying the first row of block matrices, by $(F^{11})^{-1}$ and subtracting F^{21} times this row from the second, we get

$$\left(F^{22} - F^{21}(F^{11})^{-1}F^{12}\right)A_2 = 0, \qquad (109)$$

which can be solved either by inverse iteration (Press et al., 1992) or as an eigenvalue problem since F^{22} includes a $-\varepsilon_{22}S$ term. Once A_2 is computed

Table III
Comparison of Convergence of Singular Value Decomposition Methods for $1s2p\,^1P$, $1s2s\,^3S$, $1s2s\,^1S$, and $1s^22s^2\,^1S$

i	$1s2p\,^1P$	$1s2s\,^3S$	$1s2s\,^1S$	$1s^22s^2\,^1S$
1	−2.5(−04)	−4.9(−02)	−4.9(−01)	−4.4(−01)
2	−1.4(−06)	−1.5(−03)	−3.5(−01)	−1.4(−02)
3	−5.2(−09)	−3.16(−06)	−3.0(−02)	−1.6(−04)
4	−2.1(−11)	−1.5(−09)	−5.2(−02)	−8.4(−06)
5	−4.4(−14)	−5.9(−13)	−1.0(−02)	−6.2(−07)
6	−3.0(−14)	−5.8(−15)	−4.0(−03)	−4.7(−08)
7	−	−	−1.7(−03)	−3.7(−09)
8	−	−	−8.8(−04)	−2.9(−10)
9	−	−	−5.0(−04)	−2.2(−11)
10	−	−	−2.6(−04)	−1.8(−12)

Shown are the values of $\Delta E = E_i - E_{i-1}$ as a function of the iteration i.

and normalized, A_1 can be determined through back-substitution. A process like LU-decomposition could be used, where the basic quantities are square, banded matrices.

Singular values are closely related to eigenvalues. For a square matrix A, the singular values of A are the square roots of the eigenvalues of $A^t A$. Should A be symmetric, then singular values are the absolute value of the eigenvalues and the singular vectors are the eigenvectors of $A^t A$ which, for symmetric matrices, are eigenvectors of A. When SVD is used to solve $Ax = b$, the computed value of x is such that x minimizes $|Ax - b|$. Consequently, when $b = 0$, the solution is the eigenvector of the eigenvalue nearest to zero and a routine for solving a symmetric eigenvalue problem could be used to solve the system of equations. However, practical aspects need to be considered.

Table III shows the convergence when eigenvalue methods for systems are used. For $1s2p$ the LAPACK symmetric eigenvalue solver, dsyev was used with the desired solution being the one with the smallest eigenvalue (in magnitude). Convergence was rapid and no problems were encountered. It was interesting to observe that the desired solution was the second eigenvalue. There was a solution with a lower eigenvalue, a solution in which $P_{1s}^{\text{new}} \approx P_{1s}^{\text{previous}}$ but with $P_{2p}^{\text{new}} \approx -P_{2p}^{\text{previous}}$. This change in sign in the system of equations is equivalent to changing the coefficient in the energy expression for $1s2p\,^1P$ which becomes the energy expression for $1s2p\,^3P$. For the $1s2s\,^3S$ case the desired solution is the second eigenvalue since $2s$ is the second eigenvalue of its orbital equation. With unscreened orbitals as initial estimates the desired solution was the one with the smallest eigenvalue only at the third iteration. The off-diagonal energy parameter was set to zero and no projection operators were applied.

Table IV
Eigenvalues as a Function of the Iteration Number i for the Solution of Systems of
Equations for $1s2s\,^1S$

i	E_1	E_2	E_3	E_4
1	−0.700	−0.032	−0.000	0.058
2	−1.192	−0.531	−0.441	−0.412
3	−0.890	−0.134	−0.032	0.006
4	−0.954	−0.124	−0.009	0.027
5	−0.797	−0.095	−0.001	0.025

The calculations for $1s2s\,^1S$ included both diagonal and off-diagonal energy parameters where the value for the latter was the average value computed from current estimates of ε_{ab} and ε_{ba}. The desired solution is now the eigenvector of the third eigenvalue. Iterative solutions using the `dsyev` symmetric eigenvalue solver failed to converge. This is a case where the implicit rotation of orthogonal orbitals is important and where the orthogonality for expansions in the B-spline basis required the overlap matrix S. By using the `dsygv` generalized eigenvalue solver for equations of the form

$$\left(\begin{bmatrix} H^{aa} - \varepsilon_{aa}S & H^{ab} - \varepsilon_{ab}S \\ H^{ba} - \varepsilon_{ab}S & H^{bb} - \varepsilon_{bb}S \end{bmatrix} - E \begin{bmatrix} S & 0 \\ 0 & S \end{bmatrix}\right)\begin{bmatrix} a \\ b \end{bmatrix}, \quad (110)$$

where H^{aa} represents the contributions from $I(a,a)$ and the direct Slater integrals, and H^{ab} the contributions from $G^k(a,b)$, etc. The self-consistent solution satisfies these equations with $E = 0$, but when starting even with screened hydrogenic functions as initial estimates, the first few sets of eigenvalues E_i resulting from the iterative process may be far from zero. Table IV shows the first few eigenvalues as a function of the iteration number i for $1s2s\,^1S$. Only at the fourth iteration was the desired solution the one with the smallest eigenvalue. Thus an SVD approach relying only on selecting the solution with the smallest eigenvalue would need to have initial estimates considerably more reliable than screened hydrogenic functions. For cases where orbital rotation through off-diagonal energy parameters are significant, the generalized singular value decomposition needs to be used.

When Koopmans' theorem is used to express diagonal energy parameters as the total energy minus a deviation, as was done in Eq. (95) for $1s2p\,^1P$, and where the total energy at each iteration is computed from the current orbitals, the generalized eigenvalue problem is equivalent to the method of Eq. (110) except that the eigenvalues have been shifted and the desired solution needs to be specified or known.

Table III also shows the convergence for the three cases $1s2s\,^3S$, $1s2s\,^1S$, and $1s^22s^2\,^1S$ when the generalized SVD method is used. $1s2s\,^1S$ has by far the slowest rate of convergence. Initial estimates were hydrogenic orbitals with effective nuclear charge of (2, 1) for (1s, 2s) respectively. The numerical differential equation Hartree–Fock method was able to solve this problem only with initial estimates from $1s2s\,^3S$ and the explicit rotation of orbitals (Froese Fischer, 1986). Comparing the present values with Table II, we see that convergence for $1s2p\,^1P$, $1s2s\,^3S$, and $1s^22s^2\,^1S$ is similar. One question that arises is whether orbitals should be orthogonalized when the off-diagonal energy parameters are zero. In Table III, at the end of each iteration, $2s$ was made orthogonal to $1s$. This orthogonalization can help in some cases, such as $1s2s\,^1S$, but may slow convergence slightly in others.

5.3.3. Newton–Raphson Methods

The Newton–Raphson procedure for non-linear systems of equations often exhibits quadratic convergence once an estimate is sufficiently near a solution. Consider the system of n equations $F_i(x) = 0$, $i = 1, \ldots, n$ in n unknowns, $x = (x_1, \ldots, x_n)$. Let $F(x) = 0$ be the column vector of equations and the vector x^m an estimate of a solution. Expanding in a Taylor's series,

$$F(x) = 0 = F(x^{(m)}) + (\partial F/\partial x)\Delta x + \text{higher order terms}, \tag{111}$$

where $(\partial F/\partial x)$ is the Jacobian matrix with elements $(\partial F_i/\partial x_j)$. Then the iterative process:

$$(\partial F/\partial x)\Delta x = -F(x^{(m)}),$$
$$x^{(m+1)} = x^m + \Delta x \tag{112}$$

can be shown to have quadratic convergence in many circumstances. Notice that $F(x^{(m)})$ is the residual or amount by which the equations are not satisfied by the current estimate. In the following discussion the residual for equations for orbital a will be denoted res_a, etc., and the residual for constraints will be given explicitly.

This process was applied to $1s2s\,^3S$ and ^1S (Froese Fischer et al., 1992). For the latter the unknowns were $(a, b, \varepsilon_{aa}, \varepsilon_{bb}, \varepsilon_{ab})$ and the systems of equations are the stationary conditions of Eq. (99) along with the three orthonormality conditions that are part of the energy functional. Let (a, b) be the current estimates which are used to evaluate the ε-matrix. If symmetry conditions are not satisfied, we can use the average value,

$$\varepsilon_{ab} = \varepsilon_{ba} = (b^t H^a a + a^t H^b b)/2. \tag{113}$$

Then Δa and Δb are solutions of

Table V
The Orbital Self-Consistency (ΔP) and the Deviation from the Virial Theorem ($VT+2$) as a Function the Number of Iterations, i, for $1s2s\ ^{1,3}S$ States of He with Spline Parameters $k_s = 6$ and $h = 1/8$ (Froese Fischer et al., 1992)

i	Triplet			Singlet		
	ΔP_{1s}	ΔP_{2s}	$|VT+2|$	ΔP_{1s}	ΔP_{2s}	$|VT+2|$
1	9.7(−3)	2.9(−1)		5.7(−2)	1.4(−1)	
2	8.7(−3)	1.1(−2)		6.2(−3)	1.0(−2)	
3	1.7(−5)	3.8(−5)		2.2(−4)	1.3(−4)	
4	3.9(−11)	6.4(−11)	<5(−14)	5.9(−6)	7.3(−6)	
5				3.7(−7)	4.2(−7)	4.0(−7)
6				2.1(−8)	2.4(−8)	4.5(−10)

Initial estimates were screened hydrogenic functions with $Z_{\text{eff}} = 2$ and 1, respectively, for $1s$ and $2s$, for ^3S. The latter were used as initial estimates for ^1S.

$$\begin{bmatrix} H^a - \varepsilon_{aa}S & H^{ab} - \varepsilon_{ab}S & -Sa & & -Sb \\ H^{ba} - \varepsilon_{ab}S & H^b - \varepsilon_{bb}S & & -Sb & -Sa \\ -(Sa)^t & & & & \\ & -(Sb)^t & & & \\ -(Sb)^t & -(Sa)^t & & & \end{bmatrix} \begin{bmatrix} \Delta a \\ \Delta b \\ \Delta \varepsilon_{aa} \\ \Delta \varepsilon_{bb} \\ \Delta \varepsilon_{ab} \end{bmatrix}$$

$$= \begin{bmatrix} -\text{res}_a \\ -\text{res}r_b \\ S(a,a) - 1 \\ S(b,b) - 1 \\ S(a,b) \end{bmatrix} \qquad (114)$$

where H^a and H^b in this case are the matrices of Eq. (100) and

$$H^{ab} = (H^{ba})^t = 2R^0(.,.;a,b) \pm R^0(.,a;.,b) \pm R^0(.,.;b,a).$$

With orthonormal initial estimates, the conditions for the last three equations would be zero. The performance of this method is shown in Table V where the maximum change in an orbital ΔP is shown at the end of each iteration. By the Virial theorem (VT), the ratio of the potential energy and kinetic energy should be exactly -2.0. In Table V the deviation of this ratio is also reported. Clearly, the ^3S calculations converge quadratically, with the ΔP for each orbital, effectively squared in going from one iteration to the next. Convergence for the ^1S is considerably slower, though converging without explicit rotations (Froese Fischer et al., 1992). The price for the faster convergence is the extra computation needed to generate the Jacobian matrix and the time needed to solve full systems of equations.

The Newton–Raphson method generally requires considerably more computation for setting up the system of equations than the orbital or SVD methods. Thus

Table VI
Comparison of Convergence for $2p^5 (Z = 5)$ for the Orbital
Hamiltonian Method (Orbital) and the Newton–Raphson (NR)
Method

i	Orbital	NR
1	−3.9(−01)	−3.8(−01)
2	+3.6(−01)	−7.8(−03)
3	−3.3(−01)	−2.1(−06)
4	+3.8(−01)	−1.4(−12)
5	−3.6(−01)	4.4(−14)

Shown are the values of $\Delta E = E_i - E_{i-1}$ as a function of the iteration i.

it is not always the method of choice. However, for multiply occupied shells such as $2p^5\,{}^2P$ it can be extremely effective with a minimum of overhead. The energy functional is

$$\mathcal{F}(2p^5) = 5I(a,a) + 10F^0(a,a) - 0.8F^2(a,a) \\ + \varepsilon_{aa}(1 - S(a,a)), \tag{115}$$

from which it follows that

$$H^a = 5I + 20R^0(.,a;.,a) - 1.6R^2(.,a;.,a),$$
$$\varepsilon_{aa} = a^t H^a a,$$
$$\text{res}_a = (H^a - \varepsilon_{aa} S)a,$$
$$H^{aa} = H^a + 40R^0(.,.;a,a) - 3.2R^2(.,.;a,a) - \varepsilon_{aa} S, \tag{116}$$

where H^{aa} is the second-order variation of $\mathcal{E}$ with respect to a and is a full matrix. Then the Newton–Raphson equations are

$$\begin{bmatrix} H^{aa} & -Sa \\ -(Sa)^t & 0 \end{bmatrix} \begin{bmatrix} \Delta a \\ \Delta \varepsilon_{aa} \end{bmatrix} = \begin{bmatrix} -\text{res}_a \\ S(a,a) - 1 \end{bmatrix}. \tag{117}$$

Thus with the addition of the extra $R^k(.,.;a,a)$ full matrix contributions, quadratic convergence of the iterative process is achieved. Though this process solves only a single orbital equation, it avoids the necessity of introducing damping factors for oscillations.

Table VI compares the convergence of the total energy as a function of the iteration number for $2p^5$ and $Z = 5$. In the single orbital method, the total energy oscillates and actually diverges. With the Newton–Raphson method convergence is rapid and quadratic. The important contributors are the contributions to the equations from the second-order variation of $F^k(a,a)$ integrals in the energy expression which will always be present when a shell is multiply occupied.

It needs to be remembered that the Newton–Raphson method for non-linear equations such as the HF equations, converges to the nearest energy solution. Consequently, with reasonably good initial estimates for $1s$ and $2s$, this method applied to $1s^22s$ may converge rapidly to $1s^23s$. In other words, the orbital properties of the initial estimates do not control the solution obtained by the Newton–Raphson method. This difficulty applies primarily to the application of the method to systems with singly occupied orbitals in that eigenvalues can be closely spaced.

5.3.4. Summary of Methods

In summary, three different iterative methods for solving the Hartree–Fock equations have been identified.

(1) The single-orbital Hartree–Fock Hamiltonian method is straight forward to apply and leads to a generalized eigenvalue problem for which the matrix is a full matrix except when no exchange is present. The matrix is of size $n_s \times n_s$. Projection operators can readily be incorporated for orthogonality constraints, though they do not result in orbital rotations. The index of the eigenvalue for a given orbital is known. However, without orbital rotations the final energy may not be a stationary energy. Also, the iterative process is not guaranteed to converge for multiply occupied shells, though damping factors can be introduced to eliminate oscillations in diagonal energy parameters if needed.

(2) Systems of equations for two or more orbitals can be solved by generalized singular value decomposition methods. Orbital rotations are included implicitly but initial estimates need to be sufficiently accurate so that the eigenvalue nearest to zero is associated with the desired solution. Otherwise, the index of the eigenvalue needs to be known in advance. The size of the matrix for m orbitals is $m n_s \times m n_s$ and consists of banded blocks of size $n_s \times n_s$. By rearranging the order of the matrix, it would be a banded matrix of bandwidth $m \times (2k_s - 1)$.

(3) Newton–Raphson methods may be used for quadratic convergence but become computationally intensive as the number of orbitals computed simultaneously increases. For m orbitals, a Jacobian matrix needs to be computed and a full system of equations of size $M \times M$ needs to be solved where $M = m n_s + n_c$, where n_c is the number of orthonormalization constraints. Orbital rotations are included. For the case of a single, multiply occupied orbital, the Newton–Raphson method may eliminate the need for damping factors. Accurate initial estimates may be needed to ensure that the Newton–Raphson method converges to the desired solution.

Thus a number of options are available. More study is needed for determining the most efficient combination of methods.

In this discussion we have mentioned only methods based on homogeneous equations. It is always possible to express a contribution determined from current estimates as defining a "right-hand side" for the system of equations in which case eigenvalue methods can no longer be used.

6. B-Spline MCHF Equations

6.1. With Orthonormality Constraints

Traditionally, MCHF calculations obtain a solution through an iterative process that

- improves the radial functions, then
- finds the CSF expansion coefficients.

These two phases differ in their view of angular data.

Finding the CSF expansion coefficients is a configuration-interaction (CI) problem which requires computing the matrix elements of $H = (H_{ij})$, where $H_{ij} = \langle \Phi(\gamma_i LS) | \mathcal{H} | \Phi(\gamma_j LS) \rangle$. The latter can be expressed in the form

$$H_{ij} = \sum_{ab} w_{ab}^{ij} I(a,b) + \sum_{abcd;k} v_{abcd;k}^{ij} R^k(a,b;c,d)r, \tag{118}$$

where w_{ab} and $v_{abcd;k}$ are angular coefficients obtained from integrating over all angular coordinates (Hibbert and Froese Fischer, 1991; Cowan, 1981; Gaigalas, 1999). The total energy E and the CSF expansion coefficients are eigenvalues and eigenvectors, respectively, of the real, symmetric, matrix-eigenvalue problem $(H - E)c = 0$. For large expansions, this interaction matrix is sparse and can be solved using the Davidson algorithm (Stathopoulos and Froese Fischer, 1994). In fact, generating the interaction matrix can easily be done in parallel by distributing the work by columns with excellent performance. Finding eigenvalues and eigenvectors using Davidson's algorithm is based on parallel matrix–vector multiplication with the interaction matrix distributed. Considerable communication is needed in this stage and performance is not as impressive, but solving the eigenvalue problem is a relatively small portion of the overall task (Froese Fischer et al., 1994; Stathopoulos et al., 1996).

When radial functions are improved, the energy expression is needed, namely

$$\mathcal{E}(\gamma LS) = \sum_{ab} w_{ab} I(a,b) + \sum_{abcd;k} v_{abcd;k} R^k(a,b;c,d), \tag{119}$$

where the coefficients w_{ab} and $v_{abcd;k}$ are obtained by summing over all contributions to the energy expression so that

$$w_{ab} = \sum_i^M \sum_j^M c_i c_j w_{ab}^{ij} \quad \text{and} \quad v_{abcd;k} = \sum_i^M \sum_j^M c_i c_j v_{abcd;k}^{ij}. \tag{120}$$

One obvious difference from the HF problem is the fact that not all Slater integrals are either $F^k(a, b)$ or $G^k(a, b)$ integrals. This means that the orbital equation for a single orbital cannot always be expressed as a generalized eigenvalue problem though it can be expressed as a system of equations.

Consider the case of

$$\Psi = c_1 \Phi(3p^2\,{}^1D) + c_2 \Phi(3s3d\,{}^1D), \tag{121}$$

for which (ignoring any possible core)

$$\mathcal{E}({}^1D) = c_1^2 H_{11} + 2c_1 c_2 H_{12} + c_2^2 H_{22},$$
$$\text{where } H_{11} = 2I(p, p) + F^0(p, p) + 0.04 F^2(p, p),$$
$$H_{12} = (2/\sqrt{15}) R^1(s, d; p, p),$$
$$H_{22} = I(s, s) + I(d, d) + F^0(s, d) + 0.2 G^2(s, d), \tag{122}$$

and where p, s, d are expansion vectors for $3p, 3s, 3d$ respectively. The gradient vector of $R^1(sd; pp)$ with respect to orbital s is $R^1(., d; p, p)$, a vector that cannot be written as a matrix times the vector s and so a single orbital Hamiltonian matrix is not possible for the s and d orbitals. All gradients for the variation of p, s, d can, however, be written as a system of equations in terms of banded submatrices:

$$\begin{bmatrix} H^{pp} & H^{ps} & H^{pd} \\ H^{sp} & H^{ss} & H^{sd} \\ H^{dp} & H^{ds} & H^{dd} \end{bmatrix} \begin{bmatrix} p \\ s \\ d \end{bmatrix} = 0, \tag{123}$$

where

$$H^{pp} = c_1^2 (2I + 2R^0(., p; ., p) + 0.08 R^2(., p; ., p)) - \varepsilon_{pp} S,$$
$$H^{ps} = (2c_1 c_2 / \sqrt{15}) R^1(., p; ., d),$$
$$H^{pd} = (2c_1 c_2 / \sqrt{15}) R^1(., p; ., s),$$
$$H^{ss} = c_2^2 (I + R^0(., d; ., d)) - \varepsilon_{ss} S,$$
$$H^{sd} = 0.2 c_2^2 R^2(., d; ., s),$$
$$H^{dd} = c_2^2 (I + R^0(., s; ., s)) - \varepsilon_{dd} S \tag{124}$$

and

$$H^{sp} = (H^{ps})^t, \quad H^{dp} = (H^{pd})^t, \quad H^{ds} = (H^{sd})^t.$$

This symmetric system of equations can be solved by the singular value decomposition method.

Through a generalization of Koopmans' theorem, we have

$$\varepsilon_{aa} = E(\gamma LS) - d_{aa} \tag{125}$$

where now d_{aa} includes not only the $E(\gamma; \underline{a})$ and self-energy corrections, but corrections from Slater integrals with one- or three occurrences of orbital a. Substituting for diagonal energy parameters, Eq. (123) can be cast in the form

$$(\mathbf{H} - E\mathbf{S})\mathbf{a} = 0, \tag{126}$$

where $\mathbf{H}$ and $\mathbf{S}$ are matrices of dimension $3n_s \times 3n_s$ and consist of banded blocks.

Of course, the Newton–Raphson method could also be used but overall quadratic convergence can be achieved only if the expansion coefficients c_1, c_2 are included among the set of unknowns to be determined.

In theory, the orbital expansion coefficients and the CSF expansion coefficients could be improved simultaneously but such an approach has not been tested. Certainly it is feasible for small expansions but for large MCHF CSF expansions of 1–100 thousand configuration states, sparse matrix methods for finding the eigenvalue are preferred.

6.2. WITHOUT NORMALIZATION CONSTRAINTS

6.2.1. *Correlation in the $1s^2$ Ground State*

The normalization constraint can readily be removed from the derivation of the HF equations (Froese Fischer, 1977). In fact, unconstrained forms are an easy way of interpreting the meaning of the diagonal energy parameter in an MCHF expansion (Froese Fischer, 2005).

Consider the correlation contribution of $4f^2$ to the ground state energy of $1s^2$ from a variational calculation. When orbitals and CSFs are orthonormal, the wave function has the form

$$\Psi = c_1 \left| 1s^2 \right\rangle + c_2 \left| 4f^2 \right\rangle, \quad \text{where } c_1^2 + c_2^2 = 1. \tag{127}$$

In Condon and Shortley phases (Cowan, 1981), the interaction $\langle 1s^2\,{}^1S | \mathcal{H} | 4f^2\,{}^1S \rangle$ is negative. Since correlation lowers the energy, the expansion coefficients c_1, c_2 must have the same sign and can be chosen to be positive. Consequently the square root of these coefficients can be absorbed into the definition of the radial function so that the wave function can be written in the unnormalized orbital form as

$$\Psi = \left| 1s^2 \right\rangle + \left| 4f^2 \right\rangle, \quad \text{where } \langle 1s|1s \rangle^2 + \langle 4f|4f \rangle^2 = 1. \tag{128}$$

With the constraint that Ψ be normalized, the energy functional becomes

$$\mathcal{F}(\Psi) = 2S(s,s)I(s,s) + F^0(s,s) + 2S(f,f)I(f,f) + F^0(f,f) \\ + cG^3(s,f) - E[S(s,s)^2 + S(f,f)^2], \tag{129}$$

where $c = -2/\sqrt{7}$ and only the lowest multipole of the $4f^2$ self-interaction has been included for brevity. The variational equations (in symmetric banded form) become

$$[2S(s,s)I + 2R^0(.,s;.,s)]s - [2E\,S(s,s) - 2I(s,s)]Ss \\ + cR^3(.,s;.,f)f = 0, \\ cR^3(.,f;.,s)s + [2S(f,f)I + 2R^0(.,f;.,f)]f \\ - [2E\,S(f,f) - 2I(f,f)]Sf = 0. \tag{130}$$

In this expression, $E \equiv E(\Psi)$ is the energy of the correlated state. Dividing the first set of equations by $S(s,s)$ and the second by $S(f,f)$ in order to put them in the same form as normalized equations, we can identify the coefficient of the S matrix as the diagonal energy parameter, namely

$$\varepsilon_{ss} = 2[E(\Psi) - I(s,s)/S(s,s)] = 2[E(\Psi) - I(\underline{s},\underline{s})], \\ \varepsilon_{ff} = 2[E(\Psi) - I(f,f)/S(f,f)] = 2[E(\Psi) - I(\underline{f},\underline{f})], \tag{131}$$

where $\underline{s}$ and $\underline{f}$ are the normalized s and f orbitals respectively. Along the iso-electronic sequence as Z increases, the effect of correlation decreases so that $\underline{s} \to s$. Thus, in the limit of infinite Z, $E(\Psi) = E(1s^2)$, and by definition $E(1s^2) = I(1s,1s) + F^0(1s,1s)$, from which it follows that

$$2[E(1s^2) - I(s,s)] = E(1s^2) + F^0(1s,1s), \tag{132}$$

in agreement with the definition of Eq. (24) for ε_{ss}. Associated with the $\underline{f}$ orbital is a pseudo-state, $|\underline{f}\rangle^2 F$, for which $E(\underline{f}) = I(\underline{f},\underline{f})$. Then

$$\varepsilon_{ff}/2 = [E(\Psi) - E(\underline{f})] \tag{133}$$

represents the distance of this pseudo-state from the correlated ground state and clearly is not a binding energy. Figure 6 shows that $I(\underline{f},\underline{f})$ is a positive and increasing function of Z as the nuclear charge increases. Thus the one-electron $4f$ pseudo-state lies in the continuum for all Z.

As another example, consider the natural orbital expansion of the two-electron ^1S ground state of Eq. (54), generalized to include a sum on l. With a positive coefficient for $\Phi(1s^2)$, the expansion coefficients have the sign $(-1)^l$ in Condon and Shortley phases. As a result, the wave function can be expression in terms of

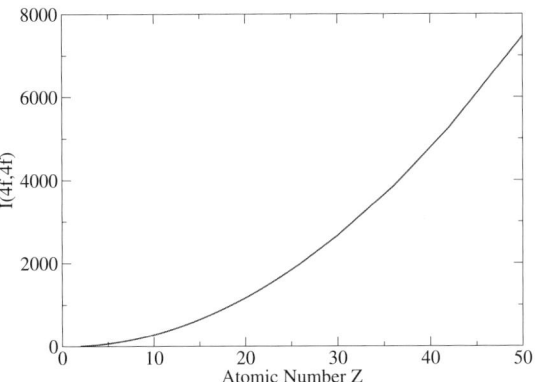

FIG. 6. The value of $I(4f, 4f)$ for the correlation orbital of the $4f^2\,{}^1S$ CSF representing correlation in the $1s^2\,{}^1S$ ground state of the He-like isoelectronic sequence.

unnormalized orbitals as

$$\Psi(1s^2\,{}^1S) = \Phi(1s^2\,{}^1S) + \sum_{n>2}\sum_{l}(-1)^{l+1}\Phi(nl^2\,{}^1S), \tag{134}$$

without the need for matrix diagonalization. If the wave function is not normalized, the variational condition needs to be applied to

$$\langle\Psi(1s^2)|\mathcal{H}|\Psi(1s^2)\rangle/\langle\Psi(1s^2)|\Psi(1s^2)\rangle,$$

augmented with off-diagonal Lagrange multipliers for orthogonality. This is equivalent to introducing the total energy E as a Lagrange multiplier for the normalization constraint on the wave function. In the simple case where only $1s^2$ and $2s^2$ are included in the expansion, the expression for the energy functional becomes (Froese Fischer et al., 1992):

$$\begin{aligned}\mathcal{F}(a,b,E,\varepsilon_{ab}) &= 2S(a,a)I(a,a) + 2S(b,b)I(b,b) \\ &\quad + F^0(a,a) + F^0(b,b) - 2G^0(a,b) \\ &\quad - E\big[S(a,a)^2 + S(b,b)^2\big] - 2\varepsilon_{ab}S(a,b).\end{aligned} \tag{135}$$

Table VII shows the rapid convergence for He and H$^-$. Initial estimates were assumed to be hydrogenic orbitals with charge $Z = 2$ for He, and $Z = 0.5$ for H$^-$. No explicit rotations were imposed. But this is a rather special case.

6.2.2. Singly Occupied Shells and Rydberg Series

The expansion coefficient of a CSF can always be absorbed into the definition of one singly occupied orbital of the CSF, provided the orbital is not occupied in any

Table VII

Convergence of the Newton–Raphson Method for $(1s^2, 2s^2)$ Approximation to the $1s^2$ Ground State of He and H$^-$ with $k_s = 6$ and $h = 1/8$ (from Froese Fischer et al., 1992)

Iteration	He		H$^-$	
	ΔP	$|VT+2|$	ΔP	$|VT+2|$
1	1.6(−1)		2.0(−1)	
2	2.1(−2)		2.6(−2)	
3	3.8(−4)	1.2(−7)	1.0(−3)	
4	7.8(−8)	2.0(−13)	5.2(−6)	
5	5.3(−15)	2.0(−13)	1.8(−10)	3.0(−13)

other CSF of the MCHF expansion. The orbital then is no longer normalized. Let us reconsider Eq. (122). Suppose we assume $3p$ and $3s$ are known, normalized orbitals and that c_2 is absorbed into the definition of d so that effectively, $c_2 = 1$. What has changed is the definition of H_{22} in that $I(s, s)$ must be replaced by $I(s, s)S(d, d)$ and, for the variational process, the normalization constraint on the d-orbital needs to be replaced by the normalization constraint on the wave function, namely $-E[c_1^2 + S(d, d)]$. In this formulation, c_1 and d can be determined from a generalized matrix-eigenvalue problem,

$$\left(\begin{bmatrix} H_{11} & (2/\sqrt{15})R^1(.,s;p,p)^t \\ (2/\sqrt{15})R^1(.,s;p,p) & H_{22} \end{bmatrix} - E\begin{bmatrix} 1 & 0 \\ 0 & S \end{bmatrix}\right)\begin{bmatrix} c_1 \\ d \end{bmatrix} = 0 \quad (136)$$

where $H_{22} = I(s, s)S + I + R^0(.,s;.,s) + 0.2R^2(.,s;s,.)$. The multiple solutions of this eigenvalue problem represent members of the $3snd$ Rydberg series that include the interaction with $3p^2$. For each energy, there will be a different c_1 and d-expansion for the orbital. It is still possible to obtain variational solutions for p, s orbitals, but it will be necessary to decide how many members of the Rydberg series are to be considered. For only one, say $d^{(1)}$, we could define $c_2 = \sqrt{S(d^{(1)}, d^{(1)})}$ and $d = d^{(1)}/c_2$ and solve the equations for p, s orbitals of Eq. (123) with d known. If the s, p orbitals are to be optimized on more than one eigenstate, it is necessary to generalize the optimization process. If in the present case, s, p are to be optimized for the first two members of the Rydberg series, then two Rydberg orbitals need to be normalized, and the energy expression assumed to be a linear combination of two energy expressions, one for the lowest state and one for the second.

This method can be generalized to wave functions with interacting Rydberg series as in

$$\Psi(^{1,3}P) = c_1\Phi(3s3p^{3\ 1,3}P) + \Phi(3s^23pns\ ^{1,3}P)$$
$$+ \Phi(3s^23pn'd\ ^{1,3}P), \tag{137}$$

where now ns and $n'd$ refer to Rydberg orbitals. Such calculations are performed with the orbitals for the $3s3p^3$ "perturber" and the $3s^2$ "target" fixed (not varied). This case differs from the previous one in that the ns Rydberg orbitals need to be orthogonal to the fixed $3s$ orbital. Applying the variational procedure of Eq. (129) and that of the approximation of Eq. (121) ignoring orthogonality initially, leads to equations of the form

$$\left(\begin{bmatrix} H_{11} & (H_{21}^s)^t & (H_{31}^d)^t \\ H_{21}^s & H_{22}^{ss} & H_{23}^{sd} \\ H_{31}^d & H_{32}^{ds} & H_{ss}^{dd} \end{bmatrix} - E \begin{bmatrix} 1 & 0 & 0 \\ 0 & S & 0 \\ 0 & 0 & S \end{bmatrix}\right) \begin{bmatrix} c_1 \\ s \\ d \end{bmatrix} = 0. \tag{138}$$

In this notation, the matrices H_{ij}^{ab} are such that $H_{ij}^{ab}b = H_{ij}^a$ where the latter is the column vector obtained by varying the matrix element $\langle i|\mathcal{H}|j\rangle$ with respect to orbital a. The orthogonality between the Rydberg orbital ns and the fixed $3s$ orbital can readily be included through the use of projection operators. Let a refer to $3s$ orbital. The Lagrange multiplier for the orthogonality condition $S(a, s) = 0$, can again be eliminated by multiplying the second "superrow" of submatrices for the s row by $(I - Saa^t)$ on the left. To symmetrize we can multiply the second "supercolumn" of submatrices by $(I - aa^tS)$ since this column is multiplied by the vector s and $(I - aa^tS)s = s$.

In general, this method assumes that there are a set of orbitals that appear in many CSFs, say the orbital set, and Rydberg "channels" which are singly occupied and appear in some subset of CSFs. Let us refer to the $(N-1)$ electron CSFs that couple to a Rydberg channel as a "target" state and the unnormalized channel function as $|n_il_i\rangle$, Then, the wave function expansion consists of N-electron CSFs or "pseudo-states" defined in terms of normalized radial functions, and targets coupled with channel functions, namely

$$\Psi(\gamma LS) = \sum_i^{M_p} c_i\Phi(\gamma_i LS) + \sum_j^{M_c} \underline{\Phi}(t_j LS), \tag{139}$$

where M_p and M_c are the number of pseudo-states and channel functions, respectively, and

$$\underline{\Phi}(t_j LS) = |(\text{target})_j \cdot |n_j l_j\rangle LS\rangle. \tag{140}$$

Our notation implies that the channel function is coupled to the target function according to the usual angular momentum rules to form an anti-symmetric eigenfunction of the total L and S. With the orbitals of the $N-1$ targets fixed, this method is referred to as the "frozen cores" approximation (Crees et al., 1978), but the term "target" is preferred here since in many calculations it is convenient to have only one common "core". In this method, a wave function typically has many targets. A description of the application of this method to the study of Rydberg series in Ca, extended to the Breit–Pauli Hamiltonian and neglecting the background interaction with the continuum, can be found in Brage and Froese Fischer (1994a, 1994b) where orthogonality constraints are imposed without symmetrization. It has also been applied by Brage et al. (1992) to the study of photodetachment and photoionization cross-sections for H^- and He for the excited states in the continuum and by Xi et al. to photodetachment in both He^- (Xi and Froese Fischer, 1999a) and Be^- (Xi and Froese Fischer, 1999b). Besides the grid which needs to be equally spaced at large r for continuum wave functions, what distinguishes the continuum problem is that there is a boundary condition only at the nucleus for a given orbital, the wave function is not normalized, and the energy is specified. By applying the Galerkin method to the Schrödinger equation

$$(\mathcal{H} - E)\Psi = 0, \tag{141}$$

equations similar to those derived by variational methods are obtained (see Froese Fischer and Idrees, 1989, where the simple case for hydrogen scattering is explained). With no boundary condition at a large radius, the coefficient matrix is not quite symmetric. Though in these early publications the systems of equations were solved by inverse iteration, they could be solved by singular value decomposition, the solution being the one for the eigenvalue nearest to zero.

6.3. WITHOUT ORTHOGONALITY CONSTRAINTS

Relaxation of the orthogonality constraint can have large benefits with a modest increase in complexity, provided non-orthogonality is used judiciously. For example, in a study of Rydberg series between the $4s\,^2S$ and $3d\,^2D$ ionization limits of calcium, six targets were identified (Brage and Froese Fischer, 1994a):

$$3d\,^2D, \quad 4p\,^2P, \quad 5s\,^2S, \quad 4d\,^2D, 5p\,^2P, \quad 4f\,^2F$$

which are the six lowest levels above $4s$ in Ca^+. The channel functions included $3dns_1$ and $5sns_2$ where $\langle ns_1|ns_2\rangle \neq 0$. Because of the different targets, much better results are obtained if ns_1 is not be required to be the same as ns_2. The 1991 version of MCHF (Froese Fischer, 1991a) allowed a limited amount of non-orthogonality for which the energy expressions never required more than two overlap integrals and was sufficient for the calcium study. Calculations were done in the Breit–Pauli approximation. Methods for computing Breit–Pauli integrals are discussed in Brage and Froese Fischer (1994b). More recently, Zatsarinny (2006) has developed determinantal methods where many constraints are removed. In particular, channel functions are not required to be orthogonal to the core with the consequence that the target functions may be MCHF wave functions for an $(N-1)$-electron state. Typically the orbitals and expansion coefficients for targets are not varied. The Slater matrix element cell methods have been extended to the Breit–Pauli integrals.

6.4. SPLINE ORBITAL BASIS METHODS: CI AND PERTURBATION THEORY

One advantage of spline orbitals is related to the fact that they form an effectively complete, orthonormal set of relatively modest size (Johnson and Saperstein, 1986). Variational methods usually find only one or possibly a few eigenvalues of an orbital Hamiltonian H^a but the equation for any singly occupied orbital can be used to generate an orthonormal spline basis. Landtman and Hansen (1993) used this basis to study the strong interactions between perturbers and Rydberg series such as between the perturber $3s3p^5\ ^3P$ and the $3s^23p^3ns\ ^3P$ and $3s^23p^3n'd\ ^3P$ Rydberg series in SI. In the Hartree–Fock approximation $3s3p^5\ ^3P$ is predicted to lie above the $3s^23p^3\ ^4S$ limit but correlation with the continuum lowers the energy to where it becomes a bound state. They claim that starting with a fixed Hartree–Fock core is preferable to a model potential.

Alternatively, the variational methods could be used to determine major components of a wave function, and then a spline orbital basis could be used to determine small corrections to first-order. Vilkas et al. (1998) use multi-reference Möller–Plesset many-body perturbation theory for relativistic calculations Vilkas and Ishikawa (2004a, 2004b) to improve variational results.

Another way of combining variational calculations with perturbation theory is to modify the CI process. In fact, this would stabilize the variational calculation where convergence slows down significantly when many small effects need to be determined. Let x be the CSF expansion vector for a variational calculation where $H^{00}x = E^0x$ defines the zero-order approximation with $x^t x = 1$. The CSFs in this expansion are the zero-order set of CSFs. Let y be the CSF expansion vector for the first-order approximation determined to first order. The CSFs in this expansion are in the first-order set. Then the CI problem for the combined set is

$$\begin{bmatrix} H^{00} & H^{01} \\ H^{10} & H^{11} \end{bmatrix} \begin{bmatrix} x \\ y \end{bmatrix} = (E^0 + E^1) \begin{bmatrix} x \\ y \end{bmatrix}, \qquad (142)$$

where H^{01} is the matrix of interaction matrix elements between the zero-order and first-order sets, and H^{11} the matrix of interaction matrix elements within the first-order set. In a first-order calculation the latter may be replaced by the diagonal matrix D for which $D_{ii} = H^{11}_{ii}$. Retaining terms to first-order we get

$$\begin{aligned} H^{00}x + H^{01}y &= (E^0 + E^1)x \quad \text{or} \quad E^1 = x^t H^{01} y, \\ H^{10}x + Dy &= E^0 y \quad \text{or} \quad y = (E^0 - D)^{-1} H^{10} x. \end{aligned} \qquad (143)$$

The CSFs in the first-order set could include all CSFs that can be constructed by the excitation of one or two orbitals from the zero-order to the first-order orbital set. Since angular data is independent of the principle quantum number, considerable simplification can be introduced as shown by Ellis et al. (2004).

Once the first-order corrections have been determined, the CSF expansions can be normalized and used in the calculation of other atomic properties.

6.5. Direct Expansions in Products of B-splines for the Many-Electron Case

Traditional multiconfiguration configuration-interaction methods expand a wave function in terms of orbitals that are orthonormal. This approach provides physical insight about a state in that often a wave function can be labelled by its dominant component and its properties predicted in qualitative (though not quantitative) manner. Orthonormality also provides mathematical convenience.

Direct expansions in a basis consisting of products of B-splines have not been carefully investigated. Such methods would have to deal with non-orthogonal basis functions, but could have advantages. Given a zero-order approximation, the first-order corrections consist of excitations $(a, b) \rightarrow (v, v')$ where (a, b) are occupied orbitals and (v, v') are virtual orbitals. The sum of CSFs for the excitation from a given pair of occupied orbitals can be expressed as symmetry adapted, two-electron, pair-correlation functions. These are linear combinations of two-electron partial waves Ψ^p characterized by the orbital angular momenta and their LS coupling, namely

$$\Psi^p(ll'LS) = \sum_v \sum_{v'} c_{vv'll'} |vlv'l'LS\rangle. \tag{144}$$

By definition,

$$|vlv'l'LS\rangle = \mathcal{N} \frac{1}{r_1 r_2} (1 - \mathcal{P}_{12}) P(vl; r_1) P(v'l'; r_2) |ll'LS\rangle, \tag{145}$$

where $\mathcal{N}$ is a normalization factor that may depend on symmetry, $\mathcal{P}_{12}$ a permutation operator, $|ll'LS\rangle$ a spin-angular factor, and $P(vl; r)$ and $P(v'l'; r)$ are orbital radial functions. Substituting into Eq. (144) and noting that each term has the same spin-angular factor, that the double sum is an expansion in a basis of a function of two variables, we get

$$\Psi^p(ll'\,LS) = \mathcal{N} \frac{1}{r_1 r_2} (1 - \mathcal{P}_{1,2}) p(r_1, r_2) |ll'LS\rangle. \tag{146}$$

We can treat $p(r_1, r_2)$ as a general two-dimensional function, usually expanded in terms of orbitals but one that could also be expanded in terms of tensor products of B-spline basis functions, namely

$$T_{ij}(r_1, r_2) = B_i(r_1) B_j(r_2). \tag{147}$$

The advantage of this basis is the local support (non-zero region) compared with products of orbitals which extend over the entire two-dimensional region.

A 1974 correlation study of the $1s^2 2s^2 \text{Be}\,^1S$ ground state used a pair-correlation approach in terms of orbitals that were non-orthogonal between different pairs (Saxena and Froese Fischer, 1974). SD-excitations from this zero-order approximation lead to four pair-correlation functions depending on the two occupied orbitals and their coupling that enter into the excitation, namely $1s^2\,^1S$, $1s2s\,^1S$, $1s2s\,^3S$, and $2s^2\,^1S$. For each expansion, a reduced form (like the natural orbital expansion for the $1s^2$ ground state of helium) was used. By this method, 99.1% of the correlation energy was accounted for in 52 CSFs yielding a total energy of -14.66587 au. The best estimate today is -14.667355748 au from an exponentially correlated Gaussian wave function (Pachuki and Komasa, 2006).

Pair functions are similar to wave functions for the 2-electron case. The feasibility of B-spline tensor-product expansions for $1s^2$, $1s2s\,^{1,3}S$, and $1s2p\,^{1,3}P$ states was studied in 1991 (Froese Fischer, 1991b). Of interest at the time was performance on parallel vector processors. Of greater importance here is the accuracy of the energy as a function of grid parameters. Shown in Fig. 7 is the log

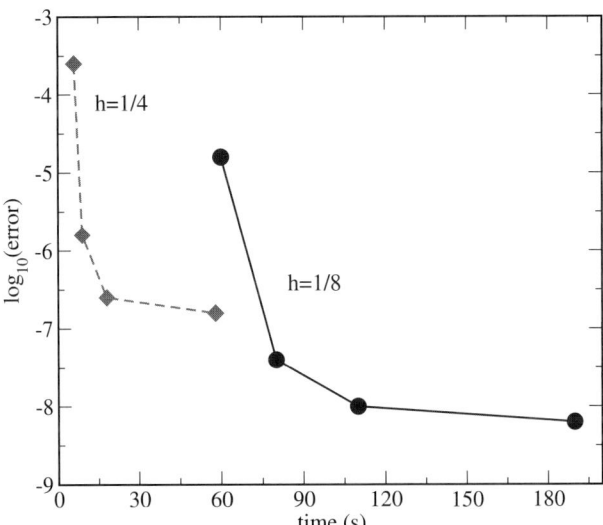

FIG. 7. The accuracy of the total energy for the $l = 0$ partial wave for the helium $1s^2$ ground state as a function of CPU time for various grid parameters. The dashed curve shows increasingly more accurate results for $k_s = 3, 4, 5, 6$ and $h = 1/4$ whereas the solid curve shows similar results for $h = 1/8$ (from Froese Fischer, 1991b).

of the error as a function of computer time for various grid parameters. Of particular importance is the size of the step-size parameter h. Notice the significant improvement in accuracy as the step-size parameter is reduced from $h = 1/4$ to $h/8$. This paper used differential equation methods for computing the B-spline Slater matrix elements which may partly explain this observation.

It is important to remember that correlation is a correction to the wave function that increases the binding of the outer electron. No increase in box-size is needed unless it increases the accuracy of the HF solution. Of particular importance is the region $r_1 = r_2$. For states like $1s^2$ where both electrons share the same region of space, exceedingly high values of l are needed because of the slow convergence with respect to l. For expansions up to $l = 7$ the errors in the total energy (in micro-Hartrees), are (Froese Fischer, 1991b):

State:	$1s^2\,^1S$	$1s2s\,^1S$	$1s2s\,^3S$	$1s2p\,^1P$	$1s2p\,^3P$
Error (mH):	50.1	3.0	0.1	2.6	0.1

Considerably more correlation remains for the $1s^2$ state than for the others and, given slow convergence with l for the energy of the ground state, computed energies are usually extrapolated. With B-splines, the expansions could be restricted to the correlation region. Consider the wave function for the $1s2p\,^1P$ state that can be expressed in terms of partial waves as

$$\Psi(^1P) = \frac{1}{2r_1 r_2} \sum_l (1-\mathcal{P}) p^{(l)}(r_1, r_2) |ll'\,^1P\rangle, \quad l' = l+1. \tag{148}$$

In orbital methods,

$$p^{(l)}(r_1, r_2) = \sum_n \sum_{n'} c_{n,n'(l)} P(nl; r_1) P(n'l'; r_2) \tag{149}$$

whereas in a direct B-spline expansion, it is

$$p^{(l)}(r_1, r_2) = \sum_i \sum_j c^{(l)}_{i,j} B_i(r_1) B_j(r_2). \tag{150}$$

These two expressions are equivalent in a complete orbital spline basis expansion, but there will be differences as soon as summations are truncated as is frequently done in spline orbital methods for a many-electron atom (see Chen et al., 2001, as an example).

Figure 8 shows the matrix of coefficients $c^{(l)}_{i,j}$ for B-spline tensor-product expansions for $l = 0, 1, 2, 3$ from left to right and top to bottom, respectively. Notice that each subgraph has a different scale with the maximum value decreasing. The modest decrease going from $l = 2$ to $l = 3$ is consistent with the slow rate of convergence with l. For $l = 0$, the matrix is approximately the cross-product of the expansions for the $1s$ and $2p$ Hartree–Fock orbitals with the maximum at

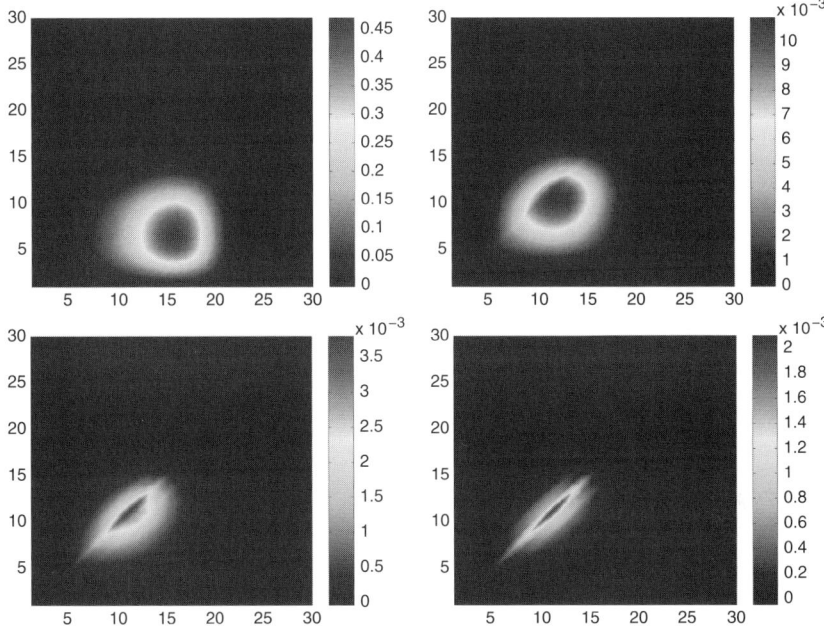

FIG. 8. A visualization of the magnitude of the matrix of expansion coefficients $c^l_{i,j}$, $\{i, j\} = 1, \ldots, 30$ for the $1s2p\,^1P$ state of helium. Shown, from left to right and top to bottom, are the expansion coefficients for the sp, pd, df, and fg partial wave functions. The maximum values are 0.45, 0.010, 0.0035, 0.0020, respectively. Note both the changing region and the decreasing maximum magnitude of each partial wave.

$r_1 = 0.707$, $r_2 = 2.05a_0$ resulting in an off-diagonal maximum for the $l = 0$ partial wave. As l increases, the maximum coefficient moves closer to $r_1 = r_2$. But more importantly, the significant components (that take into account the scale) are concentrated in a smaller and smaller region. In fact, correlation is correcting the electron–electron cusp condition first defined by Kato (1957). For two-electron systems the latter is the condition (Qiu et al., 1997)

$$\left(\frac{\partial(\Psi(\mathbf{r}_1, \mathbf{r}_2))}{\partial r_{12}}\right)_{r_{12}=0} = \frac{1}{2}\Psi(\mathbf{r}_1 = \mathbf{r}_2, \mathbf{r}_2) \tag{151}$$

and depends on what, in density functional theories (Gori-Giorgi and Savin, 2006), is referred to as "short-range correlation". This means that orbital expansions, where high-n orbitals oscillate and extend over the entire range, must deal with extensive cancellation in regions where the partial wave is near zero. At the same time, because of their local support, B-spline tensor-products can safely be restricted to the region of importance. First-order perturbation theory could be used to determine in advance whether a certain basis element should be included.

At the same time, this example shows why numerical accuracy is so important for correlation contributions. As l increases, the region around $r_1 = r_2$ contributes most to the correlation correction, and the accuracy depends more and more on the accuracy of the contribution from the diagonal cells to the Slater matrix elements. Thus the inclusion of the B-spline tensor-product basis provides a number of computational advantages that could be exploited.

7. Conclusion

This paper has reviewed variational methods in atomic structure calculations. It has been shown that differential equations with their many special cases can be eliminated and replaced by more robust linear algebra and symmetric eigenvalue problems. Through the use of B-spline expansions, Hartree–Fock orbitals can be defined as solutions of a generalized eigenvalue problem, $(H^a - \varepsilon_{aa} S)a = 0$, once orthogonality issues have been dealt with. Iterative self-consistent methods are still needed in that the "Hartree–Fock Hamiltonians" H^a depend on the other orbitals. When orthogonality conditions are significant or excited states are needed, this would be the most effective method for getting good estimates of orbitals. Once such estimates have been obtained, a number of orbitals may be updated simultaneously by expressing the equations in terms of blocks of banded submatrices for which the generation of submatrices is most efficient. Methods based on generalized singular decomposition may be used that are able to incorporate orbital rotations. For special cases, the Newton–Raphson method may be used, but with this method, the submatrices are usually dense. Thus a number of strategies are available for solving the Hartree–Fock problem.

Unlike the Hartree–Fock problem, multiconfiguration Hartree–Fock orbitals cannot always be represented as solutions of a symmetric, generalized eigenvalue problem for a given orbital, but they can be expressed as solutions of systems of banded submatrices. When normalization conditions can be relaxed, calculations for Rydberg series can be expressed as generalized eigenvalue problems, calculations that can be improved significantly through the use of non-orthogonal orbitals as implemented in BSR methods (Zatsarinny, 2006).

B-spline methods present many opportunities. Calculations can be done to low accuracy through the use of larger intervals through the parameter h and lower spline order k_s. A variety of methods may be applied. Because a spline function can be evaluated at any point in the interval, a low-order spline approximation can be redefined on a different grid and improved with a higher-order spline calculation, similar to multigrid methods for partial differential equations (Wesseling, 1992). Many options can be applied.

Though this discussion has dealt solely with the non-relativistic case, it applies also to the Dirac–Coulomb Hamiltonian (H_{DC}) and Dirac–Coulomb–Breit

Hamiltonian (H_{DCB}). Many of the convergence problems encountered in the General Relativistic Atomic Structure Package (GRASP92) (Parpia et al., 1996) are associated with initial estimates of inadequate accuracy. The simplest option in this code determines orbitals from a Thomas–Fermi potential. These are not good estimates for correlation orbitals that often may have a high principal quantum number. When instead MCHF (Froese Fischer, 1991a) orbitals are considered, converted to relativistic form to lowest-order, convergence often is achieved. Other problems are associated with node counting, something not needed for solutions of a generalized eigenvalue problem. Spline orbitals for highly excited states can be computed to considerably better accuracy than differential equation methods (Froese Fischer and Parpia, 1993).

8. Acknowledgements

This paper is dedicated to the memory of Yong-Ki Kim, whose interest in B-splines was a strong motivating factor for this review. I thank Oleg Zatsarinny for his interest in the spline project and his conviction that the Hartree–Fock problem could be formulated as one large generalized eigenvalue problem. I am grateful to Pavlas Bogdanovicius for an early reading of this manuscript.

Much of this research into spline methods for atomic structure calculations was funded during 1980–2005 through the Division of Chemical Science, Geosciences, and Bioscience, Office of Basic Energy Sciences, Office of Science, U.S. Department of Energy.

9. References

Altenbeger-Siczek, A., Gilbert, T. (1976). *J. Chem. Phys.* **64**, 432.
Bachau, H., Cormier, E., Decleva, P., Hansen, J.E., Martin, F. (2001). *Rep. Prog. Phys.* **64**, 1815.
Bentley, M. (1994). *J. Phys. B* **27**, 637.
Brage, T., Froese Fischer, C. (1994a). *Phys. Rev. A* **50**, 2937.
Brage, T., Froese Fischer, C. (1994b). *J. Phys. B* **15**, 5467.
Brage, T., Froese Fischer, C., Miecznik, G. (1992). *J. Phys. B* **25**, 5289.
Chen, M.H., Cheng, K.T., Johnson, W.R. (2001). *Phys. Rev. A* **64**. 042507.
Cowan, R.D. (1981). "The Theory of Atomic Structure and Spectra". University of California Press, Berkeley.
Crees, M., Seaton, M.J., Wilson, P.M. (1978). *Comput. Phys. Commun.* **15**, 23.
de Boor, C. (1985). "Practical Guide to Splines". Springer Verlag, New York.
Dyall, K.G., Grant, I.P., Johnson, C.T., Parpia, F.A., Plummer, E.P. (1989). *Comput. Phys. Commun.* **55**, 425.
Fletcher, C.A. (1984). "Computational Galerkin Methods". Springer-Verlag, New York.
Froese, C. (1967). *Can. J. Phys.* **45**, 7.
Froese Fischer, C. (1966). *Phys. Rev.* **150**, 1–6.

Froese Fischer, C. (1973). *J. Phys. B* **6**, 1933.
Froese Fischer, C. (1977). "The Hartree–Fock Method for Atoms". Wiley & Sons, New York.
Froese Fischer, C. (1986). *Computer Phys. Reports* **3**, 273.
Froese Fischer, C. (1987). *Comput. Phys. Commun.* **43**, 355.
Froese Fischer, C. (1991a). *Comput. Phys. Commun.* **64**, 431.
Froese Fischer, C. (1991b). *Int. J. Supercomputer Appl.* **5**, 5.
Froese Fischer, C. (2000). *Comput. Phys. Commun.* **128**, 635.
Froese Fischer, C. (2005). *Nucl. Inst. Meth. Phys. Res. B* **235**, 100.
Froese Fischer, C., Ellis, D. (2004). *Lithuanian J. Phys.* **44**, 121–134.
Froese Fischer, C., Idrees, M. (1989). *Computers in Physics* **May/June**, 53.
Froese Fischer, C., Parpia, F.A. (1993). *Physics Lett. A* **179**, 198.
Froese Fischer, C., Guo, W., Shen, Z. (1992). *Int. J. Quant. Chem.* **42**, 849.
Froese Fischer, C., Tong, M., Bentley, M., Shen, Z., Ravimohan, C. (1994). *J. Supercomputing* **8**, 117–134.
Froese Fischer, C., Brage, T., Jönsson, P. (1997). "Computational Atomic Structure". Institute of Physics Publishing, Bristol.
Gaigalas, G. (1999). *Lithuanian J. Phys.* **39**, 79.
Gori-Giorgi, P., Savin, A. (2006). *Phys. Rev. A* **73**. 032506.
Griffin, D.C., Cowan, R.D., Andrews, K.L. (1971). *Phys. Rev. A* **3**, 1233.
Hartree, D.R. (1929). *Proc. Camb. Phil. Soc.* **5**, 225.
Hartree, D.R. (1955). *Proc. Camb. Phil. Soc.* **51**, 684.
Hibbert, A., Froese Fischer, C. (1991). *Comput. Phys. Commun.* **64**, 417–430.
Johnson, W.R., Saperstein, J. (1986). *Phys. Rev. Lett.* **57**, 1126.
Kato, T. (1957). *Commun. Pure Appl. Math.* **10**, 151.
Koopmans, T.A. (1933). *Physica* **1**, 104.
Landtman, M., Hansen, J.E. (1993). *J. Phys. B* **26**, 3189.
Löwdin, P.-O. (1955). *Phys. Rev. A* **97**, 1509.
McDougall, J. (1932). *Proc. Roy. Soc. A* **138**, 550.
Pachuki, K., Komasa, J. (2006). *Phys. Rev. A* **73**. 052502.
Parpia, F.A., Froese Fischer, C., Grant, I.P. (1996). *Comput. Phys. Commun.* **94**, 249.
Press, W.H., Flannery, B.P., Teukolsky, S.A., Vetterling, W.T. (1992). "Numerical Recipes in C". Cambridge University Press, New York, USA.
Qiu, Y. (1999). "Integration by Cell Algorithm for Slater Integrals in a Spline Basis". Vanderbilt University, Nashville, TN.
Qiu, Y., Froese Fischer, C. (1999). *J. Computational Phys.* **156**, 257.
Qiu, Y., Tang, J., Burgdörfer, J., Wang, J. (1997). *J. Phys. B* **30**, L689.
Rebelo, I., Sharma, C.S. (1967). *J. Proc. Phys. B* **5**, 543.
Roothaan, C.C.J. (1951). *Rev. Mod. Phys.* **23**, 69.
Saxena, K.M.S., Froese Fischer, C. (1974). *Phys. Rev. A* **9**, 1498.
Sharma, C.S. (1967). *Proc. Phys. Soc.* **92**, 543.
Sharma, C.S., Coulson, C.A. (1962). *Proc. Phys. Soc.* **80**, 81.
Stathopoulos, A., Froese Fischer, C. (1994). *Comp. Phys. Commun.* **79**, 268–290.
Stathopoulos, A., Ynnerman, A., Froese Fischer, C. (1996). *Int. J. Supercomputers & High Performance Comput.* **10**, 41.
Szabo, A., Ostlund, N.S. (1982). "Modern Quantum Chemistry: Introduction to Advanced Electronic Structure Theory". MacMillan Publishing Co., New York.
Tachiev, G., Froese Fischer, C. (1999). *J. Phys. B* **32**, 5805–5823.
Trefftz, A., Schluter, E., Dettmar, K.-H., Jorgens, K. (1957). *Z. Astrophys.* **44**, 1.
Vilkas, M.J., Ishikawa, Y. (2004a). *J. Phys B.* **37**, 4763.
Vilkas, M.J., Ishikawa, Y. (2004b). *Phys. Rev. A* **69**. 062503.

Vilkas, M.J., Ishikawa, Y., Koc, K. (1998). *Phys. Rev. E* **58**, 5096.
Wesseling, P. (1992). "An Introduction to Multigrid Methods". John Wiley, Sons Inc.
Xi, J., Froese Fischer, C. (1999a). *Phys. Rev. A* **59**, 307.
Xi, J., Froese Fischer, C. (1999b). *J. Phys. B* **32**, 387.
Zatsarinny, O. (2006). *Comput. Phys. Commun.* **174**, 273.

ELECTRON–ION COLLISIONS: FUNDAMENTAL PROCESSES IN THE FOCUS OF APPLIED RESEARCH

ALFRED MÜLLER

Institut für Atom- und Molekülphysik, Justus-Liebig-Universität Giessen, Germany

1. Introduction 294
2. Basics of Electron–Ion Collisions 298
   2.1. The Processes 298
   2.2. Resonances and Interference Patterns 306
   2.3. Generic Energy Dependences of Cross Sections 310
   2.4. Time-Reversal Symmetry 317
3. Experimental Access to Data 319
   3.1. Plasma Techniques 327
   3.2. Traps 330
   3.3. Quasi-Free Electrons 335
   3.4. Colliding-Beams Techniques 343
4. Overview of Experimental Results on Free-Electron–Ion Collisions 355
   4.1. Electron Scattering from Ions 355
   4.2. Electron-Impact Excitation of Ions 357
   4.3. Electron–Ion Recombination 360
   4.4. Electron-Impact Ionization of Ions 384
5. Conclusions 397
6. Acknowledgements 399
7. References 399

Abstract

Electron–ion collisions are among the most important atomic processes in ionized gases. They comprise elastic scattering, excitation, ionization, and recombination. Theoretical approaches are numerous but calculating reliable results for applications is a task that has not generally been solved, although theory has progressed considerably in the last decade and can now provide reference data for electron collisions with few-electron systems. Experimental data on electron–ion collisions can be inferred from plasma observations, from ion trap measurements, from careful preparation and analysis of ion–atom collisions, and from colliding-beams experiments. The spectrum of equipment employed in the measurements ranges from table top size to large accelerator and storage-ring facilities.

This article provides an overview of the processes involved in electron–ion collisions, their cross sections, the experimental approaches and typical results. The present status of the field is discussed and characterized by numerous examples illuminating the accomplishments as well as the limitations of experimental access to accurate detailed cross sections and rate coefficients, as they are required for understanding and modelling laboratory or astrophysical plasmas.

1. Introduction

Collisions of electrons with atoms and ions are atomic interactions of fundamental nature. Studying details of these collision processes enhances our understanding of the quantum-physical basis of nature and provides knowledge about the structure and the dynamics of atomic systems. Besides their intrinsic relevance, however, electron collisions are also most important in plasma applications. They determine the charge-state balance of atoms in the plasma and the spectrum of electromagnetic radiation emitted. Understanding and diagnosing the state of a plasma, whether of astrophysical origin or man made, relies on information about cross sections and rate coefficients for electron–atom and electron–ion interactions. Accordingly, these interactions have attracted long-standing theoretical and experimental interest since the ground-breaking investigations with gas discharge tubes at the end of the 19th century (a bibliography of early work on cathode rays is provided in the Nobel Lecture by Lenard, 1906). Recent reviews on theoretical approaches appeared in this series (Moores and Reed, 1994; Burke et al., 2006; Griffin and Pindzola, 2006). Experimental work on electron–ion collisions has been reviewed from the perspective of laboratory x-ray astrophysics by Beiersdorfer (2003) and the topic is also included in a recent comprehensive review on atomic data for x-ray astrophysics by Kallman and Palmeri (2007). Brief overviews on the topic of electron–ion collisions can be found in papers by Müller (2002, 1997, 1996). Subfields of the topic were reviewed by Lindroth and Schuch (2003), Müller and Schippers (2001, 2003), Schippers (1999), Müller (1995, 1992, 1991), Chutjian (2004) and Williams (1999). Mokler and Stöhlker (1996) have addressed the role of heavy-ion storage–cooler rings in studying the physics of highly charged ions and also covered electron–ion interactions in their review. The present overview emphasizes the physics behind the large spectrum of individual processes in electron–ion collisions and concentrates on experimental approaches to the problem of determining cross sections and rate coefficients for electron-impact ionization, excitation, recombination and various scattering processes involving atomic ions.

Elastic collisions of ions with cold electrons are used to cool ion beams in storage rings (Poth, 1990). Inelastic electron interactions with atomic ions are the dominant processes in high temperature plasmas such as stellar coronae (Bryans et al., 2006). Recombination and scattering of electrons from ions are also important in low temperature plasmas, for example in photo-ionized gases (Osterbrock and Ferland, 2006) as they appear in the vicinity and the ejecta of active galactic nuclei or x-ray binaries. Thus, depending on the environment, the different electron collision processes have varying importance and the charge state distributions of atoms in the environment under consideration can span a very wide range between almost neutral and almost fully stripped (Summers et al., 2006).

Recent progress in x-ray astronomy with satellite borne spectrometers Chandra and XMM-Newton has produced high-resolution x-ray spectra of numerous astrophysical objects. Understanding those spectra requires huge amounts of data and, in particular, also knowledge about electron–ion collisions (Kallman and Palmeri, 2007). Model calculations are undertaken on the basis of existing data compilations which in turn often stem from extrapolations of theoretical calculations and of the relatively few laboratory measurements. Wrong data entering the models lead to false conclusions about the environment from which the observed spectra originate (Ferland, 2003). Really understanding astrophysical environments and their dynamics requires reliable atomic data, and electron–ion collisions are among the most important processes in those environments.

An example for the importance of basic laboratory data is the quest to understand the nature of black holes lurking in the cores of active galaxies, such as quasars and blazars. Particles are drawn into the black hole's gravitational whirlpool, emitting x-rays as they spiral down the drain. This outpouring of radiation from the galactic nucleus blasts the surrounding material, stripping away electrons from atoms and creating a gas of free electrons and ions. Understanding a black hole's environment thus requires an accurate understanding of how electrons and ions interact. The Chandra and XMM-Newton x-ray observatories have given astronomers important insights into the extreme conditions near active galactic nuclei (Paerels and Kahn, 2003). For example, spectra from both satellites reveal complex structures in the outflowing winds surrounding the central black holes. Recent x-ray observations indicate that the iron in those winds is less ionized than predicted. Figure 1 shows the measured spectrum of the active galaxy NGC 3783 together with results of two model calculations. One calculation is based on presently available (limited) atomic data sets and the other is based on the ad-hoc assumption that the recombination rates for iron ions are actually greater than the theoretically predicted rates. The agreement between the model and the observations is greatly improved by this assumption. Clearly, experiments and further development of theory are required to prove or disprove whether that assumption is valid.

A classic application of electron–ion collision data is plasma diagnostics. From the analysis of spectroscopic observations, Doppler shifts, line broadening and line intensity ratios, it is possible to infer plasma parameters such as plasma rotation velocity, ion temperature, electron temperature, and electron density. The

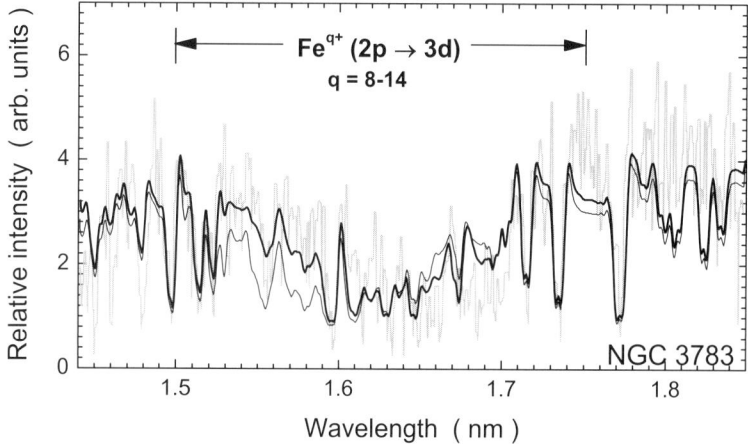

FIG. 1. Chandra x-ray spectrum of the active galaxy NGC 3783 together with laboratory astrophysics model calculations (material published by Netzer, 2004). The spectrum is rich in absorption lines from nitrogen, oxygen, neon, and other species. The observed absorption spectrum (light gray) is relatively well described by the model calculation (thin solid line) except for the range between 1.5 and 1.7 nm where $2p \to 3d$ transitions in iron ions are relevant. The model was changed by arbitrarily enhancing recombination rates of Fe^{q+} iron ions ($q = 8$–14). The result (solid black line) is then closer to the observation and indicates that the theoretical cross sections used in the original model might be too small.

field has been opened by the seminal work of Gabriel (1972) on satellites to the resonance transition in He-like ions and since then numerous studies have been devoted to the issue of plasma diagnostics employing spectra of He-like ions. In particular, dielectronic satellites, i.e., lines associated with K-shell-excited Li-like ions, turn out to be sensitive to the electron temperature in a plasma. For a quantitative derivation of electron temperatures of plasmas it is necessary to know details about electron-impact excitation and dielectronic recombination rates. The modeling of plasma ion charge state distributions requires additional knowledge about ionization rate coefficients. Considering the existing lack of reliable data on one side and the power of spectroscopic plasma diagnostics on the other, it is clear why research in the field of dielectronic satellites is very active. One of the most recent papers on the subject was published by Nahar and Pradhan (2006).

An example for a spectroscopic measurement aiming at plasma diagnostics is shown in Fig. 2. The data are from one of the many discharge shots in a tokamak fusion test device, in this case from TEXTOR in Jülich, Germany, where the plasma was seeded with argon gas for diagnostic purposes (Marchuk, 2004; Marchuk et al., 2006). From the width of the lines, an ion temperature corresponding to 1.1 keV can be inferred and the line intensity ratios yield 1.2 keV for the electron temperature. The notation of the lines associated with transitions in

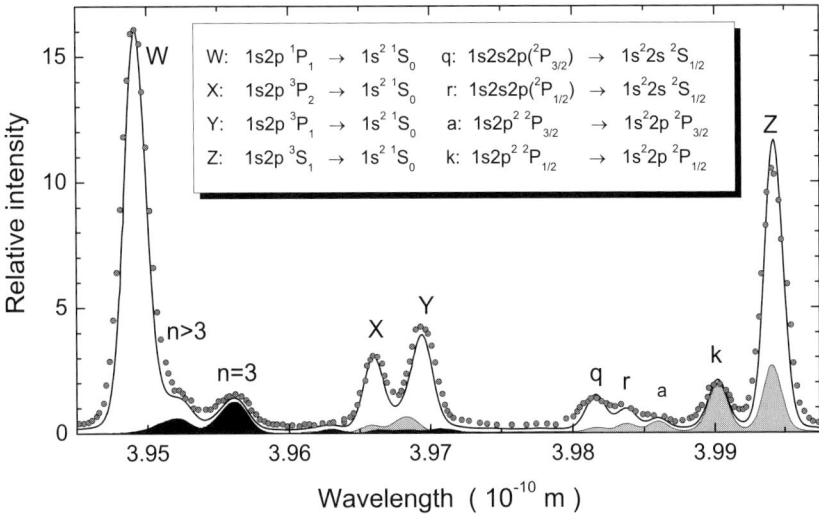

FIG. 2. Spectra of the K$_\alpha$ complex of He-like Ar XVII and its dielectronic satellites. The data points are from the TEXTOR tokamak (shot 82,069), the curves from a theoretical model calculation assuming an electron temperature T_e that corresponds to 1.2 keV (Marchuk, 2004; Marchuk et al., 2006). The notation of the lines follows the scheme introduced by Gabriel (1972). The solid line is the model calculation. The contribution of dielectronic satellites associated with $2 \to 1$ transitions in Li-like Ar XVI is gray shaded. Contributions due to $n \to 1$ transitions with $n \geqslant 3$ are shaded black.

He-like and Li-like Ar^{16+} and Ar^{15+} ions, respectively, follows the notation of Gabriel (1972). The resonance line (W) and the lines arising from intercombination transitions (X, Y, Z) are mainly due to electron-impact excitation of He-like ions. The lines with the notations $k, a, r, q, n = 3$, and $n > 3$ (dielectronic satellites) are predominantly arising from Li-like argon ions produced by dielectronic recombination of helium-like Ar XVIII.

A wide field of applications of electron–ion collisions is in research and development associated with industrial plasmas. Examples where atomic data on electron–ion rate coefficients and ion spectra are essential include the field of lighting and energy consumption (Lister, 2004) and the development of plasma sources of intense XUV radiation for 13.5 nm semi-conductor lithography (Jonkers, 2006; Krücken et al., 2004).

From a fundamental point of view, studying ions in addition to neutral atoms provides a new and welcome parameter, the ion charge state, that influences the behavior of cross sections and allows one to choose the relative strengths of electron–electron and electron–nucleus interactions in an electron–ion collision process. In particular, measurements and calculations of cross sections along isoelectronic sequences of atoms can elucidate basic relations in the structure and collisional behavior of atoms and ions. Going from neutral atomic hydrogen all the way to hydrogen-like U^{91+} adds 91 objects of similar nature to the investigation of the properties of a one-electron system. The range of possible influences of the ion charge state on such properties can be illustrated by comparing the $1s$ Lamb shift in atomic hydrogen H (33.5 µeV; Yerokhin et al., 2005) with that of hydrogen-like uranium U^{91+} (464 eV; Gumberidze et al., 2005), or, even more dramatically, the lifetime of the $2s$ state in H (0.12 s) and in U^{91+} (5.0×10^{-15} s; Jentschura, 2004) which is determined by an approximate Z^{10} increase of the one-photon M1-transition rate with the atomic number compared to a Z^6 dependence of the two-photon decay. Such properties of isoelectronic atoms in different charge states (and with different atomic numbers Z) have immediate effects also on electron-impact cross sections, thus making the investigation of electron–ion collision phenomena particularly rich compared to studies restricted to neutral atoms.

2. Basics of Electron–Ion Collisions

2.1. The Processes

Collision processes of electrons with singly or multiply charged atomic ions A^{q+} can be categorized in four groups, depending on the final situation observed in an experiment after the electron–ion interaction and after possible relaxations. The Coulomb potential supports a multitude of collision pathways which are listed in the following paragraphs. All of the processes mentioned below, even the exotic ones, can be observed in experiments.

2.1.1. Elastic Scattering

An electron scattered from a target ion may change its initial direction while the target ion remains in its original state. In the center-of-mass frame no kinetic energy is lost or gained in the elastic collision. In the laboratory frame, however, the incident electron can transfer a small amount of kinetic energy to the ion. By numerous such collisions the momentum of the ion can be substantially changed, which is employed in the cooling of ions in a storage ring where the circulating ion beam is merged in a cold beam of electrons (Poth, 1990). The process of elastic scattering can be described as

$$e + A^{q+} \rightarrow A^{q+} + e. \tag{1}$$

The simplest concept behind Eq. (1) is that of a direct elastic scattering process, in which the electron is deflected while passing the Coulomb field of the ion. However, there is a different two-step pathway by which the final state can be reached.

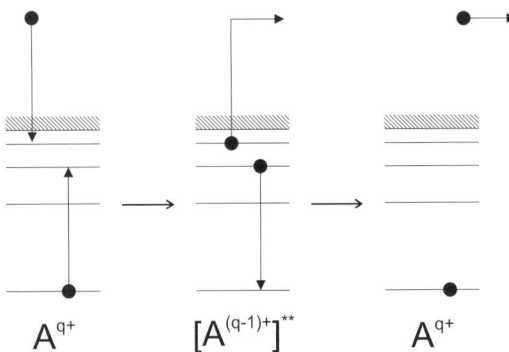

FIG. 3. Schematic energy diagram indicating the reaction steps of resonant elastic scattering. An initially free electron, moving in the continuum of ion A^{q+} with an energy E_e is captured into a subshell of the ion $A^{(q-1)+}$ and the excess energy is consumed by the excitation of a core electron of the ion to a higher level. The process is resonant, i.e., in the situation sketched here it can only happen if the energy E_e matches the difference between the ionization energy of the initial ground-state ion and the binding energy of the two active electrons in the resonant intermediate state.

The first step of this indirect elastic scattering mechanism

$$e + A^{q+} \rightarrow \left[A^{(q-1)+}\right]^{**} \quad (2)$$

is called dielectronic capture (usually abbreviated by DC) which can be treated as the time-reversed Auger process, or, more generally, as time-reversed autoionization. Dielectronic capture can only occur if the kinetic energy E_e of the projectile electron matches the difference $E_{res} = E_i - E_f$ of total binding energies of all electrons in the initial and final states of the ion considered in Eq. (2). In other words, dielectronic capture is a resonant process. The second step is relaxation of the doubly excited intermediate state $[A^{(q-1)+}]^{**}$ by emission of an electron, producing the initial state A^{q+} of the target again. This emission is associated with a characteristic lifetime, typically in the range of picoseconds or less. It involves an Auger, a Coster–Kronig or a Super Coster–Kronig transition. The whole process of resonant elastic scattering is sketched in Fig. 3.

Dielectronic capture received its name because of the interaction of two electrons in the process, the projectile and one target electron. In principle, core excitation can also involve two electrons and the associated resonant path hence leads to a triply excited ion $[A^{(q-1)+}]^{***}$ by capture of the initially free electron. A logical name for this reaction is trielectronic capture. Ejection of one electron with the energy of the original projectile would be seen as a resonance in the elastic scattering channel. This requires two-electron core transitions both in the excitation and in the subsequent relaxation which makes this 'double-trielectronic' process relatively unlikely in comparison with the regular 'double-dielectronic' interaction.

In any case, indirect elastic scattering proceeds via resonant formation and subsequent decay of an intermediate multiply excited state, called a resonance, and is thus described as

$$e + A^{q+} \rightarrow \left[A^{(q-1)+}\right]^{**} \rightarrow A^{q+} + e. \qquad (3)$$

Resonant elastic scattering [Eq. (3)] cannot be distinguished from the direct process [Eq. (1)]. As a consequence, interference between the amplitudes of the two routes of elastic scattering can occur.

Resonances are responsible for distinct features in the energy dependence of electron–ion scattering cross sections. Depending on the relaxation mechanism, dielectronic or trielectronic capture can also be the first step in inelastic electron scattering processes. Inelastic scattering of the incident electron changes the state of the target ion. Primarily that involves the occurrence of an outgoing electron which has lost some of its energy. In a more general sense, inelastic electron–ion collisions also include the formation of a stable, recombined ion, i.e., the initially free projectile electron becomes eventually bound.

2.1.2. Recombination

When dielectronic capture is followed by the emission of photons the reduced charge state of the doubly excited intermediate state $[A^{(q-1)+}]^{**}$ can be stabilized. In that case, the whole process is termed dielectronic recombination (usually abbreviated by DR). With dielectronic capture as the first step it is a resonant mechanism.

$$e + A^{q+} \rightarrow \left[A^{(q-1)+}\right]^{**} \rightarrow A^{(q-1)+} + \text{photon(s)}. \qquad (4)$$

The whole process of dielectronic recombination is illustrated by the schematic energy level diagrams in Fig. 4.

As mentioned above, resonant core excitation can also involve trielectronic capture. When the triply excited state decays by only emitting photons (and no electrons), the process of trielectronic recombination is completed (see Fig. 5).

$$e + A^{q+} \rightarrow \left[A^{(q-1)+}\right]^{***} \rightarrow A^{(q-1)+} + \text{photons}. \qquad (5)$$

Following this line of thinking, even more than two core electrons might be involved in the absorption of the energy released when an initially free electron becomes bound. Polarization of the target core electron cloud in the electron capture process associated with emission of a photon was recently discussed by Korol et al. (2006). Collective excitation of ion core electrons in general can take over the excess energy in a radiationless capture event. Subsequent radiative relaxation stabilizes the associated recombination process.

Beside resonant recombination, there is also direct or radiative recombination (usually abbreviated by RR). In this case the incident electron is captured into a

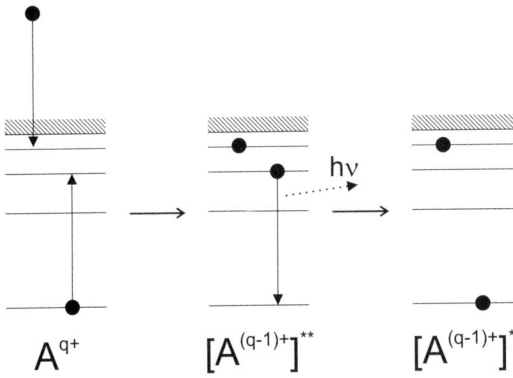

FIG. 4. The reaction pathway of dielectronic recombination [see also Eq. (4)].

bound subshell of the $A^{(q-1)+}$ ion and the excess energy is carried away by a photon.

$$e + A^{q+} \to A^{(q-1)+} + h\nu. \tag{6}$$

In sufficiently dense plasmas, a third recombination mechanism becomes effective: three-body recombination

$$e + e + A^{q+} \to A^{(q-1)+} + e. \tag{7}$$

In that case two electrons collide within the Coulomb field of an ion, one of them gains energy and leaves the scene, while the other electron, after loosing energy, finds itself trapped in the Coulomb field of the ion. This capture predominantly

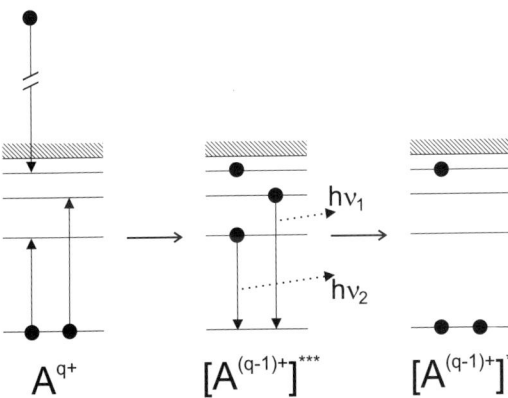

FIG. 5. The reaction pathway of trielectronic recombination [see also Eq. (5)].

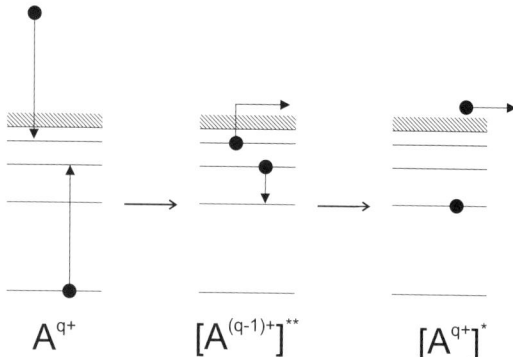

FIG. 6. Schematic energy diagram indicating the reaction steps of resonant excitation.

leads into very high Rydberg states of the ion. Such states may live for quite a long time. However, the trapped electron in the undisturbed $A^{(q-1)+}$ ion must finally end up in the ground level of the recombined ion.

2.1.3. Excitation

An electron scattered from a target ion may loose some of its initial kinetic energy, thus leaving the ion in an excited state. This process of electron-impact excitation can be described by

$$e + A^{q+} \rightarrow [A^{q+}]^* + e. \tag{8}$$

Here, as well as in elastic scattering, the final state can be reached directly or in a resonant fashion, i.e., the first step of the process is a dielectronic or trielectronic capture and the second step is autoionization to an excited state of the ion A^{q+} (see Fig. 6)

$$e + A^{q+} \rightarrow [A^{(q-1)+}]^{**} \rightarrow [A^{q+}]^* + e. \tag{9}$$

Again, as before in the case of elastic scattering, the result of the resonant excitation cannot always be distinguished from that of the direct process. As a consequence, interference between the amplitudes of the two routes of inelastic scattering is to be expected.

Excitation can also involve inner-shell electrons of the target ion A^{q+}. As a result the ion may end up in a state whose excitation energy exceeds the ionization threshold for the removal of the outermost electron. Such a state would be autoionizing and the highly excited ion can hence decay by the emission of an electron instead of a photon. The net result of that process is ionization of the parent ion (see below).

2.1.4. Ionization

Electron impact on an ion with sufficiently high collision energy may result in the removal of one or several (n) target electrons from the ion whose charge state is subsequently enhanced accordingly

$$e + A^{q+} \rightarrow A^{(q+n)+} + (n+1)e. \tag{10}$$

While being scattered off the ion, the incident electron may transfer energy to a bound electron in excess of its binding energy, so that the target electron is released. The result is observed as single ionization of the parent A^{q+} [$n = 1$ in Eq. (10)] and is referred to as direct ionization. Besides this direct knock-off process as sketched in Fig. 7, there are indirect ionization mechanisms. One of them was mentioned above already, the excitation of an inner-shell electron with subsequent autoionization. This process is usually termed excitation–autoionization (with the acronym EA)

$$e + A^{q+} \rightarrow \left[A^{q+}\right]^{**} + e \rightarrow \left[A^{(q+1)+}\right] + 2e. \tag{11}$$

It is illustrated by the three collision phases shown in Fig. 8.

As before in the excitation channel, the first step of excitation–autoionization may proceed through the formation of an intermediate resonance state. An inner-shell electron is excited and the incident electron is captured to a bound state [dielectronic capture, see Eq. (2)]. The resulting short lived recombined ion state is so highly excited that two electrons can be ejected in the relaxation process. In the end, the ion has lost one electron in the whole process. The electron emission can occur in a single (low-probability) two-electron emission process (also known as a double-Auger process, see Fig. 9 lower path) or in two successive steps (Fig. 9 upper path). The latter mechanism of resonant ionization can happen when the ion

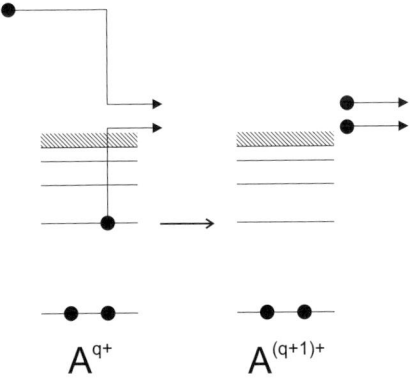

FIG. 7. Schematic energy diagram indicating the reaction steps of direct ionization.

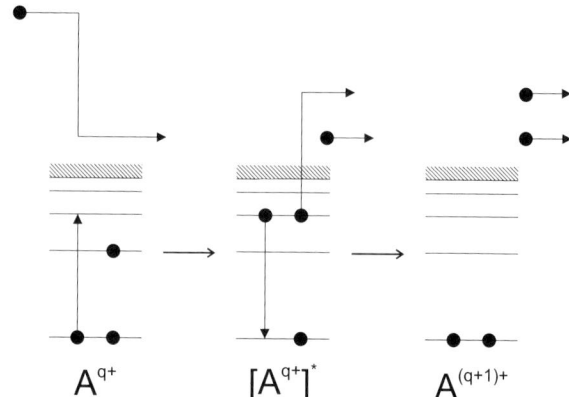

FIG. 8. Schematic energy diagram indicating the reaction steps of the indirect process of excitation–autoionization.

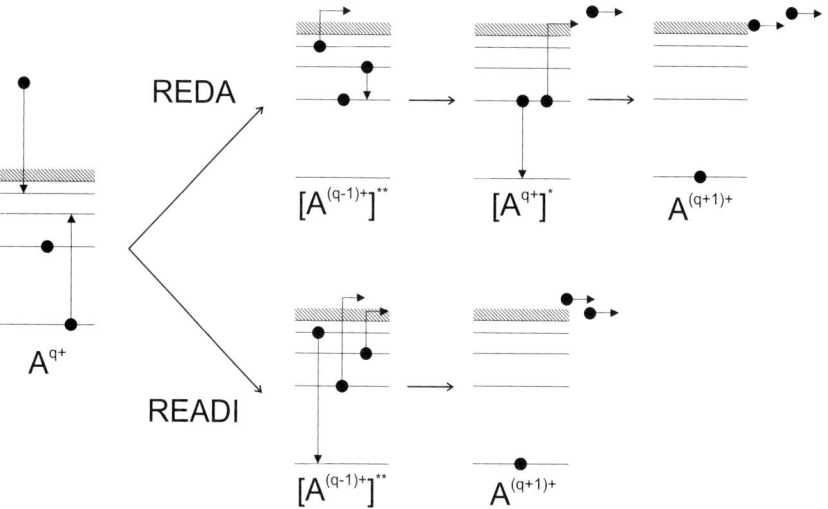

FIG. 9. Schematic energy diagram indicating the reaction steps of resonances in the electron-impact ionization of ions. REDA stands for resonant-excitation–double-autoionization, where the intermediate resonant state decays by sequential Auger or Coster–Kronig processes. READI is the resonant-excitation–auto-double-ionization process, where the intermediate resonance decays by simultaneous ejection of two electrons.

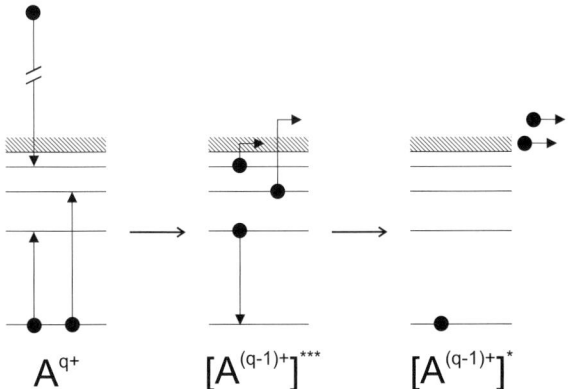

FIG. 10. Schematic energy diagram indicating the reaction steps of trielectronic capture and subsequent auto-double-ionization leading to net single ionization of the parent ion. Both individual steps of this mechanism involve multi-electron transitions.

is still in an autoionizing state, even after emission of the first electron. Accordingly, the two different mechanisms that produce resonances in single ionization of an ion or an atom are termed resonant-excitation–auto-double-ionization (with the acronym READI) and resonant-excitation–double-autoionization (with the acronym REDA). Both processes involve a radiationless (dielectronic) capture of the incident electron by the A^{q+} ion as a first step and thus are closely related to all the other resonant electron–ion interactions.

Although resonant-excitation–auto-double-ionization is already quite an exotic process involving two-electron transitions in the relaxation process, there are yet more intricate excitation processes that can lead to net single ionization of an ion by electron impact. One such possibility is direct valence-shell double excitation with subsequent autoionization formally following Eq. (11). The other one is the associated resonant double-excitation process with subsequent simultaneous or sequential emission of two electrons. The first step of this mechanism is trielectronic capture, the second is double-autoionization or auto-double-ionization. The latter is displayed in Fig. 10.

Going back to less exotic direct single ionization mechanisms, one can also find complexity. Depending on the binding energy of the electron knocked off the atom or ion, whole cascades of relaxation processes can happen. In particular, multiple ionization can be the result of several sequential Auger processes subsequent to direct single inner-shell ionization. In general, multiple ionization can also proceed via direct and indirect channels similar to the ones listed above for single ionization. In particular, direct single inner-shell ionization with subsequent electron emission often contributes considerably to net multiple ionization of ions. Also, resonant and non-resonant inner-shell excitation can deposit so much exci-

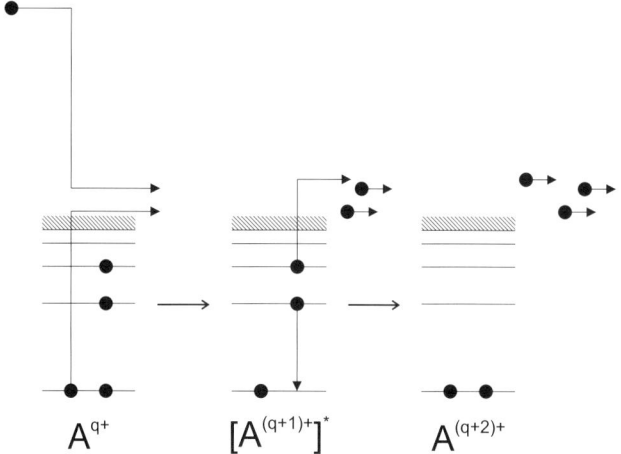

FIG. 11. Schematic energy diagram indicating the reaction steps of the often dominant indirect contributions to net multiple ionization: removal of an inner-shell electron and subsequent autoionization via an Auger process.

tation energy into the target that more than two electrons are emitted. The most important mechanism of double ionization, inner-shell ionization with subsequent autoionization, is sketched in Fig. 11.

2.2. Resonances and Interference Patterns

For all electron–ion processes associated with resonances the related cross section can be calculated by multiplying the cross section σ_d^{DC} [see Eq. (12) below] for dielectronic capture with the branching ratio for the particular decay path starting from the intermediate doubly excited state $|d\rangle$. Such indirect scattering processes may be quite complex in the case of a multi-electron target ion and hence, they cannot easily be predicted by theory.

The cross section for dielectronic capture, i.e., the first step of any resonant electron–ion collision process, can be described as

$$\sigma_d^{DC}(E_e) = 7.88 \cdot 10^{-31} \, \text{cm}^2 \, \text{eV}^2 \, \text{s} \, \frac{1}{E_e} \frac{g_d}{2g_i} \frac{A_a(d \to i) \cdot \Gamma_d}{(E_e - E_{\text{res}})^2 + \Gamma_d^2/4} \tag{12}$$

where E_e is the electron energy and $E_{\text{res}} = E(|d\rangle) - E(|i\rangle)$ the resonance energy obtained from the total energies of the resonant state $|d\rangle$ and the initial state $|i\rangle$, respectively. The quantities g_d and g_i denote the statistical weights of the state $|d\rangle$ formed by dielectronic capture and of the initial state $|i\rangle$, respectively. $A_a(d \to i)$

is the autoionization rate of $|d\rangle$ for a transition to $|i\rangle$ and Γ_d the total width of $|d\rangle$

$$\Gamma_d = \hbar \left[\sum_{k'} A_{\rm a}(d \to k') + \sum_{f'} A_{\rm r}(d \to f') \right]. \tag{13}$$

$A_{\rm r}(d \to f')$ denotes the rate for the radiative transition from $|d\rangle$ to state $|f'\rangle$. The summation indices k' and f' run over all states which can be reached from $|d\rangle$ either by autoionization or by radiative transitions, respectively.

For the calculation of the cross section for resonant elastic scattering, $\sigma_d^{\rm DC}$ has to be multiplied by the Auger yield

$$\omega_{\rm A} = \frac{A_{\rm a}(d \to i)}{\sum_{k'} A_{\rm a}(d \to k') + \sum_{f'} A_{\rm r}(d \to f')} = \hbar \frac{A_{\rm a}(d \to i)}{\Gamma_d} \tag{14}$$

of the intermediate state $|d\rangle$ leading back to the initial state $|i\rangle$. For the calculation of cross sections for dielectronic recombination, $\sigma_d^{\rm DC}$ has to be multiplied by the fluorescence yield

$$\omega_{\rm r} = \frac{\sum_f A_{\rm r}(d \to f)}{\sum_{k'} A_{\rm a}(d \to k') + \sum_{f'} A_{\rm r}(d \to f')} = \hbar \frac{\sum_f A_{\rm r}(d \to f)}{\Gamma_d} \tag{15}$$

of the intermediate state $|d\rangle$, where the summation index f runs over all states below the first ionization limit of the recombined $A^{(q-1)+}$ ion that can be reached from $|d\rangle$ by radiative transitions. Resonances in other reaction categories have to be treated accordingly. In general, however, the appearance of resonances in electron–ion collisions is usually not as simple as described above because of possible interference effects of resonant and direct channels.

There are several mechanisms in each of the collision categories introduced above which cannot always be distinguished from one another (see Fig. 12). One can imagine two reaction pathways which start from a given initial state, characterized by an electron with a defined energy and an ion, for example, in its ground state. After the collision there may be photons, electrons and a product ion in a given final electronic state. When the numbers of electrons and photons as well as their energies are identical and when the final electronic state of the target is also the same for both pathways, i.e., when the different pathways starting from a defined initial state end up in a given final state, then (and only then) the pathways cannot be distinguished and their amplitudes can interfere. This is similar to the observation of light scattered from a double-slit arrangement. The photons cannot be traced back in their pathway through one or the other slit. The result is, that the amplitudes describing the two different pathways can interfere. Such combinations can occur in elastic and inelastic scattering as well as in excitation and ionization and hence, interference in electron–ion collisions is ubiquitous. Interference patterns can be especially expected in measurements of differential cross sections where the number of final states after the collision is narrowed down by

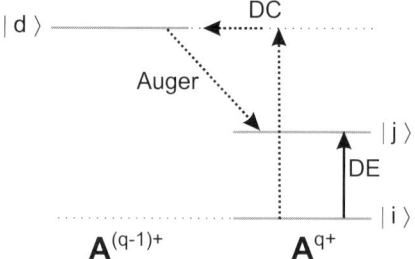

FIG. 12. Energy level scheme showing the different pathways for excitation of state $|i\rangle$ to state $|j\rangle$ in an A^{q+} ion. There is the direct excitation (DE) path [excitation energy E_{ij}, see Eq. (8)] and the indirect resonant route [see Eq. (9)] proceeding via dielectronic capture (DC) from state $|i\rangle$ to the intermediate multiply excited state $|d\rangle$ in the $A^{(q-1)+}$ ion followed by an Auger process to the excited state $|j\rangle$ in the A^{q+} ion. At an electron energy $E_e = E(|d\rangle) - E(|i\rangle)$ the initial state of the system is $e(E_e) + A^{q+}(|i\rangle)$ and the final state of the system is $e(E_e - E_{ij}) + A^{q+}(|j\rangle)$, both identical for the two different pathways. One cannot distinguish which path is taken in the excitation process and hence the amplitudes of the two pathways interfere with one another.

selective experimental conditions. For example, in an energy-loss-spectroscopy experiment, the energy of the scattered electron is selected and measured at a defined scattering angle. By that, also the target electronic state is specified and hence, interference of direct and indirect inelastic scattering can be observed with a sufficiently sensitive detector. When looking at total cross sections, however, it may be difficult to see interference patterns because they can be washed out by statistics in the sum of several individual contributions, where only one shows interference.

The best chance to see interference between different reaction pathways is in studying cross sections where, for example, a defined direct process interferes with its resonant counterpart. This is most readily seen in photoionization experiments where the dipole selection rules narrow down the number of states that can be reached. The shape of an interfering-resonance cross section as a function of energy is then no longer Lorentzian [see Eq. (12)] but becomes distorted (Fano, 1961; Fano and Cooper, 1965). The resonance acquires a shape given by

$$F(E) = \frac{A}{Q^2 \Gamma_d} \frac{2}{\pi} \left[\frac{(Q+\varepsilon)^2}{1+\varepsilon^2} - 1 \right] \quad (16)$$

with $\varepsilon = 2(E - E_{\text{res}})/\Gamma_d$, the resonance energy E_{res}, the resonance width Γ_d, and the asymmetry parameter Q. The term -1 inside the square brackets ensures that $F(E) \to 0$ for $E \to \pm\infty$. Low values of $|Q|$ correspond to strongly destructive interference and $Q = 0$ produces a symmetric dip in the cross section at the resonance energy, a so-called window resonance. In the limit $Q \to \infty$ the Fano

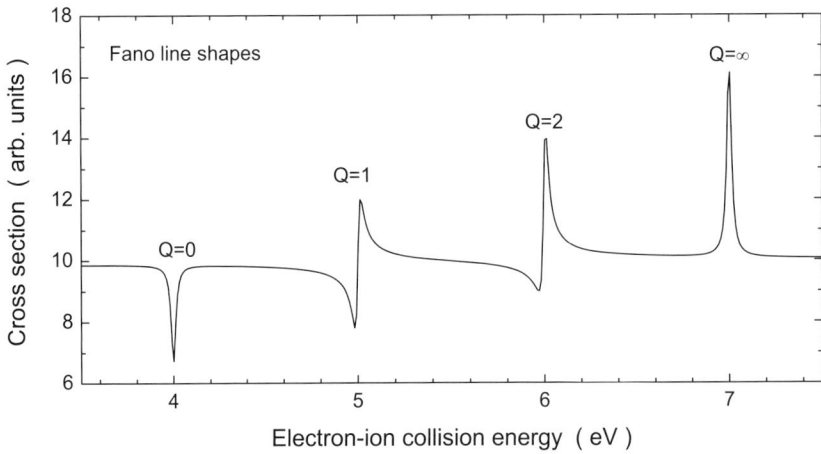

FIG. 13. Simulated resonance spectrum assuming different asymmetry parameters Q as indicated. The natural width of each resonance was set to 30 meV. Negative values of Q produce mirror images of the resonances calculated with positive Q.

profile as defined by Eq. (16) approaches the symmetric Lorentzian line profile

$$L(E) = \frac{A}{\Gamma_d} \frac{2}{\pi} \frac{1}{1+\varepsilon^2} \tag{17}$$

with the peak area A. $L(E)$ has the shape of $\sigma_d^{\mathrm{DC}}(E)$ as given by Eq. (12) with $E = E_e$.

Figure 13 shows a simulated cross section obtained with resonances at integer simulated resonance energies and different asymmetry parameters Q. For $Q = \pm\infty$ the resonances have normal Lorentzian shape. At $Q = 0$ the resonance is symmetric again, as mentioned above, but inverted. Destructive interference between different reaction channels may produce a zero cross section at the center of a window resonance. For negative values of Q the resonance shapes are mirror images (with respect to the vertical axis) of what is seen in Fig. 13.

Another scenario for interfering pathways in an electron–ion collision is of a more subtle nature. It is related to the finite widths of excited states. When two resonant states $|d_1\rangle$ and $|d_2\rangle$ of equal symmetry overlap one another, i.e., when the energy gap between the two states is smaller than the sum of the widths Γ_{d_1} and Γ_{d_2}, interference is also possible because one cannot decide from the initial and final states which reaction pathway has been taken during the collision.

2.3. Generic Energy Dependences of Cross Sections

For each electron–ion collision category listed in Section 2.1, there are characteristic energy dependences of the cross sections. In the case of elastic scattering, the total cross section, i.e., the integral over all scattering angles of the angular differential cross section, diverges if no additional differentiation between processes is introduced. At a given non-zero scattering angle, however, the differential cross section as a function of incident electron energy could have some smooth dependence with resonances interspersed in the cross section curve. This could, in principle, look like the energy dependence shown in Fig. 13.

2.3.1. Excitation

The total cross sections for direct excitation of ions [Eq. (8)] have an energy dependence that is characterized by a finite maximum at the excitation threshold and a slow, smooth decrease with increasing electron energy. The slope of the falling cross section depends on whether the transition to the excited state is optically allowed or not. Cross sections for direct excitation can be approximated for example by the so called Gaunt-factor formula of Seaton (1962) and Van Regemorter (1962)

$$\sigma = 2.36 \cdot 10^{-13} \text{ cm}^2 \text{ eV}^2 \frac{f_{ij}\bar{g}}{E_e E_{ij}}, \qquad (18)$$

where f_{ij} is the optical oscillator strength for the transition from excited state $|j\rangle$ to ground state $|i\rangle$, E_{ij} the related excitation energy, E_e the electron energy and $\bar{g}$ the effective Gaunt factor. In the vicinity of the threshold this is reasonably well approximated by $\bar{g} = 0.2$. At higher energies ($E_e \gtrsim 2E_{ij}$), the recommendation is:

$$\bar{g} = 0.28 \ln(E_e/E_{ij}). \qquad (19)$$

Together with indirect channels of excitation that proceed via formation of intermediate multiply excited resonances and neglecting the possibilities of interference between the direct and the resonant channels, the generic shape of the total cross section for electron-impact excitation of an ion to a specific excited state is of the kind shown in Fig. 14.

2.3.2. Recombination

The total cross section for radiative (i.e. direct) recombination can be conveniently calculated by using the semi-classical approximation of Kramers (1923) for hydrogenic ions (see also Bethe and Salpeter, 1957) with some modifications

$$\sigma_{\text{RR}}(E_e) = 2.1 \times 10^{-22} \text{ cm}^2 \sum_{n_{\min}}^{\infty} k_n t_n \times \frac{(Z_{\text{eff}}^2 \mathcal{R}_\infty)^2}{n E_e (Z_{\text{eff}} \mathcal{R}_\infty + n^2 E_e)}. \qquad (20)$$

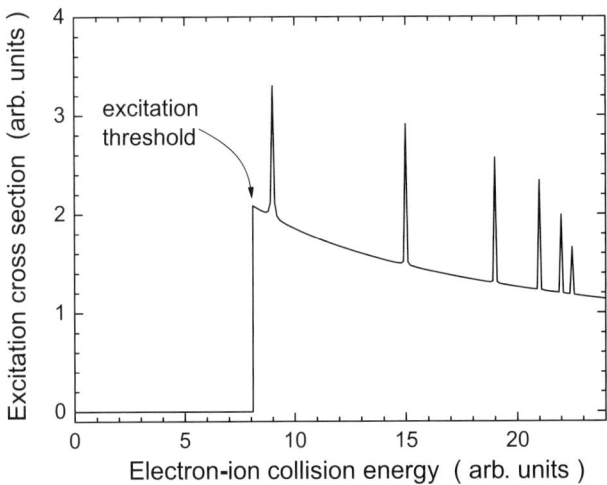

FIG. 14. Generic energy dependence of the total cross section for electron-impact excitation of an optically allowed transition in an ion A^{q+}. The resonances are assumed to show no interference with the direct excitation channel.

In Eq. (20) the constants k_n are corrections to the semi-classical cross section for low n (Andersen et al., 1990a) and the constants t_n are weight factors accounting for already partially filled n-shells. The minimum principal quantum number $n_{\min}$ denotes the lowest electron shell with at least one vacancy into which the incident electron can be captured. The effective charge can be assumed to be the ion charge q, $\mathcal{R}_\infty = E_h/2 = 13.6$ eV is half the Hartree energy. In principle, the summation extends over all Rydberg states up to $n = \infty$. However, experimental conditions, like collisions and external electric fields, limit the number of contributing Rydberg states. The sum is thus restricted to a maximum quantum number $n_{\max}$ determined by the environment in which the recombination occurs. Equation (20) shows that the recombination cross section diverges for zero electron energy. In reality, the energy of an ensemble of projectile electrons always has a finite spread which limits the recombination probability even if the average velocity of the ensemble with respect to an ion target is zero.

When the ion has at least one bound electron then dielectronic recombination also has to be considered. It is often the most important contribution to the total recombination cross section (see Fig. 15). As indicated above and on the basis of negligible interference with the direct recombination channel, the cross section for an isolated dielectronic recombination resonance $|d\rangle$ at a resonance energy $E_{\mathrm{res}} \gg \Gamma_d$ can be calculated as [see Eq. (17)]

$$\sigma_d^{\mathrm{DR}} = \frac{A_{\mathrm{DR}}}{\Gamma_d} \frac{2}{\pi} \frac{1}{1+\varepsilon^2} \qquad (21)$$

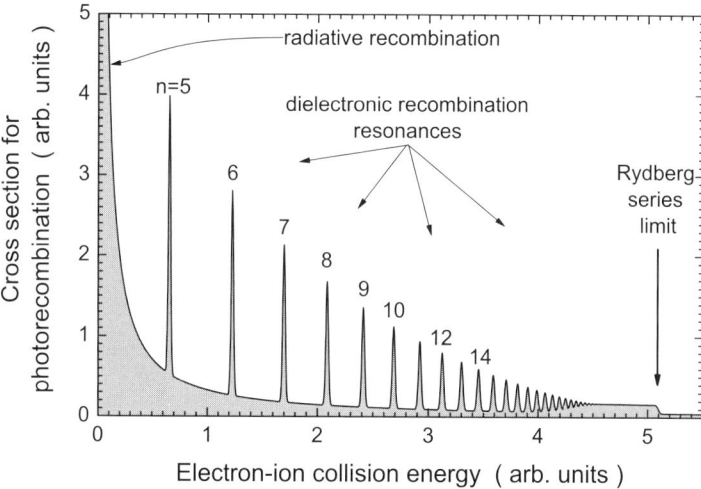

FIG. 15. Simulated total cross section for recombination of an A^{q+} ion with an electron. Radiative recombination dominates at low electron energies. Resonances associated with dielectronic recombination to Rydberg states $A^{(q-1)+}(nl)$ are marked by the principal quantum number n.

with the dielectronic recombination resonance strength in terms of the quantities introduced above (see Section 2.2)

$$A_{DR} = 4.95 \cdot 10^{-30} \text{ cm}^2 \text{ eV}^2 \text{ s}$$
$$\times \frac{1}{E_{res}} \frac{g_d}{2g_i} \frac{A_a(d \to i) \sum_f A_r(d \to f)}{\sum_k A_a(d \to k) + \sum_{f'} A_r(d \to f')}. \quad (22)$$

Hence, the dielectronic recombination cross section is

$$\sigma_d^{DR} = A_{DR} \frac{\Gamma_d}{2\pi} \frac{1}{(E_e - E_{res})^2 + \Gamma_d^2/4}. \quad (23)$$

In principle, the resonance channel can interfere with the radiative recombination process. Hence, a unified theoretical description is appropriate. Since radiative and dielectronic recombination cannot always be distinguished, the term photorecombination has been introduced to describe the unified process. As Eq. (20) makes clear, the total radiative recombination cross section is a sum of contributions of direct recombination into the ground state and numerous individual excited states which decay by the emission of one or more photons. The largest contribution to radiative recombination is related to the lowest principal quantum number n_{min} of states in the recombined ion. With increasing n the partial cross sections for direct-recombination contributions rapidly decrease with n, depending on the energy E_e of the incident electron. At sufficiently high energies, the population of a term within a given n-manifold drops like n^{-3}.

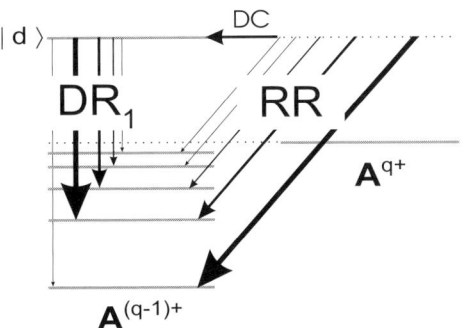

FIG. 16. Energy level diagrams of the parent A^{q+} and the recombined $A^{(q-1)+}$ ion. Possible pathways of dielectronic and radiative recombination contributing to the total recombination cross section at a given electron energy are indicated. The thickness of the different arrows schematically indicates the relative importance of the represented transition. RR stands for radiative recombination, DR_1 for the emission of the first photon from the intermediate resonant state $|d\rangle$ produced by dielectronic capture (DC).

If the incident electron energy matches the level energy of a doubly excited state $|d\rangle$ of the recombined $A^{(q-1)+}$ ion, dielectronic capture can happen and then $|d\rangle$ can decay via the same states as those involved in radiative recombination. As a result, the same photons are radiated and the identical final electronic states of the recombined ion are produced. This means that many individual resonance channels have the potential to interfere with their counterparts in the direct recombination channels. However, the relative contributions of different intermediate levels between the doubly excited state $|d\rangle$ and the ground state with $n = n_{min}$ of the recombined $A^{(q-1)+}$ ion are very different in the resonant and the direct recombination channels. While one-photon radiative recombination from the initial state $|i\rangle$ of the A^{q+} ion to the lowest available states of the recombined $A^{(q-1)+}$ ion with $n = n_{min}$ is the most likely direct recombination channel, the one-photon transition from the intermediate doubly excited state $|d\rangle$ to $n = n_{min}$ would require a two-electron transition which is very unlikely. Consequently, there is almost no dielectronic recombination for the radiative recombination to interfere with. Vice versa, the dominant one-photon–one-electron transitions in the dielectronic recombination channel proceed via (singly) excited states of the recombined $A^{(q-1)+}$ ion, where there is not much strength in the radiative recombination channel. This is illustrated by Fig. 16 where the different transition probabilities to recombined states are indicated by arrows of different thicknesses. Dielectronic recombination pathways are strong where radiative recombination is weak and vice versa. Consequently, interfering pathways are less likely than the non-interfering recombination routes.

Thus, although there are numerous singly excited intermediate states through which both dielectronic and radiative recombination can proceed and which therefore offer the possibility for interference effects, the relative contributions of each channel, direct and resonant, to a given intermediate state differ strongly. Moreover, the small contributions from the many possibly interfering pathways add to recombination with their individual strengths and possibly different asymmetry parameters, so that their interference patterns already being of low probability are likely washed out in the total sum. Therefore, it is not straightforward to see interference patterns in a total recombination cross section. On the other hand there is no doubt about the ubiquitous presence of such interference phenomena in electron–ion recombination. These issues have been discussed previously by Pindzola et al. (1992b) in a paper addressing the validity of the independent-processes and isolated-resonance approximations.

2.3.3. Single Ionization

Direct ionization cross sections (see Fig. 7) can be rather well represented by predictor formulae. Among the most often used prescriptions for calculating cross sections for the removal of a single electron from a given subshell (index ν) of an ion in a simple knock-off process is based on the formula introduced by Lotz (1967, 1968, 1969, 1970) following the quantum-mechanical results obtained by Bethe (1930) for electron-impact excitation and ionization

$$\sigma_\nu = 4.5 \cdot 10^{-14} \text{ cm}^2 \text{ eV}^2 \frac{\xi_\nu}{I_\nu E_e} \ln \frac{E_e}{I_\nu}, \tag{24}$$

where ξ_ν is the number of equivalent electrons in the νth subshell, I_ν the ionization potential of that subshell, and E_e the electron energy. The total direct single ionization cross section of an ion A^{q+} is a sum

$$\sigma_{q \to q+1} = \sum_{n_{\min}}^{\infty} \sigma_\nu \omega_r^\nu \tag{25}$$

over all subshells, that can be reached with the given electron energy E_e ($E_e \geqslant I_\nu$). The quantities ω_r^ν are averaged fluorescence yields for the associated subshells. When the vacancy produced by removal of an electron from the subshell with index ν stores enough energy to facilitate an Auger or Coster–Kronig transition, then a second electron will be ejected with high probability and the net result of the two-step process is double ionization of the parent A^{q+} ion (see Fig. 11). That is why the fluorescence yields are essential in Eq. (25).

Apart from direct single ionization there are indirect ionization mechanisms which proceed through intermediate autoionizing states produced by direct and resonant inner-shell excitation. These autoionizing states decay by the emission of electrons finally producing an ion $A^{(q+1)+}$. Direct inner-shell excitation produces distinct step features (see Fig. 14) at related threshold energies seeded

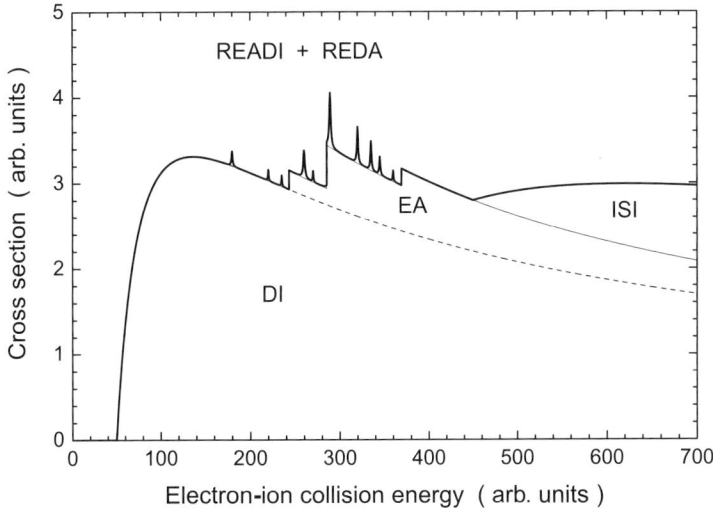

FIG. 17. Simulated total cross section for electron-impact ionization of an A^{q+} ion. The contribution from direct ionization (DI, dashed line) is simulated with the Lotz formula [Eq. (24)], the typical step features are due to excitation–autoionization (EA, thin solid line). Resonance contributions from resonant-excitation–double-autoionization (REDA) and resonant-excitation–auto-double-ionization (READI) have been simulated assuming that no interference occurs. Inner-shell ionization (ISI) can contribute if the resulting vacancy is filled via radiative processes. The sum over all contributions to single ionization is represented by the heavy solid line.

with resonances arising from the resonant-excitation–double-autoionization and resonant-excitation-auto–double-ionization mechanisms on top. Here, as well as in all other electron–ion interactions, interference of reaction channels which cannot be distinguished from one another is possible and does occur. An additional possible channel of electron-impact single ionization is inner-shell ionization (ISI) followed by radiative relaxation as mentioned above. Figure 17 shows a schematic picture of a cross section for net single ionization of an ion by electron impact.

It is interesting to note that ionization resonances can occur below the lowest inner-shell excitation threshold of the parent ion A^{q+}. The binding energy of the electron resonantly captured into an inner-shell excited state of the ion $A^{(q-1)+}$ places the resonance below the excitation threshold. Resonances in the single-ionization cross section at energies below the lowest possible excitation threshold can only be due to the resonant-excitation–auto-double-ionization process. In that process, no additional intermediate autoionizing state is required to emit the second electron (see Fig. 9). The only requirement is, that the resonance can eject two electrons simultaneously in a double-Auger process.

2.3.4. Multiple Ionization

Direct double ionization is the removal of two electrons from an atom or ion within the short time required for an electron to pass the electron cloud of the target nucleus. This time is of the order of 10^{-17} s or less. From a theoretical point of view, electron-impact double ionization is a four-body problem with three outgoing electrons in the continuum. Consequently, there is no simple formalism that would readily allow one to make reliable predictions for double ionization of a given multi-electron target.

Mechanisms that can be responsible for double ionization have been identified early on by Gryziński (1965) and Mittleman (1966a, 1966b). The incident electron can undergo two sequential ionizing collisions with the two target electrons, a process that has been termed TS2 (two-step 2) in the literature (for a review see McGuire, 1997). Alternatively TS1 (two-step 1) describes a mechanism where the projectile knocks off a target electron which, on its way out, ionizes the second target electron. Both processes have been treated theoretically by Gryziński (1965) on the basis of classical Coulomb scattering. More realistic quantum mechanical treatments followed later (see McGuire, 1997). The third mechanism results from the sudden change of charge when an electron is removed from the target atom. By the resulting change of screening the second electron may be shaken off to the continuum (Mittleman, 1966a). The shake-off mechanism is also the main contribution in particle-impact double ionization at the limit of high velocities and low projectile charge states.

For light atoms or ions a semi-empirical formula has been proposed by Shevelko et al. (2005) to calculate approximate cross sections for direct double ionization.

$$\sigma = \frac{A_{\text{dir}}}{I_{\text{th}}^{-3}} \frac{u-1}{(u+1/2)^2}, \qquad (26)$$

where A_{dir} is a constant depending on the target atomic number and charge, and I_{th} the combined binding energy of the two active electrons in the target, i.e., the threshold energy for the double ionization process. The quantity $u = E_e/I_{\text{th}}$ is the electron energy in threshold energy units. For complex targets, predictions for multiple ionization become increasingly more difficult. Shevelko et al. (2006) have suggested an extension of Eq. (26) that should accommodate a wider range of energies and target species

$$\sigma = \left(1 - \exp[-3(u-1)]\right) \frac{A_{\text{dir}}}{I_{\text{th}}^{-3}} \frac{u-1}{(u+1/2)^2} \left(1 + 0.1\ln(4u+1)\right). \qquad (27)$$

When the number of electrons bound in the target atom increases to 3 or beyond, indirect, i.e., multi-step processes become possible. Usually, multiple ionization shows substantial contributions from single ionization of an inner-shell of the target atom or ion. If an inner-shell electron is released to the continuum the resulting ion can undergo an Auger process and thus the final result of the process is a net double ionization. Such processes become possible with four or more target electrons. More complex mechanisms including inner-shell excitation or dielectronic capture with subsequent cascades of radiative and Auger decays are also possible and can contribute to multiple-ionization channels.

2.4. TIME-REVERSAL SYMMETRY

The reactions described by Eqs. (1) through (11) can all be reversed in time. It is interesting to see what the time-reversed processes are in each case. Trivially, time-reversed elastic scattering is again elastic scattering, but dielectronic capture is a non-trivial and interesting case, for which the inverse process is autoionization (Auger, Coster–Kronig or Super Coster–Kronig decay).

Time-reversed excitation is a process in which a slow electron collides with an excited ion, picks up the excitation energy and leaves the ion behind in its ground state. This process is termed super-elastic scattering.

Reading Eqs. (4)–(6) for recombination from right to left corresponds to photoionization of an ion $A^{(q-1)+}$ which, after removal of an electron, ends up as A^{q+}. This scheme works for both radiative and dielectronic recombination. However, time reversal can only be considered on a state-to-state basis.

Finally, time-reversed electron-impact single ionization is equivalent to three-body recombination.

The interactions that dominate electron–ion collisions conserve time-reversal symmetry. Hence the Hamiltonian of the system does not depend on the direction of time's arrow. As a result of this the principle of detailed balance can be derived. Cross sections for reactions proceeding in one direction ($a \to b$) can be calculated from the cross section for the time-inverse process ($b \to a$)

$$\frac{d\sigma_{a \to b}}{d\Omega} = \frac{g_b\, k_b^2}{g_a\, k_a^2} \frac{d\sigma_{b \to a}}{d\Omega}, \tag{28}$$

where g_b and g_a are the statistical weights of the final and the initial state, respectively, whereas the quantities $\hbar k_b$ and $\hbar k_a$ represent the relative momenta of the two bodies interacting with one another in the final and the initial reaction channel.

An example for the application of Eq. (28) is obvious from Eq. (12) which provides an expression for the dielectronic capture cross section. The time reversed dielectronic capture process is an Auger transition characterized by the rate $A_a(d \to i)$. Obviously, $\sigma_d^{DC}(E)$ is proportional to $A_a(d \to i)$.

The cross section $\sigma_{i \to f}^{(PR)}(E_e)$ for photorecombination

$$e + |i\rangle \to |f\rangle + \gamma + I_{\text{bind}} \tag{29}$$

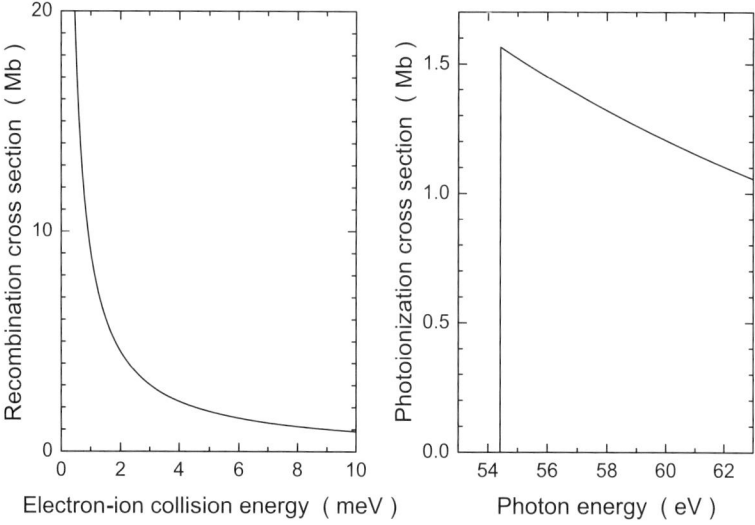

FIG. 18. Cross section for radiative recombination of He^{2+} into the 1s ground state of He$^+$ (left panel, non-relativistic quantum theory; Stobbe, 1930). From that the cross section for photoionization of ground-state He$^+$ is calculated using the principle of detailed balance [Eq. (31), right panel]. Note the different energy scales for the two cross sections. The two curves are in agreement with experiment.

of an ion in state $|i\rangle$ can be calculated from the cross section $\sigma_{f \to i}^{(PI)}(E_\gamma)$ for photoionization of an ion in state $|f\rangle$. The energies are related via the binding energy I_{bind} of the electron in state $|f\rangle$

$$E_\gamma = h\nu = E_e + I_{\text{bind}}. \tag{30}$$

E_γ is the energy of the photon and I_{bind} is the ionization potential of state $|f\rangle$. With these definitions the principle of detailed balance provides the relation between associated cross sections for photoionization and photorecombination on a state-to-state basis. For nonrelativistic photon energies $h\nu \ll m_e c^2$ the relation is

$$\sigma_{f \to i}^{(PR)}(E_e) = \frac{(h\nu)^2}{2m_e c^2 E_e} \frac{g_i}{g_f} \sigma_{i \to f}^{(PI)}(h\nu) \tag{31}$$

where E_e is the energy in the electron–ion center-of-mass frame (Flannery, 1996). The quantities g_i and g_f are the statistical weights of the ionic initial and final states, respectively.

An example for the application of the principle of detailed balance is provided in Fig. 18, where the cross section provided by Stobbe (1930) for radiative recombination of He^{2+} with an electron to the 1s state of He$^+$ is shown together with the cross section for photoionization of hydrogen-like He$^+$ as calculated from Eq. (31).

3. Experimental Access to Data

For the measurement of cross sections of a given collision process reaction rates for that process have to be observed under well-characterized experimental conditions. For electron–ion collisions the rate R of reactions resulting from a specified process that occurs in a volume τ is given by a convolution of the experimental phase space overlap between electrons and ions

$$R = \int_\tau \int_v n_e(\vec{r}) n_i(\vec{r}) \sigma(v_r) v_r f(\vec{v}_r) \, d^3v_r \, d^3r. \tag{32}$$

The electrons with a spatial number density $n_e(\vec{r})$ and the ions with a spatial number density $n_i(\vec{r})$ interact in a volume τ via the cross section $\sigma(v_r)$, where $\vec{v}_r$ is the vector of relative velocity between the collision partners and $v_r = |\vec{v}_r|$ its absolute value. The distribution function $f(\vec{v}_r)$ of the relative velocities between the colliding particles is assumed to be independent of the location $\vec{r}$. This condition can be fulfilled in an experiment by choosing a sufficiently small volume τ where the interactions occur.

The two integrations in Eq. (32) can then be carried out separately. By defining the rate coefficient

$$\alpha = \int \sigma(v_r) v_r f(\vec{v}_r) \, d^3v_r = \langle \sigma v_r \rangle, \tag{33}$$

the rate R is given by

$$R = \alpha \int_\tau n_e(\vec{r}) n_i(\vec{r}) \, d^3r. \tag{34}$$

The spatial overlap between electrons and ions depends on the particular experimental arrangement. Examples are discussed below. By observing R in an experiment, one can never get direct information about the cross section σ. Since every real experiment is restricted to velocity distributions of finite non-zero width, the only quantity that is accessible to a measurement is the rate coefficient α. Hence, a comparison of calculated cross sections with experimental data always requires an analysis of the influence of the experimental velocity (or energy) spread on the measured result. This is particularly important for resonance features in a cross section, i.e., experimental spread in the electron–ion collision energy is relevant to practically all electron–ion collision experiments. Sharp features in the cross section are smeared out by the finite experimental resolution. For any comparison of theoretical and experimental results, the experiment therefore has to provide information about the velocity distribution $f(\vec{v}_r)$ with which the measurements were carried out.

In experiments where an electron beam is employed to study electronic collisions with stationary targets the velocity distribution can be made narrow by applying appropriate techniques. The distribution f is then centered around the average velocity v_e of the electrons and can thus be appropriately expressed as $f = f(v_\mathrm{e}, \vec{v})$. When the electron velocity v_e is much greater than the velocity spread in the beam, or the width of the distribution f, then Eq. (33) can be simplified by

$$\alpha = v_\mathrm{e} \int \sigma(v) f(v_\mathrm{e}, \vec{v}) \, d^3v = v_\mathrm{e} \langle \sigma \rangle. \tag{35}$$

When $\sigma(v_\mathrm{e})$ is a smooth and slowly varying function of v_e the expectation value $\langle \sigma \rangle$ can be considered to be equal to the cross section σ. The distribution function is then assumed to be the normalized Dirac δ-function.

In a plasma where $n_\mathrm{e}(\vec{r})$ and $n_\mathrm{i}(\vec{r})$ can be assumed to be spatially constant within a sufficiently small volume, the rate of reactions is simply proportional to the volume τ. Here as well as in all other cases the rate coefficient α is determined by the distribution of relative velocities between the colliding particles. The relative velocity between a non-relativistic electron and an ion in a single collision is

$$v_\mathrm{r} = |\vec{v}_\mathrm{r}| = |\vec{v}_\mathrm{e} - \vec{v}_\mathrm{i}| = \left(v_\mathrm{e}^2 + v_\mathrm{i}^2 - 2v_\mathrm{e}v_\mathrm{i}\cos\theta\right)^{1/2}, \tag{36}$$

where $\vec{v}_\mathrm{i}$ is the velocity of the ion, $\vec{v}_\mathrm{e}$ that of the electron and θ the angle between the velocity vectors. In a plasma, the ion velocities are usually much smaller than the electron velocities and hence, the relative velocity in a collision is essentially determined by that of the electron. In a plasma one can further assume a certain temperature T for the electrons which determines their velocity distribution. In a plasma, the velocity distribution f to be inserted in Eq. (33) can often be assumed to be Maxwellian.

$$f(v) = \left(\frac{m_\mathrm{e}}{2\pi k_\mathrm{B} T}\right)^{3/2} \exp\left(-\frac{m_\mathrm{e} v^2}{2k_\mathrm{B} T}\right), \tag{37}$$

where k_B denotes Boltzmann's constant. The rate coefficient under these conditions is then expressed by

$$\alpha(T) = (k_\mathrm{B} T)^{-3/2} \sqrt{\frac{8}{m\pi}} \int_0^\infty \sigma(E) E \exp\left(-\frac{E}{k_\mathrm{B} T}\right) dE. \tag{38}$$

This $\alpha(T)$ depends on the plasma temperature (more specifically the electron temperature) and is called the plasma rate coefficient. It is the quantity that enters modeling equations for astrophysical and laboratory plasmas.

Although plasma observations with careful diagnostics of electron temperature and particle densities can be a source of information about plasma rate coefficients, there are numerous other types of experiments that provide data about rate

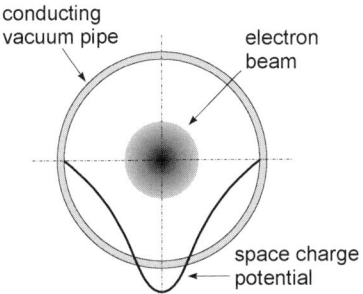

FIG. 19. Schematic of the space charge potential $V_{sc}(r)$ (solid curve) of a cylindrical electron beam within a conducting beam pipe. The space charge potential is zero at the inner surface of the beam pipe and reaches its minimum on the axis of the electron beam where, in turn, the electric field is zero.

coefficients for electron–atom or electron–ion collisions with special velocity distribution functions. One of the oldest techniques to determine cross sections for electron collisions is that of producing an electron beam with electron current I_e and letting it pass through a target of particles that are essentially at rest. The target can be a gas volume or, at higher projectile energies, a solid foil. For the sake of high reaction rates, high electron beam currents are desirable. Due to the resulting space charge, considerable electrical potential differences can result (Fig. 19).

A 1 keV electron beam of 1 cm diameter with 100 mA of electrical current at uniform electron density $n_e \approx 4 \times 10^8$ cm^{-3} produces an electric field

$$\vec{E}_{sc}(\vec{r}) = -\vec{\nabla} V_{sc}(r) = -\frac{n_e e}{2\epsilon_0} \vec{r} \tag{39}$$

that increases linearly with the distance r from the beam center and reaches its maximum, $|\vec{E}_{sc}| = 190$ V/cm, at the electron beam edge. Here, ϵ_0 is the permittivity of free space (the electric constant). In practice, only very dilute beams with electrical currents in the μA range can be produced "quasi field-free". Intense beams of electrons (and all other charged particles as well) always produce appreciable electric space-charge fields. Slow positive ions which are generated by electron-impact ionization of the residual gas are attracted by the negative potential and can partially neutralize the electron space charge.

At a sufficiently low target thickness the probability for a collision of an incident electron within the target becomes much smaller than unity. This situation would meet the so-called single-collision condition that has to be fulfilled for meaningful rate-coefficient measurements. By replacing the ion target density $n_i(\vec{r}) = n_i$ in Eq. (34) with that of the neutral target species n_0 and assuming constant target density, Eq. (34) becomes

$$R = \alpha \int_\tau n_e(\vec{r}) n_i(\vec{r}) \, d^3r = n_0 \alpha \int_\tau n_e(\vec{r}) \, d^3r = N_e n_0 \alpha \tag{40}$$

with the number N_e of electrons within τ. Given the electron velocity v_e the number of electrons is

$$N_e = \int_\tau n_e \, d^3r = \int_\tau \frac{j_e}{e v_e} \, dA \, dx = \frac{I_e}{e v_e} L, \tag{41}$$

where L is the length of the interaction region (in x-direction), $j_e = dI_e/dA = e n_e v_e$ the electrical current density in the electron beam, I_e its electrical current, and e the charge on an electron. Hence,

$$R = n_0 \alpha \frac{I_e}{e v_e} L \approx n_0 \sigma \frac{I_e}{e} L. \tag{42}$$

This equation is valid if the electron velocity v_e (and with it the energy) is well defined, which is usually the case in an electron beam. As discussed above, there is a finite velocity distribution around v_e with which the cross section σ has to be convoluted when a comparison of fine cross section features with theory is to be made. Per se the experiment can only provide an effective cross section $\langle \sigma \rangle$. The situation is demonstrated in Fig. 20. By the assumed finite resolution of the experiment the sharp features in the cross section are partly washed out. At the same time, the area under the resonance curves is conserved.

Under certain conditions electron–ion interactions can be studied by passing an ion beam through neutral gas. The gas atoms or molecules provide a target of quasi-stationary electrons which interact with the incident ions. The average relative velocity between the electrons and the ions is given by the ion velocity v_i. The bound electrons have considerable momentum which leads to substantial spread in the relative velocity between ions and the "quasi-free" electrons. Effective cross sections at $v_r = v_i$ can be inferred in a manner similar to that implied in Eq. (40).

A well-characterized target ensemble of ions exposed to an electron beam would be the method of choice to study electron–ion interactions. A stationary target of ions, however, is not as easily produced as a gas target. The space charge of the ions drives the ensemble apart. In a dilute gas target at room temperature and a pressure of $3 \cdot 10^{-6}$ hPa there are still 10^{11} particles per cubic centimeter. A cloud of ions with that density concentrated within a sphere of 1 cm radius would produce a parabolic space charge potential with a voltage of $3 \cdot 10^4$ V between the edge and the center (similar to the situation depicted in Fig. 19). Without additional forces to confine the ion cloud such a sphere would explode. This would be even more so with multiply charged ions. Stationary ion targets can thus only be realized with suitable electromagnetic traps. With the ions practically

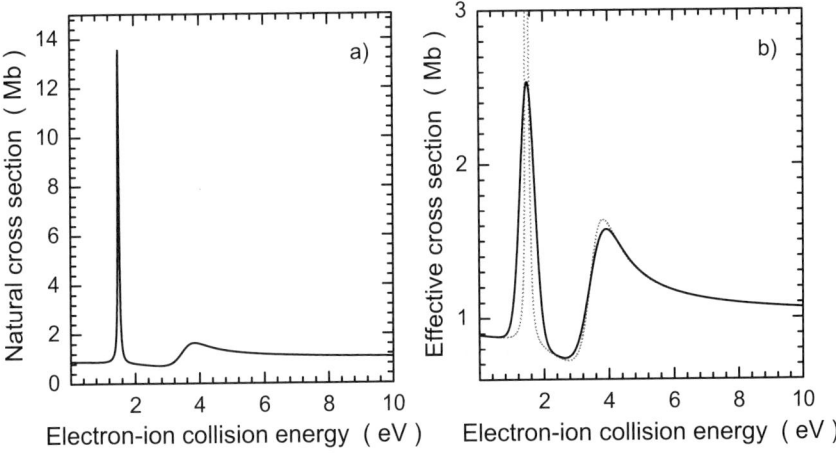

FIG. 20. Simulated "natural cross section" [panel (a), solid line] and the associated effective cross section [panel (b), solid line] measured in an experiment with a Gaussian energy distribution of 0.5 eV full width at half maximum (FWHM). For comparison panel (b) also shows the natural cross section again (dashed gray line). Note the different cross section scales.

at rest and a beam of electrons as projectiles, collision rates are again to be calculated from Eq. (40) with n_0 replaced by n_i. This scheme is effectively realized in electron beam ion trap (EBIT) devices but there are other arrangements as well suited to study electron–ion collisions.

The conceptually cleanest experimental conditions for electron–ion studies are provided in colliding-beams experiments. Crossed-, inclined- or merged-beams arrangements are possible. Each beam (assumed to be parallel in itself) can be individually produced and characterized. The average relative velocity $\vec{v}_r$ between the electrons and the ions is essentially given by Eq. (36) where the angle θ between the two beams enters. Depending on this angle the kinematics of the interacting-beams arrangement produces different distribution functions $f(v_r)$ of the relative electron–ion velocity. With respect to energy resolution, the merged-beams geometry is particularly favorable (Phaneuf et al., 1999). In that case, the two beams can interact over distances up to several meters; however, the intense electron beams used in such an arrangement have to be guided by axial magnetic fields. The principal arrangement is shown in Fig. 21. The homogeneity of the magnetic field determines the angle between the trajectories of colliding electrons with respect to the ion beam and, hence, also the relative velocity $\vec{v}_r$ and the electron–ion collision energy. Deviations of the magnetic guiding field relative to the set value $\vec{B}$ have to be reduced as much as possible, typically to the 10^{-4} level in real merged-beams experiments. Ultimately, the energy spread in such an experiment is determined by the temperatures $T_\parallel$ and $T_\perp$ associated with electron

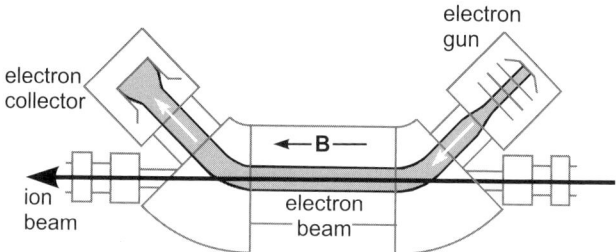

FIG. 21. Electron–ion merged-beams arrangement. The electron beam is produced within a strong axial magnetic field. As the magnetic guiding field is reduced, the electron beam expands and its transverse temperature $T_\perp$ is reduced. The figure is the schematic of an electron cooler in a storage ring.

motion parallel and perpendicular to the electron beam direction, respectively. The distribution function is generally assumed to be of the form

$$f(\vec{v}, v_r) = \left(\frac{m_e}{2\pi k_B}\right)^{3/2} \frac{1}{T_\parallel^{1/2} T_\perp} \exp\left(-\frac{m_e(v_\parallel - v_r)^2}{2k_B T_\parallel} - \frac{m_e v_\perp^2}{2k_B T_\perp}\right) \qquad (43)$$

with $v_\parallel$ and $v_\perp$ denoting the components of $\vec{v}$ perpendicular and parallel to the electron beam direction, respectively. With $T_\parallel = T_\perp$, Eq. (43) represents an isotropic Maxwellian electron velocity distribution. In a beam of accelerated electrons, however, $T_\parallel \ll T_\perp$. Therefore, Eq. (43) with $T_\parallel < T_\perp$ is termed a "flattened" Maxwellian distribution. The width of this distribution determines the experimental electron energy spread

$$\Delta E_r(\text{FWHM}) = \left[(\ln(2)k_B T_\perp)^2 + 16\ln(2)k_B T_\parallel E_r\right]^{1/2}. \qquad (44)$$

According to Eq. (44) the experimental energy spread increases at higher relative energies E_r.

Since both the electron and ion beams may be relativistic with laboratory energies E_e and E_i, respectively, the relative energy in a colliding-beams experiment has to be calculated from

$$E_r = m_i c^2 (1 + m_e/m_i) \left[\sqrt{1 + \frac{2m_e/m_i}{(1 + m_e/m_i)^2}(G - 1)} - 1\right] \qquad (45)$$

with

$$G = \gamma_i \gamma_e - \sqrt{(\gamma_i^2 - 1)(\gamma_e^2 - 1)} \cos\theta, \qquad (46)$$

where the Lorentz factors are

$$\gamma_i = 1 + \frac{E_i}{m_i c^2} \qquad (47)$$

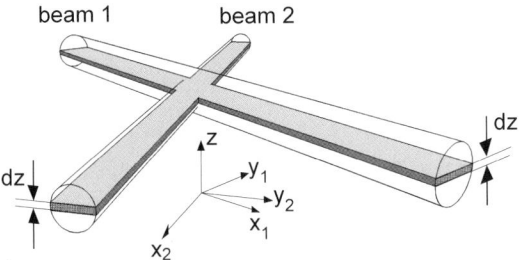

FIG. 22. Arrangement of inclined beams colliding with one another at an angle θ. The individual coordinate system of each beam is shown. The common z axis is perpendicular to the plane of the two beams which are assumed to be parallel in themselves. Slices of height dz of each beam at vertical position z completely penetrate one another. Such slices can be experimentally probed by driving a narrow horizontal slit through both beams in vertical direction. For sufficiently small dz the vertical beam profiles can be measured with good accuracy [see Eq. (51)].

and

$$\gamma_e = 1 + \frac{E_e}{m_e c^2}. \tag{48}$$

Determination of rate coefficients according to Eq. (34) can become much more complicated than measuring a cross section with a gas target of uniform pressure distribution. The particle densities n_e and n_i in Eq. (32) are related to the electron and ion flux densities J_e and J_i via the continuity equations $J_e = j_e/e = n_e v_e$ and $J_i = n_i v_i$, respectively. Accordingly, Eq. (34) can be rewritten as

$$R = \frac{\langle v_r \sigma \rangle}{v_e v_i} \Omega \tag{49}$$

with the form factor Ω accounting for the spatial overlap of the two beams

$$\Omega = \int \int \int J_e(x, y, z) J_i(x, y, z) \, dx \, dy \, dz. \tag{50}$$

Evaluating Ω would generally require the measurement of three-dimensional beam overlap factors. This is possible for example by driving narrow apertures through both beams at different locations while recording the transmitted electrical currents of electrons and ions. In many of the modern experiments the density of the electron beam is uniform and the ion beam is totally immersed in the electron beam. Under such conditions, Eq. (50) reduces to $\Omega = J_e I_i L/(qe)$ where I_i is the electrical current transported by the ion beam.

In an inclined-beams experiment (see Fig. 22) with zero-divergence beams located in the x–y plane, the form factor can be simplified since there is then no dependence of the flux densities on the coordinates in the direction of each beam. The integrations over coordinates x and y can be readily performed, producing a new expression for the beam overlap now explicitly containing the angle θ between the two beams

$$\Omega = \frac{1}{qe^2 \sin\theta} \int i_e(z) i_i(z) \, dz. \tag{51}$$

The charge state q of the ions enters Eq. (51) by the replacement of particle fluxes by electrical currents which can be determined directly by a measurement. Experimentally the integral can be evaluated by measuring differential currents $i_e(z)$ and $i_i(z)$ in each beam transmitted through a narrow horizontal slit at a number of positions z_k and replacing the integral by a sum over k. While the slit height is small and only limited by the experimental capability to measure small transmitted currents, the slit must be wide enough so that each part of the two beams can pass through it at some position z_k.

For the measurement of inclusive and partial cross sections it is important to know which states contribute to the observed product signal. This, in turn, depends on the detector system employed. For example, observing x-rays with the limited resolution of a semi-conductor detector does not necessarily provide information about the ion charge state from which the photons are emitted. Generally in electron–ion collisions, highly excited and long-lived states can be populated and it is a matter of flight time and passage of product ions through electromagnetic fields that determines whether the product can be detected or not. Resonance contributions to cross sections typically appear in the form of Rydberg series. Dielectronic capture can be viewed as core-electron excitation with simultaneous capture of the incident electron into an nl Rydberg state. External electromagnetic fields can destroy Rydberg states by field ionization.

Overlaying the Coulomb field of the nucleus in a hydrogen-like ion with a uniform electric field $\vec{E}$ produces a saddle in the resulting total potential, i.e., a low potential barrier is formed over which electrons can escape from initially bound states with high principal quantum numbers (see Fig. 23). From this concept a critical principal quantum number n_c can be derived above which all Rydberg states are field ionized. By a simple calculation not considering the change of the electron binding energy in the presence of an electric field nor the possibility of quantum tunneling through a potential barrier, one obtains (see e.g. Müller et al., 1987a) $n_c = (Z^3 E_0/16E)^{1/4}$ for a hydrogenic system with the atomic number Z and the electric field in atomic units ($E_0 = 5.142 \times 10^9$ V/cm). A more elaborate but still simplified approach (Gallagher, 1994) leads to

$$n_c = \left(Z^3 E_0/9E\right)^{1/4}. \tag{52}$$

For example, the latter equation yields $n_c = 49$ for a neutral Rydberg atom (with core charge 1e) in a field $E = 100$ V/cm.

The real situation is yet more complicated. Field ionization is not only dependent on the principal quantum number. It is different for different Stark states

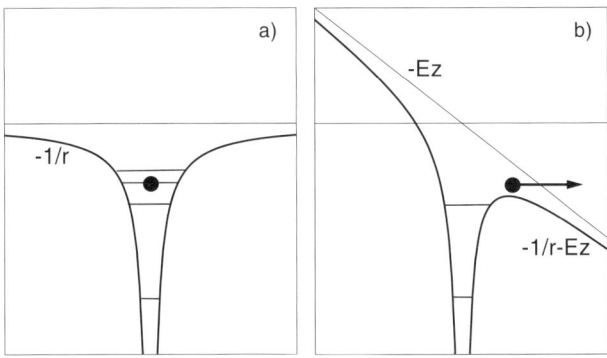

FIG. 23. Schematic of an electron in a hydrogenic atom bound by the Coulomb potential of the nucleus (a) and released when an electric field E in z direction is switched on (b).

within one n-manifold and even for different orientations characterized by the m quantum number. Hence, there is not a sharp cut-off number n_c but rather a distribution that depends on the population of the different Stark states.

3.1. PLASMA TECHNIQUES

Well-characterized Θ-pinch, tokamak and stellarator plasmas have been employed in numerous experiments to obtain plasma rate coefficients (Hinnov, 1966; Kunze et al., 1968; Kunze, 1971; Datla et al., 1975; Källne and Jones, 1977; Breton et al., 1978; Brooks et al., 1978; Isler et al., 1982; Wang et al., 1988; Griem, 1988, and references therein) for electron-impact ionization, excitation and recombination. Among the latest publications making use of plasma techniques to measure rate coefficients for electron–ion collisions are those of König et al. (1993, 1996). The concept of these measurements is to observe characteristic line emission from a plasma that is seeded with the desired atomic species, to model the observations by rate equations using measured electron densities and temperatures, and to obtain electron–ion collision rate coefficients for recombination, excitation or ionization by fitting the model to the observed quantities. The plasma has to be carefully diagnosed for this purpose and time dependences of electron densities n_e and temperatures T_e have to be recorded. Experiments with the Θ-pinch require diagnostics and line intensity measurements on a nanosecond time scale. Accordingly, the time-dependent development of densities n_q of seed ions in charge states q has to be described by a set of differential equations containing the information on collision rates of ions in different charge states q. Knowledge about the excitation of the observed line radiation from a given ion species is often taken from theoretical considerations.

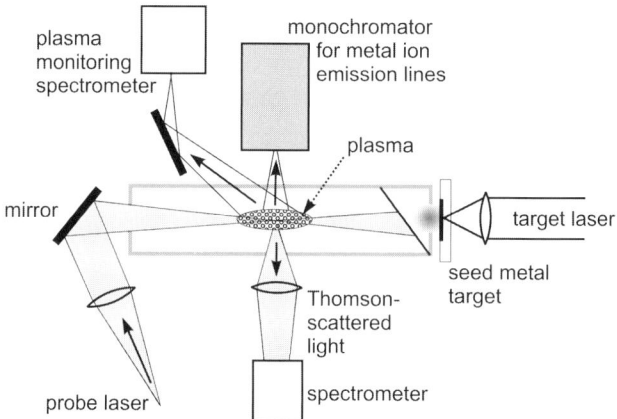

FIG. 24. Schematic of a plasma experiment to measure rate coefficients for electron-impact ionization or recombination of metal ions (Brooks et al., 1980). The plasma is produced by discharging a bank of capacitors through pinch-coils (not shown) placed around the plasma tube, thus striking a theta-pinch. The plasma is illuminated with a probe laser from which Thomson scattered light is monitored to obtain information on the electron temperature and density. Seed metal atoms are introduced by laser evaporation of a metal target. The plasma is monitored by observing visible light and short-wavelength emission lines of highly charged metal ions.

In the simplest approach a corona model can be applied to describe for example a phase of steady burn of a tokamak plasma. A set of rate equations such as

$$\frac{dn_q}{dt} = 0 = \alpha_{q-1}^{\text{ion}} n_e n_{q-1} - \alpha_q^{\text{ion}} n_e n_q + \alpha_{q+1}^{\text{rec}} n_e n_{q+1} - \alpha_q^{\text{rec}} n_e n_q \qquad (53)$$

with $q = 1, 2, 3, \ldots, Z$ relates the ionization rate coefficients α_q^{ion} and the recombination rate coefficients α_q^{rec} with the densities n_q of ions in different charge states q. These densities are inferred from the observed line radiation that is characteristic for the ion charge state of interest. All the rate coefficients can be treated like fit parameters that are determined by adjusting the model to the observations. Usually, additional assumptions have to be made in order to determine electron–ion collision rates. For example radiative recombination is usually neglected in comparison with dielectronic recombination. Although uncertainties as low as 30% have been quoted for such measurements one finds that measured plasma rate coefficients can be off by as much as a factor of 2 to 4 (Müller, 1999; Müller and Schippers, 2001). This is illustrated by Fig. 25 which compares plasma measurements with a storage ring experiment on dielectronic recombination of Fe^{15+} ions.

There are substantial differences between the plasma measurements of Isler et al. (1982) and Wang et al. (1988). Evidently the older data differ from the present expectations far beyond the quoted error bars. The data point of Wang et al. is still

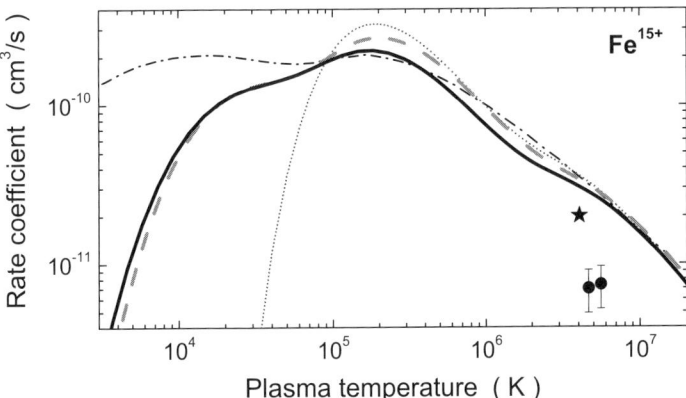

FIG. 25. Plasma rate coefficients for dielectronic recombination of Fe^{15+} ions. The star represents the tokamak measurement by Wang et al. (1988) and the solid circles with error bars are from the tokamak experiment of Isler et al. (1982). The dotted gray curve is the widely used rate coefficient recommended by Arnaud and Raymond (1992). The solid line was obtained from storage ring DR measurements (Linkemann et al., 1995a; Müller, 1999). The dashed gray curve is a theoretical result obtained by Gu (2004). The most recent calculation was published by Murakami et al. (2006) (gray chain curve).

about 50% below the storage ring experiment. Also shown in the figure are theoretical rate coefficients for dielectronic recombination of Fe^{15+}. All theories agree reasonably well with the storage ring experiment at temperatures beyond 10^5 K. The rate coefficients recommended in the widely used data compilation of Arnaud and Raymond (1992) are too small at lower temperatures. This is symptomatic of older theoretical approaches aiming at calculations for collisionally ionized plasmas, where ions like Fe^{15+} are to be expected at higher temperatures. More recent calculations by Gu (2004) are in quite good agreement with the storage ring result over the whole energy range that is displayed in Fig. 25.

While the experiment and the calculations of Gu comprise the contribution of L-shell excitation in the dielectronic recombination process, the latest theoretical results by Murakami et al. (2006) are restricted to excitations of the $3s$ valence electron. In spite of that, the new "large scale calculation", as it is referred to, is in good agreement with the other theoretical results at temperatures beyond 50 million degrees. Strong deviations from the storage ring result and the calculation of Gu are observed at temperatures below 10^5 K, a result which is not understood at this time. It is worth mentioning that Murakami et al. found discrepancy with the theory of Gu at higher temperatures. However, in the comparison they miscalculated Gu's rate coefficient from the parameter fit that Gu provided in his paper for easier use. Thus, although only selected theoretical results are shown, Fig. 25 provides a good example for the situation of data concerning plasma rate coefficients. Theory is not reliable yet at low temperatures where the exact energies of the recombination resonances are critical. Plasma techniques cannot provide data of sufficient quality. They can no longer compete with modern measurements which are based on newly developed experimental equipment that facilitated great advances in energy resolution, precision and accuracy of the data. Hence, the traditional route of plasma techniques to the determination of electron–ion collision rate coefficients is no longer pursued. The plasma technique had its justification when highly charged ions could not be produced and studied by other techniques. Of course, the research field associated with understanding, diagnosing, and modelling plasmas will continue to make good use of principal arrangements as sketched in Fig. 24. Observations of light and particles emerging from a plasma are the prime source of information about the state of the plasma.

3.2. TRAPS

Trapped-ion techniques for the determination of electron–ion collision cross sections are based on the observation of spatially confined ions that are exposed to electron impact. In the first experiments of this kind Baker and Hasted (1966) obtained information on ionization of ions by the observation of sequential ionization in an ion source within which the target ions were trapped by the space charge of an intense electron beam. By applying suitable external electric potentials additional to the "internal" space charge potential Redhead (1971) was able to obtain trapping times of the order of one second and to measure relative ionization yields in such a device. A cylindrically symmetric hollow electron beam for ion trapping and primary ionization (target preparation) and a concentric variable-energy electron beam interacting with the target ions was developed by Hasted and Awad (1972) for electron–ion collision studies (Hasted, 1983). The method was employed for the investigation of electron-impact ionization of positive ions by Hasted and Awad (1972) and Hamdan et al. (1978), for electron–ion recombination by Mathur et al. (1978), and for electron-impact dissociation of molecular ions by Mathur et al. (1979). After calibrating the apparatus to available crossed-beams data, uncertainties of measured cross sections were estimated to be in the vicinity of 30%.

The most productive use of traps for electron–ion collision studies was initiated by the development of the electron beam ion source EBIS by Donets et al. (1969) employing a very dense magnetically guided electron beam for trapping and for sequential ionization of highly charged ions. From the observation of the time history of the charge state evolution of ions extracted from an EBIS after different trapping times, cross sections were deduced for ionization of ions over a wide range of charge states by Donets and Pikin (1976), Donets and Ovsyannikov

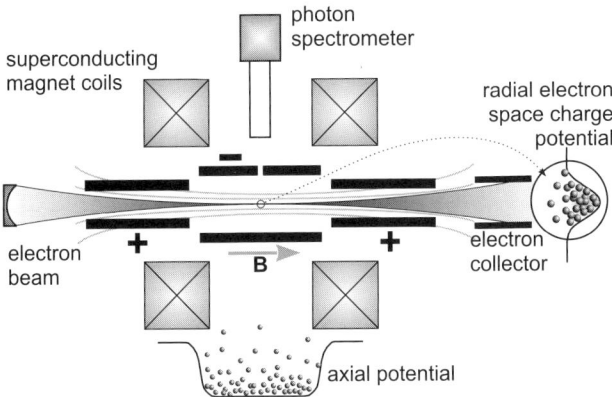

FIG. 26. Schematic of an electron beam ion trap. In a magnetic field, that is strongly increasing between the curved cathode of the electron gun and the center of the trap, an intense electron beam is compressed to diameters of about 30 to 50 μm with typical electron densities of the order of 5×10^{11} cm^{-3}. This produces a space charge potential well within the electron beam (see Fig. 19) with depths typically of the order of $q \times 10$ V within which slow A^{q+} ions can be trapped in radial direction. Axially the ions can be trapped by applying proper electric potentials to the drift tubes surrounding the electron beam. The length of the trap is typically of the order of 5 cm. Photon spectrometers covering different ranges of wavelengths at different levels of resolution are employed.

(1981) and Donets (1983). For the analysis, rate equations similar to Eq. (53) were used. While Donets et al. neglected recombination processes, this is not a good approximation when the electron energy in EBIS matches that of a recombination resonance. Ali et al. (1990) demonstrated the substantial effect of dielectronic recombination on the charge state distribution of ions extracted from an EBIS and were thus able to determine relative cross sections for dielectronic recombination.

Substantial extension of the technique became possible with the development of a modified EBIS, the electron beam ion trap EBIT by Marrs et al. (1988). The geometry of an EBIT facilitates the observation and spectroscopy of electromagnetic radiation emitted by the trapped ions. Thus, two techniques can now be used to obtain cross sections for electron collisions with highly charged ions in EBIT devices. One is associated with the extraction of ions as described above. The other is based on the observation of photons emitted by the trap inventory during electron beam exposure. In both cases the electron energy is ramped over a selected range of interest providing signatures of electron–ion collisions either in the number of extracted ions of a given charge state or in the differential photon yield at a fixed angle relative to the electron beam. The ion extraction technique has provided cross sections with experimental energy spreads of about 10 eV (DeWitt et al., 1993). Such cross sections are relative in nature and have to be normalized, for example to theory. In contrast, the x-ray technique (Knapp et al.,

1989) allows one to normalize the observed yields to the simultaneously recorded signal of radiative recombination and to make the well-founded assumption that cross sections for radiative recombination are theoretically well understood. This is especially true for radiative recombination of completely stripped ions (Ichihara and Eichler, 2000).

The electron energy spread in trap measurements is typically above 50 eV (A.J. Smith et al., 2000), but it can be made as low as 8 eV (Crespo López-Urrutia et al., 2005). The energy spread is related to the electron space charge potential required to trap slow ions in radial direction. With a given perveance of the electron gun this also limits the lowest electron energies usable in the trap experiments. At very low electron energies the space charge of the electron beam would not be sufficient for the radial trapping of the ions. It should be mentioned in this context that relative energy spreads $\Delta E/E$ of EBIT electron beams can be as low as 10^{-3}.

Ambiguity arises from the simultaneous presence of trapped ions in different charge states. All these ions (of a given element) emit similar x-ray spectra which can only be disentangled by relying on theory and by employing high-resolution spectroscopy. In his overview on laboratory x-ray astrophysics, Beiersdorfer (2003) points out that the charge state distribution in an EBIT can be influenced by proper choice of the electron energy so that the number of different charge states contributing to x-ray measurements can be as low as 2 to 3. However, 4 to 5 charge states typically contribute.

Very high charge states are accessible in traps as demonstrated by Marrs et al. (1994), but have to be produced by exposure of cold atoms to a dense electron beam of sufficiently high energy to overcome the ionization thresholds and to efficiently ionize the ions. This poses a severe challenge for the production of bare uranium requiring an electron energy of at least 200 keV (the threshold is at about 132 keV). By chopping the electron beam energy between ionization and measurement one can separate the phase of ion production from that of a cross section measurement. This phase is limited towards lower energies only by the fact that intense beams of electrons are required to obtain a measurable signal and to prevent the ions from leaving the trap region. Thus, the lower energy limit is around 100 eV. Such low-energy electron beams can only be produced in connection with strong magnetic guiding fields which are needed in such devices anyway to produce high densities in the electron beam by magnetic compression. As a consequence of the minimum requirements for the electron beam, measurements of low-energy electron collisions with highly charged ions are not accessible by the trapped-ion techniques.

By now there are numerous laboratories worldwide where EBIT devices are available. EBITs are extensively used as sources of characteristic line emission and have provided a wealth of highest quality spectroscopic information (Beiersdorfer, 2003; Beiersdorfer et al., 2005; Crespo López-Urrutia et al., 2005;

Currell, 2003, and Gillaspy, 2001). They can deliver ions in high charge states up to completely stripped uranium for external applications such as ion–atom or ion–surface interactions. Their use in the measurements of electron–ion collision cross sections is particularly valuable in the range of high electron energies where other techniques have not yet delivered many data. Recent work on electron–ion collisions studied by using EBIT devices was published for example by Fuchs et al. (1998), A.J. Smith et al. (2000), O'Rourke et al. (2001, 2004), Zhang et al. (2004) and Brown et al. (2006). At the Tokyo EBIT, KLL dielectronic recombination resonances, where a free electron is captured into the L shell and at the same time a K-shell electron is excited into the L shell, have been investigated for open-shell iodine ions by recording the detected yield of escaping ions of various charge states and modeling the charge balance in the trap (Watanabe et al., 2007). A combined experimental and theoretical investigation on KLL dielectronic recombination into He-, Li-, Be-, and B-like Hg ions has recently been presented by González Martínez et al. (2006) and Harman et al. (2006). Resonances were investigated occurring at energies around 50 keV. On the basis of the high-resolution data, effects of quantum electrodynamics as well as nuclear size effects could be studied.

As a typical example for observations of dielectronic recombination and electron-impact excitation in an EBIT, Fig. 27 shows a spectrum of Ge K_α (x-rays) at 10.2 keV as a function of electron energy taken by Zhang et al. (2004). The spectrum is composed of contributions from several different charge states as indicated by the inset. The dominant parent ion is He-like $Ge^{30+}(1s^2)$, in which one of the two K-shell electrons can be excited to the L shell at electron energies $E_e \geqslant 10.2$ keV (details of the excitation threshold are not resolved). Above the K $\rightarrow$ L threshold one expects $1s^2 \rightarrow 1s2l$ excitation with a cross-section energy dependence as discussed in Section 2.3.1: a slowly decreasing continuous direct-excitation contribution with resonances superimposed. The KMM resonance feature is then associated with doubly excited Li-like $(1s3l'3l'')$ states which can decay to He-like $(1s2l)$ states by an Auger process. The decay of the $(1s2l)$ state releases the observed K_α photon. At electron energies below 10.2 keV, dielectronic recombination produces KLL, KLM, KLO, ... resonances associated with doubly excited Li-like $(1s2l2l')$, $(1s2l3l')$, $(1s2l4l')$, ... states, respectively. All of these states emit a photon of about 10 keV when the $2l$ electron makes a transition to the K shell. Hence, the isolated resonances observed below 10.2 keV electron energy are due to dielectronic recombination (satellites to the K_α transitions in He-like Ge). Besides He-like ions there are H-like, Li-like, Be-like and B-like parent ions in the trap, producing the different contributions of recombined product states seen in the inset of Fig. 27. The whole situation can be disentangled with the support of theoretical calculations, and thus, with consideration of the angle-dependent photon emission, it was possible to pro-

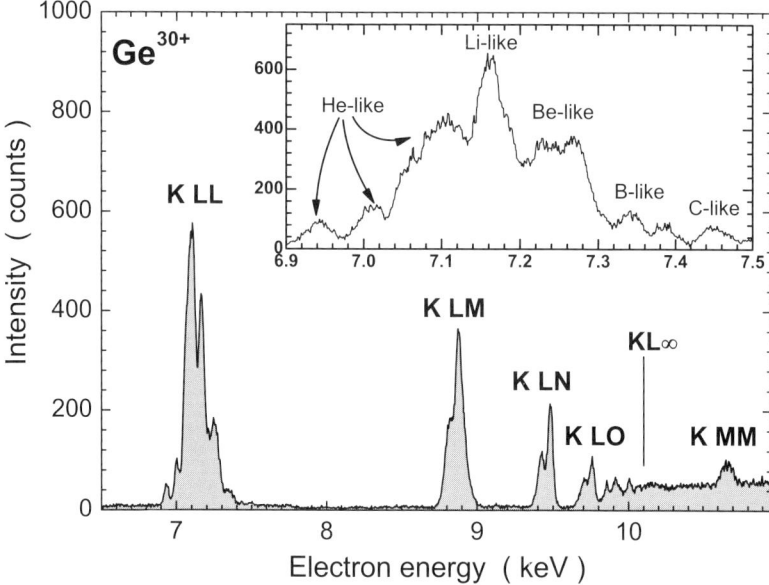

FIG. 27. Spectrum of Ge K_α transition photons observed by Zhang et al. (2004) at 10.2 keV x-ray energy at the Heidelberg EBIT while the electron energy was scanned from 6.7 to 11.1 keV. The dominant primary ion charge state in the trap was that of He-like Ge^{30+}. The inset shows a detailed scan of the KLL peak where the different charge states of the observed recombined Ge$^{(29+k)+}$ ions (with $k = 1, 0, -1, -2, -3$) are indicated. The Li-like product ions ($k = 0$) produce the dominant x-ray contribution. The vertical line marks the series limit of KLn resonances, i.e., the K → L excitation threshold.

vide resonance strengths for KLL dielectronic recombination resonances in that work.

Interference between radiative and dielectronic recombination has been a topic of special interest over many years. While theory predicts interference effects in total cross sections for recombination of ions (Badnell and Pindzola, 1992; Labzowsky and Nefiodov, 1994; Pindzola et al., 1995; Behar et al., 2004), and although interference is not an unusual phenomenon in electron–ion collisions, experimental evidence was difficult to find for the recombination channels. First evidence was reported by Knapp et al. (1995) who observed x-rays from a mixture of He-like to B-like uranium in an EBIT. In a thorough investigation of photoionization of Sc^{2+} and its time-reversed process, the photorecombination of Sc^{2+}, Schippers et al. (2002a, 2003) could demonstrate interference effects that had been predicted by Gorczyca et al. (1997). More recently, a new attempt using EBIT observations was reported by González Martínez et al. (2005). One of the results of that study is shown in Fig. 28. The asymmetric line shapes clearly indicate the presence of interference effects.

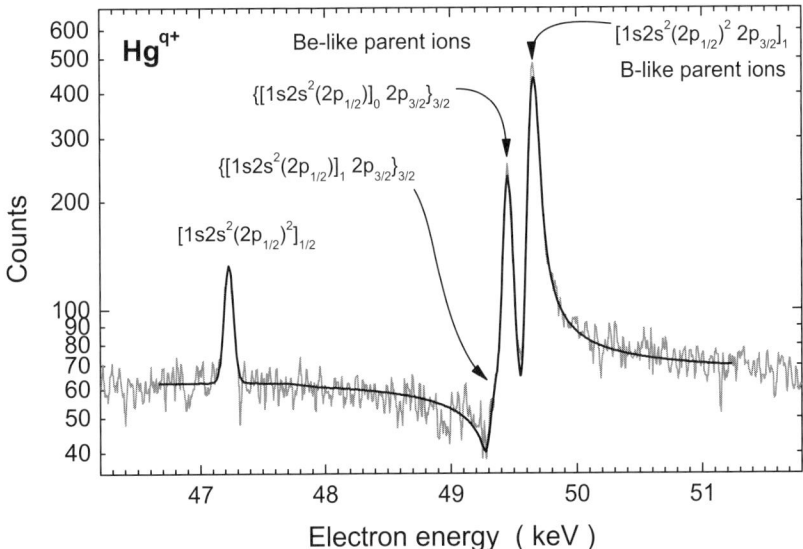

FIG. 28. EBIT measurement of x-ray emission from Be- and B-like mercury ions as a function of electron energy. The weak $[1s2s^2(2p_{1/2})^2]_{1/2}$ resonance produced from Be-like Hg^{76+} was best fitted with a symmetric type function. The $[(1s2s^2 2p_{1/2})_0 2p_{3/2}]_{3/2}$ resonance produced from Hg^{76+} and the $[1s2s^2(2p_{1/2})^2 2p_{3/2}]_1$ resonance produced from Hg^{75+} were fitted with asymmetric profiles given by Eq. (16).

3.3. Quasi-Free Electrons

Experimental studies of two-body collisions require an arrangement in which the particles of interest can interact with one another. What immediately comes to mind is a scheme in which a beam of one species interacts with a target consisting of the other species. Targets of charged particles, however, are difficult to prepare (Müller, 1989) because of space-charge limitations and, as a consequence, the target densities are very low as compared to those possible with solid or gaseous targets. Therefore the investigation of interactions between free electrons and ions often suffers from low signal rates. The density of an electron target can be enhanced by compensating the electron space charge with positive ions (Müller et al., 1987b). The most efficient scheme of this kind is the use of electrons bound in atoms or solids. This scheme is particularly obvious for the quasi-free electrons in a crystal lattice which are assumed to move freely between the positive-core ions in the three-dimensional grid. At sufficiently high projectile velocities ($v_i \gg v_0$ with v_i the velocity of the incident ion and v_0 the orbital electron velocity) also the electrons bound in target atoms can be treated as free particles, a concept that has been used with great success for almost a century to describe high-energy atomic collisions.

A number of experiments on electron-impact ionization and recombination were carried out employing the channeling technique, i.e., the electron cloud in a channel of a single crystal was used as a target (Claytor et al., 1988; Andriamonje et al., 1989; Datz et al., 1989). Numerous problems with the control of experimental parameters and conditions make the interpretation of the measured quantities difficult and lead to large uncertainties in the measured cross sections. The momentum spreads of the quasi-free electrons lead to substantial broadening of observed resonance contributions to electron–ion collisions. The topic has been reviewed by Krause and Datz (1996).

The importance of recombination and excitation measurements to test and to guide theory together with the appealing subject of observing resonance features in atomic collisions led to a very important development of accelerator-based ion–neutral-target collision studies of resonance phenomena related to electron–ion collisions (Tanis, 1992; Zouros, 1992). The interaction of highly charged ions with quasi-free electrons had already been exploited to study the radiative-recombination equivalent in ion–atom collisions, the radiative electron capture (REC; Schnopper et al., 1972). The dielectronic-capture-related process of resonant transfer with excitation (RTE) was unambiguously demonstrated for the x-ray emission channel about a decade later by Tanis et al. (1984), Schulz et al. (1987) and Mokler et al. (1990) and dielectronic recombination calculations could be directly tested by the equivalent resonant-transfer-with-excitation reaction (Tanis, 1992), provided that the momentum distribution of the quasi-free (but still bound) electrons was taken into account (Brandt, 1983). Similarly, the ion–atom equivalent to resonant elastic electron scattering from ions was found by Itoh et al. (1983) who observed the Auger decay channel of resonant-transfer-with-excitation resonances. This technique has been extended to measurements of differential cross sections for inelastic and super-elastic scattering processes (see Richard et al., 1999; Závodszky et al., 2001; Zamkov et al., 2002; Zouros et al., 2003). Resonant-transfer-with-excitation and radiative electron capture measurements were extended all the way to the highest possible charge states by Graham et al. (1990) and by Stöhlker et al. (1995, 1999). Also, measurements differential in the angle of photon emission were performed (Kandler et al., 1995; Stöhlker et al., 2001; Ma et al., 2003; Fritzsche et al., 2005) and most recently the polarization of radiative-electron-capture photons was investigated (Tashenov et al., 2006). Recently, radiative electron capture was extensively reviewed by Eichler and Stöhlker (2007) emphasizing its relation to radiative recombination and photoionization.

Behind all these studies is the so called electron scattering model or impulse approximation (Brandt, 1983; Lee et al., 1990; Eichler and Stöhlker, 2007), where the active bound target electrons as seen from the rest frame of the incident ion are considered to behave like free electrons with a given momentum distribution $f(\vec{p})$. Thus, the measured cross section for the ion–atom collision process $\sigma_{\text{ion–atom}}$ can be expressed as an average over the related electron–ion cross section $\sigma_{\text{e–ion}}$ with the momentum distribution of the electrons

$$\sigma_{\text{ion–atom}} = \int \sigma_{\text{e–ion}} f(\vec{p}) \, d^3 p. \tag{54}$$

The momentum distribution $f(\vec{p})$ results from the Fourier transform of the wave function $\Psi(\vec{r})$ describing the active bound electron

$$f(\vec{p}) = |\Phi(\vec{p})|^2 = \left| \frac{1}{(2\pi\hbar)^{3/2}} \int \exp\left(\frac{i}{\hbar} \vec{p} \cdot \vec{r}\right) \Psi(\vec{r}) \, d^3 r \right|^2. \tag{55}$$

Thus, Eq. (54) can be developed further to

$$\sigma_{\text{ion–atom}} = \int \sigma_{\text{e–ion}} |\Phi(\vec{p})|^2 d^3 p$$

$$= \int \sigma_{\text{e–ion}} \left[\int |\Phi(\vec{p})|^2 \, dp_x \, dp_y \right] dp_z. \tag{56}$$

With the definition of the Compton profile

$$J(p_z) = \int |\Phi(\vec{p})|^2 \, dp_x \, dp_y \tag{57}$$

the ion atom cross section can be written as

$$\sigma_{\text{ion–atom}} = \int \sigma_{\text{e–ion}}(p_z) J(p_z) \, dp_z. \tag{58}$$

Compton profiles $J(p_z)$ have been tabulated for all elements by Biggs et al. (1975). For one electron in a central Coulomb field, i.e., for hydrogenic atoms, the Compton profiles can be calculated in closed form. For the ground state of atomic hydrogen one obtains

$$J(p_z) = \frac{8 p_0^5}{3\pi} \frac{1}{(p_z^2 + p_0^2)^3} \tag{59}$$

with the momentum p_0 of an electron in the first Bohr orbit.

In the electron scattering model, the velocity of the impinging electron can be expressed as $\vec{v}_r = \vec{v}_i + \vec{v}$, where $\vec{v} = (v_x, v_y, v_z)$ is the velocity of the electron due to its bound motion around the target nucleus. The energy of the electron colliding with the ion is given by

$$E_r = \frac{m_e}{2} (\vec{v}_i + \vec{v})^2 - E_I \approx \frac{m_e}{2} v_i^2 + v_i p_z + \frac{p_z^2}{2 m_e} - E_I, \tag{60}$$

where E_{I} is the ionization potential of the target electron and p_z the projection of its momentum on the projectile trajectory (the z-direction is chosen along $-\vec{v}_{\mathrm{i}}$). As an approximation the transverse velocity components v_x^2 and v_y^2 are neglected in the second part of Eq. (60). The product $v_{\mathrm{i}} p_z$ is responsible for a large spread in the relative energy because of the wide distribution of bound-electron momenta.

Thus, the quasi-free electrons behave like an electron target with substantial energy spread. Even if the velocity condition $v_{\mathrm{i}} \gg v_0$ is fulfilled, the measurement of cross sections appears like a low-resolution electron–ion collision experiment. When resonances are observed as in the early resonant-transfer-with-excitation experiments then—because of the large width of $J(p_z)$—the electron–ion cross section can be approximated by $\sigma_{\mathrm{e-ion}} = A\delta(E_{\mathrm{r}} - E_{\mathrm{res}})$ with the resonance strength A, the resonance energy E_{res} and the Dirac δ-function. The impact energy [Eq. (60)] can be further approximated by neglecting the term quadratic in p_z as well as the binding energy E_{I}. By differentiating Eq. (60) one obtains then $dE_{\mathrm{r}} = v_{\mathrm{i}} dp_z$. With the substitution $J(p_z) dp_z = J(E_{\mathrm{r}}) dE_{\mathrm{r}}/v_{\mathrm{i}}$ one obtains in a non-relativistic approximation the ion–atom cross section as a function of the ion velocity v_{i}

$$\sigma_{\mathrm{ion-atom}}(v_{\mathrm{i}}) = \frac{A}{v_{\mathrm{i}}} J(p_{z,\mathrm{res}}) = \frac{A}{v_{\mathrm{i}}} J\left(\frac{E_{\mathrm{res}}}{v_{\mathrm{i}}} - \frac{m_{\mathrm{e}} v_{\mathrm{i}}}{2}\right). \tag{61}$$

The resulting peak function in the associated ion-energy dependence of the resonant cross section is centered around the resonance energy E_{res} and has a width determined by the broad Compton profile. For ground-state atomic hydrogen as a target the half-width of $J(x)$ is approximately $\Delta x = p_0$. Hence, the relative width $\Delta E_{\mathrm{i}}/E_{\mathrm{i}} \propto v_{\mathrm{i}}/E_{\mathrm{i}}$ decreases with the square root of the ion energy E_{i}. Figure 29 shows an example for a resonant-transfer-with-excitation experiment in which Li-like Ge^{29+} ions from an accelerator were passed through an H_2 gas target while the emerging Ge^{28+} ion and emitted K-x-rays were measured in coincidence (Mokler et al., 1989). There is a qualitative correspondence between the results of Fig. 29 and the EBIT measurements shown in Fig. 27. By using Eq. (60) the ion energy scale in Fig. 29 can be easily converted to the electron–ion energy of Fig. 27. The limited energy resolution of a classical resonant-transfer-with-excitation experiment becomes clearly visible.

However, this handicap can partly be overcome: if the photons or electrons emitted in the resonance process are detected with high energy resolution, detailed information can be inferred from the emission spectrum itself rather than the ion energy dependence of the total cross section. Even the moderate-resolution Si(Li) x-ray detectors used in the first resonant-transfer-with-excitation experiments made it possible to distinguish between K_α and K_β x-rays. By coincidence measurements with two Si(Li) detectors at 90° with respect to the ion beam direction and opposite to one another, details about individual groups of resonant states could be inferred (Schulz et al., 1987) and by employing the unique signature of

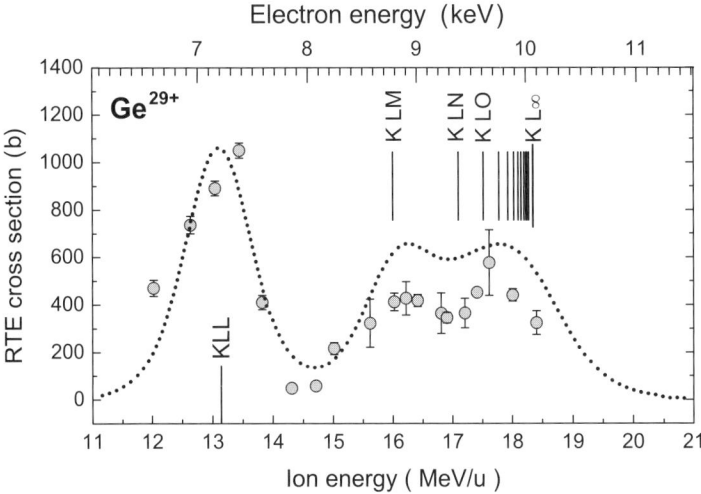

FIG. 29. Cross sections for resonant transfer with excitation in Ge^{29+}–H_2 collisions (Mokler et al., 1989). The ion energies were varied between 12 and 18.5 MeV/u corresponding to electron energies as shown on the upper horizontal axis. The contributing transitions and their resonance energies are indicated by vertical bars. The dotted line is a theoretical result obtained by Y. Hahn (unpublished; see Mokler et al., 1989).

the two-photon decay of the $1s2s\,^1S_0$ state in He-like Ge^{30+} it was possible to measure the resonance strength of a single resonance, essentially the intermediate $(2s2p\,^1P_1)$ state, in dielectronic recombination of H-like germanium (Mokler et al., 1990).

High-resolution emission spectra can exploit the luminosity advantage of the ion–neutral versus direct electron–ion experiments. Cold electron beams such as those applied for ion cooling in ion storage rings provide densities of usually less than 10^8 cm^{-3}. Targets of neutral gases can be easily produced with densities around 10^{14} cm^{-3} while preserving single-collision conditions. Solid foil targets reach electron densities of the order of 10^{22} cm^{-3}. Accordingly, ion–neutral experiments enjoy much higher counting rates than interacting-beams experiments. Moreover, the higher density of neutral targets allows one to design luminous target spots where ions and neutrals collide in a small volume from which the photons or electrons are emitted. Thus the solid angle for the observation of radiation or electron emission can be enlarged. Accordingly, the determination of differential cross sections for electron–ion collisions is relatively straightforward in ion–neutral target experiments (Kandler et al., 1995; Ma et al., 2003; Benhenni et al., 1990) and even (e, 2e) scattering on ions has become accessible (Kollmus et al., 2002). However, because of the condition $v_i \gg v_0$ that has to be fulfilled when interpreting ion–atom collisions on the basis of electron–ion scattering mechanisms, these measurements are restricted to the regime of high electron–ion collision energies. Studying low-energy electron–ion collisions by resonant-transfer-with-excitation techniques is not possible.

While the quasi-free-electron approach used to study electron–ion interactions has partly been supplanted by interacting-beams techniques, for example in the quest to study dielectronic recombination processes, there are a number of unique electron–ion interaction aspects in the experiments with fast ions incident on neutral targets that will probably never be accessible in free-electron collisions with ions just because of the target density and luminosity issues mentioned above. Among these are multiply differential measurements. There are other measurements that address electron–ion interactions by taking advantage of the high luminosity accessible in collisions of ions with neutral targets that can be complemented by experiments with free instead of quasi-free electrons, particularly in the field of elastic and inelastic electron scattering from ions.

As already discussed, the experiments employing quasi-free, i.e., loosely bound, electrons all have in common that a gas target is bombarded with fast ions of a desired charge state produced by an accelerator facility. Suitable detectors are placed around the target to determine the resulting charge state of the emerging ions and to register the angle-dependent emission of photons and electrons. While a reaction microscope (Ullrich et al., 2003) can "see" all angles and product energies at the same time, other arrangements have arrays of detectors around the target (see for example, Stöhlker et al., 2004, and Tashenov et al., 2006) observing products at different angles and energies simultaneously, though not covering anything near the full solid angle.

Compared to these arrangements a relatively simple experimental set-up is shown in Fig. 30. In this case, electron scattering at only one laboratory angle is observed: zero degrees with respect to the ion beam direction. The electron energy is analyzed by a high resolution/high transmission hemispherical electron energy analyzer. In the target-electron–projectile-ion center-of-mass frame this geometry can cover 0 or 180° electron scattering. Doppler shifts have to be considered keeping in mind that the ions have to be fast, so that the quasi-free-electron approach works.

For the interpretation of the doubly differential cross sections obtained in electron emission experiments, the equations provided above in this subsection can be employed. The ion–atom cross section given by Eq. (58) can be used in its differential form

$$\frac{d\sigma_{\text{ion–atom}}}{dp_z} = \sigma_{\text{e–ion}} J(p_z). \tag{62}$$

Differentiating Eq. (60) without neglecting terms other than those containing p_x^2 and p_y^2 yields $dE_r = (v_i + p_z/m_e)\,dp_z$ and p_z is determined by solving the

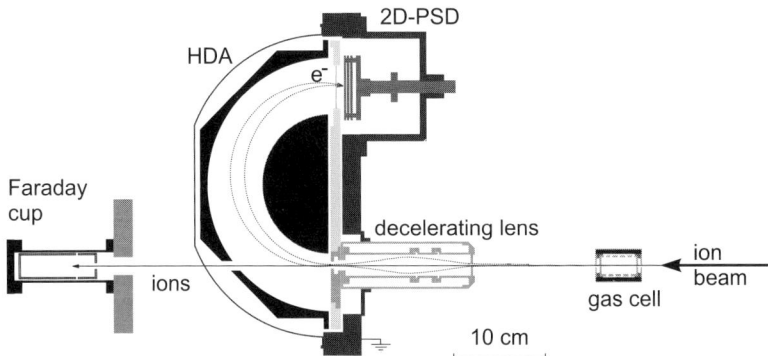

FIG. 30. Zero-degree electron spectrometer used for the measurement of elastic and inelastic electron scattering cross sections by employing the quasi-free-electron approach (Benis et al., 2004). The ion beam interacts with the gas target, confined by the gas-cell, traverses the spectrograph, and exits through a hole at the back. The ion beam current is recorded with the Faraday cup for normalization purposes. The electrons emitted from the target area in the ion-beam forward direction are decelerated by the four-element lens and focused onto the hemispherical deflector analyzer (HDA) entrance aperture. After a deflection through 180°, they are detected by the two-dimensional position sensitive detector (2D-PSD).

quadratic equation (60)

$$p_z = \sqrt{2m_e(E_r + E_I)} - m_e v_i. \qquad (63)$$

The double differential cross section for electron scattering in the ion–atom collision is then related to the single differential cross section for free-electron–ion scattering

$$\frac{d^2\sigma(E_r, \theta)}{d\Omega\, dE_r} = \frac{d\sigma(E_r, \theta)}{d\Omega} m_e \frac{J(p_z)}{m_e v_i + p_z}. \qquad (64)$$

The Compton profile $J(p_z)$ is so broad that it is almost constant within a small range of relative energies E_r in the electron–ion center-of-mass frame. At one given ion energy, a broad range of collision energies E_r is covered (similar to the irradiation of atoms by white light). Measurement of the double differential cross section for ion–atom collisions with electron emission thus yields information about the energy dependence of the cross section $d\sigma(E_r, \theta)/d\Omega$ for the related electron–ion scattering process.

An example of an experiment on elastic electron scattering from B^{4+} ions at 180° is shown in Fig. 31 (Benis et al., 2004; Zouros et al., 2003). Elastic scattering resonances within the KLL manifold are resolved and contributions from higher Rydberg states of the type KLM, KLN, KLO, ... are detected. The experimental set-up was calibrated by measuring normalized count rates for 180° scattering

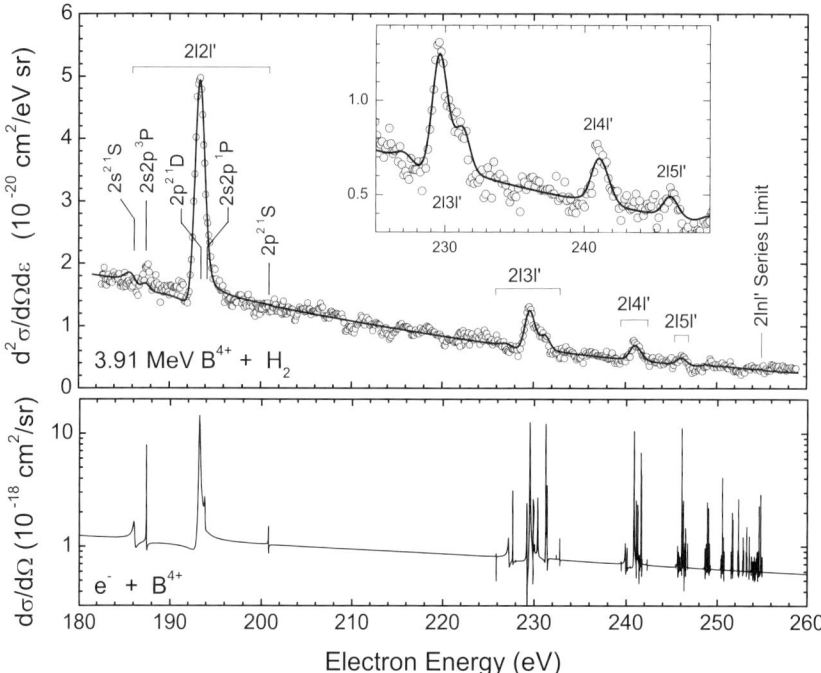

FIG. 31. Double differential cross section for 180° elastic electron scattering from B^{4+}. Experimental data (upper panel, open circles): zero-degree electron emission spectrum for 3.91 MeV B^{4+} collisions with H_2. Theory (lower panel, solid line): R-matrix calculations for elastic electron–ion scattering obtained in the framework of the electron scattering model. The theoretical results were convoluted with a 0.5 eV FWHM Gaussian (upper panel, solid line) to represent the spectrometer resolution (Zouros et al., 2003; Benis et al., 2004).

from completely stripped B^{5+}. The calibration factor was then determined by comparing the experimental rate to the Rutherford cross section. For the B^{4+} ions the identical calibration factor was used to put the measurement on an absolute scale.

With the same experimental method, inelastic electron scattering can also be studied (see e.g. Hvelplund et al., 1994; Závodszky et al., 1999). Figure 32 shows the double differential cross section for electron emission at an angle of 0° with respect to the ion velocity in the laboratory frame measured for collisions of hydrogen-like $O^{7+}(1s)$ ions with neutral H_2 (Toth et al., 1996). The cross section is displayed as a function of electron–ion center-of-mass energy. Within the electron scattering model this cross section is closely related to the single differential cross section for inelastic scattering at 180° in collisions

$$e + O^{7+}(1s) \to O^{6+}(3l3l') \to O^{7+}(2l) + e. \tag{65}$$

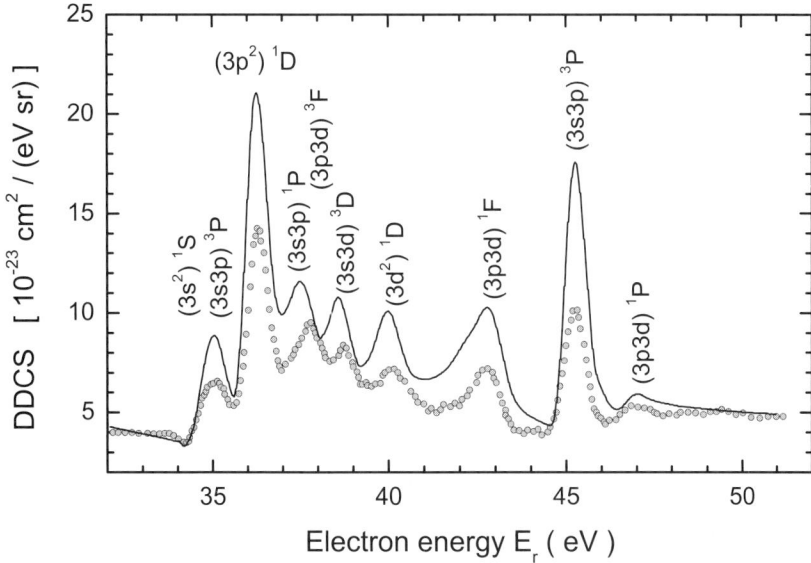

FIG. 32. Double differential cross sections for O^{7+} + H_2 collisions in the $(3l3l')$ resonance region as a function of the energy of the electron scattered at 180° in the projectile frame. Experimental double differential cross section (DDCS) data are represented by circles. Corresponding double-differential cross sections (solid line) were calculated for inelastic single differential scattering within the electron scattering model and then convoluted with a 0.4-eV FWHM Gaussian to represent the spectrometer resolution (Toth et al., 1996).

The R-matrix calculation for the process in Eq. (65) in the framework of the electron scattering model was convoluted with a 0.4 FWHM Gaussian to account for the finite resolution of the electron spectrometer.

Besides elastic scattering and electron-impact excitation, it has also been possible to observe superelastic scattering of electrons from excited metastable parent ions. Good candidates for such observations are He-like ions (Závodszky et al., 2001; Alnaser et al., 2002) with quite substantial excitation energies. The energy gain of an electron superelastically scattered from $F^{7+}(1s2s\,^3S)$ is as high as 725 eV.

3.4. Colliding-Beams Techniques

Clearly, the most direct access to unambiguous information about electron–ion collisions is the technique of interacting particle beams. The principal experimental arrangement is sketched in Fig. 33. From an accelerator, which can be as simple as a table top ion source on a static positive potential of a few kV or a major

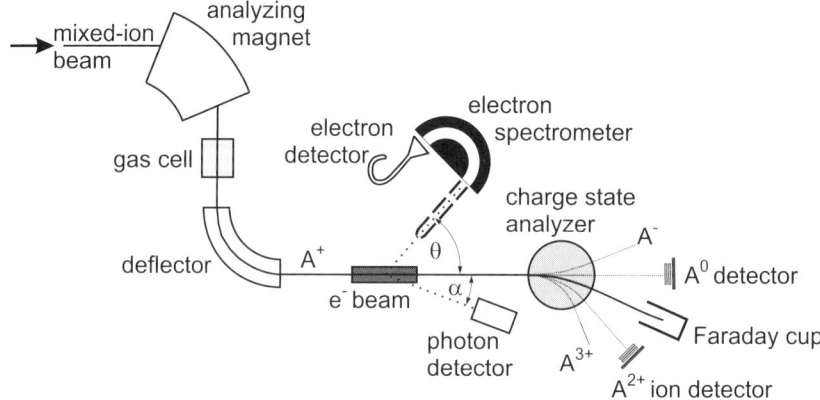

FIG. 33. Schematic of an electron–ion colliding-beams arrangement.

facility producing ions with hundreds of MeV/u, an ion beam of some unknown composition of ions is provided, typically with different masses and charge states but known energies. The first step is the selection of ions with a given mass/charge ratio at a fixed energy by employing an analyzing magnet. Some experiments allow manipulation of the selected ion beam by ion–atom interactions in a gas target, a technique that can be employed to change the fraction of long-lived excited states of ions in the beam. Just before the ions interact with the electrons their charge state spectrum is cleaned by separating ions that have changed their charge state in the gas cell or during their passage through the residual gas.

The schematic shown in Fig. 33 can be interpreted as a merged-beams setup with a cylindrically symmetric geometry or as a crossed-beams arrangement involving a ribbon shaped electron beam with a rectangular cross section. The interaction region is observed with suitable detection devices so that the collision products can be characterized and quantitatively collected. Projectile charge states are analyzed by applying static electric or magnetic fields and by using suitable single particle detectors. The same fields can be employed to detect electrons from field-ionized high Rydberg states. Short-lived excited states populated by electron–ion interactions are observed by measurements with electron-energy analyzers or energy-dispersive photon detectors. For the normalization of measured count rates it is necessary to collect the parent ion beam in a Faraday cup, to measure the electron current and to probe the beam profiles to determine the beam overlap or form factors.

The technique using colliding-beams of charged particles was first applied to electron-impact ionization of He^+ ions by Dolder et al. (1961). Only a few years later the technique was extended to the observation of photons from excited ions by Dance et al. (1966), and thus the measurement of total cross sections for electron-impact excitation of ions became possible. It took about two more decades until the recombination of ions with electrons could be addressed by the crossed-beams technique (Belić et al., 1983; Williams, 1984). Coincidences between the resonance-stabilizing photons and the recombined atoms made the measurements of cross sections possible. One reason for the experimental difficulties is background associated with the detection of recombined ions. In the experiments electron–ion recombination is in competition with electron capture of the ions from residual gas molecules. The cross sections for electron capture of an ion colliding with an atom or molecule are very high at low ion velocities and typically increase with the charge state of the ion (Müller and Salzborn, 1977; Cornelius, 2006). Hence, background arising from slow collisions of multiply charged ions becomes excessive and almost prohibits the use of the crossed-beams technique for recombination measurements with multiply charged ions (Young et al., 1994; Savin et al., 1996).

Electron capture in residual-gas collisions can be greatly reduced by employing ions with velocities v_i well above the orbital velocity v_0 of the bound target electrons since then the cross sections drop with high powers of v_i. When ions from an accelerator are used in electron–ion recombination experiments, background from residual gas collisions is greatly reduced and, moreover, fast ions can easily be stripped by passing them through solid foils so that high charge states become available for measurements. Pioneering experiments making use of accelerator-based merged beams arrangements for electron–ion recombination studies were reported by Dittner et al. (1983) and Mitchell et al. (1983). The problem with background associated with electron capture from residual gas components was partly circumvented by detecting signal from field ionization of high Rydberg states (Müller et al., 1986, 1987a) which are predominantly populated in electron–ion collisions and not as much by interactions of ions with gas molecules. The apparatus used by Müller et al. (1986, 1987a) is shown in Fig. 34.

A new observation channel in electron–ion collision studies was opened by Chutjian et al. (1983) and Zápesochnyi et al. (1986) measuring the electron-energy loss in a crossed-beams set-up thus providing the first angle-differential data for electron scattering from ions. While these measurements were successfully pursued (Williams, 1999, and references therein) producing important information on doubly differential scattering cross sections (with respect to angle and energy of the outgoing electron), the applied data needs in plasma and astrophysics called for a new access to total cross sections for electron-impact excitation of ions. An electron–ion merged beams technique employing trochoidal analyzers for merging and de-merging the interacting beams was developed for the measurement of the inelastic energy loss of electrons scattered from ions (Wåhlin et al., 1991; Smith et al., 1991). The principle of such an arrangement is shown in Fig. 35. A drawback of this method is its restriction to the threshold energy region of excitation cross sections. There are difficulties with separating

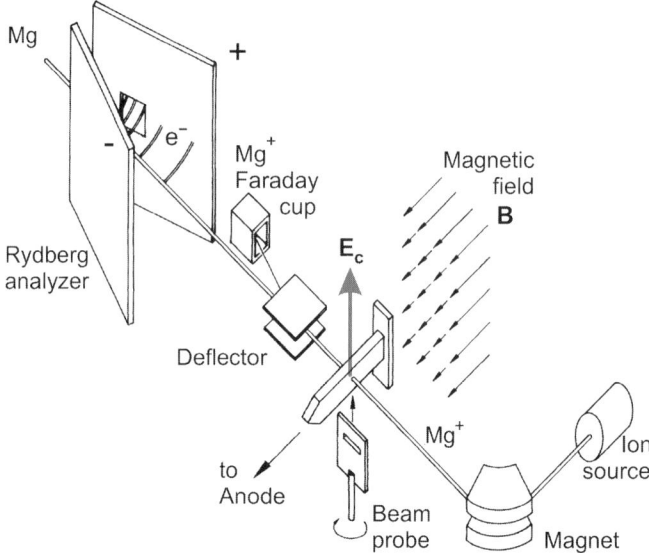

FIG. 34. Schematic of the electron–ion crossed-beams arrangement used by Müller et al. (1986, 1987a) to study e + Mg$^+$ recombination into high Rydberg states of neutral magnesium atoms. Controlled and known magnetic $\vec{B}$ and motional electric $\vec{E}_c$ fields in the collision region were employed to investigate field effects on the dielectronic recombination process. The rotatable beam probe served for measuring the beam overlap. The Rydberg analyzer with its electric field increasing in the Mg atom flight direction field-ionized Rydberg states $|n\rangle$ at different locations $z(n)$. The released electrons were measured using a position sensitive detector to identify the principal quantum number n and to measure the related resonance strength of the dielectronic recombination.

elastically and inelastically scattered electrons from one another. By introducing an electrostatic filter, an 'electrostatic aperture', Greenwood et al. (1999) were able to improve the separation of elastically from inelastically scattered electrons. Collecting all scattered electrons, even if they appear at backward angles in the electron–ion center-of-mass system, is an additional challenge in these experiments.

Low spread of the electron energy and sufficiently intense beams of electrons are natural issues in electron–ion experiments. The use of a trochoidal analyzer, i.e., a geometry with crossed homogeneous electrical and magnetic fields, for merging the two beams has the potential of selecting a narrow bandwidth from the electron energy distribution. Typical energy spreads in the electron beams used for excitation studies on the basis of electron-energy loss are of the order of 200 meV. Compared to other techniques employed in studies of electron-impact excitation, this is a remarkable achievement. As an example, Fig. 36 shows experimental (effective) cross sections for the $2s\,^2S_{1/2} \rightarrow 2p\,^2P_{1/2,3/2}$ excitation of Li-like C^{3+}

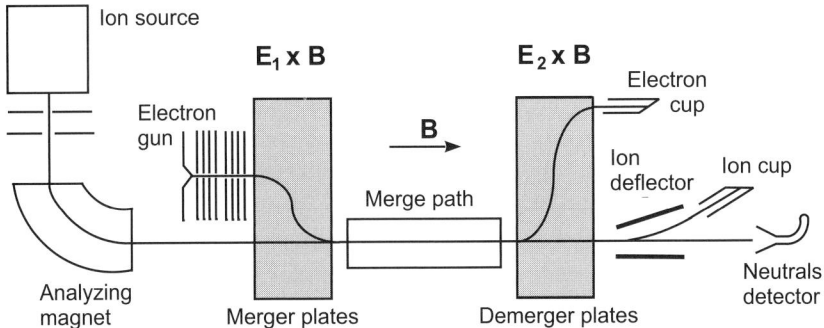

FIG. 35. Schematic of an electron–ion merged-beams set-up employing trochoidal analyzers for merging and de-merging the beams. The crossed electric and magnetic fields also separate electrons that have lost energy in the interaction region. This arrangement can therefore be employed to investigate inelastic electron scattering or electron-impact excitation (see Phaneuf et al., 1999).

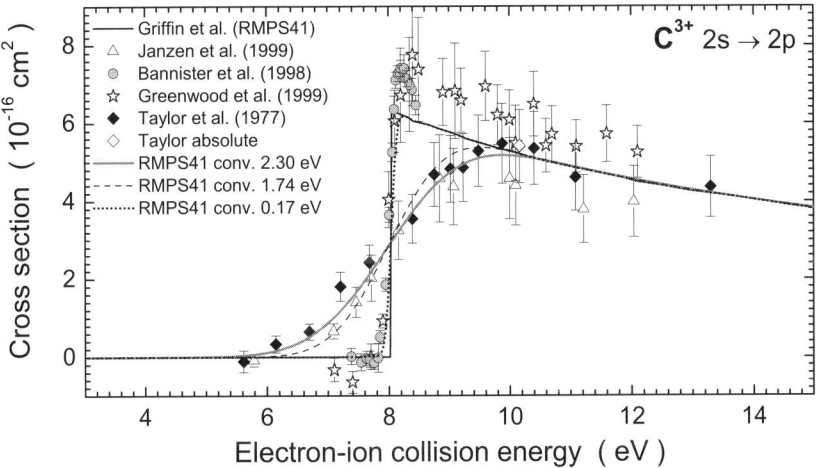

FIG. 36. Cross sections for $2s\,^2S_{1/2} \rightarrow 2p\,^2P_{1/2,3/2}$ excitation of C^{3+} ions near threshold. First experimental data were obtained by photon detection in electron–ion crossed beams arrangements (solid diamonds, electron energy spread 2.3 eV; Taylor et al., 1977; Gregory et al., 1979). A similar technique was employed by Savin et al. (1995) and the data were later renormalized (open triangles, 1.74 eV spread; Janzen et al., 1999). Trochoidal-analyzer merged-beams arrangements were used to measure the excitation cross section via the electron-energy-loss technique (gray shaded circles, 0.17 eV spread; Bannister et al., 1998; and stars, 0.17 eV spread; Greenwood et al., 1999). The error bars are total experimental uncertainties at high confidence level, except for the data of Bannister et al. which show random errors and only one total uncertainty at 8.35 eV. The solid line is a 41-state R-matrix-with-pseudo-states (RMPS41) calculation by Griffin et al. (2000). That calculation is convoluted with Gaussians with different FWHM values representing the energy spreads in the different experiments as indicated in the figure.

ions. Two sets of data were obtained in crossed-beams arrangements by observing the photons emitted after the excitation, two other sets by using the energy-loss technique in connection with trochoidal-analyzer merged-beams techniques. The improvement of energy resolution between the first and the latter data is obvious. Nevertheless, the early measurement of Taylor et al. (1977) agrees with the state-of-the-art 41-state R-matrix-with-pseudo-states (RMPS41) calculation by Griffin et al. (2000) better than any of the other measurements (Bannister et al., 1998; Greenwood et al., 1999; and Savin et al., 1995), where the latter appears to differ from the other experiments even after careful reevaluation of the data by Janzen et al. (1999). The theory was compared with other calculations with smaller sets of basis states and the results could hardly be distinguished from one another, indicating that the R-matrix calculation is converged and reliable.

For the examples discussed above for excitation of C^{3+} the energy resolution does not make a real difference for the understanding of the process, except for the different shapes of the effective cross sections. The situation would be drastically different if the energy resolution were sufficient to resolve the difference in excitations to the different ^2P fine structure levels. In that case, important additional information would be provided by the separation of individual contributions to the cross section. The value of better energy resolution also becomes obvious when there is rich structure in a cross section, e.g. by the presence of resonances. Therefore, in some electron–ion collision experiments, the electron beam was passed through an energy filter, such as a 127° electrostatic analyzer, before entering the interaction region (Peart et al., 1973, 1989; Thomason et al., 1997, and references therein). This produced energy spreads as small as about 90 meV and fine details in cross sections could thus be revealed. The electron currents that were available in these measurements were very small, sometimes in the range of only about 50 nA making it difficult to achieve sufficiently good data statistics that are necessary to really exploit the low energy spread in the electron beam.

For the measurement of cross sections with good statistics and also to access small cross sections, higher electron currents are desirable. A unique approach was introduced by Müller et al. (1980) who used a ribbon shaped electron beam that was orders of magnitude more intense than any other beam in previous electron–ion collision experiments. The electron currents produced by a space-charge limited electron gun reached up to 300 mA at 1000 eV. In subsequent experiments the design of the electron gun was improved with the goal to minimize adverse space charge effects. A gun, that has been used for more than two decades now was described by Becker et al. (1985) and Müller et al. (1985). It provides 400 to 460 mA of electron current at 1000 eV and an interaction length for the crossing ion beam of 60 mm. In the spirit of the design of this high current electron gun, a cylindrical electron beam device was constructed later by Stenke et al. (1995) to produce beams up to 6500 eV with a maximum electron

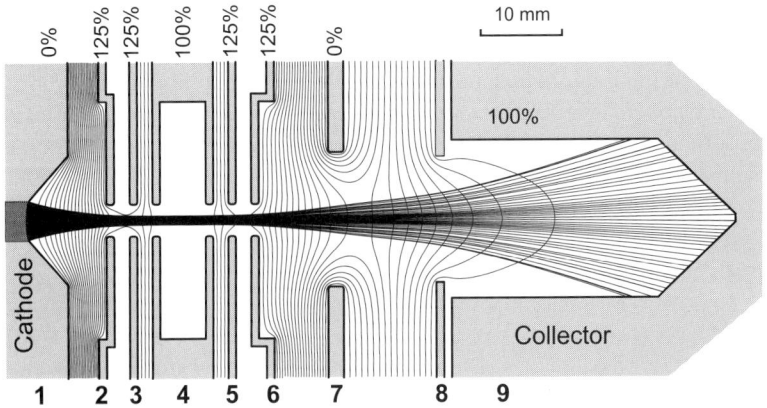

FIG. 37. Computer simulation of a versatile electron gun for electron–ion crossed-beams experiments (Shi et al., 2003). Electrodes 1 through 9 are supplied with voltages given in percent of the potential difference between the cathode (1) and the interaction region (4), i.e., the voltage that defines the laboratory energy of the electrons in collisions with ions. Electron trajectories are shown as well as equipotential lines. In the interaction region (electrode 4) the density of equipotential lines is low indicating at most small electric fields.

current of 430 mA at an interaction length (the diameter of the electron beam) of 2 mm. A more recent development (Shi et al., 2003) aimed at improved versatility of electron beam formation so that optimum conditions can be chosen at each electron energy. The new electron gun has more electrodes than its predecessors to achieve that versatility (see Fig. 37). It can enhance the electron current over that of the previous high-current gun by a factor of about 3, but also low-current modes are available. The gun was operated with 1 A of electron current at 3 keV. It provides an interaction length of 60 mm.

High electron currents inevitably produce space charge fields. The electron guns described above are designed such that the interaction region has a relatively flat effective potential distribution produced by electron space charge and the electrostatic potentials on the surrounding electrodes. At typical operation, the center of the interaction region has a potential well within which slow ions, produced from the residual gas, can be trapped and thus partially offset the space charge of the electrons. The result is that even with hundreds of mA of electron current, the relative energy resolution of the electron beam at 1 keV is $E/\Delta E \approx 300$. The minimum energy spread is $\Delta E \approx 400$ meV.

An important step in the development of the crossed beams technique was the invention of the "animated beams method" first described by Defrance et al. (1981). Rather than measuring beam overlaps [see Eq. (51)] separately, the form factors are implicitly obtained during a cross section measurement, in which signal and background are determined. The electron beam sweeps back and forth across the ion beam and the detector count rate is recorded as a function of electron beam position. From the area of the signal peak thus obtained and with proper normalization to electron and ion currents, the cross section can be directly determined. This technique can be realized by electrostatically scanning the electron beam across the ion beam (Defrance et al., 1981) or, alternatively, the whole electron gun can be moved mechanically across the ion beam (Müller et al., 1985).

Experimental approaches to study electron collisions with multiply charged ions have greatly profited from the development of powerful electron-cyclotron-resonance (ECR) ion sources (Trassl, 2003). A real experimental set-up employing a permanent-magnet ECR source (Trassl et al., 1997) that is used for studies of electron-impact ionization of ions is shown in Fig. 38. ECR sources are under continued active development. High-end sources built for injection of ions into accelerators reach mA electrical currents for multiply charged light ions. Even for ions as highly charged as U^{47+}, Leitner et al. (2007) could produce an electrical current of 4.9 µA with their forefront ECR source *VENUS*. ECR sources employ microwave heated electrons in a magnetic bottle to overcome the high ionization thresholds for highly charged ions. Bremsstrahlung from such sources reaches up to energies of about 1 MeV (Zhao et al., 2006) indicating the high energies that can be reached by the electrons. As a consequence, these sources are also capable of efficiently producing excited states of multiply charged ions and providing intense emission of short-wavelength radiation. Because of the same reason, ions extracted from an ECR source can be in long-lived excited states that do not decay before they reach the interaction region of the electron–ion collision experiment. This can be a blessing, in that it provides access to studies of collisions involving metastable excited states, or a curse, because the primary ions are not in a single well-defined initial state. In general, the presence of unknown fractions of such metastable ions in the parent beam makes it at least difficult to determine absolute cross sections with the high accuracy presently needed for the modeling of astrophysical plasmas (Savin, 2005). The problem can be overcome in most cases by storage of the ions for sufficiently long times to facilitate decay of the metastable states before the measurement starts. Such an experimental scheme for highly charged ions requires the combination of an accelerator with a heavy ion storage ring.

The accelerator based electron–ion merged-beams technique introduced to studies of recombination and later also ionization of multiply charged ions has undergone a remarkable technological development after the pioneering experiments carried out at Oak Ridge by Dittner et al. (1983). The first approach was to use a dense electron beam producing signal rates sufficiently high to allow the experimenters the separation of the desired electron–ion recombination from electron-capture background produced in collisions of the projectile ions in the residual gas. For producing a dense electron beam, magnetic compression was employed, similar in nature to the scheme used in an EBIT (see Fig. 26). The ion

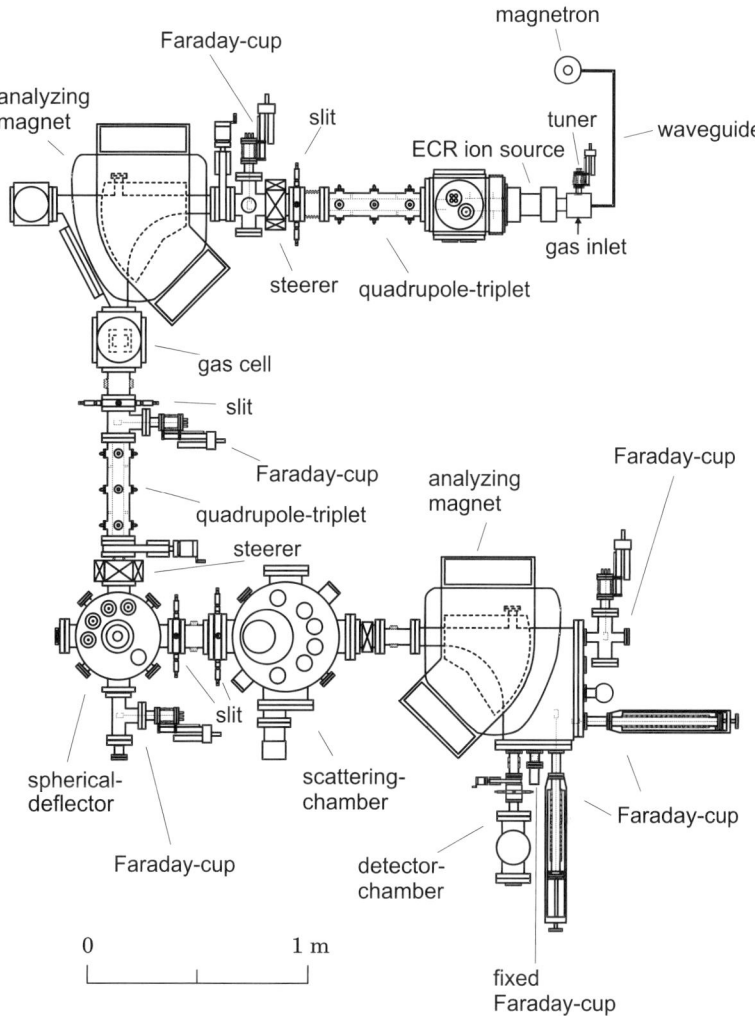

FIG. 38. Layout of an electron–ion crossed-beams set-up (Jacobi et al., 2004). The ion beam is crossed with an electron beam inside the scattering chamber. The detector chamber contains a spherical deflector bending the product ion beam out of the scattering plane to avoid background from stray photons or electrons.

beam entered the merging region through a central hole in the cathode of the electron gun and then passed axially through the electron beam. By electrostatically separating recombined from parent ions, effective cross sections [rate coefficients in the meaning of Eq. (35)] were obtained. As in all electron–ion merged beams experiments the electron velocity distribution function [Eq. (43)] applied also to the Oak Ridge experiment. By the compression of the electron beam in an increasing axial magnetic field high transverse temperatures were introduced.

By the construction of new cold electron targets the experimental resolution in electron–ion recombination measurements could be improved (see e.g. Andersen et al., 1990b; Schennach et al., 1994). The new electron beams were designed to be parallel in themselves, and by adiabatic acceleration of the electrons, substantially lower electron beam temperatures could be achieved than in the Oak Ridge device.

Another big step was accomplished by combining further optimized cold electron beams with heavy ion storage rings. By the preparation of brilliant low-emittance stored ion beams that probe the electron space charge potential (see Fig. 19) just in the center of the electron beam, the energy resolution in electron–ion experiments could again be greatly enhanced (Müller and Wolf, 1997). By adiabatic expansion rather than compression of the electron beam the transverse temperatures can be substantially reduced (Danared, 1993). Figure 39 illustrates the vast improvement in data quality, both with respect to statistics and energy resolution, that could thus be obtained. It is worth noting that the measurements obtained for C^{3+} ions at the Test Storage Ring in Heidelberg (Schippers et al., 2001) and at CRYRING in Stockholm (Mannervik et al., 1998) are in good agreement, both with respect to absolute merged-beams rate coefficients, resonance energies and energy resolution.

While in previous storage ring experiments the electron cooler had to serve two purposes, cooling of the ion beam and providing an electron target, these two functions were decoupled at the Heidelberg Test Storage Ring (TSR) by introducing an optimized electron target (Sprenger et al., 2004) additional to the existing cooling device. Thus, the ion beam can be cooled while the measurements are being taken. The target was carefully designed with respect to adiabatic acceleration of electrons and adiabatic expansion of the electron beam. Moreover, the thermal cathode can be replaced by a cooled photocathode emitting electrons with an energy spread as low as 19 meV (Orlov et al., 2006). By this new development the present record in low electron beam temperature has been set to $k_B T_\| = 0.025$ meV (after acceleration) and $k_B T_\perp = 0.5$ meV (after adiabatic magnetic expansion). Clearly, storage ring merged beams experiments have become the method of choice to study low-energy electron–ion recombination. This includes the investigation of external electric and magnetic field effects on dielectronic recombination of multiply charged ions. The Heidelberg storage ring TSR with the electron cooler and the additional electron target is schematically shown in Fig. 40.

Storage ring techniques were also applied to electron-impact ionization of highly charged ions. An important advantage of these measurements is in the accessibility of intense beams of highly charged ions, provided the ring is at-

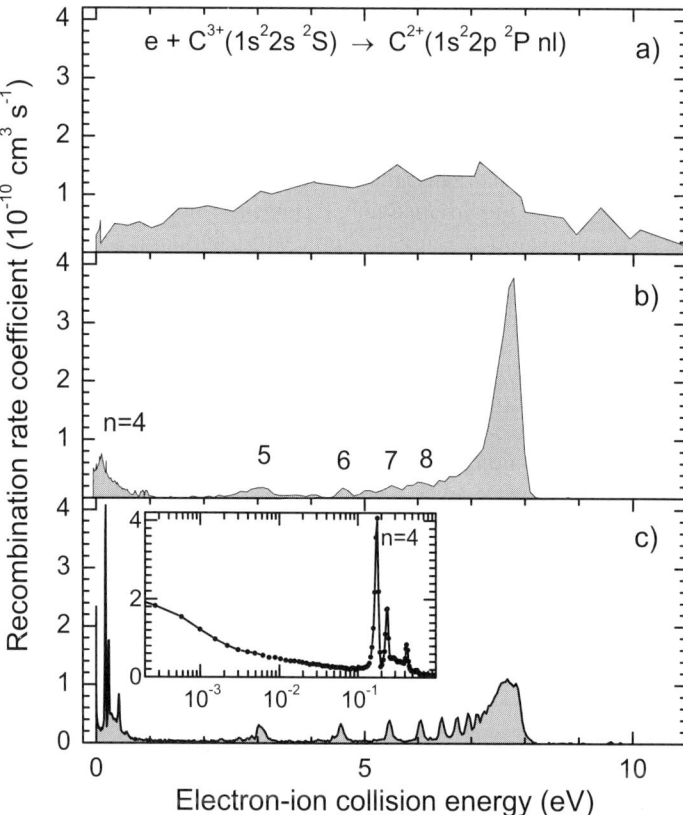

FIG. 39. Comparison of three measurements of dielectronic recombination of C^{3+} ions. From top to bottom the data are from work by (a) Dittner et al. (1983, 1987), (b) Andersen et al. (1990b), and (c) Schippers et al. (2001). In the sequence of the experiments the electron beam temperatures decreased from $k_B T_\| = 60$ meV and $k_B T_\perp = 5000$ meV to $k_B T_\| = 1$ meV and $k_B T_\perp = 135$ meV and on to $k_B T_\| = 0.15$ meV and $k_B T_\perp = 10$ meV. The inset of panel (c) emphasizes the low-energy region of the spectrum, where the storage ring experiment has its best resolution. (Note the logarithmic energy scale of the inset.)

tached to an accelerator that injects intense beam pulses of these ions which can then be accumulated and stored. With storage times in the range of seconds to hours electron–ion collision experiments can take advantage of the availability of pure ground-state ion beams for cross section measurements, thus avoiding the often faced problems of unknown fractions of metastable states in beams of highly charged ions.

A drawback of accelerator based ionization measurements is the presence of high backgrounds produced by electron-stripping collisions (Santos and DuBois,

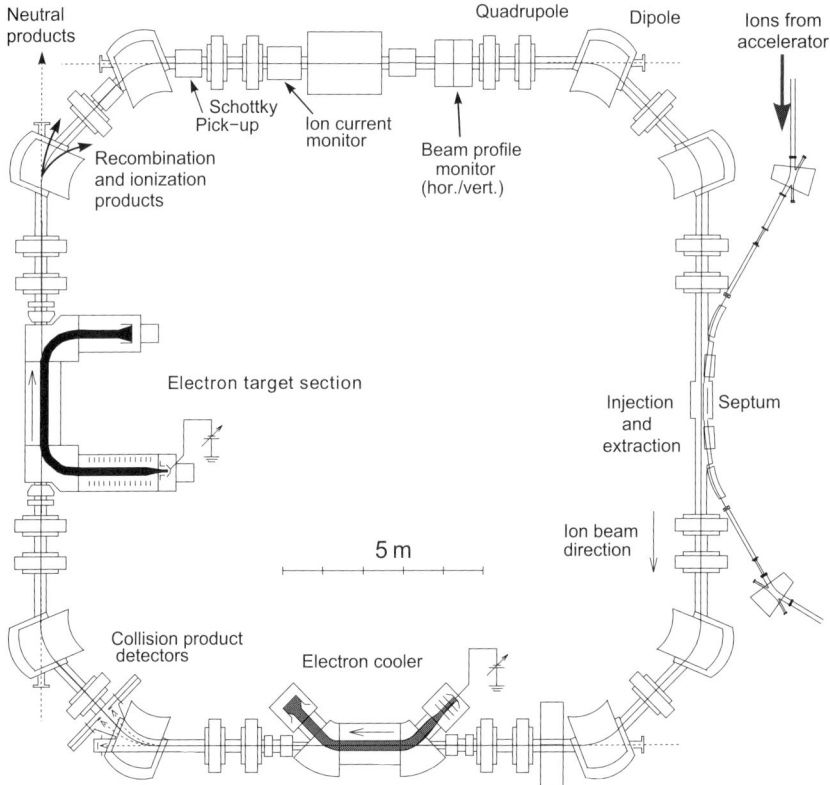

FIG. 40. Layout of the Heidelberg Test Storage Ring TSR with the electron cooling device and the additional electron target. Product ions that have changed their charge state in electron–ion collisions can be detected behind the first bending magnet following the electron target (or the electron cooler) in ion beam direction.

2004) of the fast parent ions with residual gas components—even at gas pressures in the range of 10^{-9} Pa typical for a storage ring vacuum. Therefore, experiments at low ion energies in the range of typically only 1 keV/u are still best suited for ionization studies on low to intermediate-high ion charge states that can be produced by an ion source with sufficient beam intensities. Better energy resolution and the observation of finer cross section details have been accessible in dedicated low-energy ion beam arrangements as compared to storage-ring based electron-impact ionization experiments.

Storage ring electron–ion collision experiments had previously been limited to energies below several keV. This is not a principal limitation but can be overcome e.g. by employing a dedicated electron target in addition to the electron cooler.

An alternative approach has been successfully tested at the Experimental Storage Ring ESR in Darmstadt, Germany. A U^{91+} ion beam was stochastically cooled while the electron cooler was tuned for relative energies $E_r = 60\text{--}90$ keV in the electron–ion center-of-mass system. Thus, KLL and KLM dielectronic recombination resonances which occur in that energy range could be measured (Brandau et al., 2006).

4. Overview of Experimental Results on Free-Electron–Ion Collisions

4.1. ELECTRON SCATTERING FROM IONS

In an experiment, an elastic electron scattering process can be identified by observing scattered electrons under the condition that the kinetic energy in the center-of-mass system has not changed. The electron is detected at a defined scattering angle and its energy measured at a macroscopic distance from the scattering atom or ion A^{q+}. This converts to a time span of the order of nanoseconds between the atomic process and its registration. The passage of the electron past the target A^{q+} ion is associated with deflection by the Coulomb field of the ion without any change of internal energies. This is experimentally verified also for a process where the incident electron and the target form a resonance, i.e., a short lived quasi-bound intermediate state, which then decays back to the original A^{q+} by the emission of an electron with a typical lifetime of picoseconds or less. So far, the author is not aware of any direct observation of resonant elastic electron scattering other than those using the quasi-free-electron approach in ion–atom collisions (an example is provided in Fig. 31).

Measurements of cross sections for elastic electron scattering in free-electron–ion collisions are scarce (Huber et al., 1994; Bélenger et al., 1996; Greenwood et al., 1995a, 1995b; Srigengan et al. 1996c, 1996a, 1996b; Wang et al., 2000; Brotton et al., 2002). Some of the existing data and theoretical approaches have been reviewed by Williams (1999). Elastic electron scattering from ions at low scattering angles is expected to be described by Rutherford scattering theory. At large impact parameters the scattered electron just senses the net charge of the ion. This is demonstrated by measurements of differential cross sections for elastic electron scattering from Ar^{8+} ions at $E_r = 22.46$ eV (see Fig. 41). In the angular range up to $120°$ the experimental data follow the Rutherford cross section and deviate only at larger angles. At these large angles, i.e., for small impact parameters, the electron probes the effective potential distribution inside the electron cloud of the ion and thus a more sophisticated description is required, for example by using a Hartree–Fock representation of the effective potential (Bélenger et al., 1996).

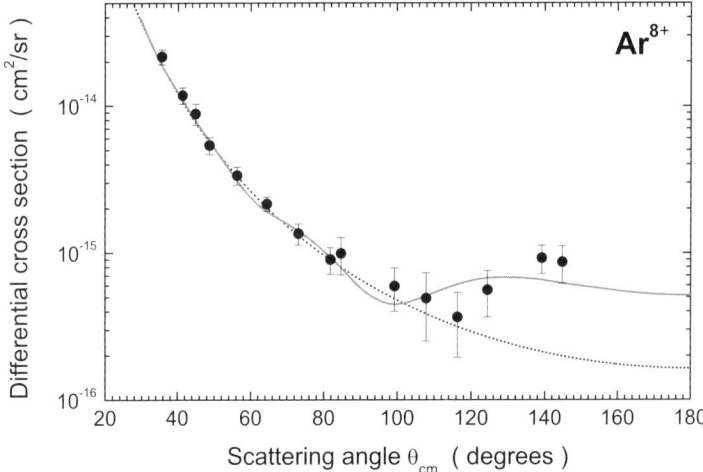

FIG. 41. Differential cross section for elastic scattering of 22.46 eV electrons from Ar^{8+} ions (Bélenger et al., 1996). The experimental data (solid circles) were normalized to relative cross sections measured for elastic electron scattering from bare helium ions in a separate experiment under identical conditions. Results for scattering of electrons from He^{2+} ions were normalized in turn to the Rutherford scattering cross section. The dotted curve is the Rutherford cross section for scattering from Ar^{8+}. The solid curve is a Hartree–Fock calculation.

With the view of the early work of Ramsauer and Kollath (1932) structure in the angular dependence of low-energy–large-angle elastic electron scattering can be expected also for ions in low charge states. This can be illustrated by results that were obtained with 16 eV electrons scattered from Ar^{1+} ions. Figure 42 shows differential cross sections for elastic scattering of 16 eV electrons from Ar^+ ions determined by Brotton et al. (2002). Normalized experimental data are displayed together with the Rutherford cross section and a so-called coupled Hartree–Fock calculation provided by the same authors. At about 75° the cross section shows a deep minimum reminiscent of the Ramsauer minima observed in the scattering of electrons from atoms and molecules. The range, where pure Coulomb scattering can represent this cross section is apparently below angles of 20°.

Beside elastic processes, inelastic electron scattering from ions has also been studied. Measurements of excitation cross sections by observing the scattered electron at a defined angle have been pioneered by Chutjian and Newell (1982), Chutjian et al. (1983) and Williams et al. (1986). Later Huber et al. (1991) and Ristori et al. (1991) started to carry out similar measurements. An interesting finding was that excitation of the target ion occurred at quite large electron scattering angles. This was particularly emphasized by experimental results of Guo et al. (1993) who found significant backscattering contributions to the total $3s - 3p$ excitation cross section of Ar^{7+} near the excitation threshold. The measurement

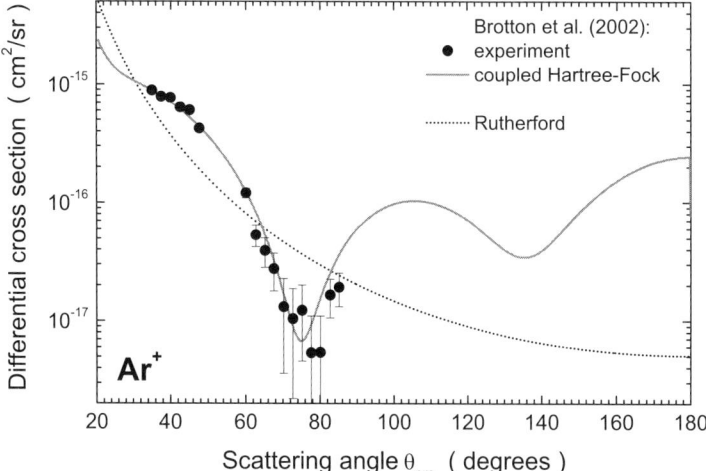

FIG. 42. Differential cross section for elastic scattering of 16 eV electrons from Ar$^+$ ions (Brotton et al., 2002). The solid points are measurements with statistical error bars. The relative experimental cross section data were normalized to coupled-Hartree–Fock calculations (solid curve) at an angle of 35°. The dotted curve is the Rutherford cross section.

was not really differential in angle but made use of the tunable kinematics in their merged-beams electron-energy-loss apparatus.

Few experimental results are available on superelastic scattering of electrons from excited ions. In collisions of electrons with metastable Ar$^{3+}(3s^23p^3\,^2D^0)$ ions, Williams et al. (1997) found evidence for superelastic scattering at an observation angle of 30°. Similar observations were later reported for scattering of quasi-free electrons from metastable F$^{7+}(1s2s\,^3S)$ ions (Závodszky et al., 2001).

4.2. ELECTRON-IMPACT EXCITATION OF IONS

While conventional electron scattering experiments with neutral atomic and molecular gases provide differential cross sections, applications in plasma and astrophysics usually require total excitation cross sections integrated over all angles. The merged-beams technique (see Fig. 35) has produced a number of interesting and important results on the excitation of optically allowed as well as optically forbidden transitions in singly and multiply charged ions (see recent work of Lozano et al., 2001; Smith et al., 2001, 2003, 2005; Popović et al., 2002; Niimura et al., 2002, and references therein). Data aspects have been reviewed by Tayal et al. (1995).

The simplest and most fundamental ionic target for electron-impact excitation is the He$^+$ ion. Simple electronic structure does not necessarily imply an easy ex-

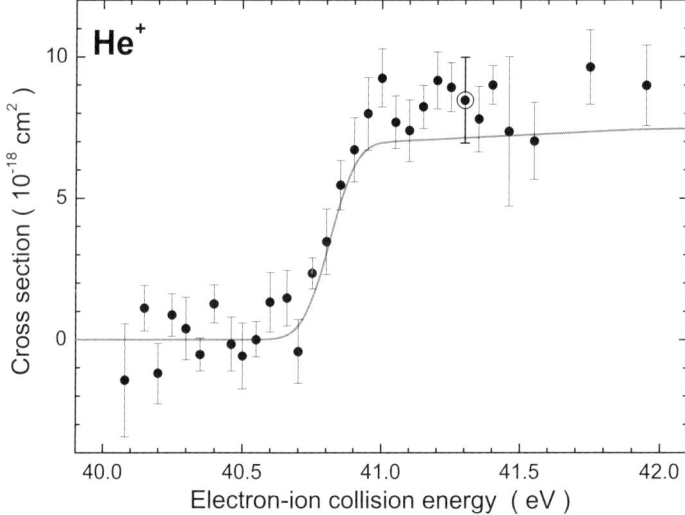

FIG. 43. Experimental cross sections for electron-impact excitation of $He^+(1s\,^2S)$ to the $2s\,^2S$ and $2p\,^2P$ states. The solid circles are experimental results (Smith et al., 2001) shown with relative error bars and an absolute total error bar on the average step size represented by the open circle at 41.3 eV. The solid curve is a convergent close coupling (CCC) calculation (Bray et al., 1993) convoluted with a 0.18 eV Gaussian characteristic of the electron energy distribution in the experiment.

periment. Cross sections for the excitation of $He^+(1s)$ to $He^+(2s)$ or $He^+(2p)$ are relatively small. With decreasing signal rates, the challenge to recover the signal from background increases quadratically. It was not until a few years ago, therefore, that the first measurement of absolute $1\,^2S \rightarrow 2\,^2S, 2\,^2P$ excitation cross sections could be reported (Smith et al., 2001). The electron-energy-loss merged-beams technique was employed (see scheme in Fig. 35). The experimental results are displayed in Fig. 43. The cross section shows the expected threshold-step behavior of direct excitation (see Fig. 14). Compared to state-of-the-art calculations, the experimental data appear high, but still agree with theory within the experimental uncertainty.

Figure 36 shows results for the excitation of C^{3+} ions obtained by using the merged-beam energy-loss technique in comparison with conventional measurements in which light emission was observed. For such collision systems, the Gaunt-factor formula [Eq. (18)] can provide a useful estimate of the cross section. In many cases, however, resonant contributions dominate over the direct process. An example is electron-impact excitation of the $3s^2\,^1S \rightarrow 3s3p\,^3P$ intercombination transition in magnesium-like Ar^{6+} ions measured by Chung et al. (1997) (see Fig. 44). Here, the cross section essentially consists of resonant contributions compared to which the direct excitation channel is almost negligible. Such reso-

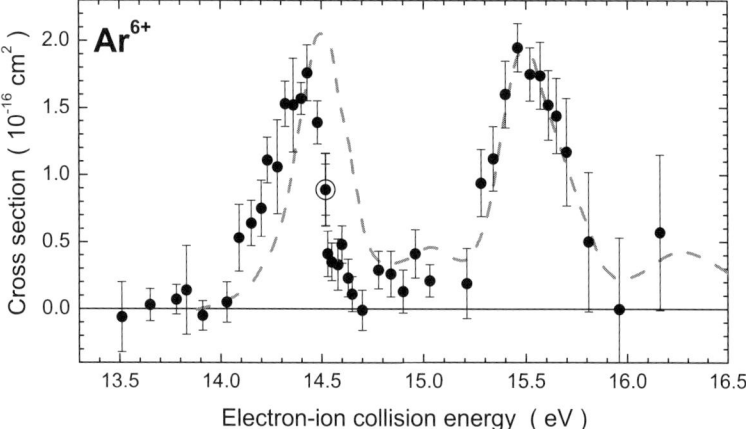

FIG. 44. Experimental cross sections for electron-impact excitation of the $3s^2\,^1S \rightarrow 3s3p\,^3P$ intercombination transition in Ar^{6+} ions. Points are experimental (Chung et al., 1997) and the curve is a theoretical result (Griffin et al., 1993) obtained from a calculation within the independent-processes and isolated-resonances approximation using distorted waves (from Fig. 5 of Chung et al., 1997). The calculation has been convoluted with a 0.24 eV Gaussian characteristic of the electron energy distribution in this experiment.

nances have been found in many of the excitation functions measured. It is worth noting that the group of A. Chutjian has succeeded in substantially extending the range of electron energies in the excitation cross section measurements above the excitation threshold. This is important for the determination of plasma rate coefficients from experimental cross section data. An example is shown in Fig. 45.

Signatures of electron-impact excitation are the emission of photons of characteristic energies and angular distributions or scattered electrons with a specific energy loss. From an excited state, also an electron with a characteristic energy can be ejected, provided the excited level is above the ionization threshold of the ion to be excited. Then, additional to a scattered electron and an emitted electron, both with characteristic energies, an ionized ion is produced. Thus, the initial excitation produces ionization as a net result of excitation plus subsequent relaxation. The process has been associated above with the excitation–autoionization contribution to the ionization of an ion. By subtracting the contribution of direct ionization from the total cross section for electron-impact single ionization, one can thus obtain an excitation cross section.

In this case, the excitation process produces one or several (often unresolved or not distinguished) multiply excited states that can autoionize. Emission of an electron from the excited state requires either excitation of an inner-shell electron or the excitation of two (outer-shell) electrons. For light ions, the fluorescence yield of doubly excited states is almost negligible as compared to the Auger yield.

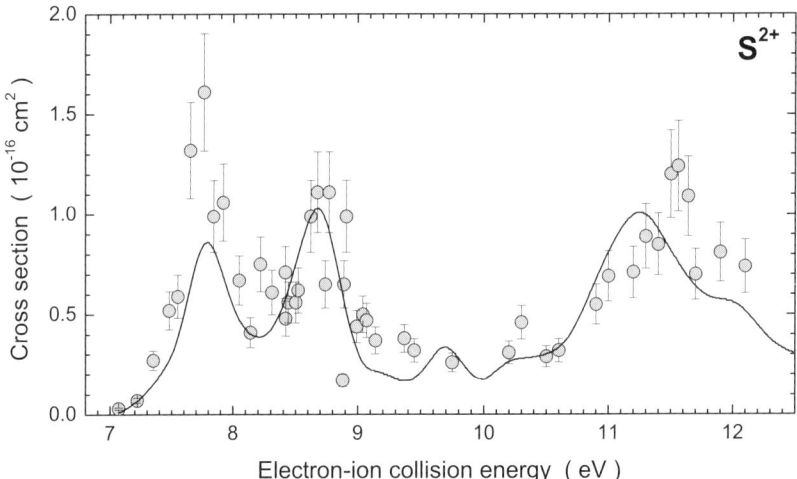

FIG. 45. Experimental cross sections for electron-impact excitation of the spin-forbidden $3s^23p^2\,^3P \to 3s3p^3\,^5S^0$ transition in S^{2+} ions. Circles are experimental and the solid curve is a close coupling calculation (both from S.J. Smith et al., 2000) convoluted with a 0.100 eV Gaussian (in the laboratory frame) characteristic of the electron energy distribution in the experiments.

Detecting the ionized ions produced in the two-step process thus provides direct access to the excitation cross section.

Figure 46 shows cross sections for electron-impact K-shell excitation of lithium-like $C^{3+}(1s^22s\,^2S)$ ions initially in the ground state. As expected for an excitation cross section, there are steps at the $1s \to 2l$ thresholds and resonance contributions are obviously also present. The dominant peak structures arise from KMM and KMN resonances.

Cross sections for electron-impact double excitation of $Li^+(1s^2)$ ions are shown in Fig. 47. As in the case of C^{3+} the experimental signature for the excitation process was taken from structure observed in the net single ionization channel, i.e., in the production of C^{4+} and Li^{2+} ions, respectively.

4.3. Electron–Ion Recombination

4.3.1. Bare Ions, Radiative Recombination, and Recombination Enhancement

For completely stripped ions the only efficient pathway for recombination with free electrons at low densities is radiative recombination. This is the normal situation with the electron beams available in colliding beams experiments, where electron densities are too low and electron beam temperatures too high to support three-body processes. Theory is expected to provide very accurate predictions for radiative recombination into hydrogen-like states (Stobbe, 1930;

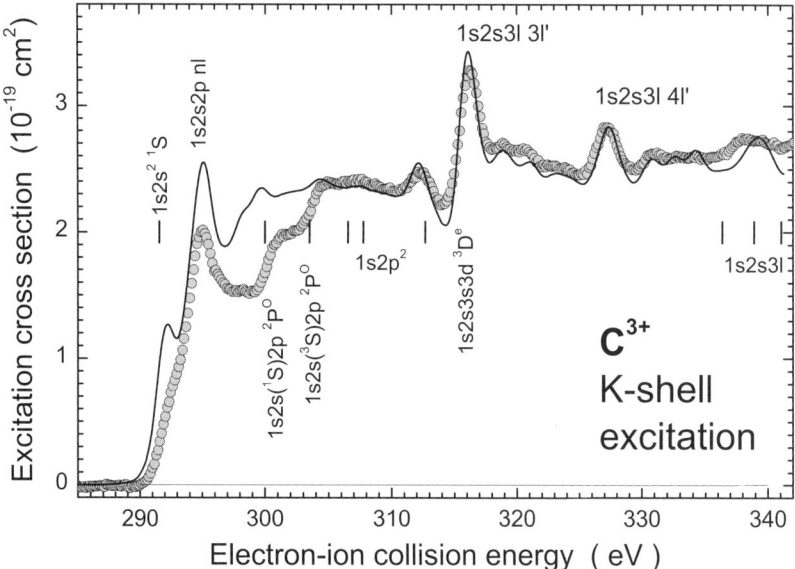

FIG. 46. Cross sections for electron-impact K-shell excitation of $C^{3+}(1s^22s)$ ions. Circles are experimental data with very low statistical uncertainty, the solid line is a 26-state R-matrix with pseudostates calculation convoluted with a 1.8-eV FWHM Gaussian to simulate the experimental energy resolution (Teng et al., 2000; artificial excitation strength calculated below the $1s2s^2$ threshold was cut off; the artifacts in such calculations have been discussed by Müller et al., 2000). Several excitation steps from the K shell to the L shell and resonance groups are indicated (characteristic energies given by vertical bars are from Mannervik et al., 1997). Note the interference of the strong $1s2s3s3d\,^3D^e$ resonance with direct K-shell excitation.

Ichihara and Eichler, 2000) and the early experiment by Andersen et al. (1990a) on bare C^{6+} agreed with the expectations. With increasing precision of merged beams experiments it was recognized that experimental rates at very low relative energies were considerably enhanced over any prediction (Müller et al., 1991). The phenomenon of enhanced recombination rates was found in all storage ring recombination experiments at relative energies in the electron–ion collision system below about 10 meV. It depends on magnetic fields in the electron target (see Fig. 21), on electron beam temperatures and on the ion charge state. Theoretical efforts have been made to understand the recombination enhancement which now appears to be due to transient-field-induced recombination due to the merging of electron and ion beams within a magnetic guiding field combined with radiative stabilization inside the central target region (Hörndl et al., 2006, and references therein). Apart from the special experimental effect of recombination rate enhancement in merged beams at very low energies, the predictions of theory for radiative recombination are generally confirmed. The situation is illustrated

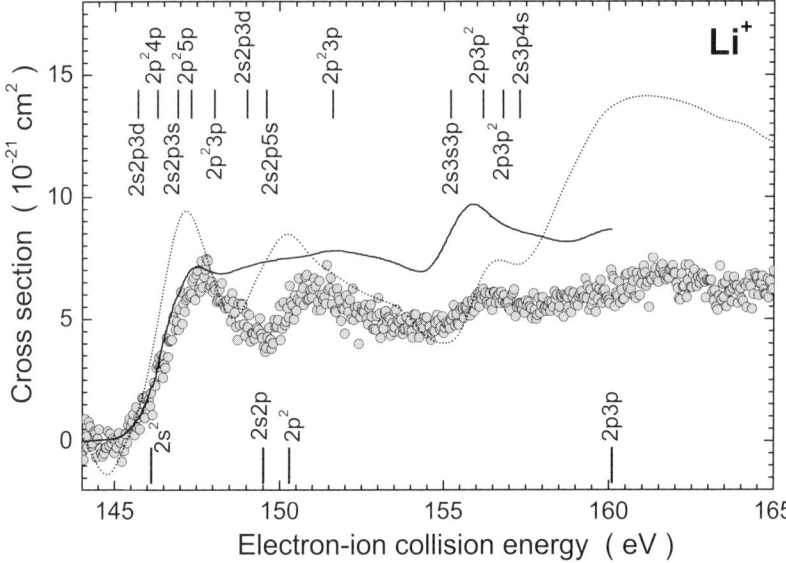

FIG. 47. Cross sections for electron-impact double excitation of $Li^+(1s^2)$ ions. Circles are experimental data of Müller et al. (1989), the dotted line is a 47-state R-matrix with pseudostates calculation convoluted with a 1.2-eV FWHM Gaussian to simulate the experimental energy resolution (Teng, 2000). The solid line is a *37-state close-coupling R-matrix calculation* by Griffin et al. (1992). Several excitation steps from the K shell to the L shell and a number of resonance energies (characterized by the dominant configuration in the excited states with level energies from Azuma et al., 1997) are indicated.

by Fig. 48 for radiative recombination of bare Cl^{17+}. Field ionization of Rydberg states in the experiment is estimated to result in a cut-off at $n_{max} = 97$. The contribution of $n = 1$ is given by the short-dashed line. Calculations with $n_{max} = 5$ and $n_{max} = 20$ show the varying contributions of different ranges of Rydberg states. At energies below 10 meV the measured rate coefficient shows the phenomenon of recombination enhancement. Similar experiments were carried out employing the storage ring merged beams technique for several species of bare ions such as D^+ (Gao et al., 1995), C^{6+} (Gwinner et al., 2000), Ne^{10+} (Gao et al., 1995), Cl^{17+} (Hoffknecht et al., 2001) and U^{92+} (Shi et al., 2001). Apart from the region of the lowest relative energies, theory and experiment are in good accord.

In principle, completely stripped ions in a dilute electron gas can also recombine in a two-step radiationless process starting with the excitation of the nucleus by the incoming free electron which thus looses so much energy that it finds itself trapped in the Coulomb field of the excited nucleus. In the subsequent relaxation the nucleus and the electron undergo radiative transitions that stabilize the recombination. This process, similar in nature to dielectronic recombination, has been

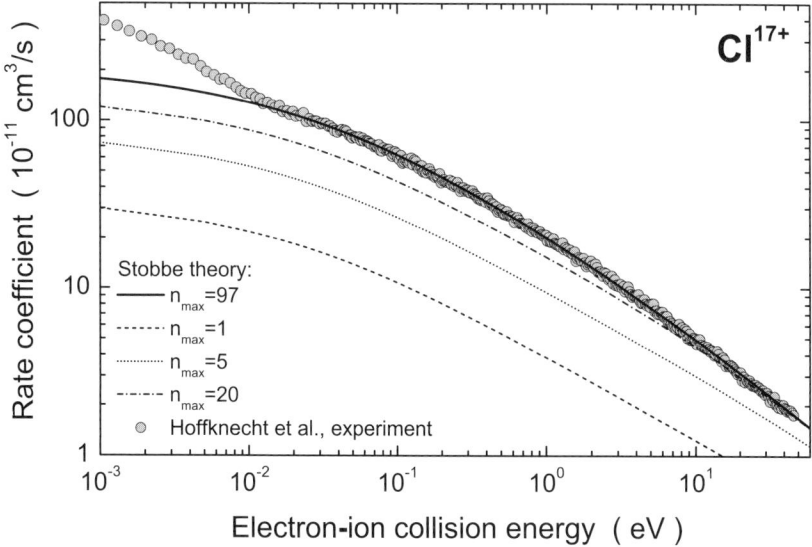

FIG. 48. Experimental and theoretical rate coefficients for radiative recombination of Cl^{17+} ions measured in an electron–ion merged-beam storage ring experiment (Hoffknecht et al., 2001). The data are compared with Stobbe-type calculations (Schippers et al., 2001; see also Eq. (20) and the associated discussion in Section 2.3.2) accounting for an experimental field-ionization cut-off at $n_{max} = 97$ (solid line). In order to illustrate the contributions of capture into various shells of the hydrogen-like product ion, results are also shown for radiative capture into $n = 1$ (short-dashed line), for capture into shells with principal quantum numbers n with $1 \leqslant n \leqslant 5$ (dotted line) and $1 \leqslant n \leqslant 20$ (dash-dotted line). The electron beam temperatures assumed for the calculation of the associated recombination rate [see Eqs. (33), (37), and (38)] are $k_B T_\parallel = 0.08$ meV and $k_B T_\perp = 38$ meV.

theoretically studied by Pálffy et al. (2006, 2007a, 2007b). While the cross sections are expected to be very small in this nuclear-excitation with electron capture, "normal" dielectronic recombination usually becomes the dominant recombination process where ever it can happen.

4.3.2. Hydrogen-Like Ions, High-Energy Resonances

The simplest atomic system supporting dielectronic recombination is a hydrogen-like ion. An example is shown in Fig. 49 presenting measured (Kilgus et al., 1990) and calculated (Pindzola et al., 1990) dielectronic recombination cross sections of hydrogen-like O^{7+} ions. Figure 50 is an enlarged plot of the KLL resonances from Fig. 49. Already in that first storage-ring dielectronic-recombination experiment it was possible to separate individual $2l2l'$ (KLL) resonances for a relatively light ion. Resonance strengths could be extracted for each of the six contributions marked in Fig. 50. According to Eq. (44) the energy resolution becomes

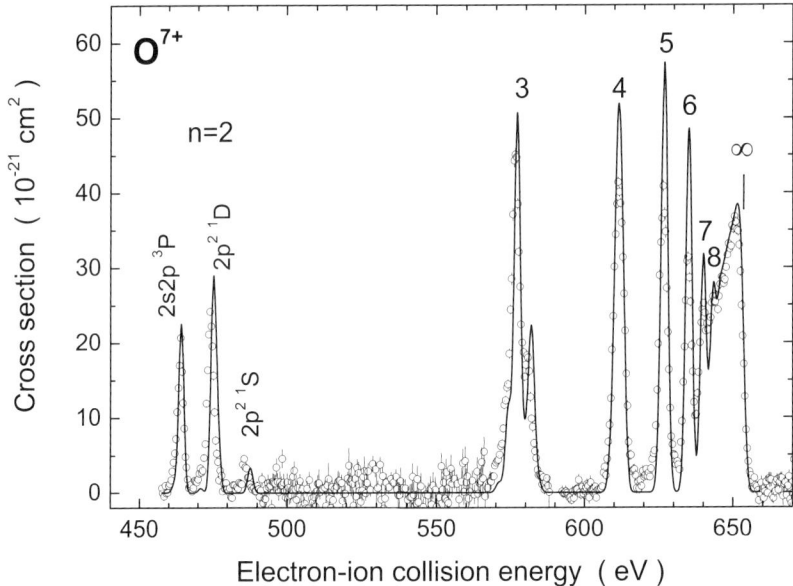

FIG. 49. Measured (open symbols; Kilgus et al., 1990) and calculated (full line; Pindzola et al., 1990) dielectronic recombination cross sections of hydrogen-like O^{7+} ions. Individual $2lnl'$ resonances can be distinguished up to $n = 8$. The theoretical cross section has been convoluted with a 2-eV FWHM Gaussian to simulate the experimental energy spread.

better with increasing relative energy E_r since $\Delta E_r/E_r \propto 1/\sqrt{E_r}$ at sufficiently high E_r. In a crude approximation the KLL resonances in dielectronic recombination of hydrogen-like ions are found near electron energies $E_r = Z^2 \cdot 7.5$ eV. Thus, one can expect to resolve features better in the more highly charged hydrogen-like ions provided the electron beam temperature stays the same. Measurements have been carried out using the heavy-ion storage ring merged-beam technique on a number of hydrogen-like ions from He^+ (DeWitt et al., 1994) to U^{91+} (Brandau et al., 2006). Energies for the investigated resonances range from about 30 eV and up in the case of He^+ to about 90 keV and below in the case of U^{91+}. Reaching such high relative energies in an electron–ion merged-beams experiment at a heavy ion storage ring was made possible by stochastically cooling the ions at a specific energy of about 400 MeV/u (corresponding to an energy near 217 keV for electron cooling) and by tuning the electron laboratory energy in the interaction region between about 24 and 42 keV.

There is an intricate additional channel for dielectronic recombination of hydrogen-like ions with electrons at low relative energies, the excitation of a hyperfine transition in the ground state of the ion associated with capture of the free electron. Measuring this effect was attempted by Brandau et al. (1999) with

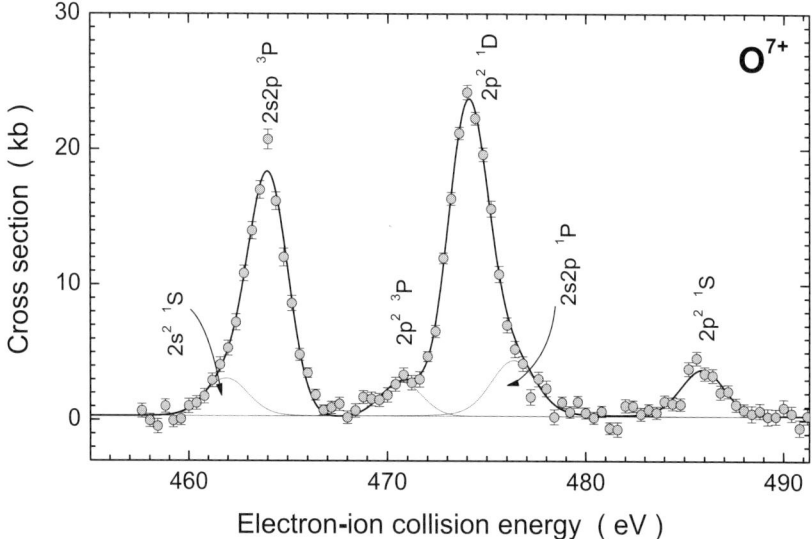

FIG. 50. The KLL resonances (2l2l' intermediate states) from Fig. 49. The circles are the experimental data of Kilgus et al. (1990). The solid line is a fit of the data with 6 Voigt profiles. The dotted lines show individual contributions of intermediate states that would otherwise not easily be seen.

hydrogen-like Bi^{82+} ions where the hyperfine splitting is $\Delta E \approx 5$ eV. However, the natural width of the high Rydberg states associated with that capture (greater than $n \approx 135$) is too large to facilitate the resolution of individual resonances (Pindzola et al., 1992a). Moreover, the resonant contribution to the observed total recombination is negligibly small compared to that of the competing process of radiative recombination.

4.3.3. Lithium-Like Ions, Plasma Applications and Spectroscopy

There are numerous results on dielectronic recombination of ions with more than one electron. Helium-like ions were investigated both starting from the ground state or metastable states (see Saghiri et al., 1999, and references therein). A detailed spectroscopic study on lithium-like carbon was performed by Mannervik et al. (1997) starting from helium-like C^{4+}. Most of the experimental efforts on understanding the process of dielectronic recombination were focussed on lithium-like ions. Available data cover more than twenty elements of the Li sequence in the range from Be^+ to U^{89+}. Under the aspect of data for plasma applications, theoretical results obtained with the AUTOSTRUCTURE code for a large number of elements and charge states are available online (Badnell, online).

For low atomic numbers Z, the dominant contributions to dielectronic recombination are from high Rydberg states. The reason is in the balance between

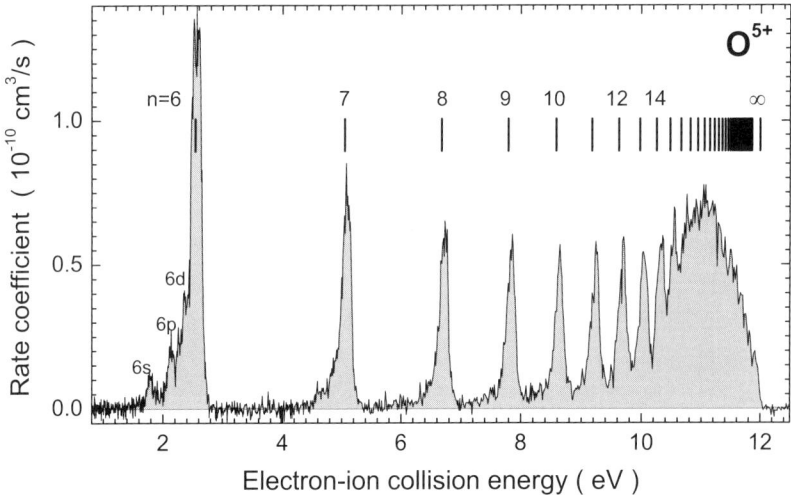

FIG. 51. Storage ring rate coefficients for dielectronic recombination of 9.4 MeV/u O^{5+} ions (Böhm et al., 2002). Principal quantum numbers n of $O^{5+}(2pnl)$ resonances are indicated. Note the splitting of the $n = 6$ resonance group with partially resolved contributions associated with configurations $(2p6s)$, $(2p6p)$, and $(2p6d)$.

radiative and Auger rates entering Eq. (22). For low Z and low n the autoionization rates are much higher than the radiative rate, and thus numerous shells with principal quantum number n can contribute to the resonance strength [Eq. (22)] with a maximum multiplicity increasing with n^2. The autoionization rate $A_a(n)$ decreases with n^{-3}. When $A_a(n)$ drops below the radiative rate A_r contributions from higher n rapidly decrease. For high Z, the radiative rates are enhanced and therefore only the lowest n-values contribute significantly. Model calculations illustrating these dependences were carried out by Müller et al. (1987b).

One of the astrophysically most important ions is O^{5+}. Storage ring experiments on dielectronic recombination have been carried out by (Böhm et al., 2002). The $2s \rightarrow 2p$ excitation energy $\Delta_{2s \rightarrow 2p}$ is 11.949 eV for the $^2P_{1/2}$ state and 12.015 eV for the $^2P_{3/2}$ state (Ralchenko et al., 2006). Hence, dielectronic recombination resonances can be expected at non-negative energies

$$E_n = \Delta_{2s \rightarrow 2p} - 13.6 \, \text{eV} \frac{Z_{\text{eff}}^2}{n^2}. \tag{66}$$

With $Z_{\text{eff}} = 5$ in the case of O^{5+}, the lowest possible principal quantum number of resonances is at $n = 6$ with an energy of about 2.5 eV. Figure 51 shows the experimental results of Böhm et al. (2002) obtained with 9.4 MeV/u O^{5+} ions. Near the $^2P_{1/2}nl$ and $^2P_{3/2}nl$ resonance-series limits at about 12 eV the contributions from Rydberg states with $n > 15$ lump together and pile up to a peak containing a considerable fraction of the total recombination strength. This peak is truncated by field ionization mainly in the motional electric field experienced by the 9.4 MeV/u radiatively pre-stabilized $O^{4+}(1snl)$ product ions passing the next ring dipole magnet (with magnetic field B_d, see layout in Fig. 40) behind the collision region. The effective electric field strength $E = v_i B_d \approx 490$ kV/cm cuts off Rydberg states with principal quantum numbers $n > n_{\text{cut}}$ where a simple model calculation on the basis of Eq. (52) gives $n_{\text{cut}} = 19$.

A more sophisticated treatment of field ionization accounts for the survival of Rydberg ions along the flight path between their production and detection. Survival probabilities in varying electromagnetic fields were calculated previously by Müller et al. (1987a) to interpret recombination experiments with Mg^+ ions. In the present case, field ionization of the recombined O^{4+} ions in the specific experimental environment of the measurements by Böhm et al. (2002) was modelled according to the procedure described by Schippers et al. (2001). The resulting survival probabilities of Rydberg states were applied to calculations of the recombination rates based on the AUTOSTRUCTURE code of Badnell (1997) such as to compare theory with the experiments. Figure 52 shows storage ring rate coefficients for dielectronic recombination of O^{5+} ions (Böhm et al., 2002) measured with different specific ion energies. When the ion energy is reduced the ion velocity v_i and the magnetic fields required to bend the ion beam on a closed orbit become smaller. The resulting smaller motional $\vec{v}_i \times \vec{B}$ electric fields and the extended flight times of the ions influence the survival of Rydberg states. The measured recombination rate coefficients piling up near the Rydberg series limit increase with decreasing ion energy and display very good agreement with the modelled AUTOSTRUCTURE calculations. When going from 9.4 to 5.0 MeV/u Rydberg states with higher principal quantum number n can survive. In the crude picture of a sharp cutoff, Rydberg states with $n = 20, 21$, and 22 can now contribute in addition to the previous $n = 6, 7, \ldots, 19$. By reducing the ion energy further to 3.3 MeV/u an additional band of Rydberg states $n = 23, 24$, and 25 enhances the recombination rate coefficient observed at the series limit.

The comparison of the detailed calculations with the experiment indicates, that the AUTOSTRUCTURE calculation is in good agreement with the experiment for Rydberg states with $n > 9$. Deviations are observed for the lower n values. This indicates the validity of the n-scaling of Rydberg resonance strengths incorporated in AUTOSTRUCTURE. Therefore, theory can be applied to infer the contributions of higher Rydberg states to the total recombination rate coefficient, which has to be taken into account in a field free plasma environment. Thus, by the measurements for low to intermediate n complemented by the theoretical results for high n (up to $n = 1000$ as a safe limit) the plasma rate coefficient can be determined by convoluting the energy dependent resonance strengths with an isotropic Maxwellian distribution according to Eq. (38). The result of that con-

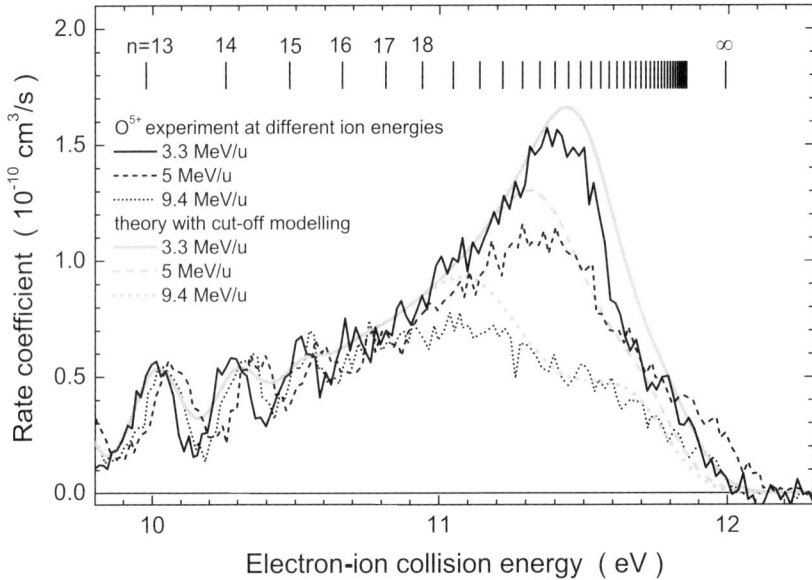

FIG. 52. Storage ring rate coefficients for dielectronic recombination of O^{5+} ions (Böhm et al., 2002) measured with different specific ion energies: 9.4 MeV/u (dotted black line), 5.0 MeV/u (dashed black line), and 3.3 MeV/u (solid black line). The gray lines (dotted, dashed and solid associated with 9.4, 5.0, and 3.3 MeV/u, respectively) show theoretical results obtained with the computer package AUTOSTRUCTURE of Badnell (1997). The theoretical curves include field ionization effects on the observed rate coefficient modelled according to the procedure described by Schippers et al. (2001).

volution for dielectronic recombination of O^{5+} ions is shown in Fig. 53 together with theoretical predictions.

The situation with the plasma rate coefficient of O^{5+} ions is quite satisfying. Theory and experiment are in accord roughly within the uncertainty of the derived rate coefficient. This is mainly due to the relatively simple structure of the lithium-like parent ion, and also, because the first resonance manifold of $(2p6l)$ states is already 2.5 eV above the threshold. Uncertainties in the calculation of the exact energy of that resonance group do not have a very significant effect on the temperature dependence of the plasma rate coefficient (as long as uncertainties in resonance energies are much less than 1 eV). This is completely different in cases when there are resonances close to zero energy. An example is dielectronic recombination of C^{3+} (see Fig. 39). Strong resonances are found between 0.1 and 1 eV. Calculating resonance energies with uncertainties of less than 0.1 eV is a challenge. Hence, a calculated resonance might be expected to be below the threshold but can, in fact, be at very small positive energy. The resonance energy

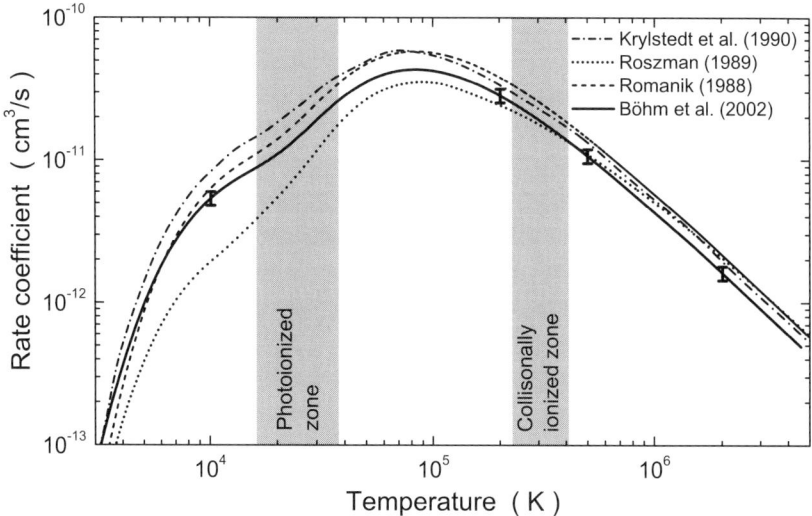

FIG. 53. Plasma rate coefficients for $2s \to 2p$ dielectronic recombination of O^{5+} ions. The solid line with error bars is the experimentally derived rate coefficient of Böhm et al. (2002). Out of numerous other published data sets, all based on theory, or on fits to theory and extrapolations of theory, only few are shown which cover the whole range of temperatures of the present plot: dotted line (Roszman, 1989), dashed line (Romanik, 1988), and dash-dotted line (Krylstedt et al., 1990). The temperature regimes where O^{5+} ions are abundant in photoionized and collisionally ionized plasmas are indicated by gray-shaded areas.

has an immediate effect on the resonance strength and of course small shifts in energy on an absolute scale have large effects on the plasma rate coefficient at low temperatures. This is an ubiquitous effect, especially with ions that have complex structure. Core electrons with many possible low excitation energies are likely to support resonances at very low energies. Examples are discussed below.

With O^{5+} ions the small fine structure splitting of the ^2P core-excited states is not resolved in the spectrum of dielectronic recombination, although the width of the $n = 6$ resonance group indicates that there are two separate manifolds of $2p\,^2Pnl$ states. With increasing atomic number the fine structure splitting rapidly increases and gives rise to two distinctly separate Rydberg series of resonances. This is illustrated by Fig. 54 showing storage ring recombination rate coefficients for $2s \to 2p$ dielectronic recombination of Ni^{25+} ions (Schippers et al., 2000). The ^2P fine structure splitting of the core is approximately 22 eV in this case. The largest energy difference between the $1s^2 2p\,^2P$ states of any Li-like ions was observed with unsurpassed accuracy in spectroscopy experiments on U^{89+} ions by Beiersdorfer et al. (2005, 1993). They found excitation energies from the $(1s^2 2s\,^2S)$ ground state of 280.645 ± 0.015 eV for

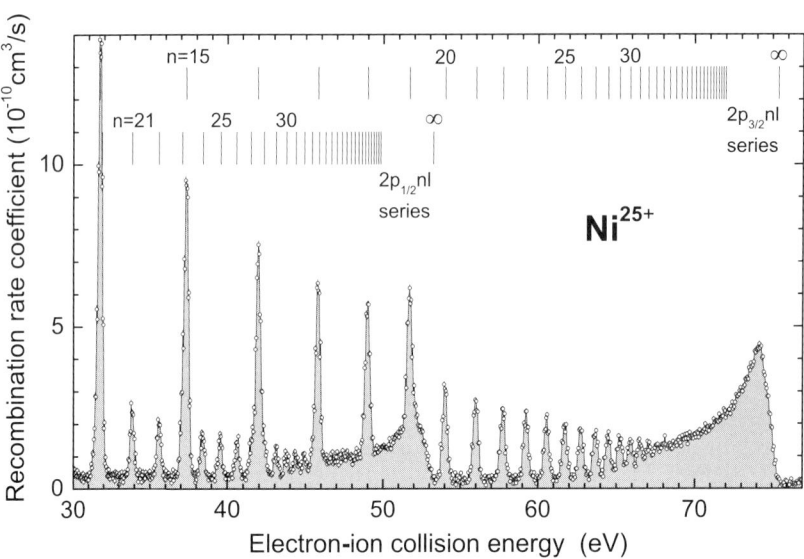

FIG. 54. Storage ring rate coefficients for $2s \rightarrow 2p$ dielectronic recombination of Ni^{25+} ions (Schippers et al., 2000).

the $^2P_{1/2}$ and 4459.37 ± 0.35 eV for the $^2P_{3/2}$ excited states. Such information can also be inferred from careful observation of dielectronic recombination resonances (Brandau et al., 2003b). However, the accuracy of the energy measurements for very heavy highly charged ions has been limited so far to about $\Delta E = 0.1$ eV. There are developments, though, that will vastly improve the precision of electron–ion recombination experiments with very heavy ions. At the future accelerator project FAIR on the site of the present GSI laboratory in Darmstadt, Germany, a high-quality electron cooler will be combined in a New Experimental Storage Ring (NESR) with a cold electron target similar to the one that is operational at the Max-Planck-Institute for Nuclear Physics in Heidelberg. The potential of this combination will be discussed below in connection with applications of electron–ion resonance spectroscopy.

The large fine structure splittings in very highly charged ions can be illustrated by the dielectronic recombination spectrum shown in Fig. 55 where storage ring rate coefficients for $2s \rightarrow 2p$ dielectronic recombination of Au^{76+} ions (Brandau et al., 2003a) are presented. Compared to previous results of Spies et al. (1992), the quality and the resolution of the measurements could be substantially improved. Nevertheless, the electron beam temperatures were still comparable to those of the first-generation storage ring experiments such as the one on O^{7+} ions (Figs. 49 and 50), leaving ample space for further progress in measurements with

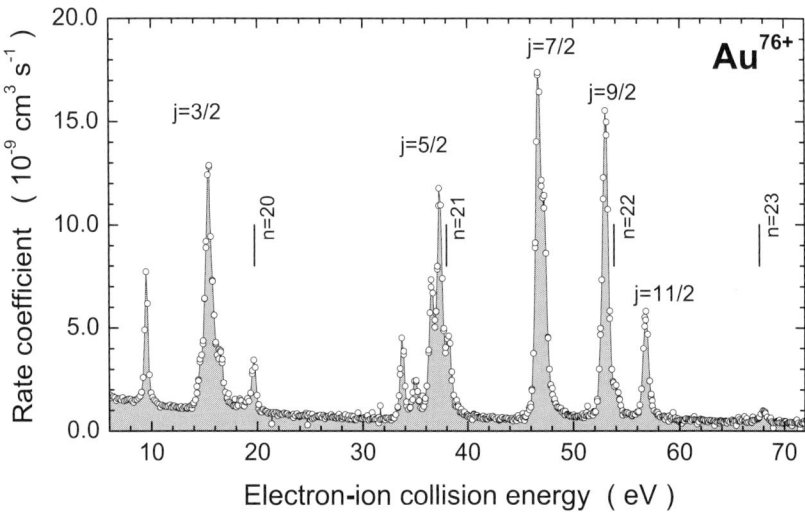

FIG. 55. Storage ring rate coefficients for $2s \to 2p$ dielectronic recombination of Au^{76+} ions (Brandau et al., 2003a). The peaks belong to two separate Rydberg sequences, one associated with Au$^{75+}(1s^2 2p\,^2P_{1/2}nl)$ resonances (the small peaks labelled with n at positions indicated by the vertical bars), the other associated with Au$^{75+}(1s^2 2p\,^2P_{3/2}6l_j)$ resonances (the dominant peaks, all with $n = 6$, labelled with the total angular momentum j of the Rydberg electron).

very heavy highly charged ions. The dominant features in the spectrum are due to Au$^{75+}(1s^2 2p\,^2P_{3/2}6l_j)$ resonances, i.e., the lowest principal quantum number in the Rydberg sequence is $n = 6$ as in the case of dielectronic recombination of O^{5+}. The individual peak groups are labelled with the total angular momentum j of the Rydberg electron, i.e., for $n = 6$ angular momentum quantum numbers $j = 1/2, 3/2, \ldots, 11/2$ are possible according to momentum coupling rules. However, the states with $j = 1/2$ are below the threshold, so their energy is negative and they can thus not be reached in collisions of free electrons with Au^{76+}. The remaining 5 states of the $n = 6$ Rydberg electron appear in the spectrum. The resonances belonging to given j quantum numbers, in particular those with $j = 3/2$ and $j = 5/2$ show substructure associated with the jj-couplings of the two active electrons. The resulting total angular momentum quantum numbers J of the Au^{75+} resonance state can take values $J = 0, 1, 2, 3$ in the $j = 3/2$ group and hence different terms have to be expected and are, indeed, observed. For example, the isolated peak at 9 eV is due to the state $[(1s^2 2p_{3/2})6p_{3/2}]_{J=2}$. The group of smaller resonance features labelled with $n = 20, 21, \ldots$ is associated with the Au$^{75+}(1s^2 2p\,^2P_{1/2}nl_j)$ Rydberg sequence. Close inspection of the experimental data shows the fine structure with different j values even for the $n = 20$ Rydberg electron. Vertical bars in Fig. 55 indicate the resonance energies

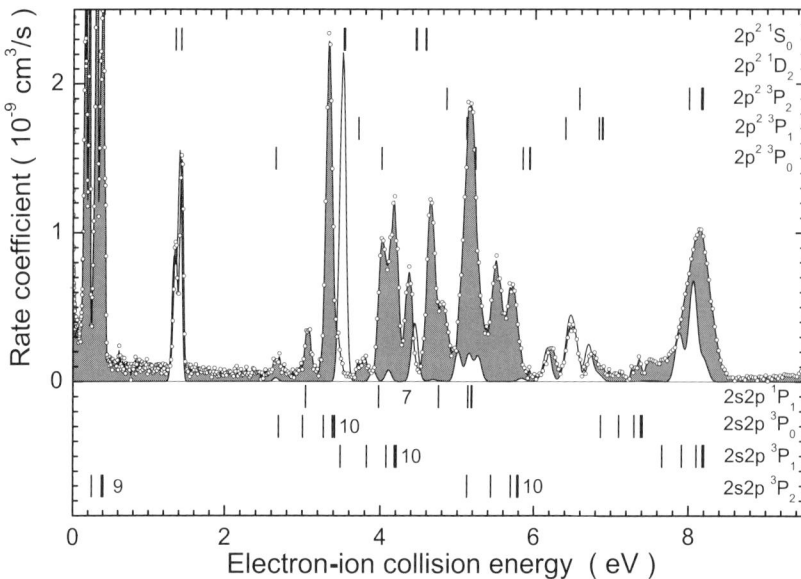

FIG. 56. Storage ring rate coefficients for recombination of Cl^{13+} ions (Schnell et al., 2003). The experimental data are the small open circles connected by a line with gray shading. AUTOSTRUCTURE calculations (see text) were analyzed for individual contributions of $2s^2 \to 2s2p$ and $2s^2 \to 2p^2$ core excitations, the latter being the trielectronic resonances. These are indicated by the smooth solid line with peak areas whited out. Resonance energies of different Rydberg sequences are indicated by the vertical bars. The bars below the data panel are associated with the possible dielectronic recombination core excitations to $2s2p\ ^1P_1$, $^3P_{0,1,2}$ states with the Rydberg electron in $n = 7$ or $n = 10, 11$ or $n = 9, 10$ depending on the specific core-excited state. The bars above the recombination peaks indicate the doubly-excited core states $2p^2\ ^1S_0$, $^1D_2$, $^3P_{0,1,2}$ which can be reached via trielectronic capture. All the bars are associated with the captured electron in Rydberg states with $n = 5$ for the $^1S_0$ core and $n = 6$ otherwise. The $^1D_2$ core state does not support resonances in the present energy range.

for the $^2P_{1/2}$ sequence simply calculated by using the Bohr formula [see Eq. (66)] with an effective atomic number $Z_{\text{eff}} = 76$ and the $2s \to 2p_{1/2}$ excitation energy from Brandau et al. (2003b).

4.3.4. Beryllium-Like Ions, Trielectronic Recombination

Adding one more electron to the Li-like parent ion provides a target that has two closed subshells in its ground state. But singly excited states, have up to three open subshells adding complexity to the recombination spectra. A number of beryllium-like ions have been studied in recombination experiments at storage rings. Recent publications address Mg^{8+} (Schippers et al., 2004), Ti^{18+} (Schip-

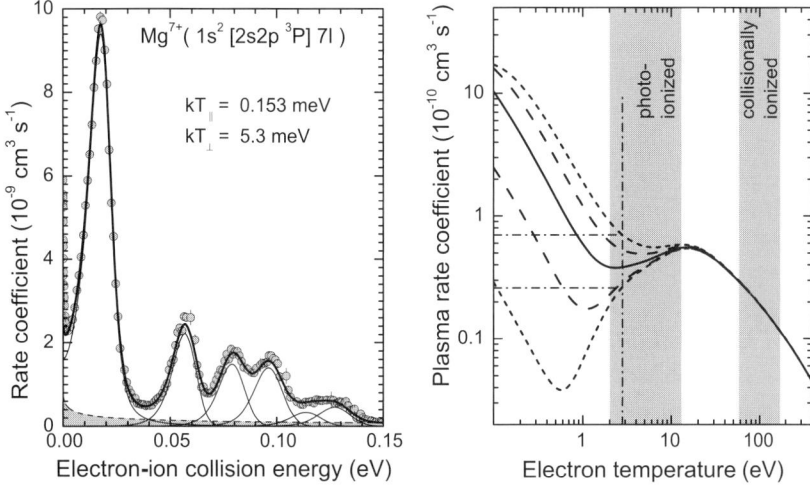

FIG. 57. Storage ring and plasma rate coefficients for recombination of Mg^{8+} ions (Schippers et al., 2004). The experimental data in the left panel are associated with $1s^2[2s2p\,^3P]7l$ dielectronic recombination resonances on top of a comparatively small contribution of radiative recombination (gray-shaded area under dashed curve). Resonance contributions shown by the thin solid lines were fitted to the data. The 6 resonances assumed for the fit add up to the heavy solid line. The fit also provided the temperatures of the electron beam as indicated in the figure. From the total measured spectrum which covers the energy range 0 to 207 eV, the plasma rate coefficient for dielectronic recombination was inferred as shown by the solid line in the right panel. The effect of small shifts of the low-energy resonances displayed in the left panel on the plasma rate coefficient is illustrated by the long-dashed and short-dashed curves corresponding to resonance shifts of $\Delta E_{\text{res}} = \pm 50$ meV and $\Delta E_{\text{res}} = \pm 100$ meV, respectively. The lower curves correspond to resonance shifts to lower energies. The vertical dash-dotted line marks the temperature 2.8 eV where the fractional abundance of Mg^{8+} peaks in a photoionized gas (Kallman and Bautista, 2001). The horizontal dash-dotted lines indicate the factor 2.7 uncertainty of the plasma rate coefficient at this temperature that would be introduced by a ± 100 meV uncertainty of the low energy resonance positions. The regions of photoionized and collisionally ionized plasmas are indicated by the shaded areas.

pers et al., 2007b), and Fe^{22+} (Savin et al., 2006). In most cases, very strong dielectronic recombination resonances are found at very low energies where theory has problems to reproduce experimental findings and, in particular, to predict the plasma rate coefficients in the photoionization regime, i.e., at relatively low plasma temperatures. One of the experiments is particularly worth mentioning. In measurements on recombination of beryllium-like Cl^{13+} Schnell et al. (2003) found the first clear evidence for substantial contributions of trielectronic recombination [see Eq. (5)]. Beside $2s^2 \to 2s2p$ core excitations, resonances due to $2s^2 \to 2p^2$ double excitations were also observed. Figure 56 shows part of the storage ring rate coefficients for recombination of Cl^{13+} ions obtained by Schnell et al. (2003). The white peaks are due to trielectronic recombination as sorted out from the output of the AUTOSTRUCTURE code (Badnell, 1997). The resonance energies to be expected for different states within $2l2l'\,nl''$ configurations are indicated. The isolated peak at about 1.4 eV can only be associated with trielectronic recombination to $2p^{2\,1}S_0 5l''$ states. It should also be noted that there are very large resonances at energies below 0.4 eV, providing the dominant contributions to the plasma rate coefficient at temperatures between 1000 and 10000 K.

Effects of low-energy dielectronic recombination resonances on plasma rate coefficients have been highlighted by Schippers et al. (2004). In the recombination of beryllium-like $Mg^{8+}(1s^2 2s^2)$ resonances associated with intermediate $1s^2[2s2p\,^3P]7l$ Rydberg configurations were found at energies extending down to below 20 meV. Figure 57 shows the strong contributions of these resonances in the range 0 to 150 meV. In a theoretical calculation of resonance contributions, the resonance energies can easily be uncertain by 100 meV and more. The effect of ± 50 and ± 100 meV shifts of the $1s^2[2s2p\,^3P]7l$ resonance group on the plasma rate coefficient is illustrated in the right panel of Fig. 57.

4.3.5. The Iron Isonuclear Sequence, Data for Astrophysics

Strong resonances near zero energy become more likely as the number of core electrons in the parent ion increases. Most of the astrophysically important iron ions support dielectronic recombination resonances at low electron–ion collision energies. As mentioned before, theory has difficulties to describe systems with several open shells. Therefore, experiments are needed particularly for the so called M-shell iron ions Fe^{8+} through Fe^{15+} with outer-shell configurations $3s^r 3p^s$, $r = 1, 2$; $s = 0, 1, 2, \ldots, 6$ and the L-shell iron ions Fe^{16+} through Fe^{23+} with outer-shell configurations $2s^r 2p^s$ $r = 1, 2$; $s = 0, 1, 2, \ldots, 6$. Experimental data are collected at the Test Storage Ring in Heidelberg, where the L-shell ions and some of the M-shell ions have already been investigated (see Schmidt et al., 2006; Savin et al., 2006, and references therein). As an example, Fig. 58 shows the low-energy region of an extended recombination spectrum taken by Schmidt et al. (2006). The resonances displayed are at least a factor of ten above the resonance features observed at higher energies. The peaks cannot easily be identified. A crude Bohr-formula estimate of Rydberg resonances attached to the $3s3p^{2\,2}P_{1/2,3/2}$ core excited states indicates that the present energy range includes resonances in the $n = 7$-manifold of the captured Rydberg electron. Also, resonances associated with ground-state fine-structure transitions fall in this energy range. Comparison with state-of-the-art calculations shows the difficulties of theory to reproduce details of the experimental data.

It is clear from this situation, that plasma rate coefficients have been very uncertain in the past. Figure 1 already provided strong hints that the dielectronic recombination rates for iron M-shell ions might have been badly underestimated

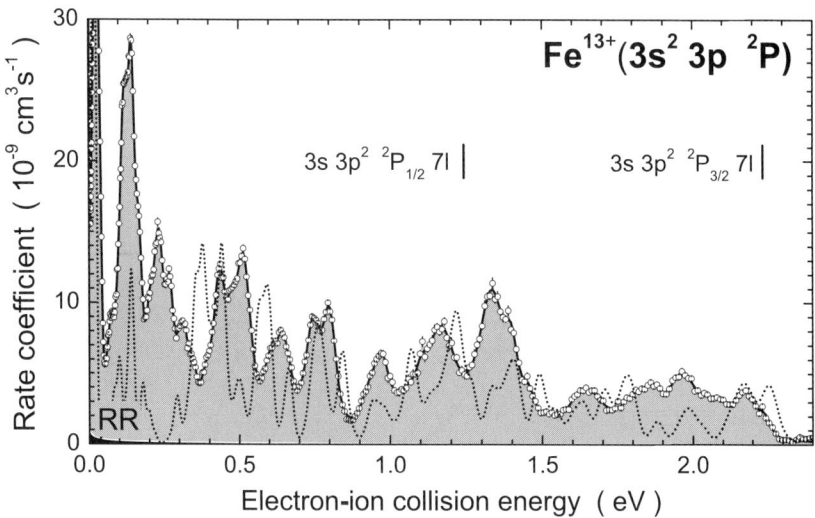

FIG. 58. Storage ring rate coefficients for dielectronic recombination of Fe^{13+} ions at low relative energies (Schmidt et al., 2006). The vertical bars indicate Rydberg energies simply calculated with the Bohr formula [according to Eq. (66)] for $3s3p^2\,{}^2P_{1/2,3/2}7l$ states without considering the splitting of levels. Radiative recombination is indicated by the black shaded area barely visible at very low energies. The figure includes one of the theoretical approaches (dotted line) recently published by Badnell (2006).

in the region of photoionized plasmas. The situation with Fe^{13+} is illustrated in Fig. 59. Radiative recombination is roughly two orders of magnitude below the dielectronic recombination rate. High Rydberg states beyond $n = 55$ do not play a significant role in radiative recombination. The extrapolated and artificially enhanced dielectronic recombination rates used by Netzer (2004) are indicated as well as the similarly obtained estimate of Kraemer et al. (2004). The experimentally inferred plasma rate coefficient of Schmidt et al. (2006) is still much greater than these estimates. It is discomforting to see that state-of-the-art theory differs by as much as 50% from the experimental data in the temperature range relevant to the modelling of photoionized plasmas.

Collisions of ions with a very complex electronic configuration can be treated theoretically by alternative approaches. Gribakin et al. (1999) and Gribakin and Sahoo (2003) used a quantum chaos approach to understand the gigantic rate coefficients observed by Hoffknecht et al. (1998) for the recombination of Au^{25+} ions with electrons, especially near zero electron–ion collision energy. Similarly large merged-beams rate coefficients in the recombination of U^{28+} were investigated by Mitnik et al. (1998) by using the AUTOSTRUCTURE and HULLAC

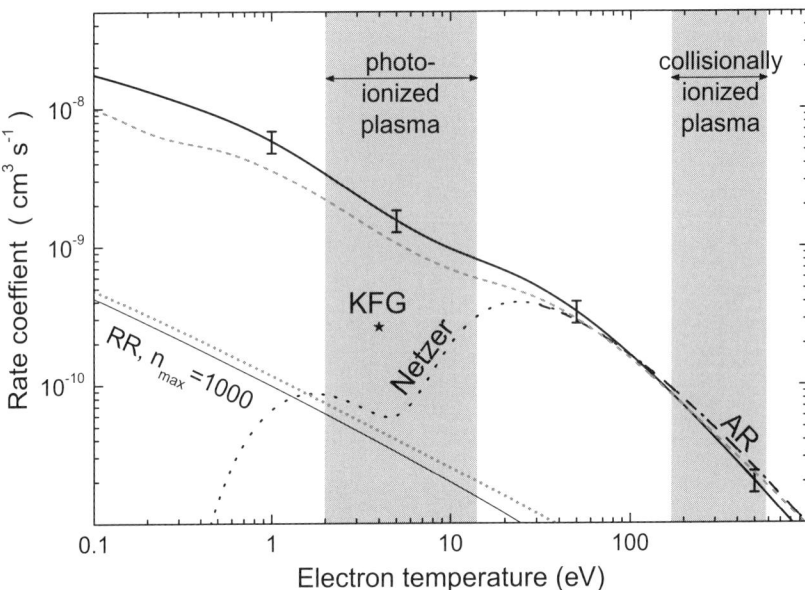

FIG. 59. Plasma rate coefficients for recombination of Fe^{13+} ions. The experimentally-derived rate coefficient (heavy full line; Schmidt et al., 2006) comprises $3l \to 3l'$ and $3l \to 4l'$ dielectronic recombination, radiative recombination, and the theoretical estimate for the unmeasured contributions of states with $n > 55$ for radiative recombination and $3l \to 3l'$ resonance contributions. The error bars denote the $\pm 18\%$ experimental uncertainty in the absolute rate coefficient. Also shown are the theoretical rate coefficient of Arnaud and Raymond (1992, heavy dashed line, labeled AR) its deliberate modification by Netzer (2004, dotted line), and the calculation of radiative recombination by Badnell (2006, short-dotted gray line). The star at 4 eV (labeled KFG) is the estimate of Kraemer et al. (2004). The curve labeled RR was calculated using Eq. (34) with $n_{max} = 1000$ (thin full line). The total recombination rate coefficient of Badnell (2006) is shown as a short-dashed gray line. The temperature ranges are highlighted where Fe^{13+} is expected to be abundant in photoionized and collisionally ionized plasmas.

(Oreg et al., 1991) computer codes. At most qualitative agreement between theory and experiment could be accomplished.

4.3.6. Effects of External Electric Fields

Soon after the establishment of dielectronic recombination as an important process (Burgess, 1964) that governs the charge state balance of ions in the solar corona, it was realized by Burgess and Summers (1969), and Jacobs et al. (1976), that cross sections for dielectronic recombination should be sensitive to external electric fields, particularly when the incident electron is captured into high Rydberg states. Such fields are present in plasma environments because of the thermal motion of charged particles in the vicinity of the recombining ion–electron pair. The resulting microfields can enhance the DR cross section contributions of those high Rydberg states. The reason is in the electric-field dependence of the autoionization rates of states involving a Rydberg electron. These rates strongly depend on the angular momentum quantum number l and the principal quantum number n of the Rydberg electron. The balance between autoionization and radiative transitions involved in dielectronic recombination [see Eq. (22)] determines the possible strength of field effects and also introduces a strong dependence of dielectronic recombination enhancement on the atomic number and the core excited states involved in the recombination process. The aspect of effects of external electromagnetic fields on dielectronic recombination has recently been reviewed by Müller and Schippers (2003).

The first experiment where external fields were applied under well-controlled conditions (see Fig. 34) was performed by Müller et al. (1986), who investigated dielectronic recombination of singly charged Mg^+ ions in the presence of known external electromagnetic fields. They observed an enhancement beyond theoretical zero-field predictions (see Fig. 60) by about an order of magnitude and found an increase of the measured n-dependent cross section by factors up to about 3 when increasing the motional $\vec{v}_i \times \vec{B}$ electric field from 7.2 to 23.5 V/cm. The agreement of these results with theoretical predictions that included effects of electric fields (LaGattuta and Hahn, 1983; Bottcher et al., 1986; Hahn and LaGattuta, 1988) was at the 20% level. The detailed measurement of cross sections for dielectronic recombination of Mg^+ ions differential in the principal quantum number of the product Mg atoms was an accomplishment that has not since been matched. Later crossed-beams experiments with multiply charged C^{3+} ions also revealed drastic enhancements of dielectronic recombination by electric fields (Young et al., 1994; Savin et al., 1996); however, the large uncertainties of these measurements left ambiguities.

The first experiment studying field effects on dielectronic recombination of highly charged ions at a storage ring was carried out with Si^{11+} ions by Bartsch et al. (1997). Motional electric fields were introduced by providing transverse magnetic field components using appropriate saddle coils on top of the electron cooler solenoid. The observed enhancement of the measured dielectronic recombination rate reached about a factor of 3 when the field was increased from 0 to 183 V/cm. The agreement with theoretical calculations was not fully satisfying. The remaining disagreement was attributed by Robicheaux et al. (1998) to additional magnetic field effects in the configuration of crossed E and B fields as they were present in the experiments (also in the experimental arrangement shown in Fig. 34).

Inspired by the theoretical predictions, Bartsch et al. (1999) carried out storage ring dielectronic recombination experiments with controlled and variable electric and magnetic fields using Li-like Cl^{14+} ions. Figure 61 shows a cas-

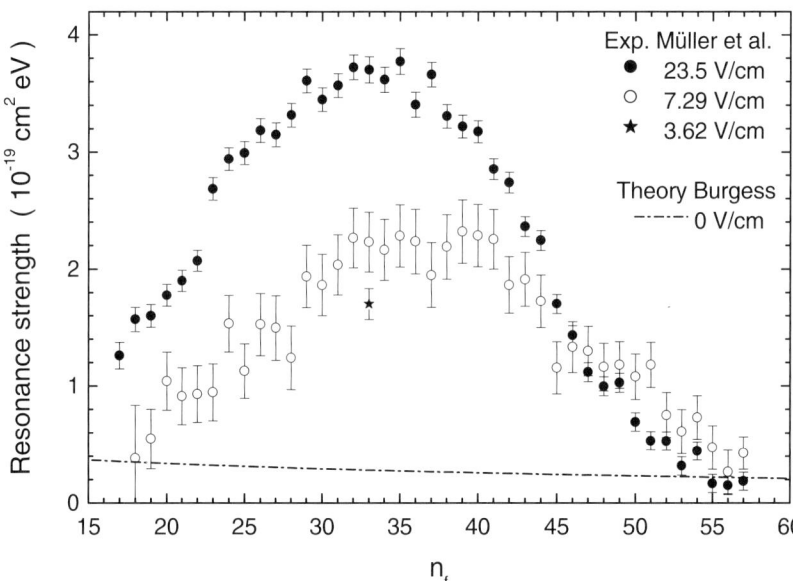

FIG. 60. Collision strengths for resonance contributions to dielectronic recombination of Mg$^+$ ions. The experimental data were obtained with the arrangement displayed in Fig. 34 (Müller, 1986, 1987a) for three motional electric fields as indicated. The dash-dotted line represents a theoretical calculation by Burgess (see Müller et al., 1987a) for zero applied field in the collision region. No effects of field ionization are included in that calculation. The horizontal axis is meant to represent the Rydberg quantum number n. In the experiment, n was determined by employing a Rydberg detector (see Fig. 34) based on field ionization. The experimental quantity n_f is a function of the electric field E applied to the Rydberg analyzer and for this plot the relation between n_f and E, $n_f = (E_0/16E)^{1/4}$, was used [Eq. (52) would give a more realistic mapping of electric field to cut-off Rydberg quantum number; however, for a comparison with theory such an approach cannot replace the consideration of individual survival probabilities of Stark states within each Rydberg manifold anyway].

cade plot of measured storage ring rate coefficients with the electric field as a parameter. Apparently the high Rydberg resonances increase in strength as the electric field increases. Moreover, a distinct effect of the magnetic field strength on the magnitude of the dielectronic recombination rate enhancement was clearly discovered by Bartsch et al. (1999) in that same series of measurements.

4.3.7. Interference Effects and Time-Reversal Symmetry

The issues of interference between dielectronic and radiative recombination and time-reversal symmetry between photorecombination and photoionization were addressed by Schippers et al. (2002b, 2003) and Müller et al. (2002). When

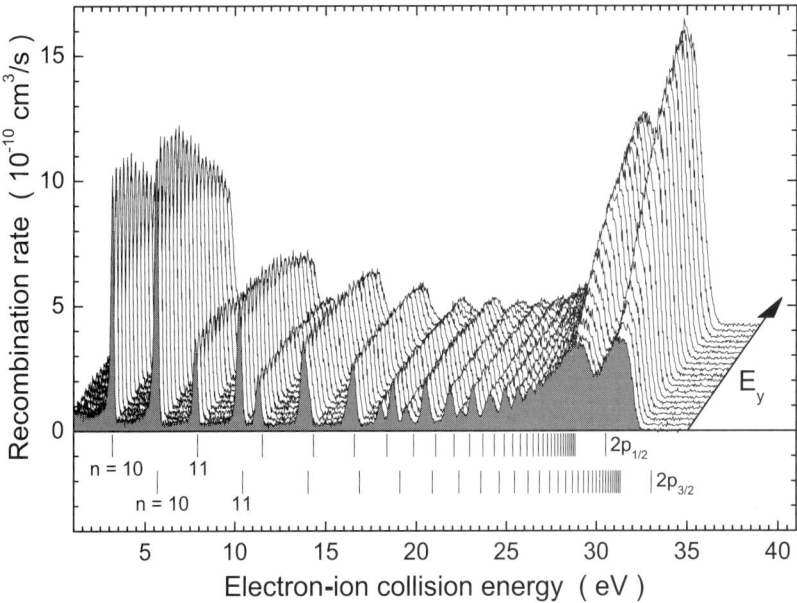

FIG. 61. Dielectronic recombination rate coefficients of Li-like Cl^{14+} (Bartsch et al., 1999) at different motional electric fields ranging from 0 to 380 V/cm. Level energies $E(n)$ of the $2p_{1/2}nl$ and $2p_{3/2}nl$ resonances are indicated.

there is interference in the photoionization channel, the principle of detailed balance [see Eq. (28)] requires that the interference also occurs in the time-reversed process, the recombination channel. It may be masked, however, by additional non-interfering recombination contributions. When the emission of a single photon after dielectronic capture can only lead directly to the ground state of the recombined ion, then chances are high to see unmasked interference with radiative recombination because the number of possible pathways in the collision process is greatly reduced as compared to the emission of two or more photons. With the same reason, the observation of clean, isolated time-reversed reaction pathways in photorecombination and photoionization is facilitated. Such a case was pointed out by Gorczyca et al. (1997) for the recombination of Sc^{3+} ions. On the basis of distorted-wave and R-matrix calculations they predicted interference effects for the particularly strong broad $3d^2(^3P)\,^2F_{5/2,7/2}$ resonances. The experimental results obtained subsequent to the prediction are shown in Fig. 62.

Photoionization of Sc^{2+} via $3p \to 3d$ excitations can be represented as

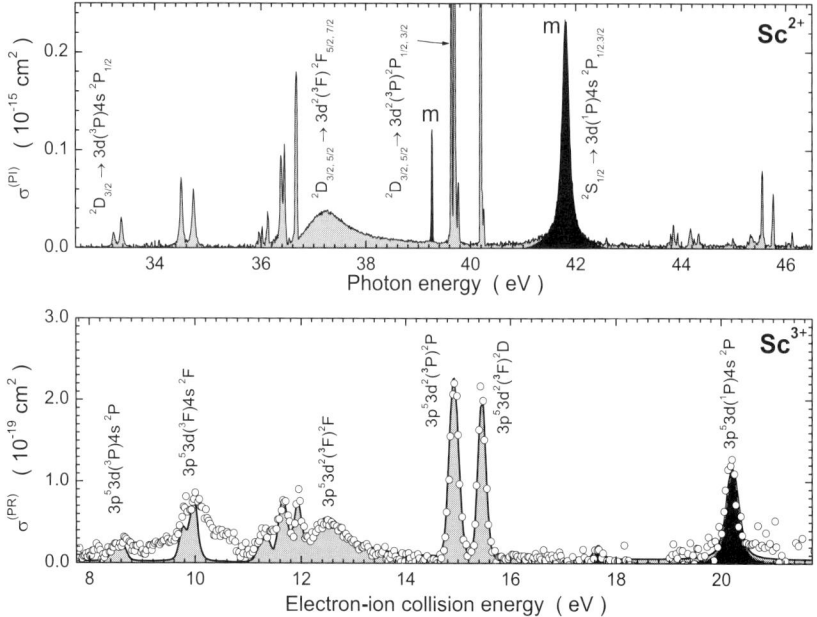

FIG. 62. Experimental cross sections for photoionization of Sc^{2+} (upper panel; Schippers et al., 2002b) and photorecombination of Sc^{3+} ions (lower panel; Schippers et al., 2002a). Resonance peak assignments are provided. The data displayed in the upper panel are cross sections for 3 different initial states of Sc^{2+}, the ground-state fine-structure doublet $^2D_{3/2,5/2}$ and the metastable $3p^64s\,^2S_{1/2}$ state. Photoionization peaks arising from the latter are black shaded and marked by "m". By applying detailed balance [Eq. (31)] to the photoionization data the solid black curve in the lower panel with gray shading is obtained after convolution with the wider energy distribution function of the recombination experiment. The black shading marks the contribution from the "m" photoionization resonances.

$$h\nu + Sc^{2+}(3s^23p^63d) \longrightarrow Sc^{2+}(3s^23p^53d^2)$$
$$\downarrow$$
$$Sc^{3+}(3s^23p^6) + e^-. \qquad (67)$$

The vertical arrow represents the decay of intermediate doubly excited $3p^53d^2$ states via Super-Coster–Kronig transitions, and the diagonal arrow represents the direct 3d photoionization channel. With the arrows reversed, Eq. (67) would represent the photorecombination of Ar-like Sc^{3+} ions. Measured Sc^{2+} photoionization and Sc^{3+} photorecombination cross sections are compared in Fig. 62. The electron energy scale (lower panel) is shifted with respect to the photon energy scale (upper panel) by the ionization potential (24.757 eV) of the $Sc^{2+}(3p^63d\,^2D_{3/2})$ ground state, such that the photoionization and photorecombination spectra can easily be compared. Apart from the lower energy resolution of the recombination experiment there are obvious similarities with respect to relative resonance positions and strengths.

The resonance at $h\nu = 41.8$ eV in the photoionization spectrum appears in the photorecombination spectrum at an energy shifted by about 3.1 eV. This shift corresponds to the $3d \to 4s$ excitation energy and is therefore a clear indication for the presence of metastable $Sc^{2+}(3p^64s)$ parent ions in the photoionization experiment. In addition to the fact that unknown fractions of metastable components were present in the primary ion beam, the measured cross section for photoionization of Sc^{2+} was initially only relative. In a first step of the analysis most of the photoionization resonances in the energy range displayed in Fig. 62 were identified and the resonance parameters were determined. With the well-founded assumption that radiative decay of the intermediate resonant states exclusively proceeds by $3d \to 3p$ transitions the principle of detailed balance was then applied to the absolute cross section for photorecombination of ground-state Sc^{3+}. The comparison of the two independent measurements provided the fractions of the $3p^63d\,^2D_{3/2}$ ground state as well as of the $3p^63d\,^2D_{5/2}$ and $3p^64s\,^2S_{1/2}$ Sc^{2+} metastable components in the parent beam used in the photoionization experiments. Absolute cross sections for photoionization of the three different initial Sc^{2+} states involved in that experiment where thus also obtained.

From the time-reversal analysis, final-state resolved contributions to photorecombination of ground-state Sc^{3+} could be inferred in addition. Their sum is indicated in the lower panel by the solid black line with the gray-shaded area under the line. Obviously, the time-reversed photoionization cross section makes up for most of the photorecombination cross section. The black shaded resonances are related to photoionization of the $3p^64s\,^2S$ metastable component of the parent ion beam and, in the lower panel, correspond to the recombination contribution via the $3p^53d(^1P)4s\,^2P$ resonance that decays almost exclusively to the metastable $3p^64s\,^2S$ state. Application of the principle of detailed balance to state-of-the-art measurements at ion storage rings and synchrotron light sources provides a powerful tool for recovering detailed information on both photoionization and photorecombination processes.

The broad $3d^2(^3P)\,^2F_{5/2,7/2}$ resonance at a photon energy of 37.1 eV clearly shows an asymmetric line shape indicating interference of the resonant with the (almost invisible) direct channel. The asymmetry parameter [see Eq. (16)] was found to be $Q = 5.02(8)$ for this resonance, which is present in both photorecombination and photoionization. By this finding the theoretical prediction by Gorczyca et al. (1997), which had initiated the experimental program on recombination of Sc^{3+} and on photoionization of Sc^{2+}, was nicely confirmed.

4.3.8. Collisional Spectroscopy

Dielectronic recombination with its resonant character has been long recognized as an ideal tool for collisional spectroscopy of ions, and particularly of ions in high charge states. Resonance energies are directly related to the core-excitation energies (and their intricate dependences, e.g. on quantum electrodynamical effects). Equation (66) shows the principal connection between resonance energies and core-excitation threshold. Resonance energies are sensitive to the charge radius of the atomic nucleus and also to its magnetic moment. In order to exploit that potential, the experimental energy resolution has to be very good.

Lindroth et al. (2001) have determined the Cu-like $Pb^{53+}(4p_{1/2} - 4s_{1/2})$ splitting with meV uncertainty by observing dielectronic recombination resonances. Brandau et al. (2003b) have determined $2s \to 2p_{1/2}$ excitation energies of very highly charged lithium-like ions by observing beryllium-like resonances populated by dielectronic recombination. Hyperfine effects were measured by Schuch et al. (2005) in studies with $^{207,208}Pb^{53+}$ ions. The technique was extended to ions with lower atomic numbers such as lithium-like Sc^{18+} by studying resonances at very low relative energies and using theoretical binding energies of high Rydberg levels (Kieslich et al., 2004). The accuracy with which the $2s \to 2p_{3/2}$ excitation energy could be determined was within an uncertainty of 1.9 meV.

The Sc^{18+} experiment was repeated by Lestinsky et al. (submitted for publication) with the ultra-cold electron beam target at the heavy-ion storage ring TSR of the Max-Planck-Institute for Nuclear Physics in Heidelberg. Exploiting the energy resolution accessible with a cooled ion beam and a cold electron target, and taking advantage of the very high precision of merged-beams experiments at low relative energies, it was possible to resolve hyperfine structures of dielectronic recombination resonances. The width of the narrowest resonance feature is no more than about 4 meV. From the measured dielectronic recombination spectrum in combination with improved theoretical calculations for the binding energies of the Rydberg electron in the recombined Sc^{17+} ions, and considering the details of hyperfine splitting of the transitions involved in the dielectronic recombination process, it is possible to extract the $2s \to 2p_{3/2}$ excitation energy and the hyperfine splitting of the $2s_{1/2}$ ground state of Sc^{18+} with an accuracy better than 0.2 meV.

In Fig. 63 the previous measurements of Kieslich et al. (2004) and the new results of Lestinsky et al. (submitted for publication) are compared in the energy range, where the lowest-energy resonances occur. The data in the upper panel were obtained by Kieslich et al. (2004) using the cooling device in the Heidelberg Test Storage Ring TSR as an electron target. The electron temperatures in that measurement corresponded to $k_B T_\perp = 7.2$ meV and $k_B T_\parallel = 180$ μeV. The sharp black-shaded peaks are calculated cross sections for the $(2p_{3/2} 10d_{5/2})_{J=4}$ and $(2p_{3/2} 10d_{3/2})_{J=2, J=3}$ resonance states using the relativistic many-body perturbation approach without accounting for hyperfine effects (Kieslich et al.,

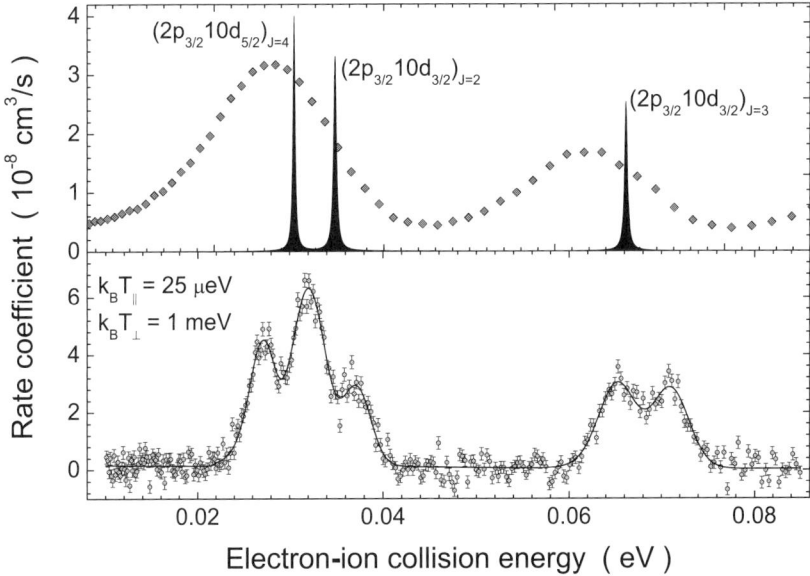

FIG. 63. Rate coefficients for dielectronic recombination of lithium-like Sc^{18+} ions. The data in the upper panel were obtained by Kieslich et al. (2004) using the cooling device in the Heidelberg Test Storage Ring TSR as an electron target. Statistical uncertainties are of the size of the symbols. The energy resolution corresponds to that of the lowest panel in Fig. 39, where measurements using the identical experimental arrangement are shown for C^{3+} ions. The sharp peaks are calculated cross sections for the indicated resonance states without accounting for hyperfine splitting (Kieslich et al., 2004). Convolution of the theoretical cross section with the experimental flattened Maxwellian distribution is in fairly good agreement with those data. The lower panel displays the experimental results (circles) of Lestinsky et al. (submitted for publication). The measurements were also performed at the TSR but with the ultra-cold electron target in addition to the cooling device. The solid line is a fit to the data considering the details of the hyperfine splitting of the transitions involved in the dielectronic recombination resonances located in the present region of interest.

2004). Convolution of the theoretical cross section with the experimental flattened Maxwellian distribution [Eq. (43)] is in fairly good agreement with those data. The lower panel shows the experimental results of Lestinsky et al. (submitted for publication). The measurements were also performed at the TSR but with the ultra-cold electron target in addition to the cooling device. The solid line is a fit to the data considering the details of the hyperfine splitting of the transitions involved in the dielectronic recombination resonances displayed. From the fit the temperatures of the electron-target beam are obtained with $k_B T_\perp = 1$ meV and $k_B T_\parallel = 25$ µeV, both roughly a factor 7 lower than those of the previous experiments with an energy-dependent improvement in resolution following from Eq. (44).

At the Experimental Storage Ring ESR in Darmstadt, Brandau et al. (2007) succeeded to measure isotope shifts by observing dielectronic recombination resonances arising from collisions of electrons with lithium-like ^{142,150}Nd^{57+}. The experiment provides an independent determination of the change in charge radius between the Nd-142 and the Nd-150 nuclei. The technique is very promising with respect to the measurement of nuclear charge radii of radioactive isotopes produced in an accelerator target, mass selected by a suitable fragment separator and stored in a heavy-ion storage ring with an electron cooler and an electron target. Such measurements are envisaged at the future Facility for Antiproton and Ion Research (FAIR) in Darmstadt.

The spectroscopic potential of dielectronic recombination has also been exploited to measure lifetimes of specific excited states. The time-dependent relative abundance of ions in a given long-lived state can be monitored in a storage ring by recording recombination count rates of resonances that can only be populated starting with the metastable excited state. Schmidt et al. (1994) used that technique to measure the magnetic-dipole decay rate of metastable helium-like carbon ions in the Heidelberg Test Storage Ring. Saghiri et al. (1999) were able to measure the very much longer lifetime of Li$^+(1s2s\,^3S)$ by using that same method. Recently, Schippers et al. (2007a) measured the rate of hyperfine-induced decay of metastable Ti$^{18+}(1s^22s2p\,^3P_0)$ by observing dielectronic recombination resonances arising from that initial long-lived state. The lifetime was determined with careful consideration of loss channels other than radiative decay.

4.4. ELECTRON-IMPACT IONIZATION OF IONS

4.4.1. Direct Single Ionization

The most fundamental direct ionization process is that of a projectile electron knocking off the bound electron from a hydrogen-like target. As one goes from the hydrogen atom to He$^+$, Li^{2+} and so forth, the interaction between the projectile and the target nucleus becomes increasingly stronger. In a non-relativistic approach, the direct single ionization cross sections σ are expected to fall on a common curve when scaled by the 4th power of the atomic number Z of the hydrogen-like target atom and displayed as a function of the electron energy in threshold units. This scaling is inherent in the Born approximation but dates back to the classical treatment of electron-impact ionization of atomic hydrogen by Thomson (1912). This approach invokes Rutherford scattering of the projectile electron from a quasi-free bound electron. The target electron is released from its bound state if the energy transferred to it in the Coulomb collision exceeds its binding energy ($= Z^2 \times 13.6$ eV in the case of a hydrogenic target).

The classical approach can easily be extended to collisions of completely stripped nuclei with hydrogenic systems. At higher energies experimental data for ionization of atomic hydrogen by impact of atomic nuclei nicely follow the predicted scaling of cross sections as shown by Müller (2005). As the discussion in Section 3.3 illustrates, the description of the initially bound electron as a quasi-free particle is a meaningful approach at higher collision energies. Experimental proof of scaling behavior in ion–ion collisions is more difficult to obtain. The only experimental data on direct ionization of hydrogen-like ions by completely stripped nuclei available for an extended energy range are for the p + He$^+$ system (Peart et al., 1983; Watts et al., 1986; Rinn et al., 1986). The cross sections from the three published experimental studies differ from one another and have large uncertainties precluding a convincing comparison with related ionization data on the basis of the scaling law. Theory does not have such restrictions. Calculations by Winter (2004) for structureless charged-particles colliding with hydrogen-like C^{5+} support the classical (and Born quantum-mechanical) scaling rule.

For the lower target-ion charge states, deviations from the Z^4 scaling behaviour are expected. Also, when the bound electron becomes relativistic the simple approach breaks down and proper wavefunctions have to be employed in a relativistic treatment of the collision. Considering a limit of $v_e = 0.2c$, i.e., an electron velocity of twenty percent of the speed of light, the limit in electron energy (ionization potential) is about 10 keV, corresponding to an atomic number of about 28. Hence, the scaling should be good for atomic numbers between roughly $Z = 4$ and $Z = 28$.

Figure 64 shows scaled experimental cross sections for electron-impact ionization of hydrogen-like ions B^{4+}, C^{5+}, N^{6+} and O^{7+} (Aichele et al., 1998) together with the simple one-parameter Lotz formula [see Eq. (24)]. The success of the formula for describing direct single ionization of ions becomes obvious from this comparison.

There are no colliding beams experiments on hydrogenic systems with higher charge states. Some data were obtained, however, in EBIT experiments which were normalized to the known cross sections for radiative recombination. Although the error bars on these data are quite large and the available data base is relatively sparse, deviations from the classical scaling begin to show for Mo^{41+} (Watanabe et al., 2002; Marrs et al., 1997), while the data for Fe^{25+} by O'Rourke et al. (2001) are still consistent with the classical expectation—at least within their error bars. For U^{91+} the cross section measured by Marrs et al. (1994) is well above the classical scaling. Relativistic treatment of both the bound electron and of the collision with the incoming electron is required and leads to higher cross sections than expected from the classical scaling.

4.4.2. Indirect Single-Ionization Contributions

Indirect processes in the ionization of ions require more than one target electron. Two target electrons in the K shell of a helium-like system can give rise

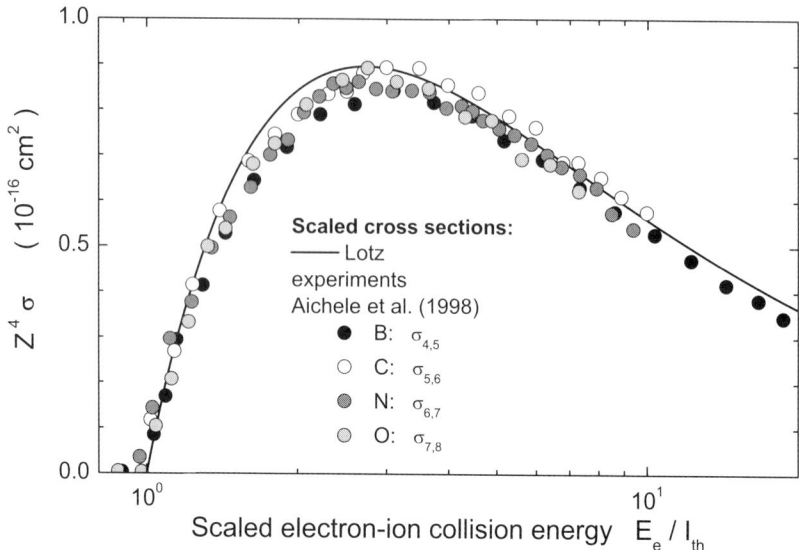

FIG. 64. Scaled experimental cross sections for electron-impact ionization of hydrogen-like ions B^{4+}, C^{5+}, N^{6+} and O^{7+} (Aichele et al., 1998). E_e is the electron–ion collision energy, I_{th} the ionization potential of the target ion with atomic number Z. The solid line is the scaled Lotz formula (Lotz, 1970).

to multi-step ionization channels if both electrons are directly or resonantly excited at a time (Müller et al., 1989; see also Fig. 47). Helium-like metastable ions can be studied in colliding-beams experiments because of their relatively long lifetime. Such metastable helium-like ions with their two electrons provide the simplest electronic configuration supporting indirect ionization mechanisms via (one-electron-)excitation–autoionization and resonant-excitation–auto-double-ionization or resonant-excitation–double-autoionization (Müller et al., 2001). The importance of excitation–autoionization was experimentally demonstrated for the first time by Peart and Dolder (1968). They found excitation–autoionization contributions to the production of Ba^{2+} from Ba^{1+} exceeding expectations for direct electron-impact ionization by factors up to 4. Indirect ionization mechanisms were also found for multiply charged lithium-like ions by Crandall et al. (1979). It soon became evident that plain direct ionization is rather the exception in electron-impact ionization of ions. Indirect processes can be much more important than the direct channel. Deviations of experimental data from theoretical predictions initiated the search for even more complex ionization contributions. Thus the resonant ionization processes were invoked and investigated by theory (LaGattuta and Hahn, 1981; and Henry and Msezane, 1982) long before their observation in experiments became feasible.

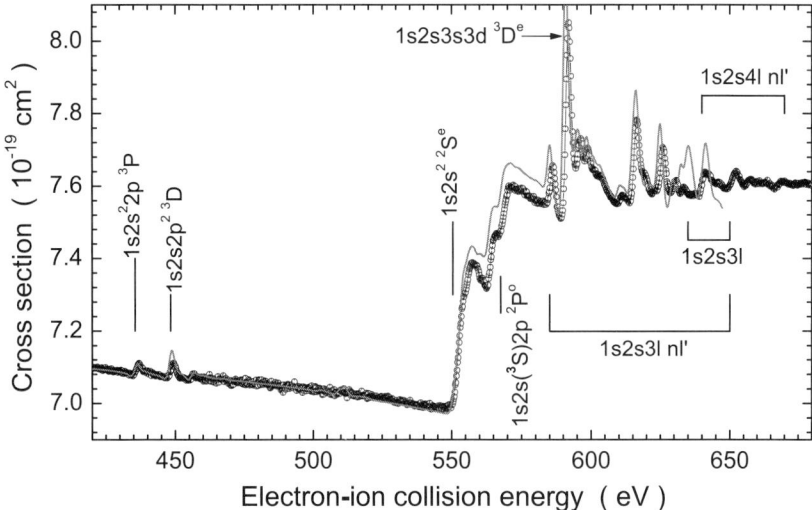

FIG. 65. Experimental cross sections for electron-impact ionization of O^{5+} ions measured by Müller et al. (2000). Intermediate multiply excited states populated during the ionization process are indicated. The gray line is based on an R-matrix calculation from the same reference.

A major leap forward in experimental precision and the capability of measuring fine details of ionization cross sections with very good statistics was facilitated by the introduction of an energy scanning technique using a high-current electron gun with low energy spread (Müller et al. 1988b, 1988a). Individual excitation thresholds as well as series of resonances contributing to single and multiple ionization became accessible to the experiments. A further step in the development was the use of the merged-beams arrangement at the Heidelberg heavy-ion storage ring to measure cross sections for electron-impact ionization of highly charged lithium-like (Kenntner et al., 1995) and sodium-like ions (Linkemann et al., 1995a), which are not easily accessible with low-energy ion beams directly from an ion source. Moreover, the Na-like ions were stored in the ring for a sufficient time to allow metastable states to decay before the actual experiment started.

There is currently a large amount of data available on electron-impact ionization of ions. The physics and technology of ionization experiments and theory has been extensively reviewed in the past (Müller, 1991; Moores and Reed, 1994; Defrance et al., 1995, and references therein). The data situation relevant to x-ray astrophysics was recently reviewed by Kallman and Palmeri (2007). Out of the many possibilities a few illustrative examples are discussed below.

Lithium-like O^{5+} has particularly attracted a great deal of experimental and theoretical interest. Figure 65 shows experimental and theoretical cross sections for electron-impact single ionization of $O^{5+}(1s^2 2s)$ ions in the region of the

K-shell excitation threshold. The shape of the cross section is similar to that displayed in Fig. 46 for C^{3+}. There the excitation aspect was discussed, whereas the contribution to ionization is emphasized here. In both cases, lithium-like ions have been exposed to electron impact. At energies near 450 eV two distinct resonance features can be seen which are associated with intermediate K-shell excited beryllium-like $1s2s^22p\,^3P$ and $1s2s2p^2\,^3D$ states, that can only decay to the observed O^{6+} channel by simultaneous emission of two electrons, i.e., by a double-Auger process. Thus, these two resonances can be unambiguously associated with resonant-excitation–auto-double-ionization. At about 550 eV the K shell can be directly excited and several distinct excited states produce threshold steps in the ionization cross section. Strong resonance features arising from $1s2s3l3l'$ configurations produce structure in the cross section around 600 eV. The dominant feature in all the lithium-like ions is the strong $1s2s3s3d\,^3D$ resonance that exhibits a marked Z-dependent relative strength. Along with the experimental results, an R-matrix calculation is shown in Fig. 65. The indirect R-matrix contributions were added to the experimentally derived direct-ionization cross section. Details of the theory curve were discussed by Müller et al. (2000). Theory and experiment both show clear evidence of interference of the $1s2s3s3d\,^3D$ resonance with the direct-excitation–autoionization background.

4.4.3. Interference Effects in Ionizing Collisions

Interference of different electron–ion collision channels of equal symmetry is an ubiquitous feature. Examples for interference patterns in electron-impact single ionization are shown in Fig. 66. From left to right the complexity of interfering channels increases. The panel labelled B^{2+} shows the cross section for single ionization of B^{2+} ions (Hofmann et al., 1990) in the vicinity of the $1s2s3s3d\,^3D$ resonance (peak A), that is found to be prominent also for the other lithium-like ions (see Fig. 65). Resonant-excitation–double-autoionization through this channel interferes with direct K-shell excitation. The fit through the data points provides the asymmetry parameter $Q = 1.82$.

The middle panel shows cross sections for single ionization of C^{3+} below the K-shell excitation threshold (Teng et al., 2000). Resonances associated with beryllium-like states are observed: $1s2s^22p\,^3P$ (peak A), $1s2s2p^2\,^3D$ (peak B), and $1s2s2p^2\,^1D$ (peak C). These can only contribute to net ionization, i.e., the production of C^{4+} ions, by double-Auger decay. There, the two outgoing electrons share the excess energy with continuous energy distributions. Thus, this channel can interfere with direct ionization with two outgoing electrons also sharing the excess energy. The fits indicate asymmetry parameters $Q_A = 2.06$, $Q_B = 3.87$, and $Q_C = 0.854$.

The right panel displays window resonances observed in electron-impact single ionization of ground-state Li^+ ions (Müller et al., 1989). The two dominant

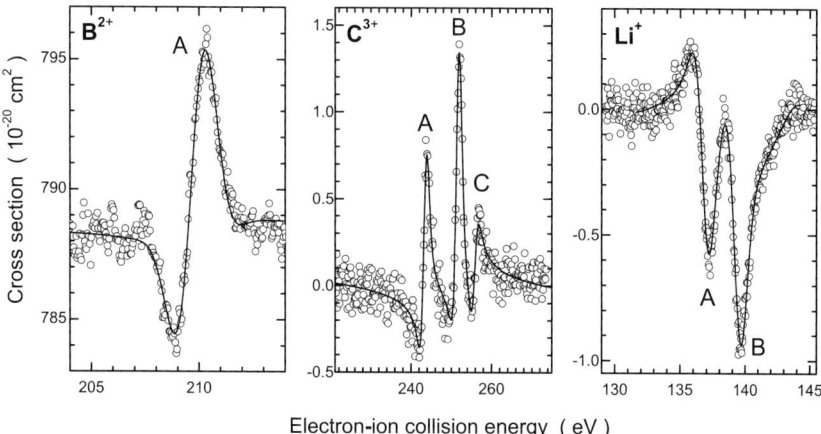

FIG. 66. Experimental cross sections for electron-impact single ionization showing obvious interference patterns. The left panel shows data for B^{2+} ions (Hofmann et al., 1990) in the vicinity of the $1s2s3s3d\,^3D$ resonance (peak A). The middle panel shows data for C^{3+} (Teng et al., 2000). The dominant resonances are $1s2s^22p\,^3P$ (peak A), $1s2s2p^2\,^3D$ (peak B), and $1s2s2p^2\,^1D$ (peak C). The right panel displays window resonances observed with ground-state Li^+ ions (Müller et al., 1989). The two dominant resonances are associated with $2s^22p\,^2P$ and $2s2p^2\,^2D$ states. The solid gray lines are Gaussian-convoluted Fano fits [see Eq. (16)] to the experimental data. Contributions of direct single ionization were subtracted from the cross sections for C^{3+} and Li^+; hence the occurrence of "negative cross sections".

features are associated with $2s^22p\,^2P$ and $2s2p^2\,^2D$ states. They are populated by resonant double excitation (trielectronic capture) and must decay by a double-Auger process in order to produce signal in the ionization channel (Li^{2+} ions were detected). When this ionization channel opens at the resonance energy, the "contribution" to the cross section is through a destructive interference with direct single ionization. The asymmetry parameters for the fit curve are $Q_A = -0.513$ and $Q_B = -0.382$. These numbers are quite uncertain; however, the destructive interference is obvious.

4.4.4. Single Ionization of Complex Ions, Metastable States, Storage Ring Approach

Electron-impact ionization of iron ions is particularly important for applications in plasmas. Iron ions are abundant both in fusion plasmas and in astrophysical environments and therefore, cross sections and rate coefficients are needed by modelers for all charge states and different electron–ion processes, ion–atom collisions, and ion–molecule interactions. As an illustration of available data for electron-impact single ionization of ions with a complex electronic structure,

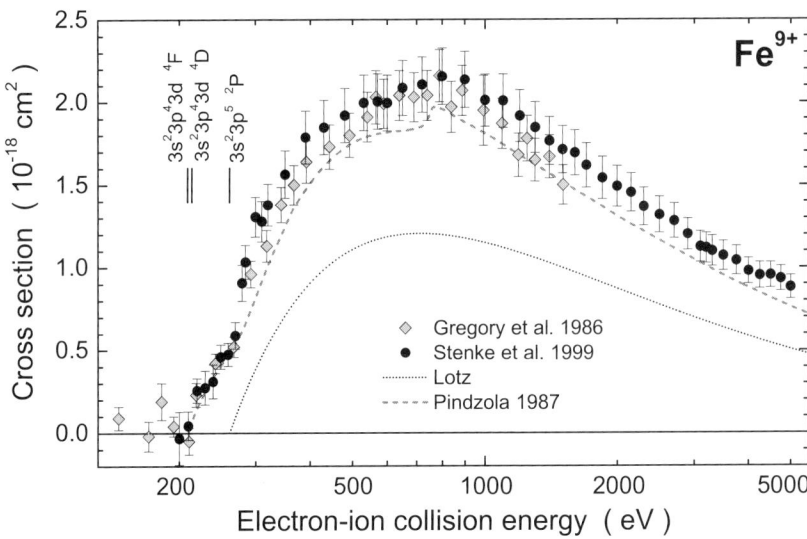

FIG. 67. Cross sections for electron-impact single ionization of Fe^{9+} ions. In addition to the experimental data of Gregory et al. (1986) and Stenke et al. (1999) the result from the Lotz formula [Eq. (24)] for ground-state ions is shown as well as a configuration-average distorted-wave calculation by Pindzola et al. (1986) for metastable $3s^23p^43d$ ions. The ionization thresholds of ground-state Fe^{9+} as well as the threshold range of metastable $3s^23p^43d\ ^4F$, ^4D states are indicated by vertical bars (energies are from Ralchenko et al., 2006).

Fig. 67 shows absolute cross sections for Fe^{9+} ions measured by Gregory et al. (1986) and by Stenke et al. (1999). The two data sets are in quite good agreement with one another. They disagree with the Lotz formula [Eq. (24)] for ground-state $Fe^{9+}(3s^22p^5\ ^2P)$ ions. Apparently, the observed ionization threshold is much lower than that of ground-state ions, indicating the presence of metastable ions. A configuration-average distorted-wave calculation by Pindzola et al. (1986) for metastable $3s^23p^43d$ ions shows good agreement with the experimental data and also fits the observed threshold. Assuming that the calculations are correct this would imply that the ion beam in both experiments consisted exclusively of $3s^23p^43d\ ^4F$, ^4D metastable states which is unlikely.

Experiments on electron-impact ionization of highly charged ions often suffer from unknown fractions of metastable ions in the parent beam. Accordingly, there are great uncertainties in the available data. Some ion species, such as lithium-like ions, do not have any low-energy metastable states; others have quite isolated and distinct metatstable states which can be recognized in the experiments. Examples are the $1s2s\ ^{1,3}S$ helium-like ions. The $1s^22s2p\ ^3P$ beryllium-like metastable ions cause problems with the determination of cross sections for either ground-state or metastable-state ions (see for example Loch et al., 2005). For most ions with a complex electronic structure, individual metastable states cannot be distinguished in the experiments and their fractions remain unknown.

An additional problem in ionization experiments can arise from the population of long lived autoionizing states. Such states can occur when an inner-shell electron is excited to a state of high spin multiplicity. This is particularly possible with ions which, in their ground state, have only one electron outside filled subshells. The presence of long-lived inner-shell excited high-spin states in the parent ion beam can easily be prohibitive for an experiment because the autoionization of ions along their flight path causes huge background count rates in the ionization detector. Such states occur in sodium-like ions (see Howald et al., 1986). In a set of challenging experiments, Lu and Phaneuf (2002) detected electrons emitted from metastable $2p^5 3s 3p\, {}^4D_{7/2}$ states in sodium-like S^{5+}, Cl^{6+}, and Ar^{7+}. Similar long-lived autoionizing states are often present in beams of potassium-like ions and can be populated also in other species, especially in higher charge states. Ion sources that can produce those multiply charged ions have sufficiently energetic electrons to also populate excited states of the ions. Whenever there is a gap in measurements along isonuclear sequences, one can be sure that background from autoionizing metastable components in the ion beam prevented measurements with acceptable statistics.

It is by far beyond the scope of this article to review all the information that is now available on electron-impact single ionization of ions. Extensive bibliographic materials have been published (see e.g. Itikawa, 2002, and references therein) and many of the available experimental data are referenced in the recent reviews of Kallman and Palmeri (2007) and Dere (2007).

It takes special efforts to get detailed information on structures in the cross sections for single ionization of many-electron ions with one electron outside filled shells in the ground state. Examples are the measurements on low-charge Na-like ions Mg^+ (Müller et al., 1990; Peart et al., 1991; Becker et al., 2004) and Al^{2+} (Thomason and Peart, 1998). In some cases, results were obtained by heroic experimental efforts. Potassium-like Ti^{3+} ions produced in an electron cyclotron resonance (ECR) ion source were observed to generate background count rates as high as 100 kHz per nA of ion beam current (van Zoest et al., 2004). By using a "cold" Penning ion source it was possible to reduce that background to about 5 kHz per nA of Ti^{3+} ions. By energy-labelling Ti^{4+} ions produced inside the electron gun region van Zoest et al. (2004) managed to reduce the background arising from the autoionizing component in the Ti^{3+} parent beam by another factor of 8. With signal to background ratios of at most 1/5, cross sections for electron-impact single ionization of Ti^{3+} ions could finally be determined in a fine energy-scan measurement with 1% counting statistics, after weeks of data taking time. Some structure could be identified in the cross section, 95% of which are due to indirect-ionization contributions. Gregory et al. (1987) measured ionization cross sections for Na-like Fe^{15+} employing an electron cyclotron resonance

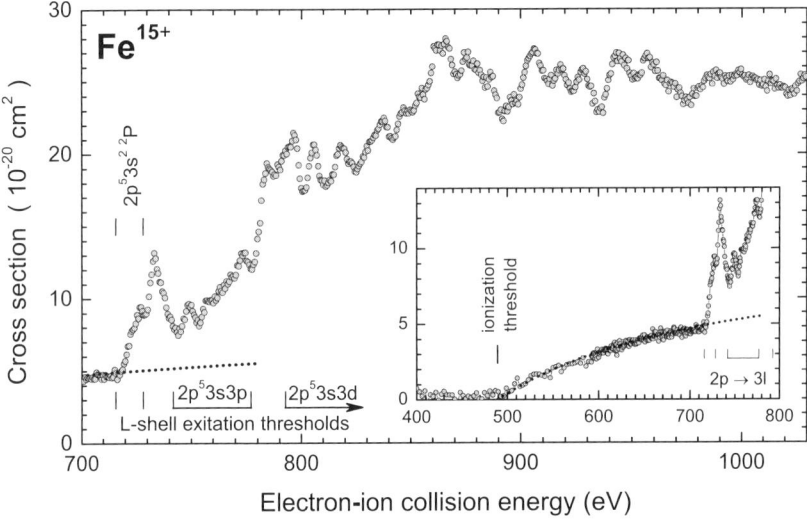

FIG. 68. Cross sections for electron-impact single ionization of Fe^{15+} ions in the ground state (Linkemann et al., 1995b). The inset shows the region from below the ionization threshold to the first onsets of L-shell excitation. Threshold energies and ranges of excited states are indicated by vertical bars (energies are from Ralchenko et al., 2006). The dotted line is the Lotz cross section for direct single ionization.

ion source. This experiment was statistically challenged by high backgrounds and low signal rates at limited ion currents. The applied as well as the fundamental importance of this quasi-one-electron ion, however, justified the large experimental effort.

A new approach to the problem came about with the advent of heavy ion storage rings that allow the measurements to be delayed until the stored ions have decayed to the ground state. Typical time spans between the production of the ions by a foil stripper and the start of electron–ion collision experiments after beam preparation and cooling in the ring are several seconds. By employing the Heidelberg Test Storage Ring, Linkemann et al. (1995b) were able to measure the cross section for electron-impact single ionization of sodium-like Fe^{15+} ([Ne] $3s\,^2S$) ions. Their experimental results are shown in Fig. 68. The relative uncertainties of the data points above the L-shell excitation threshold are about 1% (compared to 10 to 30% in the work of Gregory et al., 1987). Also, the number of data points in the investigated energy range was enhanced by more than a factor of 20. On that fine grid, details in the cross section function become visible. The experiment resolved a long standing issue concerning the relative importance of excitation–autoionization versus direct ionization.

Since the first considerations by Goldberg et al. (1965), there had been a very active theoretical development to accurately determine the excitation strengths in Fe^{15+}, and then also to include resonant processes. Figure 68 shows a resonance feature that even dominates the ionization cross section at about 730 eV. Based on the Lotz formula [Eq. (24)] for direct single ionization, the dotted line in the figure indicates the dominance of excitation–autoionization above direct single ionization at energies greater than about 720 eV. The publication of the experimental result was accompanied by theoretical calculations which are in fairly good overall agreement with the measurements but deviate in the details of resonance structures. The references provided by Linkemann et al. (1995b) outline the 30-year history of investigations on indirect contributions to the ionization of ions and, in particular, of Fe^{15+}.

4.4.5. *Direct Multiple Ionization*

Multiple ionization in electron–ion collisions has received a great deal of attention as well. In particular, for complex ions, cross sections for multiple ionization can be close to those for single ionization (Müller et al., 1984) and hence, multiple ionization cannot a priori be neglected in plasma modeling calculations (Müller, 1986).

While direct single ionization appears to be understood reasonably well on the level of total cross sections, the situation becomes very much more involved for direct double ionization processes. The fundamental importance of this two-electron process has resulted in tremendous efforts in atomic collisions research during the last two decades to understand double ionization of helium atoms exposed to electromagnetic radiation and charged-particle impact over the widest possible energy range. Special aspects of this problem can be illuminated by switching from neutral helium to electrically charged two-electron targets. Data are scarce so far because the double ionization cross sections for two-electron ions decrease very rapidly with increasing atomic number Z. Already with Li^+ the double-ionization cross-section maximum is down to about $1.5 \cdot 10^{-20}$ cm^2. Equation (26) would predict a maximum of roughly $4 \cdot 10^{-22}$ cm^2 for B^{3+} and $2 \cdot 10^{-23}$ cm^2 for O^{6+} (at about 5 keV). Such small cross sections have been measured already with the apparatus shown in Fig. 38. Work is in progress to measure cross sections for double ionization of light helium-like ions from threshold up to electron energies where the maximum occurs by using the electron gun shown in Fig. 37.

Data sets available for electron-impact double-ionization of atomic two-electron systems are displayed in Fig. 69 with a scaling slightly different from that suggested by Shevelko et al. (2005) [see Eq. (26)].

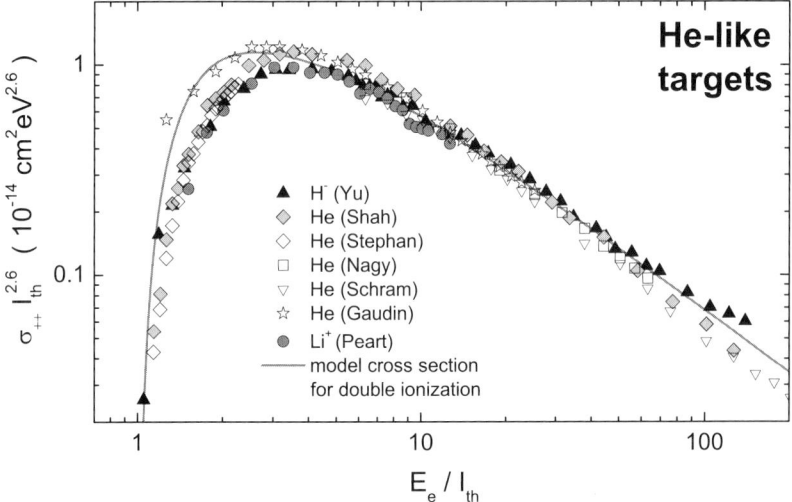

FIG. 69. Scaled cross sections $\sigma_{++} I_{th}^{2.6}$ for electron-impact direct double ionization of two-electron ions (and atoms) as a function of E_e/I_{th} where E_e is the electron–ion collision energy and I_{th} is the threshold energy. The solid line is based on a slightly rescaled version of a semi-empirical formula suggested by Shevelko et al. (2005). The data are from experiments on H$^-$ (Yu et al., 1992), He (Schram et al., 1966; Gaudin and Hagemann, 1967; Nagy et al., 1980; Stephan et al., 1980; Shah et al., 1988), and Li$^+$ (Peart and Dolder, 1969).

4.4.6. Inner-Shell Ionization with Subsequent Autoionization

When the number of electrons bound in the target atom is increased to 3 or more, indirect contributions to multiple ionization become possible. Usually, double ionization shows substantial contributions from single ionization of an inner-shell of the target atom or ion. If an inner-shell electron is released to the continuum the resulting ion can undergo an Auger process and thus the final result of the process is a net double ionization. Such processes become possible with four or more target electrons. They were first observed and interpreted in a study on Ar^{q+} ions ($q = 1, 2, 3$) by Müller and Frodl (1980), who also found signal from the emission of more than one electron by inner-shell vacancy states. This experiment was later complemented by measurements with more highly charged ions by Tinschert et al. (1989). An example for indirect processes in multiple ionization of ions involving an inner-shell ionization with subsequent emission of one, two or even three electrons is shown in Fig. 70, where cross sections for electron-impact two-fold, three- and four-fold ionization of singly charged O$^+$ ions are displayed.

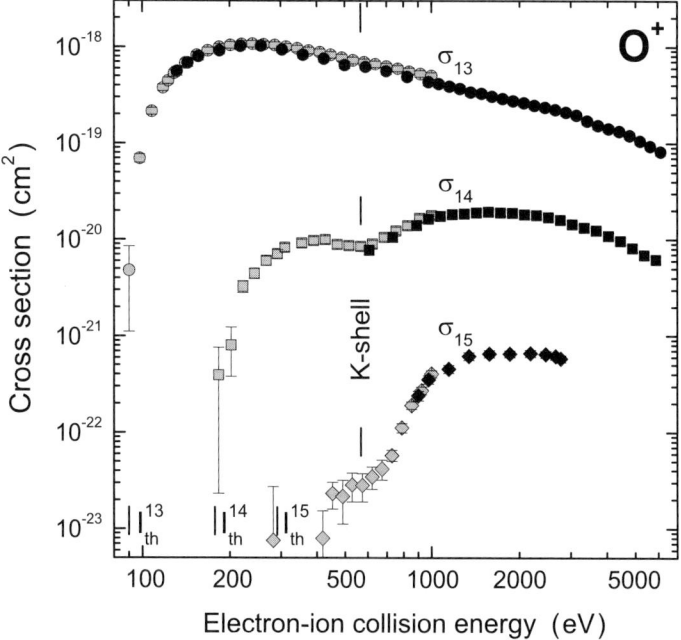

FIG. 70. Experimental cross sections for multiple ionization of O^+ ions by electron impact (Westermann et al., 1999). Threshold energies I_{th}^{1f} for multiple ionization (with the final charge state $f = 3, 4, 5$) as well as the K-shell ionization threshold are indicated by vertical bars.

4.4.7. *Deep Core Resonances and Decay Cascades*

More complex mechanisms with cascades of radiative and Auger decays are also possible and can produce signal in multiple-ionization channels, especially when deep-core excitation (Aichele et al., 2001) is involved in the process. This was demonstrated in a comprehensive study on electron-impact single and multiple ionization of Ba^{q+} ions in charge states from $q = 1$ to $q = 13$ (Knopp et al., 2003).

$$e + Ba^{q+} \rightarrow Ba^{(q+n)+} + (n+1)e. \tag{68}$$

Depending on the charge state q of the parent ion, the number n of electrons removed in a single collision ranged from $n = 1$ to $n = 7$. Absolute cross sections were measured in the energy range where the $3d$-subshell can be opened. This range changes with the parent ion charge state, and is covered by collision energies in the experiment between 600 and 1050 eV.

As one example of the measurements, the experimental cross sections for n-fold ionization of Ba^{4+} ions are shown in Fig. 71 with n ranging from 1 to 6.

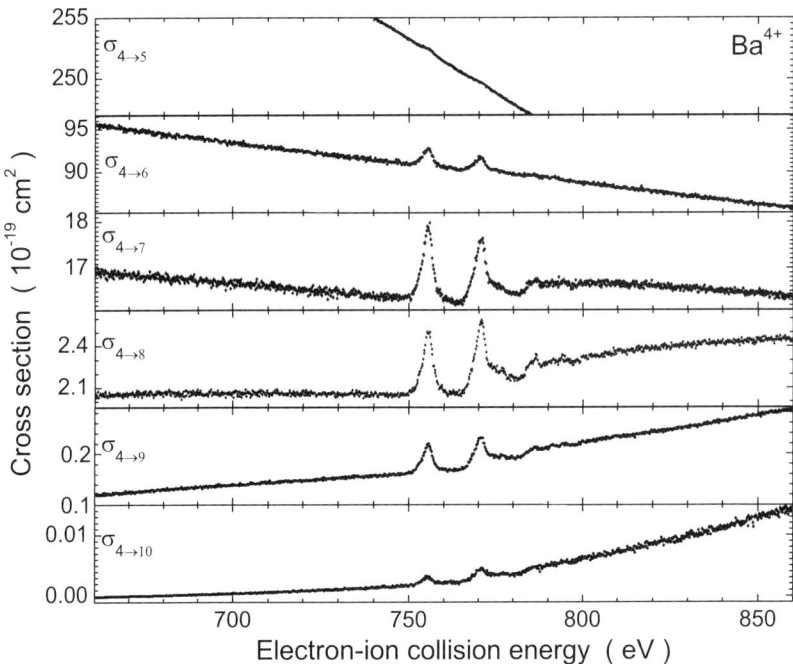

FIG. 71. Measured cross sections for n-fold ionization of Ba^{4+} ions ($n = 1, 2, \ldots, 6$) (Knopp et al., 2003).

At energies between 750 and 770 eV a pair of peaks is found in each cross section, apparently due to dielectronic capture of the incident electron by the Ba^{4+} parent ion. The resonance energy and the double-peak nature indicate the involvement of a $3d$ electron in the collision leaving a deep-core ^2D vacancy behind in an intermediate short-lived Ba^{3+} ion which subsequently decays by the emission of $(n + 1)$, i.e., up to 7 electrons. In the cross section for single ionization, the resonances are hardly visible on top of the strong non-resonant ionization. By subtracting a smooth curve from the measured cross section, however, the resonance contributions can be extracted.

On an absolute scale the strongest peak features are found for net double and triple ionization. On a relative scale, however, in comparison with the non-resonant "background" the resonances are strongest in 6-fold ionization. Here, 7 electrons are ejected from the deep-core excited intermediate Ba^{3+} ion whose excitation energy barely exceeds the threshold for 6-fold ionization. Therefore the phase space for 6-fold ionization is greatly reduced and the cross section for non-resonant processes becomes small. As a consequence the resonance contributions near the threshold of 6-fold ionization dominate the cross section. Observations like the ones described here for Ba^{4+} have been made for all investigated Ba^{q+} ions. The ions Ba^{1+} and Ba^{2+} had already previously been studied, though in a much narrower energy range (Hofmann et al., 1991). Other rare-earth elements show similar effects (Aichele et al., 2001; Mitnik et al., 2001).

5. Conclusions

Electron–ion interactions are important in all environments where matter is ionized, either by external irradiation or by collisions at sufficiently high energies. Processes involving single ions and electrons are also interesting as cases for studying relatively simple atomic systems and the implications of the laws of quantum theory on the structure and the dynamics of the building blocks of the matter that surrounds us.

Different experimental approaches have provided complementary information about different aspects of electron–ion interactions.

The plasma technique has produced a relatively limited number of electron–ion rate coefficients. Because of the inherent ambiguities and uncertainties and because of the development of the other techniques, this method has not been very actively pursued during the last decade. Observation of the radiation from plasmas and the modeling of the results on the basis of present knowledge about electron–ion collisions, however, is an active, productive and very important application.

EBIT devices have an amazing potential for producing low-velocity ions in any given charge state. Thus they are unique sources of characteristic radiation providing sufficient intensities for detailed spectroscopic studies even on very highly charged ions. Absolute cross sections for electron-impact ionization, excitation and recombination can be determined by normalizing intensities of the observed radiation to the theoretical data on radiative recombination. Uncertainties are introduced by the simultaneous presence of several ion charge states in the trap and by the necessary modelling of the trap plasma. Low-energy electron–ion collisions are hardly accessible by EBIT techniques. For collision energies beyond several keV, however, EBIT has offered almost exclusive opportunities to investigate collisions of highly charged ions with electrons. In addition, measurements on electron–ion collision systems with determination of cross sections differential in angle and photon energy are noteworthy. The spectroscopic precision reached at EBIT laboratories with respect to atomic structure of highly charged ions is breathtaking. More data can be expected from the available devices in the future.

Studying electron–ion interactions by careful observation of fast-ion–neutral-target collisions is an active field of research that produces unique information on photorecombination and on electron scattering and cross sections differential in electron energy and emission angle. Further development of detection techniques combined with the exploitation of the relatively high luminosity of these experiments will keep this field exciting and productive. While radiative and resonant electron capture have been studied in ion–atom (ion–molecule) collisions using high-energy accelerators, electron–ion scattering experiments were mainly restricted so far to small accelerators where only light ions can be studied by the ion–atom collision techniques that have been developed. The accelerator technology that is required for extending the measurements to heavier and more highly charged ions is available. Production of high-intensity ion beams up to the highest charge states is possible and can be combined with the availability of dense, cold neutral targets, providing quasi-free electrons to explore electron–ion interactions for ions of all elements in all possible charge states.

Experiments on collisions of free electrons with ions employing interacting beams have reached a very high standard. Experimental arrangements employ low-energy merged, inclined, or crossed beams of electrons and ions to study electron-impact ionization, excitation, as well as elastic and inelastic electron scattering.

Cross sections for single and multiple ionization of ions can be measured and data with very high precision and very good energy resolution can be obtained, providing information on fine details of indirect processes contributing to the ionization processes.

The theoretical calculation of cross sections for electron-impact excitation is still very difficult and predictions can easily be uncertain by substantial amounts. Experiments also continue to be difficult; however, there is steady progress in the field that is actively pursued employing the merged-beams technique. Eventually, the extension of the existing devices to accommodate more energetic ion beams will greatly enhance the ranges of collision energies and collision systems accessible to these experiments.

High-energy ion beams are now almost exclusively used in combination with ion storage rings. Merged-beams experiments predominantly concentrate on electron–ion recombination but have also been employed for ionization studies. Dielectronic recombination of highly charged ions at low center-of-mass energies remains a field of particular interest because of the difficulties that theory faces in predicting the energies and strengths of resonances near the threshold. Such information is essential for the understanding of low-temperature (photon-driven) plasmas containing highly charged ions. An aspect of special interest associated with this context is the influence of electric and magnetic fields on dielectronic recombination.

Recombination of completely stripped and very highly charged heavy ions are also receiving considerable attention. Although these data are not immediately relevant to applications in plasma physics, they provide important test cases for state-of-the-art theory describing electron–ion collision phenomena. Special appeal is in the relative energy resolution of resonances in very highly charged ions, where level splittings are maximum while the experimental energy spread is not much different from that obtained for low-charge ions.

A major challenge for electron–ion collision studies is the issue of metastable ions in the parent beams used in the experiments. Reliable information on processes with ground-state ions can usually be obtained if the parent ions are stored for at least seconds before the electron–ion collision measurement is started. This calls for the use of heavy ion storage rings to be employed in such experiments. However, not all the available experimental techniques can be transported to storage-ring devices and the use of fast beams has substantial disadvantages on the relative levels of signal and background.

Given the increasing needs for reliable atomic collision data that are required to decipher the messages from the universe conveyed to us in the form of photons and observed in expensive space missions with ever increasing spectroscopic resolution in angle and frequency, additional experimental capabilities and opportunities for carrying out laboratory astrophysics measurements are required. New, large scale accelerator projects will provide new possibilities for benchmark experiments; however, they will not be the tools to meet data needs. Dedicated laboratory astrophysics studies on electron–ion interactions are required to make effective use of the data that, for example, satellite borne high resolution x-ray observatories collect.

6. Acknowledgements

I thank Ron Phaneuf and Stefan Schippers for valuable suggestions towards this manuscript and for longstanding enjoyable cooperation. I am grateful for having had the opportunity to collaborate and interact with many of the researchers who are or were active in the field of electron–ion interactions and to participate in some of the great developments that have been made during the last 3 decades. The interaction with numerous colleagues over the years is very gratefully acknowledged. Naming everybody, who has contributed his or her share to what we know about electron–ion interactions would exceed the limitations set to an article like the present one. But a great *Thank You!* goes to all of them.

7. References

Aichele, K., Hartenfeller, U., Hathiramani, D., Hofmann, G., Schäfer, V., Steidl, M., Stenke, M., Salzborn, M., Pattard, M., Rost, J.M. (1998). Electron impact ionization of the hydrogen-like ions B^{4+}, C^{5+}, N^{6+} and O^{7+}. *J. Phys. B* **31**, 2369–2379.

Aichele, K., Hathiramani, D., Scheuermann, F., Müller, A., Salzborn, E., Mitnik, D., Colgan, J., Pindzola, M.S. (2001). Deep-core dielectronic-capture resonances in the electron-impact ionization of heavy atomic ions. *Phys. Rev. Lett.* **86**, 620.

Ali, R., Bhalla, C.P., Cocke, C.L., Stöckli, M. (1990). Dielectronic recombination of heliumlike argon. *Phys. Rev. Lett.* **64**, 633–636.

Alnaser, A.S., Landers, A.L., Pole, D.J., Hossain, S., Haija, O.A., Gorczyca, T.W., Tanis, J.A., Knutson, H. (2002). Superelastic scattering of electrons from metastable He-like C^{4+} and O^{6+} ions. *Phys. Rev. A* **65**. 042709.

Andersen, L.H., Bolko, J., Kvistgaard, P. (1990a). Radiative recombination between free electrons and bare carbon ions. *Phys. Rev. Lett.* **64**, 729–732.

Andersen, L.H., Bolko, J., Kvistgaard, P. (1990b). State-selective dielectronic-recombination measurements for He- and Li-like carbon and oxygen ions. *Phys. Rev. A* **41**, 1293–1302.

Andriamonje, S., Anne, R., de Castro Faria, N.V., Chevallier, M., Cohen, C., Dural, J., Gaillard, M.J., Genre, R., Hage-Ali, M., Kirsch, R., L'Hoir, A., Farizon-Mazuy, B., Mory, J., Mouilin, J., Poizat, C.J., Quere, Y., Remillieux, J., Schmaus, D., Toulemonde, M. (1989). Electron impact ionization and energy loss of 27-MeV/u Xe^{35+} incident ions channeled in silicon. *Phys. Rev. Lett.* **63**, 1930–1933.

Arnaud, M., Raymond, J. (1992). Iron ionization and recombination rates and ionization equilibrium. *Astrophys. J.* **398**, 394–406.

Azuma, Y., Koike, F., Cooper, J.W., Nagata, T., Kutluk, G., Shigemasa, E., Wehlitz, R., Sellin, I.A. (1997). Photoexcitation of hollow lithium with completely empty K and L shells. *Phys. Rev. Lett.* **79**, 2419–2422.

Badnell, N.R. (1997). On the effects of the two-body non-fine-structure operators of the Breit–Pauli Hamiltonian. *J. Phys. B* **30**, 1–11.

Badnell, N.R. (2006). Dielectronic recombination of Fe^{13+}: Benchmarking the M-shell. *J. Phys. B* **39**, 4825–4852.

Badnell, N.R. (online). Atomic data from AUTOSTRUCTURE. http://amdpp.phys.strath.ac.uk/tamoc/DATA/.

Badnell, N.R., Pindzola, M.S. (1992). Unified dielectronic and radiative recombination cross sections for U^{90+}. *Phys. Rev. A* **45**, 2820–2824.

Baker, F.A., Hasted, J.B. (1966). Electron collision studies with trapped positive ions. *Philos. Trans. R. Soc. London Ser. A* **261**, 33.

Bannister, M.E., Chung, Y.-S., Djurić, N., Wallbank, B., Woitke, O., Zhou, S., Dunn, G.H., Smith, A.C.H. (1998). Absolute cross sections for near-threshold electron-impact excitation of the $2s^2S \to 2p^2P$ transition in C^{3+}. *Phys. Rev. A* **57**, 278–281.

Bartsch, T., Müller, A., Spies, W., Linkemann, J., Danared, H., DeWitt, D.R., Gao, H., Zong, W., Schuch, R., Wolf, A., Dunn, G.H., Pindzola, M.S., Griffin, D.C. (1997). Field enhanced dielectronic recombination of Si^{11+} ions. *Phys. Rev. Lett.* **79**, 2233–2236.

Bartsch, T., Schippers, S., Müller, A., Brandau, C., Gwinner, G., Saghiri, A.A., Beutelspacher, M., Grieser, M., Schwalm, D., Wolf, A., Danared, H., Dunn, G.H. (1999). Experimental evidence for magnetic field effects on dielectronic recombination via high Rydberg states. *Phys. Rev. Lett.* **82**, 3779–3781.

Becker, C., Knopp, H., Jacobi, J., Teng, H., Schippers, S., Müller, A. (2004). Electron-impact single and multiple ionization of Mg^+ ions. *J. Phys. B* **37**, 1503–1518.

Becker, R., Müller, A., Achenbach, C., Tinschert, K., Salzborn, E. (1985). A dense electron target for the study of electron–ion collisions. *Nucl. Instrum. Methods B* **9**, 385–388.

Behar, E., Jacobs, V.L., Oreg, J., Bar-Shalom, A., Haan, S.L. (2004). Effects of quantum interference between radiative and dielectronic recombination on photorecombination cross-section profiles for the He-like ions Ar^{16+} and Fe^{24+}. *Phys. Rev. A* **69**. 022704.

Beiersdorfer, P. (2003). Laboratory X-ray astrophysics. *Annu. Rev. Astron. Astrophys.* **41**, 343–390.

Beiersdorfer, P., Knapp, D., Marss, R.E., Elliott, S.R., Chen, M.H. (1993). Structure and Lamb shift of $2s_{1/2} - 2p_{3/2}$ levels in lithiumlike U^{89+} through neonlike U^{82+}. *Phys. Rev. Lett.* **71**, 3939.

Beiersdorfer, P., Chen, H., Thorn, D.B., Träbert, E. (2005). Measurement of the two-loop Lamb shift in lithiumlike U^{89+}. *Phys. Rev. Lett.* **95**. 233003.

Bélenger, C., Defrance, P., Friedlein, R., Guet, C., Jalabert, D., Maurel, M., Ristori, C., Rocco, J.C., Huber, B.A. (1996). Elastic large-angle scattering of electrons by multiply charged ions. *J. Phys. B* **29**, 4443–4456.

Belić, D.S., Dunn, G.H., Morgan, T.J., Mueller, D.W., Timmer, C. (1983). Dielectronic recombination: A crossed-beams observation and measurement of cross section. *Phys. Rev. Lett.* **50**, 339–342.

Benhenni, M., Shafroth, S.M., Swenson, J.K., Schulz, M., Giese, J.P., Schöne, H., Vane, C.R., Dittner, P.F., Datz, S. (1990). Angular distribution of Auger electrons emitted through the resonant transfer and excitation process following O^{5+} + He collisions. *Phys. Rev. Lett.* **65**, 1849–1852.

Benis, E.P., Zouros, T.J.M., Gorczyca, T.W., González, A.D., Richard, P. (2004). Elastic resonant and nonresonant differential scattering of quasifree electrons from $B^{4+}(1s)$ and $B^{3+}(1s^2)$ ions. *Phys. Rev. A* **69**. 052718.

Bethe, H. (1930). Zur Theorie des Durchgangs schneller Korpuskularstrahlen durch Materie. *Ann. Phys. (Leipzig)* **397**, 325–400.

Bethe, H.A., Salpeter, E.E. (1957). "Quantum Mechanics of One- and Two-Electron Atoms". Springer, Berlin.

Biggs, F., Mendelsohn, L.B., Mann, J.B. (1975). Hartree–Fock Compton profiles for the elements. *At. Data Nucl. Data Tables* **16**, 201.

Böhm, S., Schippers, S., Shi, W., Müller, A., Eklöw, N., Schuch, R., Danared, H., Badnell, N.R., Mitnik, D., Griffin, D.C. (2002). Measurement of the field induced dielectronic recombination rate enhancement of O^{5+} ions differential in the Rydberg quantum number n. *Phys. Rev. A* **65**. 052728.

Bottcher, C., Griffin, D.C., Pindzola, M.S. (1986). Dielectronic recombination of Mg^+ in the presence of electric fields. *Phys. Rev. A* **34**, 860–865.

Brandau, C., Bartsch, T., Beckert, K., Böhme, C., Bosch, F., Franzke, B., Hoffknecht, A., Knopp, H., Kozhuharov, C., Krämer, A., Mokler, P.H., Müller, A., Nolden, F., Schippers, S., Stachura, Z., Steck, M., Stöhlker, T., Winkler, T. (1999). Recombination of highly charged ions with free electrons. *GSI Scientific Report* **99** (1), 89–90.

Brandau, C., Bartsch, T., Böhm, S., Böhme, C., Hoffknecht, A., Kieslich, S., Knopp, H., Schippers, S., Shi, W., Müller, A., Grün, N., Scheid, W., Steih, T., Bosch, F., Franzke, B., Kozhuharov, C., Krämer, A., Mokler, P.H., Nolden, F., Steck, M., Stöhlker, T., Stachura, Z. (2003a). Dielectronic recombination of very heavy lithiumlike ions. *Hyperfine Interact.* **146/147**, 41–45.

Brandau, C., Kozhuharov, C., Müller, A., Shi, W., Schippers, S., Bartsch, T., Böhm, S., Böhme, C., Hoffknecht, A., Knopp, H., Grün, N., Scheid, W., Steih, T., Bosch, F., Franzke, B., Mokler, P.H., Nolden, F., Steck, M., Stöhlker, T., Stachura, Z. (2003b). Precise determination of the $2s_{1/2} - 2p_{1/2}$ splitting in very heavy lithiumlike ions utilizing dielectronic recombination. *Phys. Rev. Lett.* **91**. 073202.

Brandau, C., Kozhuharov, C., Müller, A., Schippers, S., Beckert, K., Beller, P., Bernhardt, D., Bosch, F., Böhm, S., Currell, F., Franzke, B., Gumberidze, A., Harman, Z., Jacobi, J., Mokler, P., Nolden, F., Scheid, W., Schmidt, E.W., Spillmann, U., Stachura, Z., Steck, M., Stöhlker, T. (2006). First dielectronic recombination measurements with H-like uranium. *Rad. Phys. Chem.* **75**, 1763–1766.

Brandau, C., Kozhuharov, C., Müller, A., Beckert, K., Beller, P., Bernhardt, D., Bosch, F., Böhm, S., Currell, F.J., Gumberidze, A., Harman, Z., Jacobi, J., Kluge, H.J., Mokler, P.H., Nolden, F., Reuschl, R., Schippers, S., Schmidt, E.W., Spillmann, U., Stachura, Z., Steck, M., Stöhlker, T., Wolf, A. (2007). Photorecombination studies of highly charged ions at the storage ring ESR: A progress report. *J. Phys. Conf. Ser.* **58**, 81–86.

Brandt, D. (1983). Resonant transfer and excitation in ion–atom collisions. *Phys. Rev. A* **27**, 1314–1318.

Bray, I., McCarthy, I.E., Wigley, J., Stelbovics, A.T. (1993). Calculation of electron scattering on the He$^+$ ion. *J. Phys. B* **26**, L831–L836.

Breton, C., Michelis, C.D., Finkenthal, M., Mattioli, M. (1978). Ionization and recombination rate coefficients of highly ionized molybdenum ions from spectroscopy of tokamak plasmas. *Phys. Rev. Lett.* **41**, 110–113.

Brooks, R.L., Datla, R.U., Griem, H.R. (1978). Measurement of dielectronic recombination rates for the iron ions Fe IX–XI. *Phys. Rev. Lett.* **41**, 107–109.

Brooks, R.L., Datla, R.U., Krumbein, A.D., Griem, H.R. (1980). Measurement of effective dielectronic recombination rates for Fe IX, X, and XI. *Phys. Rev. A* **21**, 1387–1396.

Brotton, S.J., McKenna, P., Gribakin, G., Williams, I.D. (2002). Angular distribution for the elastic scattering of electrons from $Ar^+(3s^23p^5\,^2P)$ above the first inelastic threshold. *Phys. Rev. A* **66**. 062706.

Brown, G., Beiersdorfer, P., Chen, H., Scofield, J.H., Boyce, K.R., Kelley, R.L., Kilbourne, C.A., Porter, F.S., Gu, M.F., Kahn, S.M., Szymkowiak, A.E. (2006). Energy-dependent excitation cross section measurements of the diagnostic lines of Fe XVII. *Phys. Rev. Lett.* **96**. 253201.

Bryans, P., Badnell, N.R., Gorczyca, T.W., Laming, J.M., Mitthumsiri, W., Savin, D.W. (2006). Collisional ionization equilibrium for optically thin plasmas. I. Updated recombination rate coefficients for bare through sodium-like ions. *Astrophys. J. Suppl. Ser.* **167**, 343–356.

Burgess, A. (1964). Dielectronic recombination and the temperature of the solar corona. *Astrophys. J.* **139**, 776–780.

Burgess, A., Summers, H.P. (1969). The effect of electron and radiation density on dielectronic recombination. *Astrophys. J.* **157**, 1007–1021.

Burke, P., Noble, C., Burke, V. (2006). R-matrix theory of atomic, molecular and optical processes. In: Berman, P., Lin, C., Arimondo, E. (Eds.), *Adv. At. Mol. Opt. Phys.* Elsevier, pp. 237–318.

Chung, Y.-S., Djurić, N., Wallbank, B., Dunn, G.H., Bannister, M.E., Smith, A.C.H. (1997). Resonance interference and absolute cross sections in near-threshold electron-impact excitation of the $3s^2\,^1S \rightarrow 3s3p\,^3P$ and $3s^2\,^1S \rightarrow 3s3p\,^1P$ transitions in Ar^{6+}. *Phys. Rev. A* **55**, 2044–2049.

Chutjian, A. (2004). Ion collisions in the highly charged universe. *Phys. Scr. T* **110**, 203–211.

Chutjian, A., Newell, W.R. (1982). Experimental electron energy-loss spectra and cross sections for the $4\,^2S \rightarrow 4\,^2P$ transition in Zn II. *Phys. Rev. A* **26**, 2271–2273.

Chutjian, A., Msezane, A.Z., Henry, R.J.W. (1983). Angular distribution for electron excitation of the $4\,^2S \rightarrow 4\,^1P$ transition in Zn II: Comparison of experiment and theory. *Phys. Rev. Lett.* **50**, 1357–1360.

Claytor, N., Feinberg, B., Gould, H., Bemis, C.E., Campo, J.G.D., Ludemann, C.A., Vane, C.R. (1988). Electron impact ionization of $U^{88+} - U^{91+}$. *Phys. Rev. Lett.* **61**, 2081–2084.

Cornelius, K.R. (2006). Updated single-electron-capture cross-section scaling rule for $A^{q+} + H_2$ collisions. *Phys. Rev. A* **73**. 032710.

Crandall, D.H., Phaneuf, R.A., Hasselquist, B.E., Gregory, D.C. (1979). Measured cross sections for ionisation of C^{3+}, N^{4+} and O^{5+} ions with contribution due to excitation-autoionisation. *J. Phys. B* **12**, L249–L256.

Crespo López-Urrutia, J., Artemyev, A., Braun, J., Brenner, G., Bruhns, H., Draganič, I., González Martínez, A., Lapierre, A., Mironov, V., Scofield, J., Soria Orts, R., Tawara, H., Trinczek, M., Tupytsin, I., Ullrich, J. (2005). High precision measurements of forbidden transitions in highly charged ions at the Heidelberg EBIT. *Nucl. Instrum. Methods B* **235**, 85–91.

Currell, F.J. (2003). Electron beam ion traps and their use in the study of highly charged ions. In: Currell, F. (Ed.), *The Physics of Multiply and Highly Charge Ions*, vol. 1. Kluwer Academic Publishers, Dordrecht, pp. 39–75.

Danared, H. (1993). Fast electron cooling with a magnetically expanded electron beam. *Nucl. Instrum. Methods A* **335**, 397.

Dance, D.F., Harrison, M.F.A., Smith, A.C.H. (1966). A measurement of the cross section for production of $He^+(2S)$ ions by electron impact excitation of ground state helium ions. *Proc. R. Soc. A* **290**, 74–93.

Datla, R.U., Blaha, M., Kunze, H.J. (1975). Collisional rate coefficients for the iron ions Fe vii, Fe ix, and Fe x. *Phys. Rev. A* **12**, 1076–1083.

Datz, S., Vane, C.R., Dittner, P.F., Giese, J.P., Campo, J.G.D., Jones, N.L., Krause, H.F., Miller, P.D., Schulz, M., Schöne, H., Rosseel, T.M. (1989). Resonant dielectronic and direct excitation in crystal channels. *Phys. Rev. Lett.* **63**, 742–745.

Defrance, P., Brouillard, F., Claeys, W., Wassenhove, G.V. (1981). Crossed beam measurement of absolute cross sections: An alternative method and its application to the electron impact ionisation of He^+. *J. Phys. B* **14**, 103–110.

Defrance, P., Duponchelle, M., Moores, D.L. (1995). Ionization of atomic ions by electron impact. In: Janev, R.K. (Ed.), *Atomic and Molecular Processes in Fusion Edge Plasmas*. Plenum, New York, pp. 153–194.

Dere, K.P. (2007). Ionization rate coefficients for the elements hydrogen through zinc. *Astron. Astrophys.* **466**, 771–792.

DeWitt, D.R., Schneider, D., Chen, M.H., Schneider, M.B., Church, D., Weinberg, G., Sakurai, M. (1993). Dielectronic recombination cross sections of fluorinelike xenon. *Phys. Rev. A* **47**, R1597–R1600.

DeWitt, D.R., Schuch, R., Quinteros, T., Gao, H., Zong, W., Danared, H., Pajek, M., Badnell, N.R. (1994). Absolute dielectronic recombination cross sections of hydrogenlike helium. *Phys. Rev. A* **50**, 1257–1264.

Dittner, P.F., Datz, S., Miller, P.D., Moak, C.D., Stelson, P.H., Bottcher, C., Dress, W.B., Alton, G.D., Nešković, N., Fou, C.M. (1983). Cross sections for dielectronic recombination of B^{2+} and C^{3+} via 2s→2p excitation. *Phys. Rev. Lett.* **51**, 31–34.

Dittner, P.F., Datz, S., Miller, P.D., Pepmiller, P.L., Fou, C.M. (1987). Dielectronic recombination measurements for the Li-like ions B^{2+}, C^{3+}, N^{4+}, and O^{5+}. *Phys. Rev. A* **35**, 3668–3673.

Dolder, K.T., Harrison, M.F.A., Thonemann, P. (1961). A measurement of the ionization cross-section of helium ions by electron impact. *Proc. R. Soc. A* **264**, 367–379.

Donets, E.D. (1983). The electron beam method of production of highly charged ions and its applications. *Phys. Scr. T* **3**, 11.

Donets, E.D., Ovsyannikov, V.P. (1981). Investigation of ionization of positive ions by electron impact. *Sov. Phys. JETP* **53**, 466.

Donets, E.D., Pikin, A.I. (1976). Ionization of positive argon ions by electron impact. *Sov. Phys. Tech. Phys.* **43**, 1057–1062.

Donets, E.D., Ilyushchenko, V.I., Alpert, V.A. (1969). Ultra high vacuum electron beam source of highly stripped ions. In: Iere *Conférérence Internationale sur Les Sources d'Ions*. 18.–20.06.1969, Saclay, pp. 635–642.

Eichler, J., Stöhlker, T. (2007). Radiative electron capture in relativistic ion–atom collisions and the photoelectric effect in hydrogen-like high-Z systems. *Phys. Reports* **439**, 1–99.

Fano, U. (1961). Effects of configuration interaction on intensities and phase shifts. *Phys. Rev.* **124**, 1866–1878.

Fano, U., Cooper, J.W. (1965). Line profiles in the far-UV absorption spectra of the rare gases. *Phys. Rev.* **137**, A1364–A1379.

Ferland, G.J. (2003). Quantitative spectroscopy of astronomical plasmas. *Annu. Rev. Astron. Astrophys.* **41**, 517–554.

Flannery, M.R. (1996). Electron–ion and ion–ion recombination. In: Drake, G.W. (Ed.), *Atomic, Molecular & Optical Physics Handbook*. AIP Press, Woodbury, New York, pp. 605–629.

Fritzsche, S., Indelicato, P., Stöhlker, T. (2005). Relativistic quantum dynamics in strong fields: Photon emission from heavy, few-electron ions. *J. Phys. B* **38**, S707–S726.

Fuchs, T., Biedermann, C., Radtke, R., Behar, E., Doron, R. (1998). Channel-specific dielectronic recombination of highly charged krypton. *Phys. Rev. A* **58**, 4518–4525.

Gabriel, A.H. (1972). Dielectronic satellite spectra for highly-charged helium-like ion lines. *Mon. Not. R. Astr. Soc.* **160**, 99–119.

Gallagher, T.F. (1994). "Rydberg Atoms". *Cambridge Monographs on Atomic, Molecular and Chemical Physics*, vol. 3. Cambridge University Press, Cambridge, UK.

Gao, H., DeWitt, D.R., Schuch, R., Zong, W., Asp, S., Pajek, M. (1995). Observation of enhanced electron–ion recombination rates at very low energies. *Phys. Rev. Lett.* **75**, 4381–4384.

Gaudin, A., Hagemann, R.J. (1967). Absolute determination of the total and partial effective ionization cross-sections of helium, neon, argon, and acetylene for 100–2000 eV electrons. *Journal de Chimie Physique* **64**, 1209–1221.

Gillaspy, J.D. (2001). Highly charged ions. *J. Phys. B* **34**, R93–R130.

Goldberg, L., Dupree, A.K., Allen, J.W. (1965). Collisional excitation of autoionizing levels. *Annales d'Astrophysique* **28**, 589–593.

González Martínez, A.J., Crespo Lopez-Urrutia, J.R., Braun, J., Brenner, G., Bruhns, H., Lapierre, A., Mironov, V., Soria Orts, R., Tawara, H., Trinczek, M., Ullrich, J., Scofield, J.H. (2005). State-selective quantum interference observed in the recombination of highly charged $Hg^{75+-78+}$ mercury ions in an electron beam ion trap. *Phys. Rev. Lett.* **94**, 203201.

González Martínez, A.J., Crespo López-Urrutia, J.R., Braun, J., Brenner, G., Bruhns, H., Lapierre, A., Mironov, V., Soria Orts, R., Tawara, H., Trinczek, M., Ullrich, J., Artemyev, A.N., Harman, Z., Jentschura, U.D., Keitel, C.H., Scofield, J.H., Tupitsyn, I.I. (2006). Benchmarking high-field few-electron correlation and QED contributions in $Hg^{75+} \to Hg^{78+}$ ions I. Experiment. *Phys. Rev. A* **73**, 052710.

Gorczyca, T.W., Pindzola, M.S., Robicheaux, F., Badnell, N.R. (1997). Strong interference effects in the $e^- + Sc^{3+}$ recombination cross section. *Phys. Rev. A* **56**, 4742–4745.

Graham, W.G., Berkner, K.H., Bernstein, E.M., Clark, M.W., Feinberg, B., McMahan, M.A., Morgan, T.J., Rathbun, W., Schlachter, A.S., Tanis, J.A. (1990). Resonant transfer and excitation for U^{90+} projectiles in hydrogen. *Phys. Rev. Lett.* **65**, 2773–2776.

Greenwood, J.B., Williams, I.D., McGuinness, P. (1995a). Large angle elastic scattering of electrons from Ar^+. *Phys. Rev. Lett.* **75**, 1062–1065.

Greenwood, J.B., Williams, I.D., Srigengan, B., Newell, W.R., Geddes, J., O'Neill, R.W. (1995b). Low-energy elastic backscattering of electrons from Ar^+. *J. Phys. B* **28**, L307–L311.

Greenwood, J.B., Smith, S.J., Chutjian, A. (1999). Experimental cross sections for electron excitation of the $2s\,^2S \to 2p\,^2P$ transition in C^{3+}. *Phys. Rev. A* **59**, 1348–1354.

Gregory, D.C., Dunn, G.H., Phaneuf, R.A., Crandall, D.H. (1979). Absolute cross section for $2s - 2p$ excitation of N^{4+} by electron impact. *Phys. Rev. A* **20**, 410–420.

Gregory, D.C., Meyer, F.W., Müller, A., Defrance, P. (1986). Experimental cross sections for electron-impact ionization of iron ions: Fe^{5+}, Fe^{6+}, and Fe^{9+}. *Phys. Rev. A* **34**, 3657–3667.

Gregory, D.C., Wang, L.J., Meyer, F.W., Rinn, K. (1987). Electron-impact ionization of iron ions: Fe^{11+}, Fe^{13+}, and Fe^{15+}. *Phys. Rev. A* **35**, 3256–3264.

Gribakin, G.F., Sahoo, S. (2003). Mixing of dielectronic and multiply excited states in electron–ion recombination: A study of Au^{24+}. *J. Phys. B* **36**, 3349–3370.

Gribakin, G.F., Gribakina, A.A., Flambaum, V.V. (1999). Quantum chaos in multicharged ions and statistical approach to the calculation of electron–ion resonant radiative recombination. *Aust. J. Phys.* **52**, 443–457.

Griem, H.R. (1988). Electron–ion collisional rate coefficients from time-dependent plasmas. *J. Quant. Spectrosc. Radiat. Transfer* **40**, 403–420.

Griffin, D.C., Pindzola, M.S. (2006). Non-perturbative quantal methods for electron–atom scattering processes. *Adv. At. Mol. Opt. Phys.* **54**, 203–235.

Griffin, D.C., Pindzola, M.S., Badnell, N.R. (1992). The contributions of double excitation–autoionization to the electron impact ionization of Li^+. *J. Phys. B* **25**, L605–L612.

Griffin, D.C., Pindzola, M.S., Badnell, N.R. (1993). Low-energy total and differential cross sections for the electron-impact excitation of Si^{2+} and Ar^{6+}. *Phys. Rev. A* **47**, 2871–2880.

Griffin, D.C., Badnell, N.R., Pindzola, M.S. (2000). Electron-impact excitation of C^{3+} and O^{5+}: The effects of coupling to the target continuum states. *J. Phys. B* **33**, 1013–1028.

Gryziński, M. (1965). Classical theory of atomic collisions I. Theory of inelastic collisions. *Phys. Rev. A* **138**, 336.

Gu, M.F. (2004). Dielectronic recombination rate coefficients of Na-like ions from Mg II to Zn XX forming Mg-like systems. *Astrophys. J. Suppl. Ser.* **153**, 389–393.

Gumberidze, A., Stöhlker, T., Banas, D., Beckert, K., Beller, P., Beyer, H.F., Bosch, F., Hagmann, S., Kozhuharov, C., Liesen, D., Nolden, F., Ma, X., Mokler, P.H., Steck, M., Sierpowski, D., Tashenov, S. (2005). Quantum electrodynamics in strong electric fields: The ground-state Lamb shift in hydrogenlike uranium. *Phys. Rev. Lett.* **94**. 223001.

Guo, X.Q., Bell, E.W., Thompson, J.S., Dunn, G.H., Bannister, M.E., Phaneuf, R.A., Smith, A.C.H. (1993). Evidence for significant backscattering in near-threshold electron-impact excitation of $Ar^{7+}(3s \rightarrow 3p)$. *Phys. Rev. A* **47**, R9–R12.

Gwinner, G., Hoffknecht, A., Bartsch, T., Beutelspacher, M., Eklöw, N., Glans, P., Grieser, M., Krohn, S., Lindroth, E., Müller, A., Saghiri, A.A., Schippers, S., Schramm, U., Schwalm, D., Tokman, M., Wissler, G., Wolf, A. (2000). Influence of magnetic fields on electron–ion recombination at very low energies. *Phys. Rev. Lett.* **84**, 4822–4825.

Hahn, Y., LaGattuta, K.J. (1988). Dielectronic recombination and related resonance processes. *Phys. Reports* **166**, 195–268.

Hamdan, M., Birkinshaw, K., Hasted, J.B. (1978). Ionisation of positive ions by electrons in the hollow-beam trap. *J. Phys. B* **11**, 331–337.

Harman, Z., Tupitsyn, I.I., Artemyev, A.N., Jentschura, U.D., Keitel, C.H., Crespo Lopez-Urrutia, J.R., González Martínez, A.J., Tawara, H., Ullrich, J. (2006). Benchmarking high-field few-electron correlation and QED contributions in Hg^{75+} to Hg^{78+} ions. II. Theory. *Phys. Rev. A* **73**. 052711.

Hasted, J.B. (1983). Confinement of ions for collision studies. In: Brouillard, F., McGowan, J.W. (Eds.), *Physics of Ion–Ion and Electron–Ion Collisions*. In: *NATO ASI Series B: Physics*, vol. 83. Plenum Press, New York, pp. 461–500.

Hasted, J.B., Awad, G.L. (1972). Electron impact ionization of ions trapped in a hollow electron beam. *J. Phys. B* **5**, 1719–1743.

Henry, R.J.W., Msezane, A.Z. (1982). Cross sections for inner-shell excitation of Na-like ions. *Phys. Rev. A* **26**. 2545.

Hinnov, E. (1966). Excitation and ionization rates of neon ions in a stellarator discharge. *J. Opt. Soc. Am.* **56**, 1179.

Hoffknecht, A., Uwira, O., Frank, A., Schennach, S., Spies, W., Wagner, M., Schippers, S., Müller, A., Becker, R., Kleinod, M., Angert, N., Mokler, P.H. (1998). Recombination of Au^{25+} with free electrons at very low energies. *J. Phys. B* **31**, 2415–2428.

Hoffknecht, A., Schippers, S., Müller, A., Gwinner, G., Schwalm, D., Wolf, A. (2001). Recombination of bare Cl^{17+} ions in an electron cooler. *Phys. Scr. T* **92**, 402–405.

Hofmann, G., Müller, A., Tinschert, K., Salzborn, E. (1990). Indirect processes in the electron-impact ionization of Li-like ions. *Z. Phys. D* **60**, 113.

Hofmann, G., Müller, A., Weißbecker, B., Stenke, M., Tinschert, K., Salzborn, E. (1991). Resonances in the electron-impact single and multiple ionization of ions. *Z. Phys. D Supplement* **21**, S189–S191.

Hörndl, M., Yoshida, S., Wolf, A., Gwinner, G., Seliger, M., Burgdörfer, J. (2006). Classical dynamics of enhanced low-energy electron–ion recombination in storage rings. *Phys. Rev. A* **74**. 052712.

Howald, A.M., Gregory, D.C., Meyer, F.W., Phaneuf, R.A., Müller, A., Djurić, N., Dunn, G.H. (1986). Electron-impact ionization of Mg-like ions: S^{4+}, Cl^{5+}, and Ar^{6+}. *Phys. Rev. A* **33**, 3779–3786.

Huber, B.A., Ristori, C., Hervieux, P.A., Maurel, M., Guet, C., Andrä, H.J. (1991). Differential cross sections for the 3s-3p excitation of sodiumlike Ar^{7+} ions by electron impact. *Phys. Rev. Lett.* **67**, 1407–1410.

Huber, B.A., Ristori, C., Guet, C., Küchler, D., Johnson, W.R. (1994). Elastic scattering of electrons from heavy multiply charged ions (Xe^{6+}, Xe^{8+}, and Ba^{2+}). *Phys. Rev. Lett.* **73**, 2301–2303.

Hvelplund, P., Gonzáles, A.D., Dahl, P., Bhalla, C.P. (1994). Resonant inelastic scattering of quasifree electrons on C^{5+}(1s). *Phys. Rev. A* **49**, 2535–2540.

Ichihara, A., Eichler, J. (2000). Cross sections for radiative recombination and the photoelectric effect in the K, L and M shells of one-electron systems with $1 \leqslant Z \leqslant 112$ calculated within an exact relativistic description. *At. Data Nucl. Data Tables* **74**, 1–121.

Isler, R.C., Crume, E.C., Arnurius, D.E. (1982). Ionization and recombination coefficients for Fe xv–Fe xix. *Phys. Rev. A* **26**, 2105–2116.

Itikawa, Y. (2002). Annotated bibliography on electron collisions with atomic positive ions: Excitation and ionization, 1995–1999. *At. Data Nucl. Data Tables* **80**, 117–146.

Itoh, A., Schneider, T., Schiwietz, G., Roller, Z., Platten, H., Nolte, G., Scheider, D., Stolterfoht, N. (1983). Selective production of auger electrons from fast projectile ions studied by zero-degree Auger spectroscopy. *J. Phys. B* **16**, 3965–3971.

Jacobi, J., Knopp, H., Schippers, S., Müller, A., Loch, S.D., Witthoeft, M., Pindzola, M.S., Ballance, C.P. (2004). Strong contributions of indirect processes to the electron-impact ionization cross section of Sc^+ ions. *Phys. Rev. A* **70**. 042571.

Jacobs, V.L., Davies, J., Kepple, P.C. (1976). Enhancement of dielectronic recombination by plasma electric microfields. *Phys. Rev. Lett.* **37**, 1390–1393.

Janzen, P.H., Gardner, L.D., Reisenfeld, D.B., Savin, D.W., Kohl, J.L., Bartschat, K. (1999). Reevaluation of experiments and new theoretical calculations for electron-impact excitation of C^{3+}. *Phys. Rev. A* **59**, 4821–4824.

Jentschura, U.D. (2004). Self-energy correction to the two-photon decay width in hydrogenlike atoms. *Phys. Rev. A* **69**. 052118.

Jonkers, J. (2006). High power extreme ultra-violet (EUV) light sources for future lithography. *Plasma Sources Sci. Technol.* **15**, S8–S16.

Kallman, T., Bautista, M. (2001). Photoionization and high-density gas. *Astrophys. J. Suppl. Ser.* **133**, 221–253.

Kallman, T.R., Palmeri, P. (2007). Atomic data for x-ray astrophysics. *Rev. Mod. Phys.* **79**, 79–133.

Källne, E., Jones, L.A. (1977). Measurements of the ionisation rates of lithium-like ions. *J. Phys. B* **10**, 3637–3648.

Kandler, T., Stöhlker, T., Mokler, P.H., Kozhuharov, C., Geissel, H., Scheidenberger, C., Rymuza, P., Stachura, Z., Warczak, A., Dunford, R.W., Eichler, J., Ichihara, A., Shirai, T. (1995). Photon angular distribution of radiative electron capture into the M shell of He-like uranium ions at 100–140 MeV/u. *Z. Phys. D* **35**, 15–18.

Kenntner, J., Linkemann, J., Badnell, N.R., Broude, C., Habs, D., Hofmann, G., Müller, A., Pindzola, M.S., Salzborn, E., Schwalm, D., Wolf, A. (1995). Resonant electron impact ionization and recombination of Li-like Cl^{14+} and Si^{11+} at the Heidelberg Test Storage Ring. *Nucl. Instrum. Methods B* **98**, 142–145.

Kieslich, S., Schippers, S., Shi, W., Müller, A., Gwinner, G., Schnell, M., Wolf, A., Lindroth, E., Tokman, M. (2004). Determination of the 2s-2p excitation energy of lithiumlike scandium using dielectronic recombination. *Phys. Rev. A* **70**. 042714.

Kilgus, G., Berger, J., Blatt, P., Grieser, M., Habs, D., Hochadel, B., Jaeschke, E., Krämer, D., Neumann, R., Neureither, G., Ott, W., Schwalm, D., Steck, M., Stokstad, R., Szmola, E., Wolf, A., Schuch, R., Müller, A., Wagner, M. (1990). Dielectronic recombination of hydrogenlike oxygen in a heavy-ion storage ring. *Phys. Rev. Lett.* **64**, 737–740.

Knapp, D.A., Marrs, R.E., Levine, M.A., Bennet, C.L., Chen, M.H., Henderson, J.R., Schneider, M.B., Scofield, J.H. (1989). Dielectronic recombination of heliumlike nickel. *Phys. Rev. Lett.* **62**, 2104–2107.

Knapp, D.A., Beiersdorfer, P., Chen, M.H., Scofield, J.H., Schneider, D. (1995). Observation of interference between dielectronic recombination and radiative recombination in highly charged uranium ions. *Phys. Rev. Lett.* **74**, 54–57.

Knopp, H., Böhme, C., Jacobi, J., Ricz, S., Schippers, S., Müller, A. (2003). Electron-impact multiple ionization of Ba^{q+} ions ($1 \leqslant q \leqslant 13$) via resonant 3d excitation. *Nucl. Instrum. Methods B* **205**, 433–436.

Kollmus, H., Moshammer, R., Olson, R.E., Hagmann, S., Schulz, M., Ullrich, J. (2002). Simultaneous projectile-target ionization: A novel approach to (e, 2e) experiments on ions. *Phys. Rev. Lett.* **88**. 103202.

König, R., Kohl, K.-H., Kunze, H.-J. (1993). Experimental electron-impact excitation rate coefficients for lithium-like Si XII. *Phys. Scr.* **48**, 9–24.

König, R., Kolk, K.-H., Kunze, H.-J. (1996). Experimental electron-impact excitation rate coefficients for beryllium-like Si XI. *Phys. Scr.* **53**, 679–688.

Korol, A., Gribakin, G.F., Currell, F.J. (2006). Effect of target polarization in electron–ion recombination. *Phys. Rev. Lett.* **97**. 223201.

Kraemer, S.B., Ferland, G.J., Gabel, J.R. (2004). The effects of low-temperature dielectronic recombination on the relative populations of the Fe M-shell states. *Astrophys. J.* **604**, 556–561.

Kramers, H.A. (1923). On the theory of X-ray absorption and of the continuous X-ray spectrum. *Philos. Mag.* **46**, 836–871.

Krause, H.F., Datz, S. (1996). Channeling heavy ions through crystalline lattices. *Adv. At. Mol. Opt. Phys.* **37**, 139–180.

Krücken, T., Bergmann, K., Juschkin, L., Lebert, R. (2004). Fundamentals and limits for the EUV emission of pinch plasma sources for EUV lithography. *J. Phys. D* **37**, 3213–3224.

Krylstedt, P., Pindzola, M.S., Badnell, N.R. (1990). External field effects on dielectronic recombination rate coefficients for oxygen atomic ions. *Phys. Rev. A* **41**, 2506–2514.

Kunze, H.J. (1971). Collisional ionization rates for lithium- and beryllium-like ions. *Phys. Rev. A* **3**, 937–942.

Kunze, H.J., Gabriel, A.H., Griem, H.R. (1968). Measurement of collisional rate coefficients for heliumlike carbon ions in a plasma. *Phys. Rev.* **165**, 267–276.

Labzowsky, L.N., Nefiodov, A.V. (1994). Radiative interference effects in the dielectronic-recombination process of an electron with hydrogenlike uranium. *Phys. Rev. A* **49**, 236–238.

LaGattuta, K.J., Hahn, Y. (1981). Electron impact ionization of fe^{15+} by resonant double Auger ionization. *Phys. Rev. A* **24**, 2273–2276.

LaGattuta, K.J., Hahn, Y. (1983). Effect of extrinsic electric fields upon dielectronic recombination: Mg^{1+}. *Phys. Rev. Lett.* **51**, 558–561.

Lee, D.H., Richard, P., Zouros, T.J.M., Sanders, J.M., Shinpaugh, J.L., Hidmi, H. (1990). Binary-encounter electrons observed at 0° in collisions of 1–2-MeV/amu H^+, C^{6+}, N^{7+}, O^{8+}, and F^{9+} ions with H_2 and He targets. *Phys. Rev. A* **41**, 4816–4823.

Leitner, D., Todd, D.S., Galloway, M.L., Lyneis, C.M. (2007). Recent 28 GHz results with VENUS. In: *Proceedings of the 17th Int. Conf. On ECR Ion Sources (ECRIS'06)*. Lanzhou, China.

Lenard, P. (1906). On cathode rays. In: *Nobel Lectures in Physics 1901–1921*. World Scientific, Singapore, pp. 105–134.

Lestinsky, M., Lindroth, E., Orlov, D.A., Schmidt, E.W., Schippers, S., Böhm, S., Brandau, C., Sprenger, F., Terekhov, A.S., Müller, A., Wolf, A. (submitted for publication). Screened radiative corrections from hyperfine split dielectronic resonances in lithiumlike scandium. *Phys. Rev. Lett.*

Lindroth, E., Schuch, R. (2003). Recombination of cooled highly charged ions with low energy electrons. In: Currell, F.J. (Ed.), *The Physics of Multiply and Highly Charged Ions*, vol. 1. Kluwer Academic Publishers, Dordrecht, pp. 231–268.

Lindroth, E., Danared, H., Glans, P., Pesic, Z., Tokman, M., Vikor, G., Schuch, R. (2001). QED effects in Cu-like Pb recombination resonances near threshold. *Phys. Rev. Lett.* **86**, 5027–5030.

Linkemann, J., Kenntner, J., Müller, A., Wolf, A., Habs, D., Schwalm, D., Spies, W., Uwira, O., Frank, A., Liedtke, A., Hofmann, G., Salzborn, E., Badnell, N.R., Pindzola, M.S. (1995a). Electron impact ionization and dielectronic recombination of sodium-like iron ions. *Nucl. Instrum. Methods B* **98**, 154–157.

Linkemann, J., Müller, A., Kenntner, J., Habs, D., Schwalm, D., Wolf, A., Badnell, N.R., Pindzola, M.S. (1995b). Electron-impact ionization of Fe^{15+} ions: An ion storage ring cross section measurement. *Phys. Rev. Lett.* **74**, 4173–4176.

Lister, G. (2004). Keeping the lights burning: The drive for energy efficient lighting. *Optics and Photonics News* **15**, 20–25.

Loch, S.D., Witthoeft, M., Pindzola, M.S., Bray, I., Fursa, D.V., Fogle, M., Schuch, R., Glans, P., Ballance, C.P., Griffin, D.C. (2005). Influence of long-lived metastable levels on the electron-impact single ionization of C^{2+}. *Phys. Rev. A* **71**. 012716.

Lotz, W. (1967). An empirical formula for the electron-impact ionization cross-section. *Z. Phys.* **206**, 205–211.

Lotz, W. (1968). Electron-impact ionization cross-sections and ionization rate coefficients for atoms and ions from Hydrogen to Calcium. *Z. Phys.* **216**, 241–247.

Lotz, W. (1969). Electron-impact ionization cross-sections and ionization rate coefficients for atoms and ions from Scandium to Zinc. *Z. Phys.* **220**, 466–472.

Lotz, W. (1970). Electron-impact ionization cross-sections for atoms up to $Z = 108$. *Z. Phys.* **232**, 101–107.

Lozano, J.A., Niimura, M., Smith, S.J., Chutjian, A., Tayal, S.S. (2001). Experimental and theoretical cross sections for electron-impact excitation of the $2s \rightarrow 2p$ transition in O^{5+}. *Phys. Rev. A* **63**. 042713.

Lu, M., Phaneuf, R.A. (2002). Electron spectroscopy of Na-like autoionizing metastable ions. *Phys. Rev. A* **66**. 012706.

Ma, X., Mokler, P.H., Bosch, F., Gumberidze, A., Kozhuharov, C., Liesen, D., Sierpowski, D., Stachura, Z., Stohlker, T., Warczak, A. (2003). Electron–electron interaction studied in strong central fields by resonant transfer and excitation with H-like U ions. *Phys. Rev. A* **68**. 042712.

Mannervik, S., Asp, S., Broström, L., DeWitt, D.R., Lidberg, J., Schuch, R., Chung, K.T. (1997). Spectroscopic study of lithiumlike carbon by dielectronic recombination of a stored ion beam. *Phys. Rev. A* **55**, 1810–1819.

Mannervik, S., DeWitt, D., Engström, L., Lidberg, J., Lindroth, E., Schuch, R., Zong, W. (1998). Strong relativistic effects and natural linewidths observed in dielectronic recombination of lithiumlike carbon. *Phys. Rev. Lett.* **81**, 313–316.

Marchuk, O. (2004). "Modeling of He-like spectra measured at the tokamaks TEXTOR and TORE SUPRA". PhD thesis. Fakultät für Physik und Astronomie an der Ruhr-Universität Bochum.

Marchuk, O., Tokar, M.Z., Bertschinger, G., Urnov, A., Kunze, H.-J., Pilipenko, D., Loozen, X., Kalupin, D., Reiter, D., Pospieszczyk, A., Biel, W., Goto, M., Goryaev, F. (2006). Comparison of impurity transport model with measurements of He-like spectra of argon at the tokamak TEXTOR. *Plasma Phys. Control. Fusion* **48**, 1633–1646.

Marrs, R.E., Levine, M.A., Knapp, D.A., Henderson, J.R. (1988). Measurement of electron-impact–excitation cross sections for very highly charged ions. *Phys. Rev. Lett.* **60**, 1715–1718.

Marrs, R.E., Elliott, S.R., Knapp, D.A. (1994). Production and trapping of hydrogenlike and bare uranium ions in an electron beam ion trap. *Phys. Rev. Lett.* **72**, 4082–4085.

Marrs, R.E., Elliott, S.R., Scofield, J.H. (1997). Measurement of electron-impact ionization cross sections for hydrogenlike high-Z ions. *Phys. Rev. A* **56**, 1338–1345.

Mathur, D., Khan, S.U., Hasted, J.B. (1978). Dissociative recombination in low-energy e-H_2^+ and e-H_3^+ collisions. *J. Phys. B* **11**, 3615–3619.

Mathur, D., Hasted, J.B., Khan, S.U. (1979). Collision processes of electrons with molecular hydrogen ions. *J. Phys. B* **12**, 2043–2050.

McGuire, J.H. (1997). "Electron Correlation Dynamics in Atomic Collisions". Cambridge University Press, Cambridge.

Mitchell, J.B.A., Ng, C.T., Forand, J.L., Levac, D.P., Mitchell, R.E., Sen, A., Miko, D.B., McGowan, J.W. (1983). Dielectronic-recombination cross-section measurements for C^+ ions. *Phys. Rev. Lett.* **50**, 335–338.

Mitnik, D.M., Pindzola, M.S., Robicheaux, F., Badnell, N.R., Uwira, O., Müller, A., Frank, A., Linkemann, J., Spies, W., Angert, N., Mokler, P.H., Becker, R., Kleinod, M., Ricz, S., Empacher, L. (1998). Dielectronic recombination of U^{28+} atomic ions. *Phys. Rev. A* **57**, 4365–4372.

Mitnik, D.M., Griffin, D.C., Colgan, J., Pindzola, M.S., Aichele, K., Arnold, W., Hathiramani, D., Scheuermann, F., Salzborn, E. (2001). Electron-impact ionization of Sm^{12+} ions: Resonances far beyond threshold. *Phys. Rev. A* **64**. 062705.

Mittleman, M.H. (1966a). Single and double ionization of He by electrons. *Phys. Rev. Lett.* **16**, 498–499.

Mittleman, M.H. (1966b). Single and double ionization of He by electrons. *Phys. Rev. Lett.* **16**, 779.

Mokler, P.H., Stöhlker, T. (1996). The physics of highly-charged heavy ions revealed by storage/cooler rings. *Adv. At. Mol. Opt. Phys.* **37**, 297–370.

Mokler, P.H., Reusch, S., Stöhlker, T., Schuch, R., Schulz, M., Wintermeyer, G., Stachura, Z., Warczak, A., Müller, A., Awaya, Y., Kambara, T. (1989). Resonant transfer and excitation in swift, heavy few-electron projectiles. *Radiation Effects and Defects in Solids* **110**, 39.

Mokler, P.H., Reusch, S., Warczak, A., Stachura, Z., Kambara, T., Müller, A., Schuch, R., Schulz, M. (1990). Single transfer-excitation resonance observed via the two-photon decay in He-like Ge^{30+}. *Phys. Rev. Lett.* **65**, 3108–3111.

Moores, D.L., Reed, K.J. (1994). Indirect processes in electron impact ionization of positive ions. *Adv. At. Mol. Phys.* **34**, 301–426.

Müller, A. (1986). Multiple ionization and the charge state evolution of ions exposed to electron impact. *Phys. Lett. A* **113**, 415–419.

Müller, A. (1989). Targets consisting of charged particles. *Nucl. Instrum. Methods A* **282**, 80–86.

Müller, A. (1991). Ion formation processes: Ionization in ion–electron collisions. In: Mathur, D. (Ed.), *Physics of Ion-Impact Phenomena*. In: *Springer Series in Chemical Physics*, vol. 54. Springer, Berlin, Heidelberg, New York, pp. 13–90.

Müller, A. (1992). Electron-ion recombination phenomena: Formation and decay of intermediate resonant states. In: Graham, W.G., Fritsch, W., Hahn, Y., Tanis, J.A. (Eds.), *Recombination of Atomic Ions*. In: *NATO ASI Series B: Physics*, vol. 296. Plenum Press, New York, pp. 155–179.

Müller, A. (1995). Dielectronic recombination and ionization in electron–ion collisions: Data from merged-beams experiments. *Nucl. Fusion Suppl.* **6**, 59–97.

Müller, A. (1996). Fundamentals of electron–ion interaction. *Hyperfine Interactions* **99**, 31–45.

Müller, A. (1997). Dielectronic capture and subsequent relaxation in electron–ion collisions. *Nucl. Instr. and Meth.* **124**, 281–289.

Müller, A. (1999). Plasma rate coefficients for highly charged ion–electron collisions: New experimental access via ion storage rings. *Int. J. Mass Spectrom.* **192**, 9–22.

Müller, A. (2002). Experimental data for electron–ion collisions. In: Schultz, D.R., Kristić, P.S., Ownby, F. (Eds.), *Proceedings of the 3rd International Conference on Atomic and Molecular Data and Their Applications (ICAMDATA)*. In: *AIP Conference Proceedings*, vol. 636. American Institute of Physics, New York, USA, pp. 202–212.

Müller, A. (2005). Many-electron phenomena in the ionization of ions. *Nucl. Instrum. Methods B* **233**, 141–150.

Müller, A., Frodl, R. (1980). *L*-shell contributions to multiple ionization of Ar^{i+} ions ($i = 1, 2, 3$) by electron impact. *Phys. Rev. Lett.* **44**, 29–32.

Müller, A., Salzborn, E. (1977). Scaling of cross sections for multiple electron transfer to highly charged ions colliding with atoms and molecules. *Phys. Lett. A* **62**, 391–394.

Müller, A., Schippers, S. (2001). Dielectronic recombination: Experiment. In: Ferland, G., Savin, D.W. (Eds.), *Spectroscopic Challenges of Photoionized Plasmas*. In: *ASP Conference Proceedings*, vol. 247. Astronomical Society of the Pacific, San Francisco, USA, pp. 53–78.

Müller, A., Schippers, S. (2003). Dielectronic recombination in external electromagnetic fields. In: Currell, F. (Ed.), In: *The Physics of Multiply and Highly Charged Ions*, vol. 1. Kluwer Academic Publishers, Dordrecht, pp. 269–301.

Müller, A., Wolf, A. (1997). Heavy ion storage rings. In: Austin, J.C., Shafroth, S.M. (Eds.), *Accelerator-Based Atomic Physics Techniques and Applications*. AIP Press, Woodbury, p. 147.

Müller, A., Salzborn, E., Frodl, R., Becker, R., Klein, H., Winter, H. (1980). Absolute ionisation cross sections for electrons incident on O^+, Ne^+, Xe^+ and Ar^{i+} ($i = 1, \ldots, 5$) ions. *J. Phys. B* **13**, 1877–1899.

Müller, A., Achenbach, C., Salzborn, E., Becker, R. (1984). Multiple ionisation of multiply charged xenon ions by electron impact. *J. Phys. B* **17**, 1427–1444.

Müller, A., Huber, K., Tinschert, K., Becker, R., Salzborn, E. (1985). An improved crossed-beams technique for the measurement of absolute cross sections for electron impact ionisations of ions and its application to Ar^+ ions. *J. Phys. B* **18**, 2993–3009.

Müller, A., Belić, D.S., DePaola, B.D., Djurić, N., Dunn, G.H., Mueller, D.W., Timmer, C. (1986). Field effects on the Rydberg product-state distribution from dielectronic recombination. *Phys. Rev. Lett.* **56**, 127–130.

Müller, A., Belić, D.S., DePaola, B.D., Djurić, N., Dunn, G.H., Mueller, D.W., Timmer, C. (1987a). Experimental measurements of field effects on dielectronic recombination cross sections and Rydberg product-state distributions. *Phys. Rev. A* **36**, 599–613.

Müller, A., Hofmann, G., Tinschert, K., Sauer, R., Salzborn, E. (1987b). Signal enhancement in electron–ion crossed-beams experiments. *Nucl. Instrum. Methods B* **24**, 369–372.

Müller, A., Hofmann, G., Tinschert, K., Salzborn, E. (1988a). Dielectronic capture with subsequent two-electron emission in electron-impact ionization of C^{3+} ions. *Phys. Rev. Lett.* **61**, 1352–1355.

Müller, A., Tinschert, K., Hofmann, G., Salzborn, E., Dunn, G. (1988b). Resonances in electron-impact single, double and triple ionization of heavy metal ions. *Phys. Rev. Lett.* **61**, 70–73.

Müller, A., Hofmann, G., Weissbecker, B., Stenke, M., Tinschert, K., Wagner, M., Salzborn, E. (1989). Correlated two-electron transitions in electron-impact ionization of Li^+ ions. *Phys. Rev. Lett.* **63**, 758–761.

Müller, A., Hofmann, G., Tinschert, K., Weissbecker, B., Salzborn, E. (1990). Doubly autoionizing capture resonances in $e + Mg^+$ collisions. *Z. Phys. D* **15**, 145–149.

Müller, A., Schennach, S., Wagner, M., Haselbauer, J., Uwira, O., Spies, W., Jennewein, E., Becker, R., Kleinod, M., Pröbstel, U., Angert, N., Klabunde, J., Mokler, P.H., Spädtke, P., Wolf, B. (1991). Recombination of free electrons with ions. *Phys. Scr. T* **37**, 62–65.

Müller, A., Teng, H., Hofmann, G., Phaneuf, R.A., Salzborn, E. (2000). Autoionizing resonances in electron-impact ionization of O^{5+} ions. *Phys. Rev. A* **62**. 062720.

Müller, A., Böhme, C., Jacobi, J., Knopp, H., Schippers, S. (2001). Electron-impact ionization of He-like metastable ions. In: Datz, S., Bannister, M.E., Krause, L.H., Saddiq, D., Schultz, C.R., Vane, H.F. (Eds.), *Book of Abstracts of the XXIInd International Conference on Photonic, Electronic and Atomic Collisions*. Rinton Press, Princeton, New Jersey, p. 322.

Müller, A., Phaneuf, R.A., Aguilar, A., Gharaibeh, M.F., Schlachter, A.S., Alvarez, I., Cisneros, C., Hinojosa, G., McLaughlin, B.M. (2002). Photoionization of C^{2+} ions: Time-reversed recombination of C^{3+} with electrons. *J. Phys. B* **35**, L137–L143.

Murakami, I., Kato, T., Kato, D., Safronova, U.I., Cowan, T.E., Ralchenko, Y. (2006). Large-scale calculation of dielectronic recombination parameters for Mg-like Fe. *J. Phys. B* **39**, 2917–2937.

Nagy, P., Skutlartz, A., Schmidt, V. (1980). Absolute ionisation cross sections for electron impact in rare gases. *J. Phys. B* **13**, 1249–1267.

Nahar, S.N., Pradhan, A.K. (2006). Dielectronic satellite spectra of heliumlike iron and nickel from the unified recombination method. *Phys. Rev. A* **73**. 062718.

Netzer, H. (2004). The iron unresolved transition array in active galactic nuclei. *Astrophys. J.* **604**, 551–555.

Niimura, M., Smith, I.C.S.J., Chutjian, A. (2002). Measurement of absolute cross sections for excitation of the $3s^2 3p^5\,^2P^o_{3/2} - 3s^2 3p^5\,^2P^o_{1/2}$ fine-structure transition in Fe^{9+}. *Phys. Rev. Lett.* **88**. 103201.

Oreg, J., Goldstein, W.H., Klapisch, M., Bar-Shalom, A. (1991). Autoionization and radiationless electron capture in complex spectra. *Phys. Rev. A* **44**, 1750–1758.

Orlov, D.A., Lestinsky, M., Sprenger, F., Schwalm, D., Terekhov, A.S., Wolf, A. (2006). Ultra-cold electron beams for the Heidelberg TSR and CSR. In: Nagaitsev, S., Pasquinelli, R.J. (Eds.), *Beam Cooling and Related Topics: International Workshop on Beam Cooling and Related Topics—COOL05*. In: *AIP Conference Proceedings*, vol. 821. March 20, 2006, AIP, pp. 478–487. URL http://link.aip.org/link/?APC/821/478/1.

O'Rourke, B., Currell, F.J., Kuramoto, H., Li, Y.M., Ohtani, S., Tong, X.M., Watanabe, H. (2001). Electron-impact ionization of hydrogen-like iron ions. *J. Phys. B* **34**, 4003–4013.

O'Rourke, B.E., Kuramoto, H., Li, Y.M., Ohtani, S., Tong, X.M., Watanabe, H., Currell, F.J. (2004). Dielectronic recombination in He-like titanium ions. *J. Phys. B* **37**, 2343–2353.

Osterbrock, D.E., Ferland, G.J. (2006). "Astrophysics of Gaseous Nebulae and Active Galactic Nuclei", second edition. University Science Books, Sausalito, California 94965, USA.

Paerels, F.B.S., Kahn, S.M. (2003). High-resolution X-ray spectroscopy with Chandra and XMM-Newton. *Annu. Rev. Astron. Astrophys.* **41**, 291–342.

Pálffy, A., Scheid, W., Harman, Z. (2006). Theory of nuclear excitation by electron capture for heavy ions. *Phys. Rev. A* **73**. 012715.

Pálffy, A., Harman, Z., Scheid, W. (2007a). Quantum interference between nuclear excitation by electron capture and radiative recombination. *Phys. Rev. A* **75**. 012709.

Pálffy, A., Harman, Z., Surzhykov, A., Jentschura, U.D. (2007b). Photon angular distribution and nuclear-state alignment in nuclear excitation by electron capture. *Phys. Rev. A* **75**. 012712.

Peart, B., Dolder, K.T. (1968). Measurements of cross sections for the ionization of Li^+ and Ba^+ ions by electron impact. *J. Phys. B* **1**, 872–878.

Peart, B., Dolder, K.T. (1969). The ionization of Li^+ to Li^{3+} by electron impact. *J. Phys. B* **2**, 1169–1175.

Peart, B., Stevenson, J.G., Dolder, K.T. (1973). Measurements of cross sections for the ionization of Ba^+ by energy resolved electrons. *J. Phys. B* **6**, 146–149.

Peart, B., Rinn, K., Dolder, K.T. (1983). Measurements of cross sections for inelastic collisions between $^4He^+$ ions. *J. Phys. B* **16**, 2831–2835.

Peart, B., Underwood, J.R.A., Dolder, K. (1989). Autoionisation and threshold ionisation of Ba^+ by energy-resolved electrons. *J. Phys. B* **22**, 1679–1689.

Peart, B., Thomason, J.W.G., Dolder, K. (1991). Direct and indirect ionization of Mg^+ by energy-resolved electrons. *J. Phys. B* **24**, 4453–4461.

Phaneuf, R.A., Havener, C.C., Dunn, G.H., Müller, A. (1999). Merged-beams experiments in atomic and molecular physics. *Rep. Prog. Phys.* **62**, 1143–1180.

Pindzola, M.S., Griffin, D.C., Bottcher, C. (1986). Electron-impact ionization in the iron isonuclear sequence. *Phys. Rev. A* **34**, 3668–3675.

Pindzola, M.S., Badnell, N.R., Griffin, D.C. (1990). Dielectronic recombination cross sections for H-like ions. *Phys. Rev. A* **42**, 282–285.

Pindzola, M.S., Badnell, N.R., Griffin, D.C. (1992a). Dielectronic recombination between hyperfine levels of the ground state of Bi^{82+}. *Phys. Rev. A* **45**, R7659–R7662.

Pindzola, M.S., Badnell, N.R., Griffin, D.C. (1992b). Validity of the independent-processes and isolated-resonance approximations for electron–ion recombination. *Phys. Rev. A* **46**, 5725–5729.

Pindzola, M.S., Robicheaux, F.J., Badnell, N.R., Chen, M.H., Zimmermann, M. (1995). Photorecombination of highly charged uranium ions. *Phys. Rev. A* **52**, 420–425.

Popović, D.B., Bannister, M.E., Clark, R.E.H., Chung, Y.-S., Djurić, N., Meyer, F.W., Müller, A., Neau, A., Pindzola, M.S., Smith, A.C.H., Wallbank, B., Dunn, G.H. (2002). Absolute cross sections for electron-impact excitation of the $3d^2\ ^3F \to 3d4p\ ^3D,\ ^3F$ transitions in Ti^{2+}. *Phys. Rev. A* **65**. 034704.

Poth, H. (1990). Electron cooling: theory, experiment, application. *Phys. Reports* **196**, 135–297.

Ralchenko, Y., Jou, F.-C., Kelleher, D.E., Kramida, A.E., Musgrove, A., Reader, J., Wiese, W.L., Olsen, K. (2006). "NIST Atomic Spectra Database: Version 3.1.0". National Institute of Standards and Technology, Gaithersburg, MD. Available at http://physics.nist.gov/asd3.

Ramsauer, C., Kollath, R. (1932). Die Winkelverteilung bei der Streuung langsamer Elektronen an Gasmolekülen III. *Fortsetzung und Schluß. Ann. Phys. (Leipzig)* **404**, 837–848.

Redhead, P.A. (1971). Sequential impact mass spectrometry: Estimates of metastable ion cross sections. *Can. J. Phys.* **49**. 3059.

Richard, P., Bhalla, C., Hagmann, S., Zavodsky, P. (1999). Quasi-free electron–ion scattering in ion–atom collisions. *Phys. Scr. T* **80**, 87–92.

Rinn, K., Melchert, F., Rink, K., Salzborn, E. (1986). Ionisation in H^+–He^+ collisions using an improved beam-pulsing technique for measuring the formation of he^{2+}. *J. Phys. B* **19**, 3717–3726.

Ristori, C., Hervieux, P.A., Maurel, M., Andrä, H.J., Brenac, A., Crançon, J., Lamboley, G., Lamy, T., Perrin, P., Rocco, J.C., Zadworny, F., Huber, B.A. (1991). Experimental study of electron impact excitation of multiply charged ions. *Z. Phys. D* **22**, 407–409.

Robicheaux, F., Pindzola, M.S., Griffin, D.C. (1998). Effect of interacting resonances on dielectronic recombination in static fields. *Phys. Rev. Lett.* **80**, 1402–1405.

Romanik, C.J. (1988). The dielectronic recombination rate coefficient for ions in the He, Li, Be and Ne isoelectronic sequences. *Astrophys. J.* **330**, 1022–1035.

Roszman, L.J. (1989). The dielectronic recombination rate coefficient of O^{5+} and O^{2+}. *Phys. Scr. T* **28**, 36–38.

Saghiri, A.A., Linkemann, J., Schmitt, M., Schwalm, D., Wolf, A., Bartsch, T., Hoffknecht, A., Müller, A., Graham, W.G., Price, A.D., Badnell, N.R., Gorczyca, T.W., Tanis, J.A. (1999). Dielectronic recombination of ground-state and metastable Li^+-ions. *Phys. Rev. A* **60**, R3350–R3352.

Santos, A.C.F., DuBois, R.D. (2004). Scaling laws for single and multiple electron loss from projectiles in collisions with a many-electron target. *Phys. Rev. A* **69**. 042709.

Savin, D.W. (2005). Ionization and recombination with electrons: Laboratory measurements and observational consequences. In: Smith, R.K. (Ed.), *X-Ray Diagnostics of Astrophysical Plasmas: Theory, Experiment, and Observation*. In: *AIP Conference Proceedings*, vol. 774. AIP, Cambridge, Massachusetts, USA, pp. 297–304. URL http://link.aip.org/link/?APC/774/297/1.

Savin, D.W., Gardner, L.D., Reisenfeld, D.B., Young, A.R., Kohl, J.L. (1995). Absolute-rate coefficient for $C^{3+}(2s \to 2p)$ electron-impact excitation. *Phys. Rev. A* **51**, 2162–2168.

Savin, D.W., Gardner, L.D., Reisenfeld, D.B., Young, A.R., Kohl, J.L. (1996). Absolute measurement of dielectronic recombination for C^{3+} in a known external field. *Phys. Rev. A* **53**, 280–289.

Savin, D.W., Gwinner, G., Grieser, M., Repnow, R., Schnell, M., Schwalm, D., Wolf, A., Zhou, S.-G., Kieslich, S., Müller, A., Schippers, S., Colgan, J., Loch, S.D., Chen, M.H., Gu, M.F. (2006). Dielectronic recombination of Fe XXIII forming Fe XXII: Laboratory measurements and theoretical calculations. *Astrophys. J.* **642**, 1275–1285.

Schennach, S., Müller, A., Uwira, O., Haselbauer, J., Spies, W., Frank, A., Wagner, M., Becker, R., Kleinod, M., Jennewein, E., Angert, N., Mokler, P.H., Badnell, N.R., Pindzola, M.S. (1994). Dielectronic recombination of lithium-like Ar^{15+}. *Z. Phys. D* **30**, 291–306.

Schippers, S. (1999). Recombination of HCI with electrons—Fundamental atomic physics and applications. *Phys. Scr. T* **80**, 158–162.

Schippers, S., Bartsch, T., Brandau, C., Müller, A., Gwinner, G., Wissler, G., Beutelspacher, M., Grieser, M., Wolf, A., Phaneuf, R.A. (2000). Dielectronic recombination of lithiumlike Ni^{25+} ions—High resolution rate coefficients and influence of external crossed E and B fields. *Phys. Rev. A* **62**. 022708.

Schippers, S., Müller, A., Gwinner, G., Linkemann, J., Saghiri, A.A., Wolf, A. (2001). Storage ring measurement of the C iv recombination rate coefficient. *Astrophys. J.* **555**, 1027–1037.

Schippers, S., Kieslich, S., Müller, A., Gwinner, G., Schnell, M., Wolf, A., Bannister, M., Covington, A., Zhao, L.B. (2002a). Interference effects in the photorecombination of argonlike Sc^{3+} ions: Storage ring experiment and theory. *Phys. Rev. A* **65**. 042723.

Schippers, S., Müller, A., Ricz, S., Bannister, M.E., Dunn, G.H., Bozek, J.D., Schlachter, A.S., Hinojosa, G., Cisneros, C., Aguilar, A., Covington, A.M., Gharaibeh, M.F., Phaneuf, R.A. (2002b). Experimental link of photoionization of Sc^{2+} to photorecombination of Sc^{3+}: An application of detailed balance in a unique atomic system. *Phys. Rev. Lett.* **89**. 193002.

Schippers, S., Müller, A., Ricz, S., Bannister, M.E., Dunn, G.H., Schlachter, A.S., Hinojosa, G., Cisneros, C., Aguilar, A., Covington, A.M., Gharaibeh, M.F., Phaneuf, R.A. (2003). Photoionization of Sc^{2+} ions by synchrotron radiation: Measurements and absolute cross sections in the photon energy range 23–68 eV. *Phys. Rev. A* **67**. 032702.

Schippers, S., Schnell, M., Brandau, C., Kieslich, S., Müller, A., Wolf, A. (2004). Experimental Mg IX photorecombination rate coefficient. *Astron. Astrophys.* **421**, 1185–1191.

Schippers, S., Schmidt, E.W., Bernhardt, D., Yu, D., Müller, A., Lestinsky, M., Orlov, D.A., Grieser, M., Repnow, R., Wolf, A. (2007a). Storage-ring measurement of the hyperfine induced $^{47}Ti^{18+}$ $(2s2p\,^3P_0 \to 2s^2\,^1S_0)$ transition rate. *Phys. Rev. Lett.* **98**. 033001.

Schippers, S., Schmidt, E.W., Bernhardt, D., Yu, D., Müller, A., Lestinsky, M., Orlov, D.A., Grieser, M., Repnow, R., Wolf, A. (2007b). Photorecombination of berylliumlike Ti^{18+}: Hyperfine quenching of dielectronic resonances. *J. Phys. Conf. Ser.* **58**, 137–140.

Schmidt, E.W., Schippers, S., Müller, A., Lestinsky, M., Sprenger, F., Grieser, M., Repnow, R., Wolf, A., Brandau, C., Lukić, D., Schnell, M., Savin, D.W. (2006). Electron–ion recombination measurements motivated by AGN x-ray absorption features: Fe XIV forming Fe XIII. *Astrophys. J.* **641**, L157–L160.

Schmidt, H.T., Forck, P., Grieser, M., Habs, D., Kenntner, J., Miersch, G., Repnow, R., Schramm, U., Schüssler, T., Schwalm, D., Wolf, A. (1994). High-precision measurement of the magnetic-dipole decay rate of metastable heliumlike carbon ions in a storage ring. *Phys. Rev. Lett.* **72**, 1616–1619.

Schnell, M., Gwinner, G., Badnell, N.R., Bannister, M.E., Böhm, S., Colgan, J., Kieslich, S., Loch, S.D., Mitnik, D., Müller, A., Pindzola, M.S., Schippers, S., Schwalm, D., Shi, W., Wolf, A., Zhou, S.-G. (2003). Observation of trielectronic recombination in Be-like Cl ions. *Phys. Rev. Lett.* **91**. 043001.

Schnopper, H.W., Betz, H.D., Delvaille, J.P., Kalata, K., Sohval, A.R., Jones, K.W., Wegner, H.E. (1972). Evidence for radiative electron capture by fast, highly stripped heavy ions. *Phys. Rev. Lett.* **29**, 898–901.

Schram, B.L., Boerboom, A.J.H., Kistemaker, J. (1966). Partial ionization cross sections of noble gases for electrons with energy 0.5–16 keV. I. Helium and neon. *Physica* **32**, 185–196.

Schuch, R., Lindroth, E., Madzunkov, S., Fogle, M., Mohamed, T., Indelicato, P. (2005). Dielectronic resonance method for measuring isotope shifts. *Phys. Rev. Lett.* **95**. 183003.

Schulz, M., Justiniano, E., Schuch, R., Mokler, P.H., Reusch, S. (1987). Separated resonances in simultaneous capture and excitation of S^{15+} in H_2 observed by K-x-ray–K-x-ray coincidences. *Phys. Rev. Lett.* **58**, 1734–1737.

Seaton, M.J. (1962). The theory of excitation and ionization by electron impact. In: Bates, D.R. (Ed.), *Atomic and Molecular Processes*. Academic Press, New York, p. 374 (Ch. 11).

Shah, M.B., Elliott, D.S., McCallion, P., Gilbody, H.B. (1988). Single and double ionisation of helium by electron impact. *J. Phys. B* **21**, 2751–2761.

Shevelko, V.P., Tawara, H., Scheuermann, F., Fabian, B., Müller, A., Salzborn, E. (2005). Semiempirical formulae for electron-impact double-ionization cross sections of light positive ions. *J. Phys. B* **38**, 525–545.

Shevelko, V.P., Tawara, H., Tolstikhina, I.Y., Scheuermann, F., Fabian, B., Müller, A., Salzborn, E. (2006). Double ionization of heavy positive ions by electron impact: Empirical formula and fitting parameters for ionization cross sections. *J. Phys. B* **39**, 1499–1516.

Shi, W., Böhm, S., Böhme, C., Brandau, C., Hoffknecht, A., Kieslich, S., Schippers, S., Müller, A., Kozhuharov, C., Bosch, F., Franzke, B., Mokler, P.H., Steck, M., Stöhlker, T., Stachura, Z. (2001). Recombination of U^{92+} ions with electrons. *Eur. Phys. J. D* **15**, 145–154.

Shi, W., Jacobi, J., Knopp, H., Schippers, S., Müller, A. (2003). A high-current electron gun for electron–ion collision physics. *Nucl. Instrum. Methods B* **205**, 201–206.

Smith, A.C.H., Bannister, M.E., Chung, Y.-S., Djuric, N., Dunn, G.H., Neau, A., Popovic, D., Stepanovic, M., Wallbank, B. (2001). Excitation of He^+ to the $2\,^2S$ and $2\,^2P$ states by electron impact. *J. Phys. B* **34**, L571–L577.

Smith, A.J., Beiersdorfer, P., Widmann, K., Chen, M.H., Scofield, J.H. (2000). Measurement of resonant strengths for dielectronic recombination in heliumlike Ar^{16+}. *Phys. Rev. A* **62**. 052717.

Smith, S.J., Man, K.-F., Mawhorter, R.J., Williams, I.D., Chutjian, A. (1991). Absolute, cascade-free cross sections for the $^2S \rightarrow {}^2P$ transition in Zn^+ using electron-energy-loss and merged-beams methods. *Phys. Rev. Lett.* **67**, 30–33.

Smith, S.J., Greenwood, J.B., Chutjian, A., Tayal, S.S. (2000). Electron excitation cross sections for the $3s^2 3p^2\,^3P \rightarrow 3s3p^3\,^5S^o$ transition in S^{2+}. *Astrophys. J.* **541**, 501–505.

Smith, S.J., Lozano, J.A., Tayal, S.S., Chutjian, A. (2003). Electron excitation cross sections for the $2s^2 2p\,^2P^o \rightarrow 2s2p^2\,^4P$ and $2s2p^2\,^2D$ transitions in O^{3+}. *Phys. Rev. A* **68** (6). 062708.

Smith, S.J., Djurić, N., Lozano, J.A., Berrington, K.A., Chutjian, A. (2005). Measurement of absolute cross sections for excitation of the $2s^2\,^1S \rightarrow 2s2p\,^1P^o$ transition in O^{+4}. *Astrophys. J.* **630**, 1213–1216.

Spies, W., Müller, A., Linkemann, J., Frank, A., Wagner, M., Kozhuharov, C., Franzke, B., Bosch, F., Eickhoff, H., Jung, M., Klepper, O., König, W., Mokler, P.H., Moshammer, R., Nolden, F., Schaaf, U., Spädtke, P., Steck, M., Zimmerer, P., Grün, N., Scheid, W., Pindzola, M.S., Badnell, N.R. (1992). Dielectronic and radiative recombination of lithiumlike gold. *Phys. Rev. Lett.* **69**, 2768–2771.

Sprenger, F., Lestinsky, M., Orlov, D.A., Schwalm, D., Wolf, A. (2004). The high-resolution electron–ion collision facility at TSR. *Nucl. Instrum. Methods A* **532**, 298–302.

Srigengan, B., Williams, I.D., Newell, W.R. (1996a). Angular distribution for the elastic scattering of electrons from Na^+ ions. *Phys. Rev. A* **54**, R2540–R2542.

Srigengan, B., Williams, I.D., Newell, W.R. (1996b). Dynamic effects in low-energy electron—ion scattering. *J. Phys. B* **29**, L897–L900.

Srigengan, B., Williams, I.D., Newell, W.R. (1996c). Experimental measurements of low-energy elastic scattering of electrons from Cs^+ ions. *J. Phys. B* **29**, L605–L610.

Stenke, M., Aichele, K., Hathiramani, D., Hofmann, G., Steidl, M., Völpel, R., Salzborn, E. (1995). A high-current electron gun for crossed-beams electron–ion collision studies at keV energies. *Nucl. Instrum. Methods B* **98**, 573–576.

Stenke, M., Aichele, K., Hartenfeller, U., Hathiramani, D., Steidl, M., Salzborn, E. (1999). Electron impact single ionization of multiply charged iron ions. *J. Phys. B* **32**, 3627–3639.

Stephan, K., Helm, H., Märk, T.D. (1980). Mass spectrometric determination of partial electron impact ionization cross sections of He, Ne, Ar and Kr from threshold up to 180 eV. *J. Chem. Phys.* **73**, 3763–3778.

Stobbe, M. (1930). Zur Quantenmechanik photoelektrischer Prozesse. *Ann. Phys. (Leipzig)* **7**, 661–715.

Stöhlker, T., Kozhuharov, C., Mokler, P.H., Warczak, A., Bosch, F., Geissel, H., Moshammer, R., Scheidenberger, C., Eichler, J., Ichihara, A., Schirai, T., Stachura, Z., Rymuza, P. (1995). Radiative electron capture studied in relativistic heavy-ion–atom collisions. *Phys. Rev. A* **51**, 2098–2111.

Stöhlker, T., Ludziejewski, T., Bosch, F., Dunford, R.W., Kozhuharov, C., Mokler, P.H., Beyer, H.F., Brinzanescu, O., Franzke, B., Eichler, J., Griegal, A., Hagmann, S., Ichihara, A., Krämer, A., Lekki, J., Liesen, D., Nolden, F., Reich, H., Rymuza, P., Stachura, Z., Steck, M., Swiat, P., Warczak, A. (1999). Angular distribution studies for the time-reversed photoionization process in hydrogenlike uranium: The identification of spin-flip transitions. *Phys. Rev. Lett.* **82**, 3232–3235. Erratum: *Phys. Rev. Lett.* **84** (2000) 1360.

Stöhlker, T., Ma, X., Ludziejewski, T., Beyer, H.F., Bosch, F., Brinzanescu, O., Dunford, R.W., Eichler, J., Hagmann, S., Ichihara, A., Kozhuharov, C., Krämer, A., Liesen, D., Mokler, P.H., Stachura, Z., Swiat, P., Warczak, A. (2001). Near-threshold photoionization of hydrogenlike uranium studied in ion–atom collisions via the time-reversed process. *Phys. Rev. Lett.* **86**, 983–986.

Stöhlker, T., Banas, D., Fritzsche, S., Gumberidze, A., Kozhuharov, C., Ma, X., Orsic-Muthig, A., Spillmann, U., Sierpowski, D., Surzhykov, A., Tachenov, S., Warczak, A. (2004). Angular correlation and polarization studies for radiative electron capture into high-Z ions. *Phys. Scr. T* **110**, 384–388.

Summers, H.P., Dickson, W.J., O'Mullane, M.G., Badnell, N.R., Whiteford, A.D., Brooks, D.H., Lang, J., Loch, S.D., Griffin, D.C. (2006). Ionization state, excited populations and emission of impurities in dynamic finite density plasmas: I. The generalized collisional-radiative model for light elements. *Plasma Phys. Control. Fusion* **48**, 263–293.

Tanis, J.A. (1992). Resonant transfer excitation associated with single X-ray emission. In: Graham, W.G., Fritsch, W., Hahn, Y., Tanis, J.A. (Eds.), *Recombination of Atomic Ions*. In: *NATO ASI Series B: Physics*, vol. 296. Plenum Press, New York, pp. 241–257.

Tanis, J.A., Bernstein, E.M., Graham, W.G., Stockli, M.P., Clark, M., McFarland, R.H., Morgan, T.J., Berkner, K.H., Schlachter, A.S., Stearns, J.W. (1984). Resonant electron transfer and excitation in two-, three-, and four-electron $_{20}Ca^{q+}$ and $_{23}V^{q+}$ ions colliding with helium. *Phys. Rev. Lett.* **53**, 2551–2554.

Tashenov, S., Stöhlker, T., Banaś, D., Beckert, K., Beller, P., Beyer, H.F., Bosch, F., Fritzsche, S., Gumberidze, A., Hagmann, S., Kozhuharov, C., Krings, T., Liesen, D., Nolden, F., Protic, D., Sierpowski, D., Spillmann, U., Steck, M., Surzhykov, A. (2006). First measurement of the linear polarization of radiative electron capture transitions. *Phys. Rev. Lett.* **97**. 223202.

Tayal, S.S., Pradhan, A.K., Pindzola, M.S. (1995). Excitation of atomic ions by electron impact. In: Janev, R.K. (Ed.), *Atomic and Molecular Processes in Fusion Edge Plasmas*. Plenum, New York, pp. 119–152.

Taylor, P.O., Gregory, D., Dunn, G.H., Phaneuf, R.A., Crandall, D.H. (1977). Absolute cross sections for $2s - 2p$ excitation of C^{3+} by electron impact. *Phys. Rev. Lett.* **39**, 1256–1259.

Teng, H. (2000). Indirect processes in electron-impact ionization of Li^+. *J. Phys. B* **33**, L227–L232.

Teng, H., Knopp, H., Ricz, S., Schippers, S., Berrington, K.A., Müller, A. (2000). Interference of direct and resonance channels in electron impact ionization of C^{3+} ions: Unified R-matrix calculation and experiment. *Phys. Rev. A* **61**. 060704.

Thomason, J.W.G., Peart, B. (1998). The electron-impact ionization of Al^{2+} ions. *J. Phys. B* **31**, L201–L207.

Thomason, J.W.G., Peart, B., Hayton, S.J.T. (1997). The double ionization of Cs^+ and Sr^+ by energy-resolved electrons. *J. Phys. B* **30**, 749–756.

Thomson, S.J.J. (1912). Ionization by moving electrified particles. *Philos. Mag.* **23**, 449–457.

Tinschert, K., Müller, A., Phaneuf, R.A., Hofmann, G., Salzborn, E. (1989). Electron impact double ionisation of Ar^{q+} ions ($q = 1, 2, \ldots, 7$): Two-electron processes compared with inner-shell contributions. *J. Phys. B* **22**, 1241–1248.

Toth, G., Grabbe, S., Richard, P., Bhalla, C.P. (1996). Inelastic scattering of quasifree electrons on O^{7+} projectiles. *Phys. Rev. A* **54**, R4613–R4616.

Trassl, R. (2003). ECR ion sources. In: Currell, F. (Ed.) *The Physics of Multiply and Highly Charged Ions*, vol. 1. Kluwer Academic Publishers, Dordrecht, pp. 3–37.

Trassl, R., Hathiramani, P., Broetz, F., Greenwood, J.B., McCullough, R.W., Schlapp, M., Salzborn, E. (1997). Characterization and recent modifications of a compact 10GHz electron cyclotron resonance (ECR) ion source for atomic physics experiments. *Phys. Scr. T* **73**, 380–381.

Ullrich, J., Moshammer, R., Dorn, A., Dörner, R., Schmidt, L., Schmidt-Böcking, H. (2003). Recoil-ion and electron momentum spectroscopy: reaction-microscopes. *Rep. Prog. Phys.* **66**, 1463–1545.

Van Regemorter, H. (1962). Rate of collisional excitation in stellar atmospheres. *Astrophys. J.* **136**, 906.

van Zoest, T., Knopp, H., Jacobi, J., Schippers, S., Phaneuf, R.A., Müller, A. (2004). Electron-impact ionization of Ti^{3+} ions. *J. Phys. B* **37**, 4387–4395.

Wåhlin, E.K., Thompson, J.S., Dunn, G.H., Phaneuf, R.A., Gregory, D.C., Smith, A.C.H. (1991). Electron-impact excitation of $Si^{3+}(3s \rightarrow 3p)$ using a merged-beam electron-energy-loss technique. *Phys. Rev. Lett.* **66**, 157–160.

Wang, J.S., Griem, H.R., Hess, R., Rowan, W.L. (1988). Measurement of ionization and recombination rates for Fe xvi–Fe xxii from time-resolved spectroscopy of tokamak plasmas. *Phys. Rev. A* **38**, 4761–4766.

Wang, Z., Matsumoto, J., Tanuma, H., Danjo, A., Yoshino, M., Kobayashi, N. (2000). Differential cross section measurements for elastic scattering of electrons by highly charged argon ions. *J. Phys. B* **33**, 2629–2640.

Watanabe, H., Currell, F.J., Kuramoto, H., Ohtani, S., O'Rourke, B.E., Tong, X.M. (2002). Electron impact ionization of hydrogen-like molybdenum ions. *J. Phys. B* **35**, 5095–5103.

Watanabe, H., Tobiyama, H., Kavanagh, A.P., Li, Y.M., Nakamura, N., Sakaue, H.A., Currell, F.J., Ohtani, S. (2007). Dielectronic recombination of He-like to C-like iodine ions. *Phys. Rev. A* **75**. 012702.

Watts, M.F., Dunn, K.F., Gilbody, H.B. (1986). Redetermination of cross sections for charge transfer and ionisation in H^+-He^+ collisions. *J. Phys. B* **19**, L355–L359.

Westermann, M., Scheuermann, F., Aichele, K., Hathiramani, D., Steidl, M., Salzborn, E. (1999). Multiple ionization of C^{q+}, N^{q+} and O^{q+} ions by electron impact. *Phys. Scr. T* **80**, 285–286.

Williams, I.D. (1999). Electron-ion scattering. *Rep. Prog. Phys.* **62**, 1431–1469.

Williams, I.D., Chutjian, A., Mawhorter, R.J. (1986). Differential electron scattering cross sections for the first optically forbidden and resonance transitions in Mg II, Zn II and Cd II. *J. Phys. B* **19**, 2189–2198.

Williams, I.D., Srigengan, B., Platzer, A., Greenwood, J.B., Newell, W.R., O'Hagan, L. (1997). Super-elastic scattering of electrons from Ar^{3+}. *Phys. Scr. T* **73**, 121–122.

Williams, J.F. (1984). Dielectronic recombination for Ca^+ via $4s \rightarrow 4p$ excitation. *Phys. Rev. A* **29**, 2936–2938.

Winter, T.G. (2004). Ionization, excitation, and electron transfer in Mev-energy collisions between light nuclei and C^{5+}(1s) ions studied with a Sturmian basis. *Phys. Rev. A* **69**. 042711.

Yerokhin, V.A., Indelicato, P., Shabaev, V.M. (2005). Two-loop self-energy contribution to the Lamb shift in H-like ions. *Phys. Rev. A* **71**. 040101(R).

Young, A.R., Gardner, L.D., Savin, D.W., Lafyatis, G.P., Chutjian, A., Bliman, S., Kohl, J.L. (1994). Measurement of C^{3+} dielectronic recombination in a known external field. *Phys. Rev. A* **49**, 357–362.

Yu, D.J., Rachafi, S., Jureta, J., Defrance, P. (1992). The formation of H^+ from H^- by electron impact. *J. Phys. B* **25**, 4593–4600.

Zamkov, M., Aliabadi, H., Benis, E.P., Richard, P., Tawara, H., Zouros, T.J.M. (2002). Absolute cross sections and decay rates for the triply excited $B^{2+}(2s2p^2\,^2D)$ resonance in electron–metastable-ion collisions. *Phys. Rev. A* **65**. 032705.

Zápesochnyi, A.P., Imre, A.I., Aleksakhin, I.S., Zapesochnyi, I.P., Zatsarinnyi, O.I. (1986). Resonance effects in the inelastic interactions of slow electrons with alkali metal ions. *Sov. Phys. JETP* **63**, 1155–1160.

Závodszky, P.A., Tóth, G., Grabbe, S.R., Zouros, T.J.M., Richard, P., Bhalla, C.P., Tanis, J.A. (1999). Forward-backward asymmetry in the inelastic scattering of electrons from highly charged ions. *J. Phys. B* **32**, 4425–4435.

Závodszky, P.A., Aliabadi, H., Bhalla, C., Richard, P., Tóth, G., Tanis, J.A. (2001). Superelastic scattering of electrons from highly charged ions with inner shell vacancies. *Phys. Rev. Lett.* **87**. 033202.

Zhang, X., Crespo López-Urrutia, J.R., Guo, P., Mironov, V., Shi, X., González Martínez, A.J., Tawara, H., Ullrich, J. (2004). Experimental study of the deep-lying dielectronic recombination resonances of He-like germanium ions. *J. Phys. B* **37**, 2277–2284.

Zhao, H.Y., Zhao, H.W., Ma, X.W., Zhang, S.F., Feng, W.T., Zhu, X.L., Zhang, Z.M., He, W., Sun, L.T., Feng, Y.C., Cao, Y., Li, J.Y., Li, X.X., Wang, H., Ma, B.H. (2006). Measurements of bremsstrahlung spectra of Lanzhou ECR ion source no. 3 (LECR3). *Rev. Sci. Instrum.* **77**. 03A312.

Zouros, T.J.M. (1992). Resonant transfer excitation associated with Auger electron emission. In: Graham, W.G., Fritsch, W., Hahn, Y., Tanis, J.A. (Eds.), *Recombination of Atomic Ions*. In: *NATO ASI Series B: Physics*, vol. 296. Plenum Press, New York, pp. 271–300.

Zouros, T.J.M., Benis, E.P., Gorczyca, T.W., González, A.D., Zamkov, M., Richard, P. (2003). Differential electron scattering from positive ions measured by zero-degree ion–atom spectroscopy. *Nucl. Instrum. Methods B* **205**, 508–516.

ROBUST PROBABILISTIC QUANTUM INFORMATION PROCESSING WITH ATOMS, PHOTONS, AND ATOMIC ENSEMBLES

L.-M. DUAN and C. MONROE

FOCUS, MCTP, and Department of Physics, University of Michigan,
Ann Arbor, MI 48109-1040, USA

1. Introduction 420
2. Quantum Communication with Atomic Ensembles 421
   2.1. Quantum Repeaters for Scalable Communication 421
   2.2. Collective Enhancement in Interaction between Light and Atomic Ensembles 423
   2.3. Entanglement Generation, Connection, and Entanglement-Based Communication Schemes 425
   2.4. Built-in Entanglement Purification and Scaling of the Communication Efficiency . . 429
   2.5. Experimental Quantum Communication with Atomic Ensembles 432
3. Quantum State Engineering with Realistic Linear Optics 434
   3.1. Linear Optics and Quantum State Engineering 434
   3.2. Preparation of Arbitrary Graph States 436
   3.3. Efficient Generation and Scaling of Tree Graph States 439
4. Quantum Computation through Probabilistic Atom–Photon Operations 442
   4.1. Robust Probabilistic Gates 442
   4.2. Probabilistic Gates from Free-Space Atom–Photon Coupling 443
   4.3. Scalable Quantum Computation with Probabilistic Gates 447
   4.4. Experiments towards Probabilistic Ion Gates 452
5. Summary 459
6. Acknowledgements 460
7. References 460

Abstract

In this article, we review several new approaches to scalable and robust quantum communication, state engineering, and quantum computation. We consider the use of atomic ensembles, linear optical elements, and trapped ions for this purpose, all having significant experimental simplifications compared to conventional systems. These new approaches are based on probabilistic entanglement of quantum bits, where the dominant source of error is the (typically small) probability of entanglement success per attempt. By exploiting the properties of this particular noise process, we can design scalable quantum network schemes that are inherently insensitive to the noise, resulting in error-correction thresholds that are much more forgiving than any conventional threshold requirement. We review several such types of schemes in different contexts, and show their close relations with the current experimental implementations of scalable quantum information processing. Experimental progress along these approaches will be briefly remarked, especially in a system of trapped atomic ions.

1. Introduction

Quantum information systems hold great promise for superfast computation and secure communication (Nielsen and Chuang, 2000). However, practical quantum hardware can be highly susceptible to external noise and decoherence, leading to quantum information errors. This is especially true during quantum logic operations, where interactions between quantum bits must be controlled with great precision. Quantum error correction of some form will therefore be essential for reliable quantum information processing (Shor, 1995; Steane, 1997). The fundamental quantum error threshold theorem (Preskill, 1998) states that the error per quantum operation must be less than a particular (small) threshold value for effective error correction. However, noise levels in current experimental systems are typically orders of magnitude larger than fault-tolerant thresholds for arbitrary quantum errors. An alternative approach to general error correction is to exploit the properties of the noise itself to design schemes that either automatically correct the dominant source of noise, or lead to specific types of errors that can be more easily corrected later.

In this article, we review several schemes for reliable quantum communication, state engineering, and quantum computation, that can tolerate very high levels of experimental noise. The experimental systems in our consideration range from atomic ensembles, to photonic systems (such as the spontaneous parametric down conversion or cavity QED systems), and to individual trapped atoms or ions. The precise physical sources of noise in these different systems can be quite different, and there are a number of system-specific properties which are discussed. However, in spite of these differences, the dominant noise in these systems share very similar properties. All of these schemes involve photons, and the dominant source of noise is always some type of photon loss, related to "quantum leakage" errors, where the state of the physical system continuously leaks outside of the logical Hilbert space that carries the quantum information. Because of this similarity in the noise properties, we can identify very efficient error-correction methods for all of these systems and achieve scalability in quantum communication, state engineering, and quantum computation in the face of high noise levels.

In the next section, we review the scalable quantum communication scheme using atomic ensembles first proposed in Duan et al. (2001). This scheme implements a quantum repeater architecture (Briegel et al., 1998), although does not require the dominant noise to be below a stringent threshold value. The recent remarkable experimental progress using this approach is discussed briefly (Kuzmich et al., 2003; Van der Wal et al., 2003; Chou et al. 2004, 2005; Matsukevich and Kuzmich, 2004; Blinov et al., 2004; Balic et al., 2005; Chaneliere et al., 2005; Eisaman et al., 2005; Black et al., 2005; Manz et al., 2006; Riedmatten et al., 2006; Matsukevich et al., 2006). In Section 3, we review a recent scheme for quantum state engineering with linear optics (Bodiya and Duan, 2006), where entangled "graph states" can be created in a scalable fashion with realistic linear optical systems under current technology. Finally, in the last section, we review approaches to scalable quantum computation and networking based on probabilistic entangling gates (Duan et al., 2005; Barrett and Kok, 2005), and the implementation of these types of gates using trapped ions (Duan et al., 2006) and cavity QED systems (Duan et al., 2005). Some recent trapped ion experiments are described that represent an initial step in achieving such gate operations (Madsen et al., 2006; Maunz et al., 2006).

2. Quantum Communication with Atomic Ensembles

2.1. QUANTUM REPEATERS FOR SCALABLE COMMUNICATION

The communication of quantum information over remote distances is essential for realizing quantum networks and secretly transferring messages by means of quantum cryptography. The key resource in quantum communication is the generation of nearly perfect entangled states between distant sites. Such states can be used then to implement secure quantum cryptography (Ekert, 1991) or to transfer arbitrary quantum messages (Bennett et al., 1993). Realistic schemes for quantum communication are based on photonic channels, as photons are the only viable particles that can be transmitted with high speeds over long distances. To overcome the inevitable signal attenuation in the channel, the concept of entanglement purification was invented (Bennett et al., 1997). However, entanglement purification does not fully solve the problem for long-distance quantum communication. Due to the exponential decay of the entanglement with the channel length, one needs an exponentially large number of partially entangled states to obtain one highly entangled state, which means that for a sufficiently long distance the task becomes nearly impossible.

The idea of a quantum repeater was proposed by Briegel et al. (1998) to mitigate the exponential decay of fidelity with distance. In principle, quantum repeaters allow the overall communication fidelity to approach unity, with the communication time growing only polynomially with the transmission distance. In analogy with fault-tolerant quantum computing (Preskill, 1998), the quantum repeater is a concatenated (nested) entanglement purification protocol for communication systems. The idea is to divide the transmission channel into many segments, with the length of each segment comparable to the channel attenuation length. Entanglement is first generated and purified for each segment, then the purified entanglement is extended to longer lengths by connecting two adjacent segments through the entanglement swapping protocol (Bennett et al., 1993). Following this, the overall entanglement decreases and must be purified again. These rounds of entanglement swapping and purification can be repeated until nearly perfect entangled states are created between two distant sites. Similar to fault tolerant quantum computation, the conventional quantum repeater protocol requires the noise per quantum operation (such as a local gate operation for the entanglement purification or a quantum transmission operation) to be below a certain threshold value. However, this threshold value is considerably less stringent than qubit error-correction thresholds for quantum computing. Error thresholds for quantum repeaters are typically estimated to be at the $\sim 1\%$ level.

In this section, we review the implementation scheme of quantum repeaters proposed by Duan, Lukin, Cirac, and Zoller (2001) (DLCZ). This scheme has a nested architecture similar to the original quantum repeater protocol (Briegel et al., 1998), but it uses a very different noise reduction method. Instead of explicit entanglement purifications at each step, the DLCZ scheme features inherent fault tolerance. In this scheme, the dominant noise is first classified in the proposed experimental system (e.g., atomic ensembles). Based on the properties of this noise, the entanglement generation, connection (swapping), and application schemes are designed so that each step of this protocol has some function of built-in entanglement purification. Through this built-in entanglement purification, each step partially removes the noise from all of the previous steps, and as a result, the noise accumulates much more slowly. The effect of the remaining noise is removed by the last step of entanglement application, and we can then prove that one has an overall efficient scaling. Because there is no need for additional steps of explicit entanglement purification, this scheme overcomes stringent error threshold requirements, and can tolerate important experimental noise at much higher levels. For instance, with the DLCZ scheme, photon detector efficiencies can be around 50% or even lower without significantly influencing the overall scaling of the protocol. In the conventional approach (Briegel et al., 1998), even with a forgiving error-correction threshold at the percent level, photon detector efficiencies (and any other sources of photon loss) must still be larger than 99% which is still considered challenging with current technology. So, inherent fault tolerance to high levels of noise is the essential feature of this DLCZ scheme. Such a property is critical for the recent remarkable experimental progress using this approach (Kuzmich et al., 2003; Van der Wal et al., 2003; Chou et al., 2004, 2005; Matsukevich and Kuzmich, 2004; Blinov et al., 2004; Balic et al., 2005; Chaneliere et al., 2005; Eisaman et al., 2005; Black et al., 2005; Manz et al., 2006; Riedmatten et al., 2006; Matsukevich et al., 2006).

2.2. Collective Enhancement in Interaction between Light and Atomic Ensembles

The DLCZ scheme proposes atomic ensembles as nodes for quantum repeaters. The atomic ensemble contains a large number of identical neutral atoms, which might consist of either laser-cooled atoms (Kuzmich et al., 2003; Chou et al., 2004; Black et al., 2005; Manz et al., 2006; Matsukevich et al., 2006), or a room-temperature vapor (Van der Wal et al., 2003; Julsgaard et al., 2001). The motivation of using atomic ensembles instead of single-particles for quantum information processing is two-fold: first, laser manipulation of atomic ensembles without separate addressing of individual atoms is simpler than the laser manipulation of single particles; and second, the use of the atomic ensembles allows for collective effects resulting from many-atom coherences to enhance the signal-to-noise ratio, which is critical for increasing efficiencies of some quantum information protocols (Fleischhauer and Lukin, 2000; Duan et al., 2000, 2001; Julsgaard et al., 2001). The collective enhancement in atomic ensembles and its applications in quantum information has been reviewed in several recent articles (Cirac et al., 2001; Lukin, 2003).

In the DLCZ scheme, a different level configuration is used, and the collective enhancement effect for this configuration becomes more subtle, as shown in detail in Duan et al. (2002). The atomic ensemble consists of a cloud of N_a identical atoms with the relevant level structure shown in Fig. 1A. A pair of stable lower states $|g\rangle$ and $|s\rangle$ can correspond, for instance, to hyperfine or Zeeman sublevels of electronic ground state of alkali atoms. The relevant coherence between the levels $|g\rangle$ and $|s\rangle$ can be maintained for a sufficiently long time, which provides the desired quantum memory. All the atoms are initially prepared in the ground state $|g\rangle$ through optical pumping.

The ensemble is then illuminated by a weak pumping laser pulse which drives the transition $|g\rangle \to |e\rangle$ with a large detuning Δ, and we look at the spontaneous emission light from the transition $|e\rangle \to |s\rangle$, whose polarization and/or frequency are assumed to be different from that of the pumping laser. The pumping laser is directed onto all the atoms in the focusing area so that each atom has nearly the same small probability to be excited into the state $|s\rangle$ through the Raman transition. (It is a simple matter to extend the analysis to inhomogeneous couplings where different atoms have different excitation probabilities by appropriately redefining the collective atomic mode.)

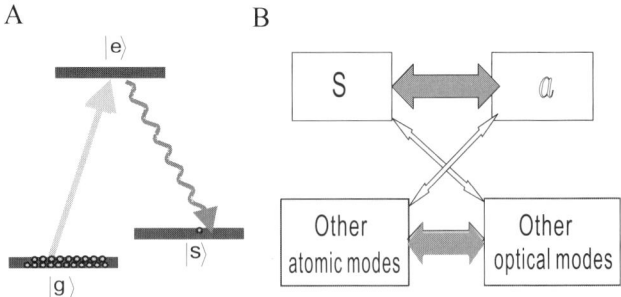

FIG. 1. (A) The relevant atomic level structure with ground and metastable states $|g\rangle$ and $|s\rangle$, and excited state $|e\rangle$. (B) Correlations between the atomic modes of an atomic ensemble (left), and the free-space optical modes (right). The symmetric atomic mode S is predominantly correlated with the forward scattered optical mode a, with only weak correlations to other optical modes that account for spontaneous emission noise. Likewise, the forward scattered mode a is only weakly correlated with other (nonsymmetric) atomic modes (see Duan et al., 2002).

With the above Raman transition, the corresponding spontaneous photon from each atom can be emitted in any direction given by the dipole emission pattern, and without any selection on the photonic or atomic state, there will be no collective enhancement effect that gives rise to directional emission. (In the weak-pumping regime considered here, we neglect effects of superradiant emission.) However, considering the atomic gas with many atoms, we can define collective atomic modes. If we look at a particular atomic mode, it gets correlated with a well-defined photon mode that is directional in space. For instance, the symmetric atomic mode S (resulting from homogeneous excitation) is defined as

$$S \equiv (1/\sqrt{N_a}) \sum_{i=1}^{N_a} |g\rangle_i \langle s|. \tag{1}$$

This atomic mode is correlated with a special optical spontaneous emission mode a (called the signal mode) which is essentially collinear with the pumping laser. The signal mode a can be written by expanding the spontaneous emission field in plane wave modes:

$$a = \int f_{\mathbf{k}}^* a_{\mathbf{k}} d^3 \mathbf{k}, \tag{2}$$

where $a_{\mathbf{k}}$ represents the plane wave mode with the wave vector $\mathbf{k}$. The operators $a_{\mathbf{k}}$ satisfy the standard commutation relations $[a_{\mathbf{k}}, a_{\mathbf{k}'}^{\dagger}] = \delta(\mathbf{k} - \mathbf{k}')$, and $f_{\mathbf{k}}^*$ is the normalized signal mode function whose explicit form depends on the pumping laser profile and the geometry of the atomic ensemble, as specified in Duan et al. (2002). The modes S and a are correlated with each other: if the atom collection is excited into the symmetric collective mode S, the accompanying spontaneous emission photon is be emitted into the signal mode a, and vice versa. There are many other atomic modes in the ensemble and optical modes in the spontaneous emission field, and these background modes can be correlated with each other in a complicated pattern that depends on details concerning the thermal motion of the atoms. However, the correlation between the particular modes S and a is maintained even with a large degree of atomic thermal motion. In experiments with cold atomic ensembles, the correlation pattern between the atomic and the optical modes gets simplified as the atomic thermal motional effect is minimized, and correlations between other atomic and optical modes can be used, where the spontaneous emission photon is non-collinear with the pumping pulse (Balic et al., 2005). Due to density fluctuations in the atomic gas, the modes S and a can still get weakly correlated with the other atomic and optical modes, inducing so-called spontaneous emission noise. This noise vanishes as the atomic gas becomes optically thick. The explicit characterization of the noise can be found in Duan et al. (2002). The correlation picture between different atomic and optical modes is shown schematically in Fig. 1B.

The application of this system to quantum communication comes from the correlation between the modes S and a, which can both be selectively detected. The mode a is directional in space with a well-defined mode structure, so it can be coupled into a single-mode optical fiber and then detected by a single-photon detector. The excitation in the atomic mode S can be subsequently transferred into a directional optical photon with a repumping laser pulse (Fleischhauer and Lukin, 2000; Cirac et al., 2001; Liu et al., 2001; Phillips et al., 2001) and then can be similarly detected by a single-photon detector coupled through a fiber. With this detection method, we can neglect the other atomic and optical modes as they have no influence on the above measurements, and we are left with an effective two-mode problem. The pure correlation between the modes S and a can then be used to generate entanglement between two distant atomic ensembles, which eventually leads to realization of quantum repeaters as described in the next section.

2.3. Entanglement Generation, Connection, and Entanglement-Based Communication Schemes

To realize long-distance quantum communication, first we need to entangle two atomic ensembles within the channel attenuation length, and then connect different segments of entanglement to generate a long-distance entangled state. We follow the same approach as in Duan et al. (2001).

The entanglement generation scheme is based on photon interference at photodetectors, which critically uses the fault-tolerance property of photon detection. With a weak pumping laser pulse, the state of the atomic and the optical modes S and a can be written in the form (Duan et al., 2002)

$$|\phi\rangle = |0_a\rangle|0_p\rangle + \sqrt{p_c} S^\dagger a^\dagger |0_a\rangle|0_p\rangle + O(p_c), \tag{3}$$

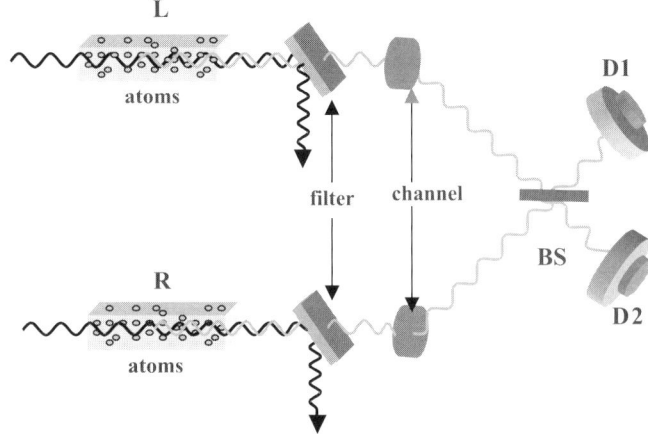

FIG. 2. Schematic setup for generating entanglement between the two atomic ensembles L and R (see Duan et al., 2001). The two ensembles are pencil shaped and illuminated by the synchronized classical laser pulses. The forward-scattered Stokes pulses are collected after the filters (polarization and frequency selective) and interfered at a 50%–50% beam splitter (BS) after the transmission channels, with the outputs detected respectively by two single-photon detectors D_1 and D_2. If there is a click in D_1 or D_2, the process is finished and we successfully generate entanglement between the ensembles L and R. Otherwise, we first apply a repumping pulse to drive the transition $|2\rangle \to |3\rangle$ on the ensembles L and R to reset the state of the ensembles back to the ground state $|0\rangle_a^L \otimes |0\rangle_a^R$, then the same classical laser pulses as the first round drive the transition $|1\rangle \to |3\rangle$ and we detect again the forward-scattering Stokes pulses after the beam splitter. This process is repeated until finally we have a click in the D_1 or D_2 detector.

where $p_c \ll 1$ is a small excitation probability, and $|0_a\rangle$ and $|0_p\rangle$ denote the vacuum (no excitation) state of the modes S and a_s, respectively.

Now we explain how to use this setup to generate entanglement between two distant ensembles L and R using the configuration shown in Fig. 2. Here, two laser pulses excite both ensembles simultaneously, and the whole system is described by the state $|\phi\rangle_L \otimes |\phi\rangle_R$, where $|\phi\rangle_L$ and $|\phi\rangle_R$ are given by Eq. (3) with all the operators and states distinguished by the subscript L or R. The forward scattered Stokes signal from both ensembles is combined at the beam splitter and a photodetector click in either D_1 or D_2 measures the combined radiation from two samples, $a_+^\dagger a_+$ or $a_-^\dagger a_-$ with $a_\pm = (a_L \pm e^{i\varphi} a_R)/\sqrt{2}$. Here, φ denotes an unknown difference of the phase shifts in the two-side channels. We can also assume that φ has an imaginary part to account for the possible asymmetry of the setup, which will also be corrected automatically in the scheme. But the setup asymmetry can be easily made very small, and for simplicity of expressions we assume that φ is real in the following. Conditional on the detector click, we should apply a_+ or a_- to the whole state $|\phi\rangle_L \otimes |\phi\rangle_R$, and the projected state of the ensembles L and R is nearly maximally entangled with the form (neglecting the high-order terms of order p_c),

$$|\Psi_\varphi\rangle_{LR}^\pm = \left(S_L^\dagger \pm e^{i\varphi} S_R^\dagger\right)/\sqrt{2}\,|0_a\rangle_L |0_a\rangle_R. \tag{4}$$

The probability for getting a click is given by p_c for each round, so we need repeat the process about $1/p_c$ times for a successful entanglement preparation, and the average preparation time is approximately given by t_Δ/p_c, where t_Δ is the duration of each pumping cycle. The states $|\Psi_r\rangle_{LR}^+$ and $|\Psi_r\rangle_{LR}^-$ can be easily transformed to each other by a simple local phase shift. Without loss of generality, we assume in the following that the entangled state $|\Psi_r\rangle_{LR}^+$ is generated.

The presence of noise modifies the projected state of the ensemble to

$$\rho_{LR}(c_0,\varphi) = \frac{1}{c_0+1}\left(c_0 |0_a 0_a\rangle_{LR}\langle 0_a 0_a| + |\Psi_\varphi\rangle_{LR}^+ \langle\Psi_\varphi|\right), \tag{5}$$

where the "vacuum" coefficient c_0 is determined by the dark count rates of the photon detectors. It is seen below that any state in the form of Eq. (5) will be purified automatically to a maximally entangled state in the entanglement-based communication schemes. We therefore call this state an effective maximally entangled (EME) state with the vacuum coefficient c_0 determining the purification efficiency.

After successful generation of entanglement within the attenuation length, we now extend the quantum communication distance. This is done through entanglement swapping with the configuration shown in Fig. 3. Suppose that we start with two pairs of the entangled ensembles described by the state $\rho_{LI_1} \otimes \rho_{I_2R}$, where ρ_{LI_1} and ρ_{I_2R} are given by Eq. (5). In the ideal case, the setup shown in Fig. 3 measures the quantities corresponding to operators $S_\pm^\dagger S_\pm$ with $S_\pm = (S_{I_1} \pm S_{I_2})/\sqrt{2}$. If the measurement is successful (i.e., one of the detectors registers one photon), we prepare the ensembles L and R into another EME state. The new φ-parameter is given by $\varphi_1+\varphi_2$, where φ_1 and φ_2 denote the old φ-parameters for the two segment EME states. Even in the presence of realistic noise such as photon loss, an EME state is still created after a detector click. The noise only influences the success probability to get a click and the new vacuum coefficient in the EME state. The above method for connecting entanglement can be continued to extend the communication over an arbitrary distance.

After an EME state has been established between two distant sites, we would like to use it in the communication protocols, such as for quantum teleportation, cryptography, or Bell inequality detection. It is not obvious that the EME state [Eq. (5)], which is entangled in the Fock basis, is useful for these tasks since in the Fock basis it is experimentally hard to do certain single-bit operations. In the following we show how the EME states can indeed be used to realize all these protocols with simple experimental configurations.

Quantum cryptography and the Bell inequality detection are achieved with the setup shown by Fig. 4. The state of the two pairs of ensembles is expressed as

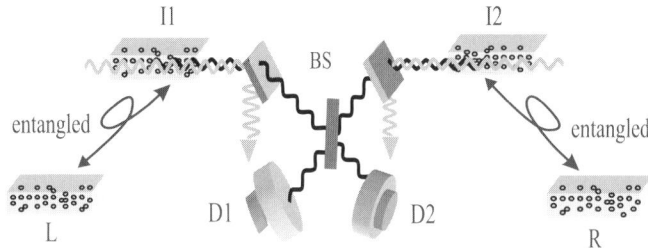

FIG. 3. Illustration for the entanglement connection (swapping) (see Duan et al., 2001). We have two pairs of ensembles L, I_1 and I_2, R distributed at three sites L, I and R. Each of the ensemble-pairs L, I_1 and I_2, R is prepared in an EME state in the form of Eq. (3). The excitations in the collective modes of the ensembles I_1 and I_2 are transferred simultaneously to the optical excitations by the repumping pulses applied to the atomic transition $|2\rangle \rightarrow |3\rangle$, and the stimulated optical excitations, after a 50%–50% beam splitter, are detected by single-photon detectors D_1 and D_2. If either D_1 or D_2 clicks, the protocol is successful and an EME state in the form of Eq. (3) is established between the ensembles L and R with a doubled communication distance. Otherwise, the process fails, and we need to repeat the previous entanglement generation and swapping until finally we have a click in D_1 or D_2, that is, until the protocol finally succeeds.

$\rho_{L_1 R_1} \otimes \rho_{L_2 R_2}$, where $\rho_{L_i R_i}$ ($i = 1, 2$) denote the same EME state with the vacuum coefficient c_n if we have carried out the entanglement connection n times. The φ-parameters in $\rho_{L_1 R_1}$ and $\rho_{L_2 R_2}$ are the same provided that the two states are established over the same stationary channels. We register only the coincidences of the two-side detectors, so the protocol is successful only if there is a click on each side. Under this condition, the vacuum components in the EME states, together with the state components $S_{L_1}^\dagger S_{L_2}^\dagger |vac\rangle$ and $S_{R_1}^\dagger S_{R_2}^\dagger |vac\rangle$, where $|vac\rangle$ denotes the ensemble state $|0_a 0_a 0_a 0_a\rangle_{L_1 R_1 L_2 R_2}$, have no contributions to the experimental results. So, for the measurement scheme shown by Fig. 4, the

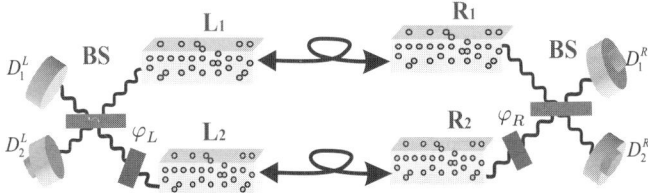

FIG. 4. Schematic setup for the realization of quantum cryptography and Bell inequality detection (see Duan et al., 2001). Two pairs of ensembles L_1, R_1 and L_2, R_2 have been prepared in the EME states. The collective atomic excitations on each side are transferred to the optical excitations, which, respectively after a relative phase shift φ_L or φ_R and a 50%–50% beam splitter, are detected by the single-photon detectors D_1^L, D_2^L and D_1^R, D_2^R. We look at the four possible coincidences of D_1^R, D_2^R with D_1^L, D_2^L, which are functions of the phase difference $\varphi_L - \varphi_R$. Depending on the choice of φ_L and φ_R, this setup can realize both the quantum cryptography and the Bell inequality detection.

ensemble state $\rho_{L_1R_1} \otimes \rho_{L_2R_2}$ is effectively equivalent to the following "polarization" maximally entangled (PME) state (the terminology of "polarization" comes from an analogy to the optical case)

$$|\Psi\rangle_{PME} = (S_{L_1}^\dagger S_{R_2}^\dagger + S_{L_2}^\dagger S_{R_1}^\dagger)/\sqrt{2}|vac\rangle. \tag{6}$$

The success probability for the projection from $\rho_{L_1R_1} \otimes \rho_{L_2R_2}$ to $|\Psi\rangle_{PME}$ (i.e., the probability to get a click on each side) is given by $1/[2(c_n+1)^2]$. One can also check that in Fig. 4, the phase shift φ_Λ ($\Lambda = L$ or R) together with the corresponding beam splitter operation are equivalent to a single-bit rotation in the basis $\{|0\rangle_\Lambda \equiv S_{\Lambda_1}^\dagger |0_a 0_a\rangle_{\Lambda_1 \Lambda_2}, |1\rangle_\Lambda \equiv S_{\Lambda_2}^\dagger |0_a 0_a\rangle_{\Lambda_1 \Lambda_2}\}$ with the rotation angle $\theta = \varphi_\Lambda/2$. Since we have the effective PME state and we can perform the desired single-bit rotations in the corresponding basis, it is clear how to use this facility to realize quantum cryptography, Bell inequality detection, as well as teleportation (see Duan et al., 2001 for details).

2.4. BUILT-IN ENTANGLEMENT PURIFICATION AND SCALING OF THE COMMUNICATION EFFICIENCY

It is remarkable that all the steps of entanglement generation, connection, and application schemes described above are robust to practical noise. Now we analyze the built-in entanglement purification in each step, which, combined together, makes the whole scheme noise resilient.

In the entanglement generation, the dominant noise is photon loss, which includes the contributions from the channel attenuation, spontaneous emission in the atomic ensembles (which results in the population of the collective atomic mode correlated with the accompanying photon going to other directions), the coupling inefficiency of the Stokes light into and out of the channel, and the inefficiency of the single-photon detectors. The loss probability is denoted by $1 - \eta_p$ with the overall efficiency $\eta_p = \eta_p' e^{-L_0/L_{att}}$, where we have separated the exponential channel attenuation factor (with channel attenuation length L_{att}) from other noise contributions η_p' that are independent of the communication distance L_0. The photon loss decreases the success probability for getting a detector click from p_c to $\eta_p p_c$, but it has no influence on the resulting EME state. Due to this noise, the entanglement preparation time is now written as $T_0 \sim t_\Delta/(\eta_p p_c)$. The second source of noise comes from the dark counts of the single-photon detectors. The dark count gives a detector click, but without population of the collective atomic mode, so it contributes to the vacuum coefficient in the EME state. If the dark count comes with a probability p_{dc} over the time interval t_Δ, the vacuum coefficient is given by $c_0 = p_{dc}/(\eta_p p_c)$, which is typically much smaller than unity since the Raman transition rate is much larger than the dark count rate. The final source of noise, which influences the fidelity to get the EME state, is caused by the event that more than one atom are excited to the collective mode S whereas there is only one click in D_1 or D_2. The conditional probability for that event is given by p_c, so we can estimate the infidelity $\bar{F}_0 \equiv 1 - F_0$ for the entanglement generation by

$$\bar{F}_0 \sim p_c. \tag{7}$$

Note that by decreasing the excitation probability p_c, the infidelity can be made closer to zero with the price of a longer entanglement preparation time T_0. This is the basic idea of the entanglement purification. So, in this scheme, the confirmation of the click from the single-photon detector generates and purifies entanglement at the same time.

In the entanglement swapping step, the dominant noise remains photon loss, which include the contributions from the detector inefficiency, the inefficiency of the excitation transfer from the collective atomic mode to the optical mode, and the small decay of the atomic excitation during the storage. (Note that by introducing the detector inefficiency, we have automatically taken into account the imperfection that the detectors cannot distinguish between one and two photons.) With all these losses, the overall efficiency in the entanglement swapping is denoted by η_s. The loss in the entanglement swapping gives contributions to the vacuum coefficient in the connected EME state, since in the presence of loss a single detector click might result from two collective excitations in the ensembles I_1 and I_2, and in this case, the collective modes in the ensembles L and R have to be in a vacuum state. After taking into account the realistic noise, we can specify the success probability and the new vacuum coefficient for the ith entanglement connection by the recursion relations

$$p_i \equiv f_1(c_{i-1}) = \frac{\eta_s \left[1 - \frac{\eta_s}{2(c_{i-1}+1)}\right]}{c_{i-1}+1}, \tag{8}$$

$$c_i \equiv f_2(c_{i-1}) = 2c_{i-1} + 1 - \eta_s. \tag{9}$$

The coefficient c_0 for the entanglement preparation is typically much smaller than $1 - \eta_s$, so $c_i \approx (2^i - 1)(1 - \eta_s) = (L_i/L_0 - 1)(1 - \eta_s)$, where L_i denotes the communication distance after i entanglement connections. With this expression for the c_i, we can easily evaluate the probability p_i and the communication time T_n for establishing a EME state over the distance $L_n = 2^n L_0$. After the entanglement connection, the fidelity of the EME state also decreases, and after n entanglement connections, the overall fidelity imperfection $\bar{F}_n \sim 2^n \bar{F}_0 \sim (L_n/L_0)\bar{F}_0$. This infidelity $\bar{F}_n$ can be small by simply decreasing the excitation probability p_c, from Eq. (7).

It is important to point out that this entanglement connection scheme also has built-in entanglement purification. This can be understood as follows: each time we connect entanglement, the imperfections of the setup decrease the entanglement fraction $1/(c_i + 1)$ in the EME state. However, the entanglement fraction decays only linearly with the distance (the number of segments), which is in contrast to the exponential decay of the entanglement for the connection schemes without entanglement purification. The reason for the slow decay is that in each time of the entanglement connection, we need repeat the protocol only until there is a detector click. The confirmation of a click removes part of the added vacuum noise, since a larger vacuum component of the EME state results in more repetition. The built-in entanglement purification in the connection scheme is essential for the polynomial scaling law of the communication efficiency.

As in the entanglement generation and connection schemes, the entanglement application schemes also have built-in entanglement purification which makes them resilient to practical noise. First, we have seen that the vacuum components in the EME states are removed from the confirmation of the detector clicks and thus have no influence on the fidelity of all the application schemes. Second, if the single-photon detectors and the atom-to-light excitation transitions in the application schemes are imperfect with the overall efficiency denoted by η_a, these imperfections only influence the efficiency to get the detector clicks with the success probability now given by $p_a = \eta_a/[2(c_n + 1)^2]$, and have no effects on the communication fidelity. Finally, we have seen that the phase shifts in the stationary channels and the small asymmetry of the stationary setup are removed automatically when we project the EME state to the PME state, and thus have no influence on the communication fidelity.

As a result of the built-in entanglement purification in each step of the DLCZ scheme, we can fix the communication fidelity to be nearly perfect, and at the same time keep the communication time to increase only polynomially with the distance. Assume that we want to communicate over a distance $L = L_n = 2^n L_0$. By fixing the overall fidelity imperfection to be a desired small value $\bar{F}$, the entanglement preparation time becomes $T_0 \sim t_\Delta/(\eta_p \bar{F}_0) \sim (L_n/L_0) t_\Delta/(\eta_p \bar{F})$. For an effective generation of the PME state (6), the total communication time is $T_{tot} \sim T_n/p_a$, with $T_n \sim T_0/\prod_{i=1}^{n} p_i$. So the total communication time scales with the distance by

$$T_{tot} \sim 2t_\Delta \left(\frac{L}{L_0}\right)^2 \frac{1}{\eta_p p_a \bar{F} \prod_{i=1}^{n} p_i}, \tag{10}$$

where the success probabilities p_i, p_a for the ith entanglement connection and for the entanglement application have been specified earlier. Equation (10) confirms that the communication time T_{tot} increases with the distance L only polynomially, with the understanding that the number of segments n itself depends logarithmically on L.

We illustrate this polynomial scaling explicitly in two limiting cases.

- For high efficiency entanglement swapping $(1-\eta_s \ll 1)$, the communication time in Eq. (10) is

$$T_{tot} \sim \tau_C \left(\frac{L}{L_0}\right)^2 e^{L_0/L_{att}}, \tag{11}$$

with the prefactor $\tau_C \equiv 2t_\Delta/(\eta'_p \eta_a \bar{F})$ independent of the segment and the total distances L_0 and L. In this case, the communication time T_{tot} increases quadratically with L.

- For low efficiency entanglement swapping with a significant inefficiency $1-\eta_s$, the communication time is approximated by

$$T_{tot} \sim \tau_C \left(\frac{L}{L_0}\right)^{\frac{1}{2}\log_2 \frac{L}{L_0} + \log_2(\frac{1}{\eta_s}-1) + \frac{5}{2}} e^{L_0/L_{att}}, \tag{12}$$

also exhibiting a polynomial increases with L. (More rigorously, this scaling is considered sub-exponential, but there is no practical difference because the polynomial order $\log_2(L/L_0)$ is well bounded from above for any reasonably long distance.) In general, when T_{tot} increases with L/L_0 by an mth-order power law $(L/L_0)^m$, choosing a segment length $L_0 = mL_{att}$ minimizes the net communication time T_{tot}. As a simple estimate of the improvement in the communication efficiency, we assume that the total distance L is about $100L_{att}$, for a choice of the parameter $\eta_s \approx 2/3$, the communication time $T_{tot}/\tau_c \sim 10^6$, using the optimal segment length $L_0 \sim 5.7L_{att}$. This result is a dramatic improvement compared with the direct communication case, where the communication time T_{tot} for getting a PME state increases with the distance L exponentially: $T_{tot} \sim T_0 e^{L/L_{att}}$. For the same distance $L \sim 100L_{att}$, this requires $T_{tot}/T_0 \sim 10^{43}$ for direct communication, which implies that for this example, the DLCZ scheme becomes 10^{37} times more efficient through implementation of the quantum repeater architecture.

2.5. Experimental Quantum Communication with Atomic Ensembles

Remarkable experimental advances in implementing the DLCZ scheme have been reported in recent years (Kuzmich et al., 2003; Van der Wal et al., 2003; Chou et al., 2004, 2005; Matsukevich and Kuzmich, 2004; Blinov et al., 2004; Balic et al., 2005; Chaneliere et al., 2005; Eisaman et al., 2005; Manz et al., 2006; Riedmatten et al., 2006; Matsukevich et al., 2006), targeted to demonstration of scalable quantum communication with atomic ensembles. The first two experiments were reported in 2003 by the Caltech and the Harvard groups (Kuzmich et al., 2003; Van der Wal et al., 2003), where the non-classical correlation between the symmetric atomic mode S and the signal light mode a was observed. Demonstration of this correlation is fundamental to the entanglement generation scheme in the DLCZ approach, and also shows the collective enhancement effect in the atomic ensemble for the associated level configuration.

In these initial demonstrations, the symmetric atomic mode S is transferred to another photon mode b, and the correlation is actually measured between the photonic modes a and b. The Caltech experiment (Kuzmich et al., 2003) operates in the weak pumping region as assumed in the DLCZ scheme, and it introduces a useful quantity from the Cauchy–Schwarz inequality to characterize the nonclassical correlation between the modes a and b (and thus S and a). From the photon counting measurements of these two photonic modes, the self-correlation functions $\tilde{g}_{a,a}$, $\tilde{g}_{b,b}$ and the cross-correlation function $\tilde{g}_{a,b}$ can be defined. For any two classical light fields (as in quantum optics, "classical" here means that the fields can be described with a positive P-representation), these correlations satisfy the so-called Cauchy–Schwarz inequality $[\tilde{g}_{a,b}]^2 \leqslant \tilde{g}_{a,a}\tilde{g}_{b,b}$, while this inequality could be violated for fields with non-classical correlation. So, if one defines a quantity $\eta \equiv [\tilde{g}_{a,b}]^2/(\tilde{g}_{a,a}\tilde{g}_{b,b})$, $\eta > 1$ is a clear experimental signature for nonclassical correlation. This quantity η has been measured to be about 1.84 in the first experiment (Kuzmich et al., 2003), and now its value can be pushed above hundreds (Riedmatten et al., 2006; Matsukevich et al., 2006), signaling a much better signal-to-noise ratio. The Cauchy–Schwarz inequality and the associated ratio η is widely used in experiments to characterize noise and quantify entry into the quantum region.

More recently, several experiments have coherently manipulated two separate atomic ensembles (Matsukevich and Kuzmich, 2004; Chou et al., 2005; Chaneliere et al., 2005; Eisaman et al., 2005; Matsukevich et al., 2006). The coherence or entanglement of the two ensembles using the above entanglement generation scheme has been shown (Matsukevich and Kuzmich, 2004; Chou et al., 2005; Matsukevich et al., 2006), and quantum states have been coherently transferred from one ensemble to the other (Chaneliere et al., 2005; Eisaman et al., 2005).

The level configuration investigated in Duan et al. (2001) also provides a new scheme to generate entangled photon pairs, with one photon of the pair able to be stored in the atomic ensemble with a controllable delay time. This ability, together with the projection measurements by the photon counts, have a number of interesting applications for engineering the states of the photon pulses (Chou et al., 2004) and for proof-of-principle demonstration of storage of the single photon pulses (Chaneliere et al., 2005; Eisaman et al., 2005).

In addition to atomic ensembles, various other physical systems have been proposed to replace the atomic ensembles in the DLCZ scheme. For instance, ensembles of electron spins or nuclear spins in quantum dots (or other solid-state systems) can be considered (Childress et al., 2006), given that the electron or nuclear spins are sufficiently identical (have a small inhomogeneous shift), and exhibit a sufficiently long coherence time.

Single trapped atoms or ions can also replace the atomic ensemble (Duan et al., 2004). In this case, there is of course no collective enhancement effect for the coupling to light, so the entanglement connection efficiency can be significantly reduced. However, for local ions at the same node, the Coulomb interaction can be exploited to perform local quantum gates and entangle different segments of the system. This provides an alternative way for efficient entanglement connections and the construction of quantum repeaters or quantum networks (Duan et al., 2004). The advantage with the trapped ion system is that single ions behave as near-ideal quantum memories, having coherence times far exceeding that of atomic ensembles. Moreover, the ion trap system is a leading candidate for the implementation of quantum computers (Monroe, 2002). In this context, it may be desirable to wire together remotely-located ions through photons, an important step towards distributed quantum computation and the "quantum internet". Along these lines, an initial experiment has been reported in 2004 (Blinov et al., 2004), where entanglement was observed between a single trapped ion qubit and a spontaneous emission photon, therefore demonstrating entanglement between an ideal quantum memory and a flying qubit. Further experiment has confirmed the violation of the Bell inequality corresponding to this entanglement (Moehring et al., 2004). Similar entanglement has also been measured subsequently between a single neutral atom and a photon (Volz et al., 2006).

3. Quantum State Engineering with Realistic Linear Optics

3.1. Linear Optics and Quantum State Engineering

In this section, we review some protocols to generate many-body entangled states with linear optics elements. It is closely related to topics in the previous section in that these schemes are inherently robust to the dominant noise in the corresponding experimental system. For quantum communication, we would like to establish an entangled EPR-like or Bell state over a long distance, and are interested in the scaling of total communication time and the communication distance. For quantum state engineering, we would like to generate more diverse and complex quantum states of many qubits, and are interested in the scaling of total preparation time and the number of qubits.

Linear optics, combined with practical single-photon sources and detectors, has provided a powerful tool to test a number of quantum information protocols (Kwiat et al., 1995; Bouwmeester et al., 1997; Boschi et al., 1998; Pan et al., 2001; Walther et al., 2005). In linear optics implementations of quantum information processing, the post-selection technique of photon detection typically plays a critical role. However, as the system is scaled up, this post-selection naïvely leads to an exponential scaling of the overall efficiency (or success probability) with the size of the system. A remarkable linear optics quantum computation scheme was proposed by Knill et al. (2001), which in principle can be used to overcome this scaling problem. But the implementation of the KLM scheme requires photon detectors with a very high efficiency, far beyond than the efficiency of the state-of-the-art photon detectors. In the past few years, there have been a number of proposals to improve upon the original linear optics computation scheme (Yoran and Reznik, 2003; Nielsen, 2004; Browne and Rudolph, 2005), some of which employ a graph state approach to quantum computation (Raussendorf and Briegel, 2001; Hein et al., 2006). The threshold efficiency for the photon detectors has also been improved considerably, with the most recent estimate about 99.7% (Dawson et al., 2006) (although it is still significantly beyond the efficiency of practical photon detectors).

We now review quantum state engineering schemes that do not require a high threshold efficiency on the photon detectors. Earlier, it was shown that GHZ types of entangled states can be prepared with linear optical devices and low efficiency photon detectors (Duan, 2002). Recently, this scheme has been extended to generate any "graph" state, which can be used for quantum information protocols and universal quantum computation (Raussendorf and Briegel, 2001; Hein et al., 2006). Moreover, a particular class of entangled states represent by "tree" graphs can be prepared efficiently with photon detectors of *any* efficiency. This approach overcomes the inefficient scaling through the "divide-and-conquer" method, similar to the quantum repeater protocol. While tree graph states are not universal for quantum computation (Shi et al., 2006), they can still be used for implementation of a number of other quantum information protocols, including quantum communication, networking, and fundamental test of quantum mechanics (Hein et al., 2006).

In this section, we review this scheme for engineering graph state entanglement following the approach in Bodiya and Duan (2006). We analyze the effect of a polarization beam splitter (PBS) in the Hilbert subspace post-selected by the photon detections, and show that a single PBS actually represents a powerful gate for generating graph states of arbitrary shapes. This PBS gate is more efficient than other linear optical quantum gates (Knill et al., 2001; Nielsen, 2004; Browne and Rudolph, 2005), not wasting any ancilla photons within each gate operation. Finally, we review a method for scalable generation and detection of many-qubit entangled states represented by tree graph states. Here, by "scalable", we mean the overall efficiency for preparation of a large-scale entanglement with a tree-graph structure scales nearly polynomially with the number of qubits. This efficient scaling persists no matter how small the efficiencies of the photon sources or detectors.

3.2. Preparation of Arbitrary Graph States

We assume to have an imperfect source of entangled photon pairs, which generates states of the following form

$$\rho_s = (1 - \eta_s)\rho_{vac} + \eta_s |\Psi\rangle_{12}\langle\Psi|, \qquad (13)$$

where $|\Psi\rangle_{12} = (|HH\rangle_{12} + |VV\rangle_{12})/\sqrt{2}$ represents single photons in two distinct (spatial) modes (1 and 2) entangled through their (internal) polarization states $|H\rangle$ and $|V\rangle$. The state ρ_{vac} represents the vacuum component with zero photons in modes 1 and 2; and η_s is the source efficiency for producing the entangled photon pair. In experiments, the entangled photon source is typically provided through the process of spontaneous parametric down conversion (SPDC), where the source efficiency $\eta_s \ll 1$ (Kwiat et al., 1995; Bouwmeester et al., 1997; Boschi et al., 1998). The pair state of Eq. (13) can also be generated from other experimental setups, such as from decay of a single dipole (which could be a single atom, ion, or a quantum dot) in free space or in a cavity (Blinov et al., 2004), or from decay of an collective excitation in an atomic ensemble (Duan et al., 2001; Chaneliere et al., 2005; Eisaman et al., 2005). In these cases, one mode of the entangled pair is typically represented by a matter qubit (see the previous section), which can be transferred later to a photon qubit after a controllable delay.

Now we show any graph state in the Hilbert subspace post-selected by the photon detection can be generated from the pair states [Eq. (13)] through a series of PBS gates. An n-qubit graph state is defined as the co-eigenstate of n independent stabilizer operators $S_i = X_i \prod_j Z_j$, where i denotes qubit i (each qubit is associated with a vertex of the graph), j runs over all the neighbors of the qubit i, and X_i, Z_i are the Pauli operators σ_x and σ_z for qubit i (Raussendorf and Briegel, 2001; Hein et al., 2006). In a graph, the qubits i and j are called neighbors if they are connected with an edge. The graph state reduces to a cluster state if the corresponding graph is a periodic lattice (Hein et al., 2006).

To show construction of the graph states, first we must consider the effect of a PBS in the subspace post-selected by the photon detection. For linear optics quantum information, all the photon modes are measured eventually in an appropriate polarization basis by a single-photon detector. We are interested only in the measurement outcomes with one photon registered from each mode (its polarization can be arbitrary). So, by this final measurement, one post-selects a Hilbert subspace, which we denote as S. We need only determine the state evolution in this "physical" subspace S, as the state component outside S has no influence on the final measurement of the polarization qubits.

A PBS lets the photon through if it is in the H polarization state, and reflects it if it is in the V polarization state. After the PBS, photons from the two incoming modes therefore emerge along different modes if and only if both photons have the same polarization (either HH or VV). Otherwise, the photons emerge from the

PBS in the same spatial mode with the other mode in the vacuum state—outside of the "physical" subspace S. So, within the subspace S, the effect of a PBS is to perform a projection on the input state, described by the projector

$$P = |HH\rangle_{12}\langle HH| + |VV\rangle_{12}\langle VV|. \tag{14}$$

This projection is equivalent to a measurement of the operator $Z_1 Z_2$ on the two input qubits 1 and 2, with the final state undisturbed only for measurement outcome "+1" ($|HH\rangle_{12}$ and $|VV\rangle_{12}$ are eigenstates of $Z_1 Z_2$ with an eigenvalue "+1"). So in the physical subspace S, a single PBS performs an effective $Z_1 Z_2$ measurement gate with a success probability of 1/2 (the probability to stay in the "physical" space S after the PBS).

We start with two entangled pairs 1, 2 and 3, 4, each pair described by the state of Eq. (13). In the subspace S, the effective state is then given by $|\Psi\rangle_{12}$, which can be transferred to a two-bit graph state with a straightforward Hadamard gate on one of the qubits. Consequently, for the pairs 1, 2 and 3, 4, we can assume them to have the stabilizer operators $X_1 Z_2$, $X_2 Z_1$ and $X_3 Z_4$, $X_4 Z_3$, respectively. If the qubits 2 and 3 pass through a PBS, the effective output state in the subspace S is then stabilized by the operators $Z_2 Z_3$, $X_2 X_3 Z_1 Z_4$, $X_2 Z_1$, and $X_4 Z_3$. (The second operator is just a product of the previous stabilizers $X_2 Z_1$ and $X_3 Z_4$, and it remains unchanged after the PBS because it commutes with the effective measurement gate $Z_2 Z_3$.) With a straightforward Hadamard gate $X_3 \leftrightarrow Z_3$, implemented with a half-wave plate, the above four stabilizers transform to the standard stabilizers for the 4-bit star-shape graph state as shown in Fig. 5.

For convenience, we label the combination of the PBS and single-bit Hadamard operation the PBS gate (see Fig. 5). An extension of the above construction yields the following important result: the PBS gate always joins two pieces of graphs, independent of the shapes of the initial pieces. This result can be proven generally as follows. We start with two pieces of graph states G_1 and G_2, with n and m qubits, respectively. The stabilizers associated with the qubits i_1 and i_2 are given by $S_{i_l} = X_{i_l} \prod_{j_l \in N(i_l)} Z_{j_l}$ ($l = 1, 2$), where i_l is an arbitrary vertex of the graph G_l and $N(i_l)$ denotes all the neighbors of the qubit i_l in the graph G_l. After a PBS gate on the qubits i_1 and i_2, the stabilizers S_{i_2} and S_{i_1} are replaced by

$$S'_{i_2} = X_{i_2} Z_{i_1} \quad \text{and} \quad S'_{i_1} = X_{i_1} Z_{i_2} \prod_{j_1 \in N(i_1)} Z_{j_1} \prod_{j_2 \in N(i_2)} Z_{j_2}.$$

All the other stabilizers of the initial graphs G_1 and G_2 remain unchanged after the gate. One can immediately see that the effective output state of the PBS gate is still a graph state which combines the two initial graphs G_1 and G_2, with i_2 attached to i_1, and i_1 attached to i_2 and all their initial neighbors in the graphs G_1 and G_2 [see Fig. 5(c)].

Given the above result, it becomes possible to construct graph states of any shape with a series of PBS gates. In Figs. 5 and 6, we illustrate this by constructing

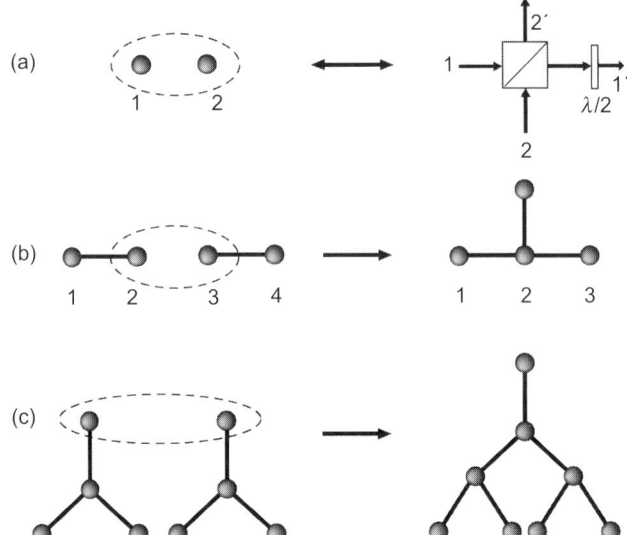

FIG. 5. (a) Representation of the PBS gate, which consists of a polarization beam splitter and a half-wave plate (for a Hadamard operation on one mode). (b) and (c): Illustration in which PBS gates generate tree graph states. It is obvious that tree graphs of any shapes can be generated with this method (see Bodiya and Duan, 2006).

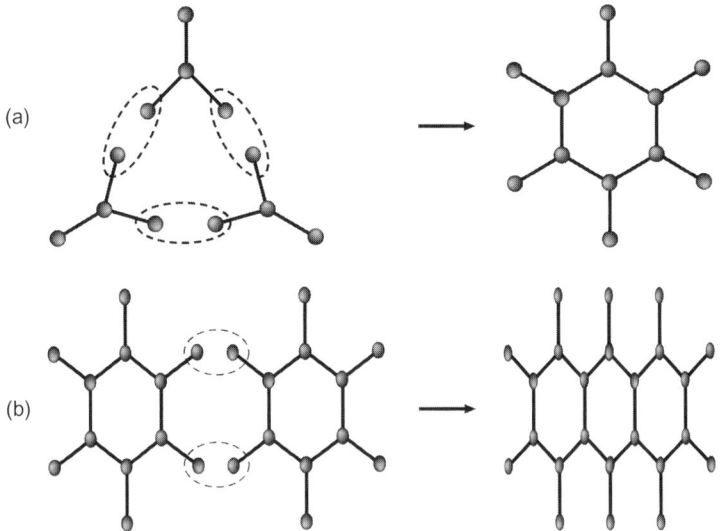

FIG. 6. (a) and (b): Illustration of using the PBS gates to generate 2-dimensional graphs states (see Bodiya and Duan, 2006).

graph states representing a tree graph and a two-dimensional graph with many loops. This construction method is efficient, as no photons are wasted during the state preparation. Starting with n entangled pairs, we can build graph states of $2n$ qubits with various shapes.

3.3. EFFICIENT GENERATION AND SCALING OF TREE GRAPH STATES

If we assume both the photon sources and detectors have large inefficiencies, we still have inefficient (exponential) scaling for construction of large-scale graph states even with the above PBS gate. In order to generate an n-qubit entangled graph-state, we must consume $n/2$ imperfect entangled pairs represented by the state [Eq. (13)] and detect n photon modes at the end. So, there is a factor of $\eta_d^n \eta_s^{n/2}$ in the preparation efficiency, where η_d is the efficiency for each individual detector. If we require $m \leqslant n/2$ PBS gates to arrive at such a graph state, there is an additional factor of $(1/2)^m$ in the preparation efficiency associated with the intrinsic gate success probability to stay in the subspace S. In the case of a small source efficiency η_s (such as for the SPDC experiments), the preparation efficiency degrades rapidly with the size of the state, limiting the current implementation to only a few qubits (Kwiat et al., 1995; Bouwmeester et al., 1997; Boschi et al., 1998; Pan et al., 2001; Walther et al., 2005). In the following, we show that an important subclass of graph states—"tree" states—can be prepared and detected efficiently with a number of operations that scales polynomially in the state size. As the name implies, tree states are defined as graph states where any two vertices are connected by exactly one path.

This efficient scaling method is based on a combination of the ideas of the divide-and-conquer (quantum repeater) protocol and the post-selection measurements. We note that for applications of graph states in linear optics quantum information, each photon mode needs to be eventually measured in some polarization basis. This suggests that the whole protocol can be divided into two logical steps: the graph state preparation and the application measurement. For the second step, measurement of each photon mode has a finite failure probability, where instead of getting the photon's polarization, one does not register any photon. To boost the efficiency of the whole protocol, it is better to sort out and discard these failure events as soon as possible. In this spirit, we can try to apply the application measurements on some individual qubits before we finish the first logic step of the graph-state preparation. We measure the qubits as soon as we do not need to apply the PBS gates on those qubits any more. When we register a failure event, we immediately discard the qubits that are influenced by the failure event, and restart the state preparation for that segment.

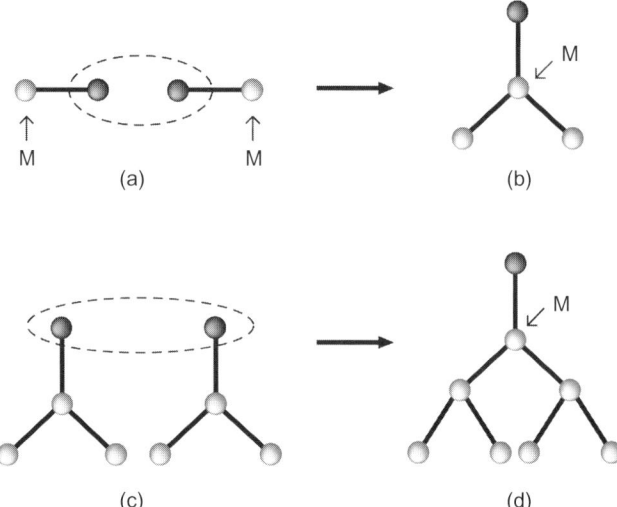

FIG. 7. Efficient construction of tree-graph states. White circles represent qubits that have been measured in appropriate polarization bases, and black circles represent the connection qubits (unmeasured) that enable the next-step connection. (a) Before connection of the two center qubits, the two edge qubits have been measured. (b) After connection, we immediately measure one of the connection qubits, and leave the other one for the next step connection as shown in (c). (c) and (d): Repetition of the process of connection/measurement for construction of larger graphs (see Bodiya and Duan, 2006).

Figure 7 illustrates how such an idea works for preparation and detection of a tree-graph state. We start with two pairs $(1, 2)$ and $(3, 4)$, with the pair state described by Eq. (13). As we do not need to apply the PBS gates on the qubits 1 and 4 in the following steps, we immediately measure them in the polarization basis chosen according to the targeted application protocol. The measurement on the qubit 1 (or 4) succeeds with a probability $p_0 = \eta_s \eta_d$, and upon success, the vacuum component in the imperfect state [Eq. (13)] is eliminated. If we fail in the measurement of qubit 1, we simply prepare the state for the pair $(1, 2)$ once again, with the pair $(3, 4)$ intact. After an average number of trials $\sim 1/p_0$, we succeed in entangling the qubit pair $(1, 2)$ and eliminating their vacuum component. Simultaneously, we can do the same to qubit pair $(3, 4)$ after $\sim 1/p_0$ trials. Then, we continue with the connection of the qubits 2 and 3 through a PBS gate, and after the connection, we immediately measure the qubit 2 as we only need keep qubit 3 for the next step of connection. This process is continued until we get an effective tree-graph state with a desired number of qubits, as outlined in the figure.

To determine the overall efficiency for generation of this graph-state entanglement, we specify recursion relations for each step of connection. For each connection, the number of qubits is doubled. For the mth connection, the effective state before connection can be written as $\rho_{2n}^a = \rho_n \otimes \rho_n$, where ρ_n is the state of a segment which has $n = 2^m$ qubits. The segment state can be expressed as $\rho_n = a_{m-1}\rho_g + (1 - a_{m-1})\rho_{vac}$, where ρ_g denotes the effective n-qubit tree graph state and ρ_{vac} represents the vacuum component where the connection qubit of the graph is in the vacuum state. For the 1st connection [of pairs (1, 2) with (3, 4)], $a_0 = 1$ because the vacuum component has been eliminated by the measurement on qubits 1 and 4. After the mth connection, we immediately measure one of the two connection qubits (the other one is kept as the connection qubit for the next step). The success probability for this measurement is given by

$$p_m = \eta_d \left[\frac{a_{m-1}^2}{2} + \frac{a_{m-1}^2(2 - \eta_d)}{4} + a_{m-1}(1 - a_{m-1}) \right], \tag{15}$$

where we have assumed the detector cannot distinguish the single-photon and two-photon counts, as is the case in practice. Upon a success of this measurement, the effective state for the $2n$ qubits becomes $\rho_{2n} = a_m \rho_g + (1 - a_m)\rho_{vac}$, where ρ_g and ρ_{vac} have the same meaning as before except that they are for $2n$ qubits now, and the coefficient a_m is given by the recursion relation $a_m = 2a_{m-1}/(4 - \eta_d a_{m-1})$. Together with $a_0 = 1$, this recursion relation yields

$$a_m = \frac{2}{2^m(2 - \eta_d) + \eta_d}. \tag{16}$$

To prepare and confirm an $n = 2^m$ qubit entanglement represented by the tree graph state, the overall efficiency of the scheme can be characterized by the total preparation time T. From the above recursion relations, we find that

$$T = \frac{t_0}{\eta_d a_{m-1}} \prod_{i=0}^{m-1} \frac{1}{p_i}$$

$$\approx \left(\frac{t_0}{\eta_s \eta_d} \right) n^{\frac{1}{2}(\log_2 n - 1) + \log_2(\frac{1}{\eta_d} - \frac{1}{2})}, \tag{17}$$

where the approximation is valid when $\eta_d/2 \ll n$, and we have assumed that the two segments of graphs states before each connection can be prepared in parallel simultaneously. Note that overall preparation time scales linearly with t_0, the time to generate an imperfect pair [Eq. (13)]. In the case of SPDC photon sources, this is roughly the inverse of the pulse repetition rate in these systems (Kwiat et al., 1995; Bouwmeester et al., 1997; Boschi et al., 1998; Pan et al., 2001; Walther et al., 2005). We find that T also scales nearly polynomially with the size n of the final graph state, and such a scaling holds for any source efficiency η_s and detector efficiency η_d.

The following example illustrates the dramatic improvement in scaling compared with conventional (sequential) entanglement techniques. If we take the source efficiency $\eta_s \sim 1\%$ and the detector efficiency $\eta_d \sim 33\%$, we find

$T/t_0 \sim 3.0 \times 10^7$ (3.6×10^4) for preparation of a graph state of 32 (8) qubits. If the pulse repetition rate is 80 MHz, typical for mode-locked laser sources (Kwiat et al., 1995; Bouwmeester et al., 1997), the total preparation time is $T \sim 0.37$ s (0.45 ms), which is still reasonable. If we do not use this divide-and-conquer technique, the total time required to produce a 32-qubit graph state is $T/t_0 = \eta_s^{-32/2} \eta_d^{-32} 2^{32/2-1} \sim 10^{52}$, or $T \sim 10^{44}$ s.

We point out that a very recent experiment has reported demonstration of six photon graph states (Lu et al., 2006), using exactly the above PBS gate.

4. Quantum Computation through Probabilistic Atom–Photon Operations

4.1. Robust Probabilistic Gates

In this section, we review some schemes and experimental progress towards scalable quantum computing with probabilistic type of gates on atoms or ions. In the previous two sections, we have reviewed methods to achieve inherently robust quantum communication and state engineering. Similarly, we also strive to achieve inherent fault tolerance to the dominant experimental noise in quantum computing applications. This is the main motivation for using probabilistic quantum gates, where the dominant noise merely contributes to the gate inefficiency and does not necessarily lead to infidelity of quantum gates. With noise, the entangling gate succeeds with a finite (even small) probability, but we know with near-certainty when it does succeed. It becomes much easier to correct these probabilistic type of errors. We show that universal quantum computers can be constructed efficiently even if the entangling gate only succeeds with an arbitrarily small probability. Thus, through design of probabilistic gates, we can tolerate the dominant experimental noise at very high levels.

The probabilistic gate covered in this section is different from the probabilistic entangling operations and the post-selected PBS gate that we have reviewed in the last two sections. Although all of these operations share the property of inherent insensitivity to noise, only the probabilistic gate can lead to scalable quantum computation. Compared with deterministic gates, the additional overhead in resources (such as the number of qubit manipulations) for quantum computing with probabilistic gates scales only polynomially with both the size of the computation and the inverse of the gate success probability.

In the next section, following the approach in Duan et al. (2006), we review a scheme for probabilistic gates on remote ions or atoms through interference of photonic qubits stored in the frequency (color) of the photons. There are other proposals for implementation of probabilistic gates using atomic qubits (Barrett and

Kok, 2005; Duan et al., 2005; Lim et al., 2006). However, the scheme reviewed here is considerably simpler with regard to experimental requirements:

- The gate operates on atoms in free space without the need for optical cavities, and uses ideal atomic ground state hyperfine energy levels as matter qubits.
- Optical *frequency* qubits are used to connect and entangle matter qubits at distant locations. The two states comprising this optical qubit have the same polarization, but differ in frequency by the atomic hyperfine splitting (typically in the microwave region). These closely-spaced frequency components have basically zero dispersion in typical optical paths, thus this optical qubit is highly insensitive to phase jitter inherent in optical interferometers.
- The gate scheme does not require interferometric stabilization of the optical path lengths to near or within an optical wavelength.
- The motion of the atomic qubits need not be confined to within an optical wavelength (the Lamb–Dicke regime).

After a review of this probabilistic gate scheme, we then show how efficient universal quantum computation can be accomplished with probabilistic entangling quantum gates, following the approach in Duan and Raussendorf (2005). Finally, we review recent experimental progress towards demonstration of the probabilistic quantum gates in the trapped ion system (Madsen et al., 2006; Maunz et al., 2006).

4.2. Probabilistic Gates from Free-Space Atom–Photon Coupling

The scheme for probabilistic gates between remote atoms is illustrated in Fig. 8. The qubit is represented by two $S_{1/2}$ ground state hyperfine levels of an alkali-like atom (ion), with $|0\rangle \equiv |F, m = 0\rangle$, and $|1\rangle \equiv |F + 1, m = 0\rangle$. These "clock" states are particularly insensitive to stray magnetic fields. In the figure, for simplicity, we take $F = 0$, which is the case for ions such as ^{111}Cd$^+$, but the scheme works for any value of F. To perform a probabilistic gate on two remote atoms 1 and 2, we first excite both of the atoms to the $P_{1/2}$ excited electronic state with a π-polarized ultrafast laser pulse. We assume the laser has a bandwidth which is larger than the ground state hyperfine splitting (14 GHz for ^{111}Cd$^+$), but smaller than the fine structure splitting between $P_{1/2}$ and $P_{3/2}$ (74 THz for ^{111}Cd$^+$). Typical picosecond pulses used in experiments (bandwidth $\sim$ 500 GHz) satisfy these requirements (Madsen et al., 2006). Under the above condition, we can assume the pulse drives only the $D1$ transition from the ground state $S_{1/2}$ to the excited state $P_{1/2}$. [For any atoms with nuclear spin $I = 1/2$ such as ^{111}Cd$^+$, one can also drive the $D2$ line $S_{1/2} \to P_{3/2}$, where the two corresponding hyperfine transitions are given by $|F, m = 0\rangle \to |F', m = 0\rangle$ and $|F + 1, m = 0\rangle \to |F' + 1, m = 0\rangle$ with $F' = F + 1$, see Madsen et al. (2006).] Due

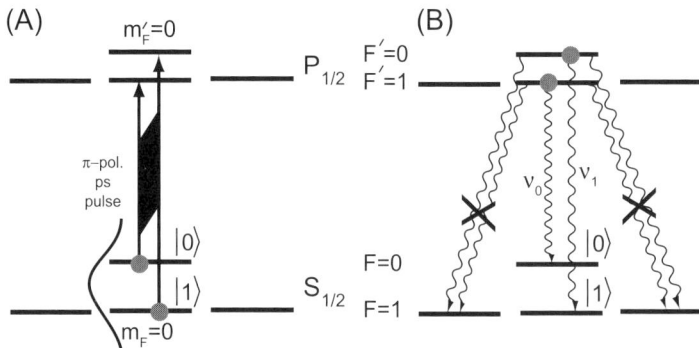

FIG. 8. The atomic level configuration and the laser excitation scheme. (A) An ultrafast laser pulse transfers the atomic qubit state from the ground levels to the excited levels. (B) The atom decays back to the ground levels, with the frequency of the spontaneously emitted photon correlated with the atomic qubit state (marked as the signal mode ν_0 and ν_1 in the figure). The photon from the $\sigma^{\pm}$ decay channels is filtered through polarization selection (see Duan et al., 2006).

to dipole selection rules, for a π-polarized pulse, only the hyperfine transitions $|F, m = 0\rangle \to |F' + 1, m = 0\rangle$ and $|F + 1, m = 0\rangle \to |F', m = 0\rangle$ are allowed, where the upper hyperfine spin $F' = F$. Thanks to the selection rules, each qubit state is transferred to a unique excited hyperfine level after the pulsed laser excitation. This point is critical for successful gate operation.

After this laser excitation, the atoms eventually decay back to their ground $S_{1/2}$ states. There are several decay channels, denoted as π or $\sigma^{\pm}$ in Fig. 8. The spontaneous emission photons from the π and the $\sigma^{\pm}$ decay channels have orthogonal polarizations along the observation direction. We can distinguish them and can block any photon from the $\sigma^{\pm}$ decay channels through a polarization filter. We then consider the π decay channels. In this case, the excited levels $|F' + 1, m = 0\rangle$ and $|F', m = 0\rangle$ can only decay back to the ground states $|F, m = 0\rangle$ and $|F + 1, m = 0\rangle$, respectively. While photons from these two decay channels have the same polarization, they have slightly different frequencies. The frequency difference is given by $\Delta_{HF}^{S} + \Delta_{HF}^{P}$, the sum of the hyperfine splittings of the ground $S_{1/2}$ and excited $P_{1/2}$ states. This frequency difference is typically much larger than the natural linewidth of the excited level. For instance, for ^{133}Cs atoms or ^{111}Cd^{+} ions, the hyperfine splitting is about 9 GHz (14 GHz), while the natural linewidth of the excited level (the inverse of the lifetime) is around 5 MHz (60 MHz). In both cases, the condition is well satisfied. So the corresponding photons from the two π-decay channels are well resolved in frequency. This defines two frequency modes for the emitted photon field, and we call them ν_0 and ν_1 modes, respectively. If the atom is initially in the qubit state $|\Psi_a\rangle = c_0|0\rangle + c_1|1\rangle$, then after this excitation-decay process the atom–photon system evolves to an entangled state

$$|\Psi_{ap}\rangle = c_0|0\rangle|v_0\rangle + c_1|1\rangle|v_1\rangle \qquad (18)$$

if we only collect the photon from the π decay channels, where $|v_0\rangle$ and $|v_1\rangle$ represent a single photon state in the frequency modes v_0 and v_1, respectively. This result is somewhat similar to the experiment of the atom–photon entanglement (Blinov et al., 2004), but there are important differences. First, the final state $|\Psi_{ap}\rangle$ keeps track of the information c_0, c_1 of the initial qubit state. Thus, the scheme here is not just an entangling protocol, but is instead an entangling *gate* with the final quantum state depending on the initial state. As we see later, this type of gate can form the basis for scalable quantum computation, and is therefore more powerful than merely an entangling operation. Second, the spontaneous emission photon with either frequency v_0 or v_1 has the same spatial mode, so good spatial mode-matching of this photonic qubit is possible even if we increase the solid angle of collection. In the previous entangling protocol (Duan et al., 2004; Blinov et al., 2004), the quantum information is carried by different polarization modes of the photon, which have different spatial emission patterns. This requires small collection solid angles in order to both maintain orthogonality and ensure adequate spatial matching of the photonic qubit states.

To perform a gate on two remote atoms, the spontaneous emission photons from the decay channels in each atom are collected in a certain solid angle, and directed onto a beam splitter for interference (see Fig. 9). The output of the beam splitter is measured by two single-photon detectors. We keep the resulting outcome atomic state only when we register a photon from each detector. In this case, what we have performed is a "measurement gate" on the atoms 1 and 2. It corresponds to a quantum non-demolition measurement of the operator $Z_1 Z_2$, where Z_i (or X_i) stands for the z (or x) component of the Pauli matrix associated with atomic qubit i. After the coincidence measurement of photons on both detectors, the atomic state is projected to the eigenspace of $Z_1 Z_2$ with -1 eigenvalue. To see this, we note that before the measurement, the state of both atom–photon systems can be written as $|\Psi_{ap}\rangle_1 \otimes |\Psi_{ap}\rangle_2$, where $|\Psi_{ap}\rangle_1$ has the form of Eq. (14), and $|\Psi_{ap}\rangle_2$ can be written as $|\Psi_{ap}\rangle_2 = d_0|0\rangle_2|v_0\rangle_2 + d_1|1\rangle_2|v_1\rangle_2$. To register a photon from each detector, the two photons before the beam splitter need to go to different sides, which means they should be in the anti-symmetric component $|\Phi_{AS}\rangle = (|v_0\rangle_1|v_1\rangle_2 - |v_1\rangle_1|v_0\rangle_2)/\sqrt{2}$ (for photons in the symmetric states, they always go to the same detector). So, given that the photons take separate paths after the beam splitter, the state of the atoms 1, 2 is given by the projection

$$\begin{aligned}|\Psi_{12}\rangle &\propto \langle\Phi_{AS}||\Psi_{ap}\rangle_1 \otimes |\Psi_{ap}\rangle_2 \\ &\propto c_0 d_1|0\rangle_1|1\rangle_2 - c_1 d_0|1\rangle_1|0\rangle_2 \\ &\propto Z_1(I - Z_1 Z_2)|\Psi_a\rangle_1 \otimes |\Psi_a\rangle_2,\end{aligned} \qquad (19)$$

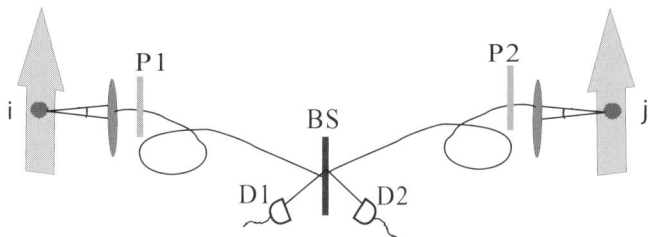

FIG. 9. The ZZ measurement gate on the atoms i and j. The spontaneous emission photons from the π decay channels of these two atoms are collected, interfered at the beam splitter (BS), and then detected by two single-photon detectors (D_1 and D_2). If each detector registers a photon, the atomic state is projected onto the eigenspace of the $Z_i Z_j$ operator (see Duan et al., 2006).

where $I - Z_1 Z_2$ is the corresponding projector, and Z_1 is a trivial additional single-bit gate on atom 1 which we neglect in the following. This measurement gate, of course, only succeeds with a finite probability. The overall success probability is given by $p_s = \eta_d^2 \eta_c^2 \eta_b^2 / 4$, where η_d is the quantum efficiency of each detector, η_c is the photon collection efficiency (proportional to the solid angle), and η_b is the branching ratio for the atom to decay along the π channel. We have an additional factor of $1/4$ in p_s describing the average probability for the two spontaneous emission photons to go to different detectors (averaged over all the possible initial atomic states). In the above contributions to the success probability, the collection efficiency is typically the smallest and thus dominates the overall efficiency. That is why it is important to increase the collection solid angle as much as possible. Alternatively, one can also increase this efficiency with the use of optical cavities surrounding the atoms (McKeever et al., 2003). The overall success probability p_s is typically small. For instance, for a ^{111}Cd$^+$ ion decaying from the $P_{3/2}$ state, the branching ratio into the π-channel $\eta_b = 2/3$, the photon collection efficiency $\eta_c \sim 2\%$ in free space (Blinov et al., 2004), and the detector efficiency $\eta_d \sim 20\%$. This leads to an overall success probability $p_s \sim 2 \times 10^{-6}$. If the collection efficiency is increased by a factor of 10 with a larger collection solid angle or with a surrounding cavity (not necessarily high-finesse), the success probability will be significantly increased with a factor of 100.

The above measurement gate is robust to noise. We do not require that the atoms be localized to the Lamb–Dicke limit. In general, atomic motion occurs with a time scale of the trap frequency ν_t, typically much smaller than the decay rate γ of the excited atomic level. Thus, for each spontaneous emission pulse, we can safely assume the atom to be in a fixed but random position $\mathbf{r}$. In this case, both of the frequency components $|\nu_0\rangle$ and $|\nu_1\rangle$ will acquire the same random phase factor proportional to $e^{i\mathbf{k}\cdot\mathbf{r}}$, where $\mathbf{k}$ is the wave vector associated with the spontaneous emission photon. This overall phase therefore has no effect on the resultant measurement gate as shown in Eq. (19). If we take into account the motion of the atom within the pulse duration, the pulse from this moving atom also has a slight Doppler shift $\delta\omega = \mathbf{k} \cdot \mathbf{v} \sim |\mathbf{k}|v_t l_s$ in its frequency, where $\mathbf{v}$ is the random atom velocity at that moment, and l_s is the characteristic length scale for the atom oscillation. We need this random Doppler shift to be significantly smaller than the bandwidth of the pulse in order to have a good shape matching of the spontaneous emission pulses from different atoms. So, there is a further requirement $|\mathbf{k}|v_t l_s \ll \gamma$, which is consistent with the assumption $v_t \ll \gamma$. Finally, this gate is also very insensitive to the birefringence and the phase drift in the optical interferometer. Both of the components $|v_0\rangle$ and $|v_1\rangle$ have the same polarization, and they are very close in frequency. So, they essentially experience the same noisy phase shift under fluctuation of the optical path length, again canceling.

4.3. SCALABLE QUANTUM COMPUTATION WITH PROBABILISTIC GATES

In the previous section, we have shown how to perform probabilistic entangling gates on remote atoms. Such gates only succeed with a small probability, but they are very robust to noise. Naively, if the entangling gates only succeed with a certain probability p, one cannot have efficient computation as the overall success probability (efficiency) scales down exponentially as p^n with the number n of gates. However, in this section we review a method which shows that efficient quantum computation can be constructed with the required computational overhead (such as the computation time or the repetition number of the entangling gates) scaling up slowly (polynomially) with both n and $1/p$. We will follow the approach in Duan and Raussendorf (2005). (Another method was proposed in Barrett and Kok (2005), which applies to a more restricted noise model, where one assumes that the qubits are subject only to Z type of errors when the gate fails. The implementation scheme we reviewed in the previous section does not satisfy this restriction.) The demonstration of this result combines the ideas from the divide-and-conquer method used in the DLCZ scheme (Duan et al., 2001), the cluster state approach to quantum computation (Raussendorf and Briegel, 2001), and the repetitive error correction important for construction of the two-dimensional clusters.

To be more specific, we assume that one can reliably perform the above ZZ measurement gates with a small success probability p. We neglect the noise for all the single-bit operations, which is well justified for typical atomic or optical experiments. Then, to realize efficient quantum computation, our task reduces to how to efficiently construct large scale two-dimensional (2D) cluster states, as 2D clusters, combined with free single-bit operations, realize universal quantum computation. Our basic steps are: first we show how to efficiently prepare a 1D cluster state from the probabilistic ZZ gates using the divide-and-conquer method, then we give a construction to efficiently generate 2D cluster states from 1D chains.

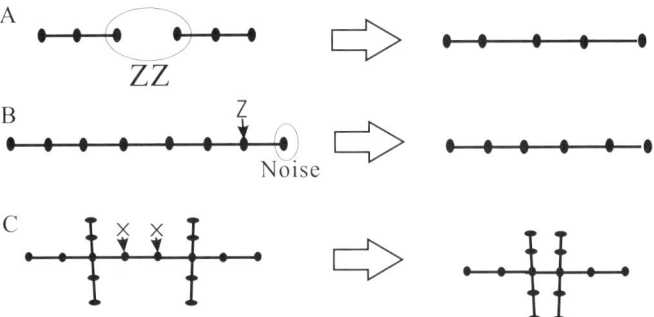

FIG. 10. Illustration of the three properties of the cluster states which are important for our construction of such states with the probabilistic entangling gates: (A) extend cluster states with ZZ measurement gates; (B) recover cluster states by removing bad qubits; (C) shrink cluster states for more complicated links (see Duan and Raussendorf, 2005).

With respect to a given lattice geometry, the cluster state is defined as co-eigenstates of all the operators $A_i = X_i \prod_j Z_j$, where i denotes an arbitrary lattice site and j runs over all the nearest neighbors of the site i. The X_i and Z_j denote respectively the Pauli spin and phase flip operators on the qubits at the sites i, j. In our construction of lattice cluster states with probabilistic ZZ gates, we will make use of the following properties of the ZZ gate or the cluster states: (1) If one starts with two qubits (atoms) in the co-eigenstate of X_1 and X_2 (a product state), the final state after a ZZ measurement is projected to a co-eigenstate of the stabilizer operators $Z_1 Z_2$ and $X_1 X_2$, which is equivalent to the two-qubit cluster state under single-bit rotations (Barrett and Kok, 2005); (2) Assume that one has prepared two 1D cluster chains, each of n qubits. The stabilizer operators for the boundary qubits n and $n+1$ of the two chains are denoted by $X_n Z_{n-1}$ and $X_{n+1} Z_{n+2}$, respectively. A ZZ measurement of these two boundary qubits generates the new stabilizer operators $Z_n Z_{n+1}$ and $X_n X_{n+1} Z_{n-1} Z_{n+2}$. This operation actually connects the two chains into a cluster state of $2n - 1$ qubits (the central qubits n and $n+1$ together represent one logic qubit with the encoded $X_L = X_n X_{n+1}$ and $Z_L = Z_n$ or Z_{n+1}. One can also measure the single-bit operator X_{n+1} to reduce the encode operators X_L and Z_L to X_n and Z_n); (3) If we destroy the state of an end qubit of an n-qubit cluster chain, for instance, through an unsuccessful attempt of the ZZ gate, we can remove this bad qubit by performing a Z measurement on its neighboring qubit, and recover a cluster state of $n - 2$ qubits. (4) We can shrink a cluster state by performing X measurements on all the connecting qubits. The last three properties of the cluster states, illustrated in Fig. 10, can be conveniently explained from their above definition (Hein et al., 2006).

If we have generated two sufficiently long cluster chains each of n_0 qubits, we can just try to connect them through a probabilistic ZZ gate. If this attempt fails, through the property (ii), we can recover two $(n_0 - 2)$-qubit cluster chains through a Z measurement, and try to connect them again. As one continues with this process, the average number of qubits in the connected chain is then given by

$$n_1 = \sum_{i=0}^{n_0/2} (2n_0 - 1 - 4i) p(1-p)^i \simeq 2n_0 - 1 - 4(1-p)/p,$$

where the last approximation is valid when $e^{-n_0 p/2} \ll 1$. As a result the average chain length goes up if $n_0 > n_c \equiv 4(1-p)/p + 1$. We can iterate these connections to see how the computation overhead scales with the qubit number n. We measure the computation overhead in terms of the total computation time and the total number of attempts for the ZZ gates. For the rth ($r \geq 1$) round of successful connections, the chain length n_r, the total preparation time T_r, and the total number of attempts M_r scale in a manner that can be obtained from the recursion relations

$$n_r = 2n_{r-1} - n_c, \quad T_r = T_{r-1} + t_a/p, \quad \text{and} \quad M_r = 2M_{r-1} + 1/p,$$

respectively. In writing the recursion relation for T_r, we have assumed that two cluster chains for each connection are prepared in parallel, and we neglect the time for single-bit operations (t_a denotes the time for each attempt of the ZZ gate). From the above recursion relations, we conclude that if we can prepare cluster chains of n_0 ($n_0 > n_c$) qubits in time T_0 with M_0 attempts of the probabilistic gates, for a large cluster state, the preparation time T and the number of attempts M scale with the chain length n as

$$T(n) = T_0 + (t_a/p) \log_2\bigl[(n - n_c)/(n_0 - n_c)\bigr], \quad \text{and}$$
$$M(n) = (M_0 + 1/p)(n - n_c)/(n_0 - n_c) - 1/p.$$

In the above, we have shown that if one can prepare cluster chains longer than some critical length n_c, one can generate large scale 1D cluster states very efficiently. The problem then reduces to how to efficiently prepare cluster chains up to the critical length n_c. If one wants to prepare an n-qubit cluster chain, we propose to use a repeater protocol which divides the task into $m = \log_2 n$ steps: for the ith ($i = 1, 2, \ldots, m$) step we attempt to build a 2^i-bit cluster state by connecting two 2^{i-1}-bit cluster chains through a probabilistic ZZ gate. If such an attempt fails, we discard all the qubits and restart from the beginning. For the ith step, the recursion relations for the preparation time T_i and the number of attempts M_i are given by $T_i = (1/p)(T_{i-1} + t_a)$ and $M_i = (1/p)(2M_{i-1} + 1)$, which, together with $T_1 = t_a/p$ and $M_1 = 1/p$, give the scaling rules $T(n) \simeq t_a(1/p)^{\log_2 n}$ and $M(n) \simeq (2/p)^{\log_2 n}/2$. The cost is more significant, but it is still a polynomial function of n. To construct a n-qubit cluster chain, in total we need $n - 1$

successful ZZ gates. In a direct protocol, we need all these attempts succeed simultaneously, which gives the scaling $T(n) \propto M(n) \propto (1/p)^{n-1}$. By dividing the task into a series of independent pieces, we improve the scaling with n from exponential to polynomial (for $n \leqslant n_c$).

To generate a cluster chain of a length $n > n_c$, we simply combine the above two protocols. First, we use the repeater protocol to generate n_0-qubit chains with $n_0 > n_c$. Then it is straightforward to use the connect-and-repair protocol to further increase its length. For instance, with $n_0 = n_c + 1$ (which is a reasonable close-to-optimal choice), the overall scaling rules for T and M are (for $n > n_c$),

$$T(n) \simeq t_a(1/p)^{\log_2(n_c+1)} + (t_a/p)\log_2(n - n_c), \qquad (20)$$

$$M(n) \simeq (2/p)^{\log_2(n_c+1)}(n - n_c)/2. \qquad (21)$$

As the critical length is $n_c \simeq 4/p$, T and M in our protocol scale with $1/p$ as $(1/p)^{\log_2(4/p)}$, which is much more efficient than the super-exponential scaling $(1/p)^{4/p}$ in the previous work.

We have shown that for any success probability p of the probabilistic entangling gate, 1D cluster states of arbitrary length can be created efficiently. For universal quantum computation, however, such 1D cluster states are not sufficient. They need to be first connected and transformed into 2D cluster states (for instance, with a square lattice geometry). It is not obvious that such a connection can be done *efficiently*. First, in the connect-and-repair protocol, when an attempt fails, we need to remove the end qubits and all of their neighbors. This means that in a 2D geometry, the lattice shrinks much faster to an irregular shape in the events of failure. Furthermore, a more important obstacle is that we need to connect many more boundary qubits if we want to join two 2D cluster states. For instance, for a square lattice of n qubits, the number of boundary qubits scales as $\sqrt{n}$ (which is distinct from a 1D chain). If we need to connect all the corresponding boundary qubits of the two parts, the overall success probability is exponentially small.

To overcome this problem, we introduce a method which enables efficient connection by attaching a long leg (a 1D cluster chain) to each boundary qubit of the 2D lattice. The protocol is divided into the following steps: First, we try to build a "+" shape cluster state by probabilistically connecting two cluster chains each of length $2n_l + 1$ (the value of n_l is specified below). This can be done through the probabilistic ZZ gate together with a simple Hadamard gate H and an X measurement, as shown in Fig. 2A and explained in its caption. With on average $1/p$ repetitions, we get a "+" shape state with the length of each of the four legs given by n_l. We use the "+" shape state as the basic building blocks of large scale 2D cluster states. In the "+" shape state, we have attached four long legs to the center qubit. The leg qubits serve as ancilla to generate near-deterministic connection from the probabilistic ZZ gates. The critical idea here is that if we want to connect two center qubits, we always start the connection along the end qubits of one of

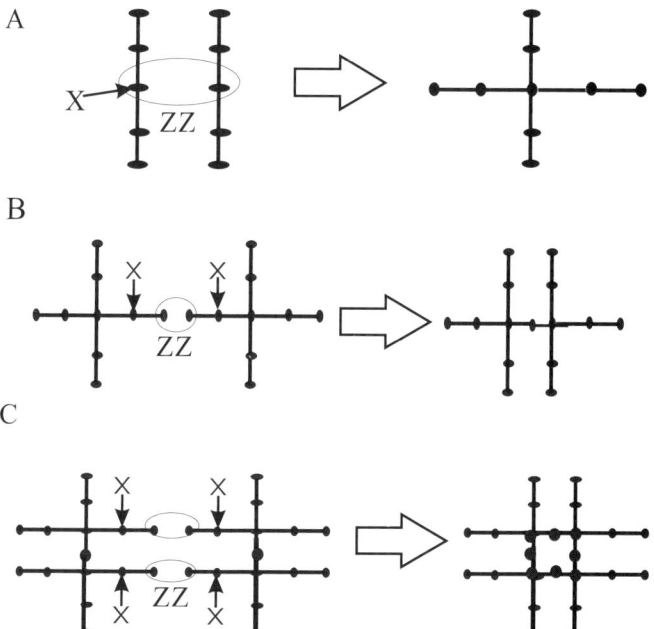

FIG. 11. Illustration of the steps for construction of the two-dimensional square lattice cluster states from a set of cluster chains. (A) Construction of the basic "+" shape states from cluster chains by applying a ZZ gate to connect the two middle qubits, and a X measurement on one middle qubit to remove it. (B) and (C): Construction of the square lattice cluster state from the "+" shape states through probabilistic ZZ gates along the legs and X measurements to remove the remaining redundant qubits. See Duan and Raussendorf (2005) for a similar construction with the controlled phase flip gates.

the legs (see illustration in Fig. 11). If such an attempt fails, we can delete two end qubits and try the connection again along the same legs. If the leg is sufficiently long, we can almost certainly succeed before we reach (destroy) the center qubits. When we succeed, and if there are still redundant leg qubits between the two center ones, we can delete the intermediate leg qubits by performing simple single-bit X measurements on all of them (see Figs. 11 and 10C for the third property of the cluster state). With such a procedure, we can continuously connect the center qubits and form any complex lattice geometry (see the illustration for construction of the square lattice state in Figs. 11B and C). What is important here is that after each time of connection of the center qubits, in the formed new shape, we still have the same length of ancillary legs on all the boundary qubits, which enables the succeeding near-deterministic connection of these new shapes.

Now we investigate for the 2D case how the computational overhead scales with the size of the cluster state. If the ancillary legs have length n_l, for each connection of two center qubits, we can try at most $n_l/2$ times of the probabilistic ZZ gates, and the overall success probability is given by $p_c = 1-(1-p)^{n_l/2}$. If we want to build a square lattice cluster state of N qubits, we need about $2N$ times of connections of the center qubits (there are about $2N$ edges in an N-vertex square lattice). The probability for all these connections to be successful is given by p_c^{2N}. We require this overall success probability is sufficiently large with $p_c^{2N} \geqslant 1 - \epsilon$, where ϵ is a small number characterizing the overall failure probability. From that requirement, we find $n_l \simeq (2/p)\ln(2N/\epsilon)$. To construct a square lattice cluster state of N qubits, we need to consume N "+" shape states, and each of the latter requires on average $2/p$ cluster chains with a length of $2n_l + 1$ qubits. So we need in total $2N/p$ $(2n_l + 1)$-bit cluster chains, which can be prepared in parallel with $(2N/p)M(2n_l+1)$ ZZ attempts within a time period $T(2n_l+1)$ [see Eqs. (1) and (2) for expressions of the $M(n)$ and $T(n)$]. This gives the resources for preparation of all the basic building blocks (the chains). Then we need to connect these blocks to form the square lattice. We assume that the connection of all the building blocks are done in parallel. The whole connection takes on average $2N/p$ CPF attempts, and consumes a time at most $t_a/p \ln(2N/\epsilon)$. Summarizing these results, the temporal and the operational resources for preparation of an N-bit square lattice cluster state are approximately given by

$$T(N) \simeq t_a(1/p)^{\log_2(4/p-3)} + \frac{t_a}{p}\ln(2N/\epsilon)$$

$$+ \frac{t_a}{p}\log_2\left(\frac{4}{p}[\ln(2N/\epsilon) - 1]\right), \quad (22)$$

$$M(N) \simeq (2/p)^{2+\log_2(4/p-3)} N[\ln(2N/\epsilon) - 1] + 2N/p. \quad (23)$$

In the 2D case, the temporal and the operational overhead still have very efficient scaling with the qubit number N, logarithmically for $T(N)$ and $N\ln(N)$ for $M(N)$. Their scalings with $1/p$ are almost the same as in the 1D case except for an additional factor of $1/p^2$ for $M(N)$. Through some straightforward variations of the above method, it is also possible to efficiently prepare any complicated graph state using probabilistic ZZ gates. This shows that in principle we do not need to impose any threshold on the success probability of the ZZ gates for construction of efficient quantum computation. Although the probabilistic gate that we mentioned in the previous section succeeds only with a small probability in practice, it still provides a way to scalable quantum computation.

4.4. Experiments towards Probabilistic Ion Gates

4.4.1. Ion–Photon Entanglement

While matter–light entanglement has been implicit in many experimental systems over the past few decades (Haroche et al., 2002; Kuhn and Rempe, 2002;

McKeever et al., 2003; Freedman and Clauser, 1972; Aspect et al., 1982; Eichmann et al., 1993; DeVoe and Brewer, 1996; Kuzmich et al., 2000, 2003; Julsgaard et al., 2001; Van der Wal et al., 2003; Blinov et al., 2004; Moehring et al., 2004; Matsukevich and Kuzmich, 2004; Chou et al., 2005; Volz et al., 2006), atom–light entanglement is particularly clean in the ion trap system, where both atomic and photonic qubits are under great control. Early experiments involved the entanglement between the polarization of the photon with the hyperfine ground state of a trapped ion (Blinov et al., 2004; Moehring et al., 2004). Polarization qubits are more sensitive to decoherence from birefringence, and typically are useful only along certain directions of atomic emission. We therefore concentrate on the use of photonic frequency qubits, where the central frequency of the photon is entangled with the internal qubits state, as described above.

In a recent experiment, indirect evidence of the entanglement between an atomic qubit and a photon frequency qubit was demonstrated in the ^{111}Cd$^+$ system (Madsen et al., 2006). A diagram of the relevant energy levels and a description of the experiment are given in Fig. 12. First, the ion is optically pumped to $|0, 0\rangle \equiv |\uparrow\rangle$, and a microwave pulse prepares the ion in the state $(|\downarrow\rangle + |\uparrow\rangle)/\sqrt{2}$ [Fig. 12(a)]. Next, a single π-polarized ultrafast laser pulse coherently drives the superposition to the clock states in the $^2P_{3/2}$ manifold with near unit probability. The coherence in this excitation scheme is demonstrated using a microwave Ramsey experiment. In the absence of ultrafast laser pulses, the Ramsey contrast is essentially perfect. Following the application of the ultrafast laser pulse the atom is driven to the excited state. The excited atom then spontaneously decays, and without precise measurement of the photon polarization, frequency, and emission time, the coherence is lost, as seen in Fig. 12(e). The uncontrolled measurement of the photon results in tracing over the photon portion of the density matrix and the resulting loss in contrast is consistent with prior ion–photon entanglement.

To show that the excitation pulse is indeed coherently driving the superposition to the excited state, the Ramsey coherence is recovered by driving the ion back down to the ground state before spontaneous emission occurs [Fig. 12(c)]. With a pair of picosecond laser pulses incident on the ion between the microwave pulses, the contrast reappears with a phase shift proportional to the time Δt spent in the excited state and the hyperfine frequency difference between the ground and excited state levels: $\Delta t(v_0 - v_1) = (680 \text{ ps})(13.9 \text{ GHz}) = 18.9\pi$ [Fig. 12(e)]. The observed contrast is only 40% of the contrast without ultrafast laser pulses due to limited laser power in the second pulse and spontaneous decay ($\sim 23\%$) during the delay time between the ultrafast pulses.

4.4.2. Two Photon Interference

As mentioned previously, once two atoms are entangled with their respective photons, the next step for remote atom entanglement is the interference of the photon

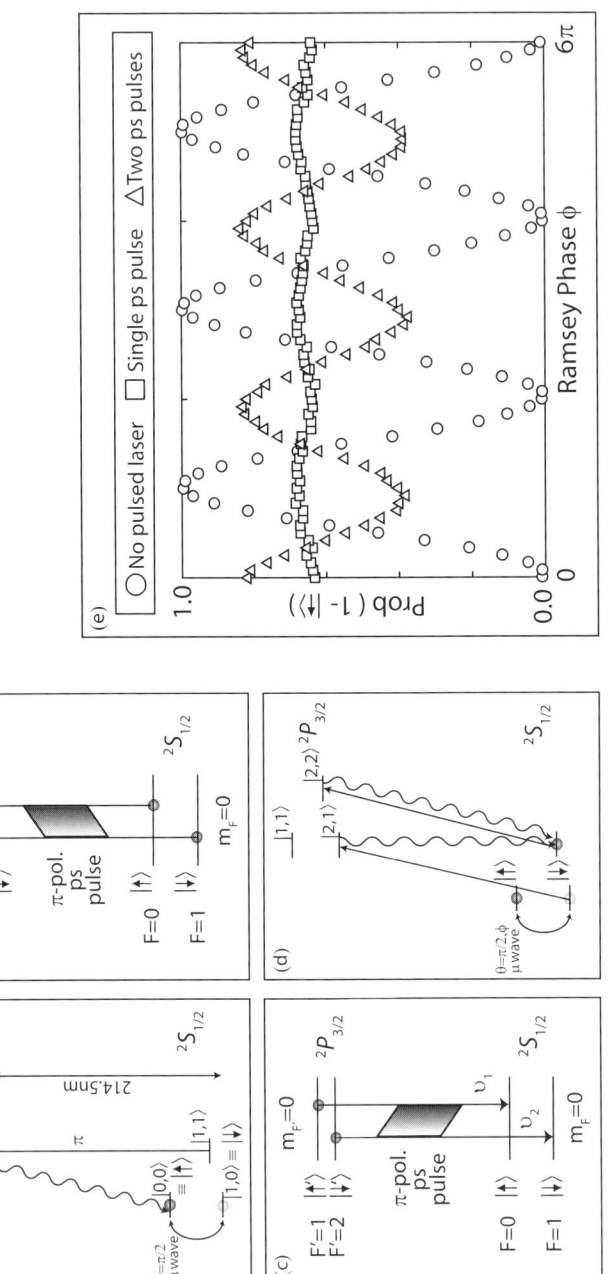

FIG. 12. Experimental procedure for atom–photon entanglement with photon frequency qubits (Madsen et al., 2006). (a) The ion is initialized in the state $(|\downarrow\rangle + |\uparrow\rangle)/\sqrt{2}$ via optical pumping to the $(0, 0)$ state and a microwave $\pi/2$ pulse. (b) The superposition of atomic qubit states is coherently driven to the $^2P_{3/2}$ excited state via a resonantly tuned π-polarized ultrafast laser pulse. (c) A second pulse drives the qubit back to the ground state a short time later. (d) A second $\pi/2$ microwave pulse with variable phase completes the Ramsey experiment and the atomic state is measured using a resonance fluorescence technique. (e) Results from the microwave Ramsey experiment. Circles show the near perfect Ramsey fringes for the case with no ultrafast laser pulse. With a single ultrafast laser pulse, the coherence is lost due to the spontaneous emission of a photon that is not measured in a controlled, precisely timed fashion (squares). The average population in the bright state is above 0.5 due to the fluorescence branching ratios [Fig. 16(inset)]. Upon application of a second ultrafast laser pulse, the coherence in the ion is maintained by driving the qubit states back down to the ground states (triangles).

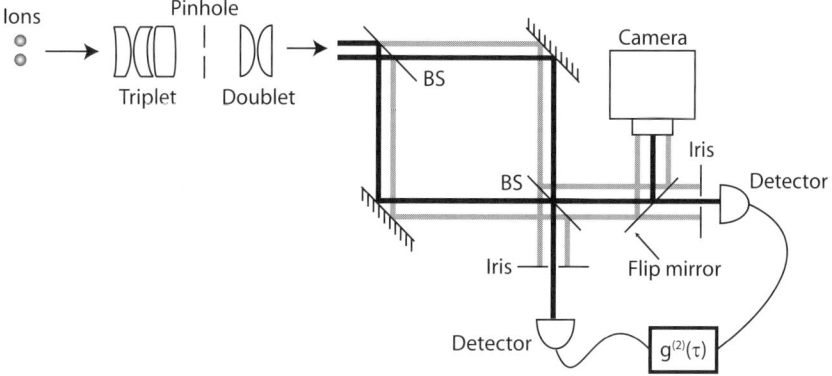

FIG. 13. Detection system for the two-photon interference experiment. The light from the two ions is separated on a beam splitter (*BS*) and mode-matched on the second *BS*. The photons are detected on single photon sensitive photomultiplier tubes (PMTs). A camera is used for coarse alignment, and the non-overlapping photon modes are blocked by irises.

modes from each atom on a beam splitter. Progress toward this end has been recently demonstrated (Beugnon et al., 2006; Maunz et al., 2006). In the cadmium ion system, two ions are placed in a trap and a beam-splitter setup is used to interfere the emitted photons (Fig. 13). In this setup, light scattered by the two ions is collected using an $f/2.1$ objective lens with a working distance of 13 mm. A pinhole is placed at the intermediate image for suppression of background photons and the intermediate image is re-imaged by a doublet lens. The image is then broken up into two paths by a beam splitter, and the transmitted and reflected beam pairs are directed to a second beam splitter where the light from each ion is superimposed. Irises are used to block the unwanted beams and the overlapping beams are directed to photomultiplier tubes (PMTs) with a time resolution of about 1 ns (Moehring et al., 2006). The equal path lengths of the transmitted and reflected beams ensure that the photons emitted by two ions are mode-matched in size and divergence. Coarse alignment is performed by imaging the light after the second beam splitter on a single photon sensitive camera, where the overall magnification of the imaging system is about 1000 and the diffraction-limited image of the two ions are separated by 2 mm, each with a spot size of 0.5 mm.

To demonstrate two photon interference, first the photon statistics of a single ion excited by a σ^+-polarized cw laser is investigated (dashed line in Fig. 14). In this case, the $g^{(2)}$ autocorrelation function shows the expected damped Rabi oscillations (Diedrich and Walther, 1987; Itano et al., 1988) between the $^2S_{1/2}|1, 1\rangle$ and $^2P_{3/2}|2, 2\rangle$ levels. It is unlikely that two photons are emitted from one ion in close proximity since after emission of a single photon, the ion is assured to be in the ground state. The maximum observed antibunching for the single ion

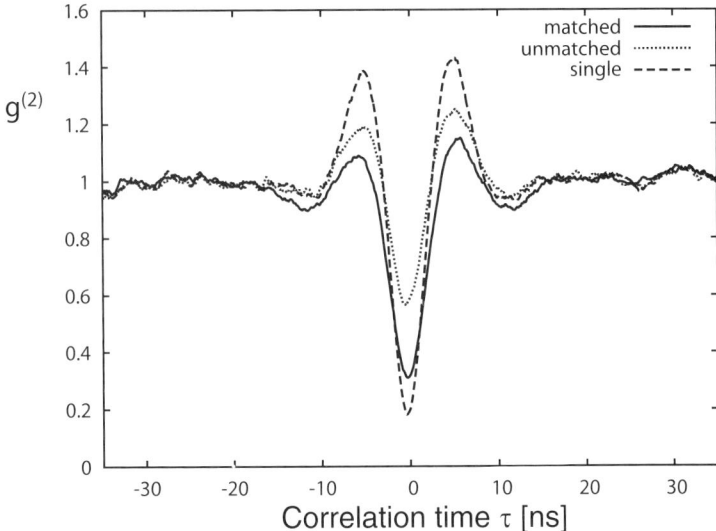

FIG. 14. Intensity autocorrelation for cw-excitation. The dashed line shows strong antibunching for a single ion with $g_1^{(2)}(0) = 0.18$, limited by the resolution of the detection system. With this value, the expected antibunching of light from two non-overlapping ions is expected to be $g_{2,um}^{(2)}(0) = 0.59$ in good agreement with the experimental value (dotted line). If the two photon modes are matched, the interference leads to a significant reduction of coincidence detections (solid line). The measured antibunching was $g_{2,m}^{(2)}(0) = 0.31$, corresponding to a mode overlap of about 57%.

is $g_1^{(2)}(0) = 0.18$ as expected for the time resolution of the PMTs (Maunz et al., 2006).

Next, two ions are illuminated equally and purposefully not mode-matched on the beam splitter. In this case, half of the signal results from two photons from the same ion, and the other half result from one photon from each ion. Since these photon modes are not matched on the beam splitter, the detected photons are uncorrelated. We therefore expect a reduced antibunching,

$$g_{2,um}^{(2)}(0) = \frac{1}{2}\left(1 + g_1^{(2)}(0)\right) \approx 0.59,$$

in agreement with the measurement (dotted line in Fig. 14).

If the photon modes from each ion are matched on the beam splitter, then the photons always leave on the same output port, and thus no coincident detections are observed (Mandel, 1999). The suppression of coincidence events is clearly visible in the autocorrelation signal of the mode-matched ions (solid line in Fig. 14) and has a measured $g_{2,m}^{(2)}(0)$ of 0.31. This corresponds to an interference signal of about 57% (amplitude matching of 75%), and compares well to the results observed in reference (Beugnon et al., 2006). This mode overlap is not ideal and is likely due to phase front distortions from the two atomic sources as they sample different parts of many optical surfaces before finally interfering on the beam splitter.

In order to entangle two remotely located atoms, it is likely necessary to use single mode optical fibers. This is because interfering the two photon modes requires very high stability of the atom and collection optics with respect to the beam splitter, as well as good spatial mode matching from the two imaging systems. With free-space mode-matching, any relative motion of the trapped atoms and the imaging optics can ruin the entanglement fidelity by producing false positive detection events, while in the fiber coupled case, effects such as mechanical vibrations and thermal drifts simply lower the rate of coincidence counts (ignoring dark counts). In the cadmium system, however, the spontaneously emitted photons are deep in the ultraviolet at 214.5 nm where it is very difficult to use optical fibers. Very recently, single photons emitted from two remotely-located Yb^+ ions have been interfered using optical fiber, resulting in better than 80% contrast in the two-photon coincidence rate.

4.4.3. Single Photon Sources

It is important for the atoms to emit only a single photon during an entanglement trial, especially with remote-atom entanglement. Such a single photon source was demonstrated recently using optical excitation of a single cadmium ion with a picosecond mode-locked Ti:sapphire laser (Maunz et al., 2006). This laser is tuned to 858 nm and is sent through a pulse picker to reduce the repetition rate from 81 MHz to 27 MHz with an extinction ratio of better than 100:1 in the infrared. The pulses are frequency quadrupled through single pass nonlinear crystals and the resulting 214.5 nm laser pulses have a pulse extinction ratio near 10^{-8} and a transform-limited pulse width of about 1 ps. This allows excitation of the ion on a timescale much faster than the 2.65 ns excited state lifetime.

The single ion is repeatedly excited with the pulsed laser resulting in a periodic emission of photons at the laser pulse separation time of 37.5 ns, and the intensity autocorrelation function of the photons is recorded using a multi-channel scaler (Fig. 15). The half width of each peak is given by the excited state lifetime and the peak at zero time delay corresponding to coincidentally detected photons is almost entirely suppressed. This near-perfect antibunching is highly non-classical and demonstrates that at most one photon is emitted from the ion following an excitation pulse (limited by the possibility of emitting and detecting a photon *during* the excitation pulse $\approx 10^{-6}$). The residual peak at zero time delay has a height of about 2% of the other peaks, originating from diffuse scattered light from the pulsed laser. With fast electronics, this residual peak could be identically zero by vetoing photons emitted during the picosecond laser pulse.

The use of ultrafast lasers also allows unit-probability excitation ($p_e \sim 1$) while maintaining a single photon source. This corresponds to performing a Rabi π

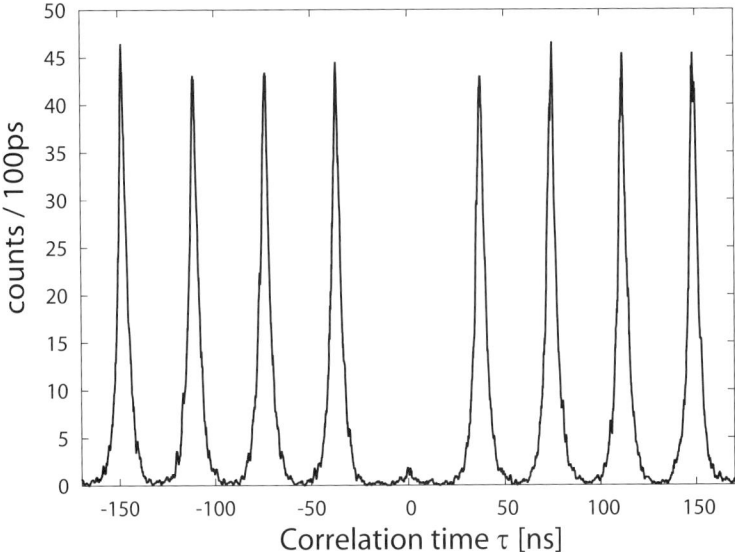

FIG. 15. Intensity autocorrelation of the light emitted by a single ion excited by an ultrafast laser. The near perfect antibunching at $t = 0$ shows that at most one photon is emitted from an excitation pulse.

pulse on the optical S–P transition. In a recent experiment, the Rabi angle was measured by preparing the ion in a known initial ground state and applying a single excitation pulse of known polarization (Madsen et al., 2006). With knowledge of the fluorescence branching ratios and the ability to perform efficient state detection, Rabi flopping with the pulsed laser can be detected using every laser pulse with a high signal to noise ratio (Fig. 16). An alternative method would be to detect the photon scattering rate from an ion as a function of the pulse energy where Rabi angles with an odd (even) multiple of π would have a maximum (minimum) of scattered photons as the ion would be left in the excited (ground) state at the end of each pulse (Darquie et al., 2005).

In the experiment, the ion is prepared in the $|0, 0\rangle$ ground state through optical pumping (Lee et al., 2003). A single linearly polarized ps laser pulse excites the ion to the $P_{3/2}|1, 0\rangle$ state. After a time (10 μs) much longer than the excited state lifetime, the ion has decayed back to the $S_{1/2}$ ground state levels via spontaneous emission following the fluorescence branching ratios. The atomic ground states are then measured using resonance fluorescence detection where all three $F = 1$ states are equally bright, while the $F = 0$ state is dark (Blatt and Zoller, 1988; Lee et al., 2005; Acton et al., 2006), with the results shown in Fig. 16. The available power from the pulsed laser limits the Rabi rotation angle to roughly π, and the

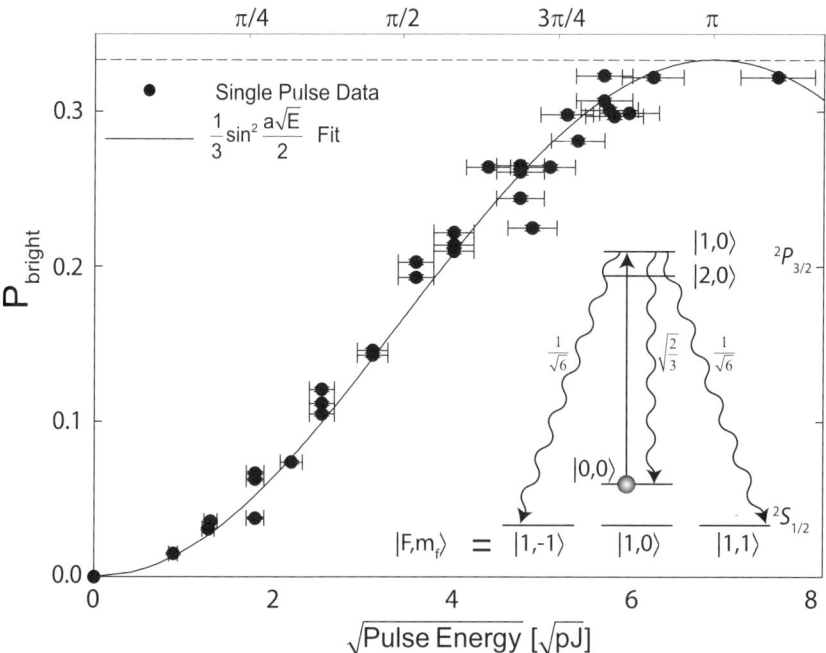

FIG. 16. The ion bright state population as a function of pulse energy. Each point represents a collection of 60,000 runs. As the population in the excited P state is driven to unity, the bright state population approaches $1/3$ (horizontal dashed line), determined by the spontaneous emission branching ratio. The data are fit to a single parameter giving a value $a = 0.42$ $\text{pJ}^{-1/2}$. Inset: Relevant energy levels for the S–P Rabi oscillation experiment. A π-polarized ultrafast laser pulse excited the ion from the ground state to the excited state with variable energy. The three possible decay channels are shown with their respective fluorescence branching ratios. After a time (10 µs) following the excitation pulse, the bright state population of the ion was measured using resonance fluorescence detection.

data agree well with the estimates based on the beam waist, pulse length and pulse shape (Madsen et al., 2006). The probability of measuring the bright state is equal to $1/3$ the probability of excitation to the excited state, as follows from the Clebsch–Gordan coefficients [Fig. 16(inset)]. Hence, we have shown that unit excitation and single photon emission can be achieved with ultrafast laser pulses.

5. Summary

In this article, we have reviewed several schemes towards the goal to realize scalable quantum communication, state engineering, and quantum computation with different physical systems, including atomic ensembles, photons under linear optical devices, and trapped atoms or ions. For all of these schemes, we can reduce the dominant experimental noise to particular types of errors, and then correct such errors at a arbitrarily high level. As a result, these schemes are inherently immune to these special sources of error. This inherent insensitivity to noise opens up a practical route for realization of scalable quantum information with realistic physical devices. Examples of experimental progress along these lines have been described.

6. Acknowledgements

It is a pleasure to thank our collaborators for the works described in this review. The quantum communication scheme with atomic ensembles was done in collaboration with Ignacio Cirac, Mikhail Lukin and Peter Zoller. The state engineering scheme with linear optics was done in collaboration with Tim Bodiya (then an undergraduate student at UM). The scaling method with probabilistic gates was developed in collaboration with Robert Raussendorf. The ideas and the experiments for probabilistic ion gates rely on the collaboration with Boris Blinov, Rudy Kohn, Martin Madsen, Dzmitry Matsukevich, David Moehring, Peter Maunz, Steve Olmschenk, and Kelly Younge. We finally gratefully acknowledge fruitful interactions and collaborations with Jeff Kimble, Alex Kuzmich, Eugene Polzik, Alex Andre, Sean Barrett, Paul Berman, Hans Briegel, Michael Fleischhauer, Peter Kok, Jianwei Pan, Anders Sorensen, and Bin Wang. This work was supported by National Science Foundation award 0431476, the NSF ITR and PIF programs, the National Security Agency and the Disruptive Technology Organization under Army Research Office contracts, the Michigan Center for Theoretical Physics, and the A.P. Sloan Foundation.

7. References

Acton, M., Brickman, K.-A., Haljan, P., Lee, P.J., Deslauriers, L., Monroe, C. (2006). Near-perfect simultaneous measurement of a qubit register. *Quant. Inf. Comp.* **6**, 465–482.
Aspect, A., Grangier, P., Roger, G. (1982). Experimental realization of Einstein–Podolsky–Rosen–Bohm *Gedankenexperiment*: A new violation of Bell's inequalities. *Phys. Rev. Lett.* **49**, 91–94.
Balic, V., Braje, D.A., Kolchin, P., Yin, G.Y., Harris, S.E. (2005). Generation of paired photons with controllable waveforms. *Phys. Rev. Lett.* **94**, 183601–183604.
Barrett, S.D., Kok, P. (2005). Efficient high-fidelity quantum computation using matter qubits and linear optics. *Phys. Rev. A* **71R**. 060310.
Bennett, C.H., et al. (1993). Teleporting an unknown quantum state via dual classical and Einstein–Podolsky–Rosen channels. *Phys. Rev. Lett.* **73**, 3081–3084.
Bennett, C.H., et al. (1997). Purification of noisy entanglement and faithful teleportation via noisy channels. *Phys. Rev. Lett.* **76**, 722–725.

Beugnon, J., Jones, M.P.A., Dingjan, J., Darquie, B., Messin, G., Browaeys, A., Grangier, P. (2006). Quantum interference between two single photons emitted by independently trapped atoms. *Nature* **440**, 779–782.

Black, A.T., Thompson, J.K., Vuletic, V. (2005). On-demand superradiant conversion of atomic spin gratings into single photons with high efficiency. *Phys. Rev. Lett.* **95**. 133601.

Blatt, R., Zoller, P. (1988). Quantum jumps in atomic systems. *Eur. J. Phys.* **9**, 250–256.

Blinov, B.B., Moehring, D.L., Duan, L.-M., Monroe, C. (2004). Observation of entanglement between a single trapped atom and a single photon. *Nature* **428**, 153–157.

Bodiya, T.P., Duan, L.-M. (2006). Scalable generation of graph-state entanglement through realistic linear optics. *Phys. Rev. Lett.* **97**. 143601-1-4.

Boschi, D., Branca1, S., De Martini, F., Hardy, L., Popescu, S. (1998). Experimental realization of teleporting an unknown pure quantum state via dual classical and Einstein–Podolsky–Rosen channels. *Phys. Rev. Lett.* **80**, 1121–1124.

Bouwmeester, D., et al. (1997). Experimental quantum teleportation. *Nature* **390**, 575–579.

Briegel, H.-J., Duer, W., Cirac, J.I., Zoller, P. (1998). Quantum repeaters: The role of imperfect local operations in quantum communication. *Phys. Rev. Lett.* **81**, 5932–5935.

Browne, D.E., Rudolph, T. (2005). Resource-efficient linear optical quantum computation. *Phys. Rev. Lett.* **95**, 010501–010504.

Chaneliere, T., Matsukevich, D.N., Jenkins, S.D., Lan, S.-Y., Kennedy, T.A.B., Kuzmich, A. (2005). Storage and retrieval of single photons transmitted between remote quantum memories. *Nature* **438**, 833–836.

Childress, L., Taylor, J.M., Sorensen, A.S., Lukin, M.D. (2006). Fault-tolerant quantum communication based on solid-state photon emitters. *Phys. Rev. Lett.* **96**. 070504.

Chou, C.W., Polyakov, S.V., Kuzmich, A., Kimble, H.J. (2004). Single-photon generation from stored excitation in an atomic ensemble. *Phys. Rev. Lett.* **92**, 213601–213604.

Chou, C.W., de Riedmatten, H., Felinto, D., Polyakov, S.V., van Enk, S.J., Kimble, H.J. (2005). Measurement-induced entanglement for excitation stored in remote atomic ensembles. *Nature* **438**, 828–832.

Cirac, J.I., Duan, L.-M., Zoller, P. (2001). Quantum optical implementation of quantum information processing. In: *Proceedings of International School of Physics "Enrico Fermi" Courses*. See also quant-ph/0405030.

Darquie, B., Jones, M.P.A., Dingjan, J., Beugnon, J., Bergamini, S., Sortais, Y., Messin, G., Browaeys, A., Grangier, P. (2005). Controlled single-photon emission from a single trapped two-level atom. *Science* **309**, 454–456.

Dawson, C.M., Haselgrove, H.L., Nielsen, M.A. (2006). Noise thresholds for optical quantum computers. *Phys. Rev. Lett.* **96**, 020501–020504.

DeVoe, R.G., Brewer, R.G. (1996). Observation of superradiant and subradiant spontaneous emission of two trapped ions. *Phys. Rev. Lett.* **76**, 2049–2052.

Diedrich, F., Walther, H. (1987). Nonclassical radiation of a single stored ion. *Phys. Rev. Lett.* **58**, 203–206.

Duan, L.-M. (2002). Entangling many atomic ensembles through laser manipulation. *Phys. Rev. Lett.* **88**, 170402–170405.

Duan, L.-M., Raussendorf, R. (2005). Efficient quantum computation with probabilistic quantum gates. *Phys. Rev. Lett.* **95**, 080503–080506.

Duan, L.-M., Cirac, J.I., Zoller, P., Polzik, E.S. (2000). Quantum communication between atomic ensembles using coherent light. *Phys. Rev. Lett.* **85**, 5643–5646.

Duan, L.-M., Lukin, M.D., Cirac, J.I., Zoller, P. (2001). Long distance quantum communication with atomic ensembles and linear optics. *Nature* **414**, 413–418.

Duan, L.-M., Cirac, J.I., Zoller, P. (2002). Three-dimensional theory for interaction between atomic ensembles and free-space light. *Phys. Rev. A* **66**. 023818. 13 pp.

Duan, L.-M., Blinov, B.B., Moehring, D.L., Monroe, C. (2004). Scalable trapped ion quantum computation with a probabilistic ion–photon mapping. *Quant. Inf. Comp.* **4**, 165–173.

Duan, L.-M., Wang, B., Kimble, J. (2005). Robust quantum gates on neutral atoms with cavity-assisted photon-scattering. *Phys. Rev. A* **72**. 032333.

Duan, L.-M., Madsen, M.J., Moehring, D.L., Maunz, P., Kohn, R.N., Monroe, C. (2006). Probabilistic quantum gates between remote atoms through interference of optical frequency qubits. *Phys. Rev. A* **73**. 062324.

Eichmann, U., Bergquist, J.C., Bollinger, J.J., Gilligan, J.M., Itano, W.M., Wineland, D.J., Raizen, M.G. (1993). Young's interference experiment with light scattered from two atoms. *Phys. Rev. Lett.* **70**, 2359–2362.

Eisaman, M.D., Andre, A., Massou, F., Fleischhauer, M., Zibrov, A.S., Lukin, M.D. (2005). Electromagnetically induced transparency with tunable single-photon pulses. *Nature* **438**, 837–841.

Ekert, A. (1991). Quantum cryptography based on Bell's theorem. *Phys. Rev. Lett.* **67**, 661–663.

Fleischhauer, M., Lukin, M.D. (2000). Dark-state polaritons in electromagnetically induced transparency. *Phys. Rev. Lett.* **84**, 5094–5097.

Freedman, S.J., Clauser, J.F. (1972). Experimental test of local hidden-variable theories. *Phys. Rev. Lett.* **28**, 938–941.

Hein, M., Dür, W., Eisert, J., Raussendorf, R., Van den Nest, M., Briegel, H.-J. (2006). Entanglement in graph states and its applications. quant-ph/0602096.

Haroche, S., Raimond, J.M., Brune, M. (2002). In: *Experimental Quantum Computation and Information*. IOS Press, Amsterdam, pp. 3–36.

Itano, W.M., Bergquist, J.C., Wineland, D.J. (1988). Photon antibunching and sub-Poissonian statistics from quantum jumps in one and two atoms. *Phys. Rev. A* **38**, 559–562.

Julsgaard, B., Kozhekin, A., Polzik, E.S. (2001). Experimental long-lived entanglement of two macroscopic objects. *Nature* **413**, 400–403.

Knill, E., Laflamme, R., Milburn, G.J. (2001). A scheme for efficient quantum computation with linear optics. *Nature* **409**, 46–52.

Kuhn, A., Rempe, G. (2002). In: *Experimental Quantum Computation and Information*. IOS Press, Amsterdam, pp. 37–66.

Kuzmich, A., Mandel, L., Bigelow, N.P. (2000). Generation of spin squeezing via continuous quantum nondemolition measurement. *Phys. Rev. Lett.* **85**, 1594–1597.

Kuzmich, A., Bowen, W.P., Boozer, A.D., Boca, A., Chou, C.W., Duan, L.-M., Kimble, H.J. (2003). Generation of nonclassical photon pairs for scalable quantum communication with atomic ensembles. *Nature* **423**, 731–734.

Kwiat, P.G., Mattle, K., Weinfurter, H., Zeilinger, A., Sergienko, A.V., Shih, Y.H. (1995). New high-intensity source of polarization-entangled photon pairs. *Phys. Rev. Lett.* **75**, 4337.

Lee, P.J., Blinov, B.B., Brickman, K.-A., Deslauriers, L., Madsen, M.J., Miller, R., Moehring, D.L., Stick, D., Monroe, C. (2003). Atomic qubit manipulations with an electro–optic modulator. *Optics Letters* **28**, 1582–1584.

Lee, P.J., Brickman, K.-A., Deslauriers, L., Haljan, P.C., Duan, L.-M., Monroe, C. (2005). Phase control of trapped ion quantum gates. *Journal of Optics B* **7**, S371–S383.

Lim, Y.L., Barrett, S.D., Beige, A., Kok, P., Kwek, L.C. (2006). Repeat-until-success quantum computing using stationary and flying qubits. *Phys. Rev. A* **73**. 012304.

Liu, C., Dutton, Z., Behroozi, C.H., Hau, L.V. (2001). Observation of coherent optical information storage in an atomic medium using halted light pulses. *Nature* **409**, 490–493.

Lu, C.-Y., et al. (2006). Experimental entanglement of six photons in graph states. quant-ph/0609130.

Lukin, M.D. (2003). Colloquium: Trapping and manipulating photon states in atomic ensembles. *Rev. Mod. Phys.* **75**, 457–472.

McKeever, J., Buck, J.R., Boozer, A.D., Kuzmich, A., Nagerl, H.-C., Stamper-Kurn, D.M., Kimble, H.J. (2003). State-insensitive cooling and trapping of single atoms in an optical cavity. *Phys. Rev. Lett.* **90**, 133602–133605.

Madsen, M.J., Moehring, D.L., Maunz, P., Kohn Jr., R.N., Duan, L.-M., Monroe, C. (2006). Ultrafast coherent coupling of atomic hyperfine and photon frequency qubits. *Phys. Rev. Lett.* **97**, 040505–040508.

Mandel, L. (1999). Quantum effects in one-photon and two-photon interference. *Rev. Mod. Phys.* **71**, S274–S282.

Manz, S., Fernholz, T., Schmiedmayer, J., Pan, J.-W. (2006). Collisional decoherence during writing and reading quantum states. quant-ph/0608159.

Matsukevich, D.N., Kuzmich, A. (2004). Quantum state transfer between matter and light. *Science* **306**, 663.

Matsukevich, D.N., Chaneliere, T., Jenkins, S.D., Lan, S.-Y., Kennedy, T.A.B., Kuzmich, A. (2006). Entanglement of remote atomic qubits. *Phys. Rev. Lett.* **96**, 030405–030408.

Maunz, P., Moehring, D.L., Madsen, M.J., Kohn Jr., R.N., Younge, K.C., Monroe, C. (2006). Quantum interference of photon pairs from two trapped atomic ions. quant-ph/0608047.

Moehring, D.L., Madsen, M.J., Blinov, B.B., Monroe, C. (2004). Experimental Bell inequality violation with an atom and a photon. *Phys. Rev. Lett.* **93**, 090410–090413.

Moehring, D.L., Blinov, B.B., Gidley, D.W., Kohn Jr., R.N., Madsen, M.J., Sanderson, T.B., Vallery, R.S., Monroe, C. (2006). Precision lifetime measurement of a single trapped ion with ultrafast laser pulses. *Phys. Rev. A* **73**. 023413.

Monroe, C. (2002). Quantum information processing with atoms and photons. *Nature* **416**, 238–246.

Nielsen, M.A. (2004). Optical quantum computation using cluster states. *Phys. Rev. Lett.* **93**, 040503–040506.

Nielsen, M.A., Chuang, I.L. (2000). "Quantum Computation and Quantum Information". Cambridge University Press.

Pan, J.-W., Daniell, M., Gasparoni, S., Weihs, G., Zeilinger, A. (2001). Experimental demonstration of four-photon entanglement and high-fidelity teleportation. *Phys. Rev. Lett.* **86**, 4435–4438.

Phillips, D.F., et al. (2001). Storage of light in atomic vapor. *Phys. Rev. Lett.* **86**, 783–786.

Preskill, J. (1998). Reliable quantum computers. *Proc. R. Soc. Lond. A* **454**, 385–410.

Raussendorf, R., Briegel, H.J. (2001). A one-way quantum computer. *Phys. Rev. Lett.* **86**, 5188–5191.

Riedmatten, H., Laurat, J., Chou, C.W., Schomburg, E.W., Felinto, D., Kimble, H.J. (2006). Direct measurement of decoherence for entanglement between a photon and stored atomic excitation. *Phys. Rev. Lett.* **97**, 113603–113606.

Shi, Y.-Y., Duan, L.-M., Vidal, G. (2006). Classical simulation of quantum many-body systems with a tree tensor network. *Phys. Rev. A* **74**. 022320.

Shor, P.W. (1995). Scheme for reducing decoherence in quantum computer memory. *Phys. Rev. A* **52**. R2493.

Steane, A.M. (1997). Error correcting codes in quantum theory. *Phys. Rev. Lett.* **77**, 793.

Van der Wal, C.H., Eisaman, M.D., Andr, A., Walsworth, R.L., Phillips, D.F., Zibrov, A.S., Lukin, M.D. (2003). Atomic memory for correlated photon states. *Science* **301**, 196.

Volz, J., Weber, M., Schlenk, D., Rosenfeld, W., Vrana, J., Saucke, K., Kurtsiefer, C., Weinfurter, H. (2006). Observation of entanglement of a single photon with a trapped atom. *Phys. Rev. Lett.* **96**, 030404–030407.

Walther, P., et al. (2005). Experimental one-way quantum computing. *Nature* **434**, 169–176.

Yoran, N., Reznik, B. (2003). Deterministic linear optics quantum computation with single photon qubits. *Phys. Rev. Lett.* **91**, 037903–037906.

Index

AC Stark shift, 19, 26
Acousto–optical modulator (AOM), 19, 149, 174
Alignment, 183
Alkali ion–alkali atom collisions, 162–70
Anharmonic waveguides, 80
Animated beams, 349–50
Anisotropic targets, 142, 183–4
Annihilation operators, 99
Asymmetry parameters, 388
Atom chip, 65, 109
Atom–molecule systems, 99, 101, 108, 118, 124–5
Atom–photon coupling, 443–7
Atom waveguides, 64–7
Atomic beam, 19
Atomic clocks, 45–6, 208
Atomic ensembles, 423
– experimental quantum communication with, 432–4
Atomic modes, 424–5, 429–30, 433
Atomic orbital (AO) calculations, 162, 164, 165, 167
Atomic parity nonconservation, 193, 202, 207, 219–22, 224, 227
Auger processes, 304, 305, 317
Auger yield, 307, 359
Autoionization, 302, 366, 394
AUTOSTRUCTURE, 365, 367, 374, 375
Axicon MOT, 144

B-spline MCHF equations, 275–88
B-spline methods, 251–61
– integration methods, 254–6
– for many-electron Hartree–Fock problem, 261–75
– for solution of differential equations, 253
– spline grid for radial functions, 254
B-spline R-matrix (BSR), 238, 256, 259, 260, 288
Basis generator method (BGM), 151
BBR shift, 225
Bell inequality detection, 427–9

Beryllium-like ions, 372–4
Beryllium-like states, 388
Beta decay processes, 183
Bethe ansatz, 89–98
Bethe–Peierls boundary condition, 69
Binding energy, 303
Black holes, 295
Blackbody radiation (BBR) shift, 225
Bogoliubov–Born–Green–Kirkwood–Yvon (BBGKY) equations, 99
Bogoliubov–de Gennes equations, 100, 118
Bosons, short-range-interacting 1D, 103–13
Bound states
– confinement-induced, 75–8
– – composition, 76–8
– – energies, 75
– many-body, 90, 94, 96, 122
– three-body, 125
Boundary conditions
– Bethe–Peierls, 69
– energy-dependent, 70
– multichannel, 69
– periodic, 86, 88, 92, 94, 96, 103, 123
Bragg reflectors, 35
Break-points, 251
Breit interaction, 192, 195–6, 213, 221
Breit–Pauli integrals, 260, 282, 283
Brillouin's theorem, 242, 250
Brueckner orbitals (BO), 201, 202, 220

C_3 coefficient, 225–7
C_6 coefficient, 225–7
Calogero–Sutherland model, 96
Carrier–envelope offset phase (ϕ_{ce}), 5, 7
Carrier offset frequency (f_o), 34
Cauchy–Schwarz inequality, 433
Cavity enhanced DFCS, 32–8
Cavity ring-down spectroscopy (CRDS), 33, 37, 38
CC (coupled-cluster), 194, 198, 200, 203, 205, 209–12, 214, 218
CCSD (coupled-cluster single–double), 193–4, 200, 203, 204, 221

Center-of-mass frame, 298, 340
Cesium (Cs)
– multi-photon DFCS, 27
– single-photon DFCS, 18–20
Chandra, 295
Channel functions, 281–2
Channeling technique, 336
Charge-state balance, 294
Chemistry, one-dimensional, 117–22
CI (configuration interaction), 194, 214–7, 219, 284
– and perturbation theory, 283–4
CI-MBPT (configuration-interaction–MBPT), 216–7
Classical trajectory Monte Carlo (CTMS) calculations, 156
Clock states, 443, 453
Cluster states, 436, 447–52
Coherence time, 40
Coherent accumulation, 14, 23, 24, 40
Coherent anti-Stokes Raman spectroscopy (CARS), 11
Coherent control, 38–45
Cold electrons, 294
Collective enhancement, 423–5
Collective excitation, 300
Colliding-beams experiments, 323–6, 340, 343–55
Collisional spectroscopy, 382–4
Collisionally ionized plasmas, 329
Collisions
– in atom waveguides, 71, 79–81
– one-dimensional two-channel, 73
– resonant, 68–71
COLTRIMS, 140, 141, 144, 146
Compton profile, 337, 338, 341
Configuration interaction (CI), 194, 214–7, 219, 284
– and perturbation theory, 283–4
Configuration-interaction–MBPT, 216–7
Configuration state function (CSF), 238–9
Confinement-induced resonance (CIR), 63, 74, 75
Constrained thermodynamics, 116–7
Corona model, 328
Correlation, 277, 285, 286
– short-range, 287
Correlation functions, 104–8
– experimental tests, 109–13
Coster–Kronig processes, 304

Coulomb self-energy correction, 242
Coupled-cluster (CC), 194, 198, 200, 203, 205, 209–12, 214, 218
Coupled-cluster single–double (CCSD), 193–4, 200, 203, 204, 221
CRDS, 33, 37, 38
Cross sections, generic energy dependencies, 310–7
Crossed beams, 323, 344, 345, 348, 351
CRYRING, 352
Cut-off number (n_c), 327

Davidson algorithm, 275
Deactivation
– of broad molecules, 121–2
– resonance-enhanced, 70
– in waveguides, 79–80
Decay cascades, 395–7
Deep core resonances, 395–7
Delay line detectors, 145
Demkov coupling, 159
Destructive interference, 308–9, 389
Detailed balance, 317, 318, 379, 381
Diagonal energy parameter, 238, 242, 245, 267, 270, 278
Dielectronic capture (DC), 299, 300, 302, 306, 317
Dielectronic recombination (DR), 300, 307, 311–3, 329, 366, 371, 377–8
– in collisional spectroscopy, 382–4
– KLL, 333–4, 355, 363–5
– KLM, 355
Dielectronic satellites, 296
Difference-frequency generation (DFG), 52
Differential cross sections, 143, 161, 164–72, 180, 341
– double, 340–2
Dirac–Coulomb–Breit Hamiltonian, 288–9
Dirac–Coulomb Hamiltonian, 288
Dirac equation, 194, 202, 219
Dirac–Hartree–Fock (DHF) equations, 192–3, 260
Direct density matrices, 257–8
Direct double ionization, 316
Direct elastic scattering, 298
Direct frequency comb spectroscopy (DFCS), 12–30
– cavity enhanced, 32–8
– extension to strong field coherent control, 40–5

– multi-photon, 22–9
– short wavelength, 29–30
– single-photon, 15–22
Direct inner-shell excitation, 314
Direct ionization (DI), 303, 314–6, 384–5, 393–4
Direct multiple ionization, 316, 393–4
Direct single ionization, 314, 384–5, 393
DLCZ scheme, 422–3, 431–3
Doppler shift, 18, 27, 447
Doppler width, 30
Double-Auger process, 303, 315, 388, 389
Double differential cross sections, 340–2
Double excitation, 360
Double ionization, 306, 393–4
– direct, 316

EBIS, 330–1
EBIT, 323, 331–5, 385
ECR, 350, 391
Effective maximally entangled (EME) states, 427–31
Elastic collision, 298
Elastic electron scattering, 341–2, 355
Elastic scattering, 298–300, 307
Electric space-charge, 321
Electro–optical phase modulator, 8
Electromagnetic induced transparency (EIT), 18
Electron beam ion source (EBIS), 330–1
Electron beam ion trap (EBIT), 323, 331–5, 385
Electron capture, 345
Electron capture to continuum (ECC), 150–2, 154
Electron collisions, 181
Electron cooler, 354
Electron-cyclotron-resonance (ECR) ion sources, 350, 391
Electron–electron cusp condition, 287
Electron energy, 340
Electron-energy loss, 357
Electron excitation to continuum (EEC), 152, 154
Electron guns, 348–9
Electron-impact excitation, 302, 357–60
Electron-impact ionization, 344, 352, 384–97
Electron-impact single ionization, 359
Electron–ion recombination *see* Recombination

Electron–ion scattering, 300
Electron scattering, 300, 336, 337, 341, 342, 355–7
Electron temperature, 296, 320
Electron transfer, 140
– anisotropic targets, 183–4
– case studies of differential cross sections, 161–72
– case studies in total collisions, 149–61
– population dynamics measurements, 176, 178
– relative cross sections, 147–9
EME states, 427–31
Energy expression, 240, 245, 248, 263, 264, 275
Energy filters, 348
Energy functional, 243, 262–4, 278, 279
Energy-loss spectroscopy, 308
Energy spreads, 331–2, 346, 348
Entanglement connection, 427–8, 430–1
Entanglement generation, 425–7
Entanglement purification, 421, 429–30
Entangling gates, 445, 447
ESR, 355, 384
Exchange models, 97
Excitation, 302, 310, 347
Excitation–autoionization (EA), 303, 315, 359, 386, 393
Excitation threshold, 356
Excited fractions, 172–6
Experimental Storage Ring (ESR), 355, 384
External electric fields, 376–8
External electromagnetic fields, 376–8

Faddeev–Lovelace equations, 119–20
FAIR, 370, 384
Fano profile, 308–9
Fermi–Bose mapping, 84, 87, 97–8
Fermi–Huang pseudopotential, 69, 71
Feshback resonance, 63, 65–8
Field ionization, 326–7, 362, 367
Field shift (FS), 227–30
Finite-range interactions, 81, 88, 96
"Flattened" Maxwellian distribution, 324
Fluorescence yield, 307, 359
Fock operator, 245
Forbidden transitions, 20–2
Form factor, 325
Fourier transform, 337
Fraunhofer diffraction, 162–4, 170

Free spectral range (FSR), 31, 34–6
Frozen cores approximation, 281
FS, 227–30
FSR, 31, 34–6

Galerkin method, 236, 253, 256, 282
Gaunt-factor formula, 310, 358
Gaussian integration, 255, 259
Generalized eigenvalue problem, 237, 263, 264, 266
Gradient vector, 268, 276
Graph states
– preparation, 436–9
– tree, 435, 439–42
GRASP, 250, 289
Grating spectrometer, 32
Grid, in Dirac–Hartree–Fock calculations, 260
Gross–Pitaevskii equation, 99, 102, 103, 122, 123
Gross–Pitaevskii regime, 104–7
Group delay dispersion (GDD), 8

Hadamard operation, 437
Harmonic waveguide, 67, 71, 73, 82
Hartree–Fock approximation, 238–48
– B-spline MCHF equations, 275–88
– B-spline methods for, 261–75
– multiconfiguration (MCHF), 249–51
– with no orthogonality constraints, 240–2
– with orthogonality constraints, 242–8
Hartree–Fock–Bogoliubov (HFB) method, 101
Hartree–Fock Hamiltonians, 244
Hartree–Fock operator, 241
Heavy ion storage rings, 352, 392
Helium-like ions, 296, 297, 365, 386
Helium-like system, 385
High-harmonic generation (HHG), 11, 47–9
Highly charged ions, 390
HULLAC, 375
Hurwitz zeta function, 72, 82, 126–7
Hydrogen (H), 49
Hydrogen-like atom, 384
Hydrogen-like ions, 298, 363–5, 385
Hydrogen maser, 19
Hyperbolic model, 96–7
Hyperfine constants, 193, 194, 203–7, 212–13, 216–8
Hyperfine splitting, 365, 383

Impulse approximation, 336–7
Inclined beams, 323, 325–6
Indirect elastic scattering, 299, 300
Indirect ionization, 314, 385–8, 391
Inelastic electron scattering, 300, 342, 356
Inelastic scattering, 302
Infidelity, 430
Inner-shell excitation, 305, 314
Inner-shell ionization (ISI), 306, 315, 394
Integrability, 113
– physical manifestations, 113–6
Intensity fluctuations, 44
Interacting-beams techniques, 323–6, 340, 343–55
Interference, 302, 307–9, 388–9
– between radiative and dielectronic recombination, 314, 334, 378–81
– destructive, 308–9, 389
– quantum, 14, 29, 38
– two-photon, 453–7
Ioffe–Pritchard trap, 64, 108
Ion–photon entanglement, 452–3
Ion storage rings *see* Storage rings
Ionization, 303–6
Ionization cross sections, 151–2
Iron ions, 389–90
Iron isonuclear sequence, 374
Isoelectronic sequences, 297
Isotope shift, 227–30

jj-couplings, 371

K-shell excitation, 360, 388
KLL resonances, 333–4, 355, 363–5
KLM resonances, 355
KLM scheme, 435
Knot sequence, 251, 252
Koopmans' theorem, 242, 245, 265, 270, 277

Laboratory energies, 324
Lagrange multipliers, 240–3, 279
– off-diagonal, 265, 267–8
Lambda transition, 3–4
Lieb–Liniger equation, 103–4
– experimental tests, 109–13
Lieb–Liniger–McGuire model, 93–5, 104, 121–2
Lifetimes, 205, 209, 217, 218, 224, 384
Linear optics, 434–6, 439

Lithium-like ions, 365–72, 386–8
Lithography, 297
Logarithmic grid, 254, 259
Long-lived autoionizing states, 391
Longitudinal extraction, 144, 145
Lorentz factors, 324
Lorentzian line profile, 309
Lotz formula, 314, 385, 390, 393
Luttinger liquid model, 86

M matrix, 68
Magnetic bottle, 350
Magnetic field gradient, 144, 175
Magnetic guiding field, 323, 361
Magneto–optical trap (MOT), 17
Many-body 1D problems, 85–6
Many-body perturbation theory (MBPT), 193–4, 196–8, 200–1, 202–3, 205
– for atoms with two valence electrons, 213–4
– C_3 coefficient, 226
– CI-MBPT, 216–7
– isotope shift, 228–9
Measurement gates, 445–52
Merged beams, 323, 344, 345, 350–2
Metastable ions, 350, 386, 390
Metastable states, 350, 390
Micro-structured fiber, 7
Microwave heated electrons, 350
Molecular dynamics, 40–1, 43
Molecular orbital (MO) calculations, 162, 164–6
MOTRIMS, 141
– advantages, 141–2
– apparatus, 142–7
Multicomponent models, 95–6
Multiconfiguration Hartree–Fock (MCHF) approximation, 249–51
– B-spline MCHF equations, 275–88
Multiconfiguration Hartree–Fock (MCHF) program, 247, 248
Multi-photon absorption, 13
Multi-photon DFCS, 22–9
Multi-photon transitions, 22
Multiple ionization, 306, 316–7, 393–7
– direct, 316, 393–4
Multiply charged ions, 298, 357
Multi-reference Möller–Plesset (MR-MP) methods, 250, 283
Multi-reference set, 250–1

New Experimental Storage Ring (NESR), 370
Newton–Raphson method, 271–4, 277, 288
No-pair Hamiltonian, 195, 196
Nonlinear conversion, 29
Normal mass shift (NMS), 227–8
Normalization constraint, 277, 280
Nuclear charge radii, 384

Off-diagonal energy parameter, 242, 248, 249, 266, 269–70
Off-diagonal Lagrange multipliers, 265, 267, 268
Offset frequency (f_o), 5
Optical cavity, 8–12, 33, 34
Optical frequency comb, 5–12
– control via optical cavity, 8–12
– degrees of freedom, 5–8
Optical frequency synthesis, 51–2
Optical lattices, 65, 109
– blue-detuned, 65, 110, 114
– red-detuned, 65, 110
Optical modes, 424, 425
Optically forbidden transitions, 357
Orbital basis set methods, 237
Orbital Hamiltonian, 283
Orbital Hartree–Fock Hamiltonian matrix, 266, 283, 288
Orbital Hartree–Fock Hamiltonian method, 274
Orbital rotations, 244, 248
Orientation, 183
Orthogonal transformations, 244, 245, 248
Orthogonality conditions, 242, 249, 250, 264, 265, 271

p-wave scattering, 81–4
Pair-correlation functions, 284, 285
Parallel spectroscopy, 30–9
Parametric approximation, 100–1
Parity nonconservation, 193, 202, 207, 219–22, 224, 227
Partial waves, 284, 286
PBS gates, 435–40, 442
Photo-associative ionization, 181–3
Photocathode, 352
Photodetachment, 282
Photoionization, 181–2, 185, 238, 282, 318, 378–81
Photoionization regime, 373

Photon emission spectroscopy (PES), 156
Photon frequency cubits, 453, 454
Photon loss, 429, 430
Photorecombination, 312, 317–18, 378–81
Piezo-electric transducer (PZT), 5
Plasma, 320, 327–30
Plasma diagnostics, 295–6
Plasma rate coefficient, 320, 327, 328, 368–9, 373–4
Plasma temperature, 320
PME state, 429, 431
PNC (parity nonconservation), 193, 202, 207, 219–22, 224, 227
Polarizability, 209, 222–5
Polarization beam splitter (PBS) gates, 435–40, 442
Polarization maximally entangled (PME) state, 429, 431
Polynomial scaling, 431–2
Population dynamics, 176–81
Positron sensitive detector (PSD), 142, 143, 147
Potassium-like ions, 391
Power broadening, 24–5
Probabilistic gates, 442–3
– experiments towards, 452–9
– from atom–photon coupling, 443–7
– scalable quantum computation with, 447–52
Projection operators, 266, 269, 281
Proton–sodium collisions, 150–5
Pseudopotential energy-dependent, 70
– Fermi–Huang, 69, 71
– one-dimensional
– – effective, 74
– – even, 73
– – odd, 82
Pulse area, 42
Pulse shaping, 13, 39
Pump–dump, 41–4
Pyramid MOT, 144

Q-value, 140, 145, 147–9, 176
Quantum cryptography, 427–9
Quantum cummulants, 101–2
Quantum dots, 433
Quantum electrodynamical effects, 382
Quantum error threshold theorem, 420
Quantum interference, 14, 29, 38
Quantum Newton's cradle, 114–16

Quantum repeaters, 421–2
Quantum state engineering, 434–42
Quasi-free electrons, 335–43, 355
Quasicondensate, 102, 108–10

Rabi frequencies, 21
Rabi oscillations, 455, 457–9
Radial equation, 236, 237, 256
Radiative electron capture (REC), 336
Radiative recombination (RR), 300–1, 312–13, 318, 360–3, 375
Raman transition, 40–3, 423–4, 429
Ramsauer minima, 356
Ramsey–Bordé fringes, 21
Ramsey spectroscopy, 2, 13, 29
Random-phase approximation (RPA), 201, 220, 221, 223
Rate coefficient, 319–21, 327, 328, 368–9, 373–4
Rate equations, 328
Reaction microscope, 144, 146–7, 340
READI, 304, 305, 315, 386
REC, 336
Recoil spectrometer, 143
Recombination, 300–2, 310–4, 317–8, 352, 360–84
– dielectronic see Dielectronic recombination
– radiative, 300–1, 312–13, 318, 360–3, 375
– resonant, 300
– three-body, 301
– trielectronic, 372–4
Rectangular waveguides, 80–1
REDA, 304, 305, 315, 386, 388
Relative energy, 324
Relativistic many-body perturbation theory (RMBPT) methods, 238
Repetition frequency, 34
Repetition rate (f_r), 5
Resonance strength, 312
Resonances, 306–9
Resonant comb mode index, 16
Resonant double excitation, 389
Resonant elastic scattering, 299, 300, 307
Resonant excitation, 302
Resonant-excitation–auto-double-ionization (READI), 304, 305, 315, 386
Resonant-excitation–double-autoionization (REDA), 304, 305, 315, 386, 388
Resonant ionization, 386
Resonant recombination, 300

Resonant transfer with excitation (RTE), 336, 338–40
RIMS, 141–3
RPA, 201, 220–3
RTE, 336, 338–40
Rubidium (Rb)
– multi-photon DFCS, 23–7
– single-photon DFCS, 16–18
Rutherford cross section, 342, 355, 356
Rydberg constant, 49–50
Rydberg series, 237, 238, 279–83, 288
Rydberg states, 367

s-wave scattering, 71–81
Saddle point (SP) mechanism, 152, 154, 155
Saturation intensity, 175
Saturation parameter, 176
Scaling rule, 385
Scattering amplitude
– one-dimensional
– – even, 74
– – odd, 83
– – of transmission and reflection, 90
– resonant, 69
– – p-wave, 81
SCF iterations, 248, 266
SCF theory, 192
SD (linearized coupled-cluster single–double), 194, 198–206, 209–10
– all-order calculations, 217–22, 225–7, 229, 230
– limitations, 206–7
– motivation for further development, 206–9
– non-linear term inclusion, 210–12
SDpT (single–double–partial triple), 194, 198, 204–6, 222, 229
– limitations, 206
Self-consistent field (SCF) iterations, 248, 266
Self-consistent field (SCF) theory, 192
Self-referencing, 7
Shake-off mechanism, 316
Signal mode, 424–5, 433
SIMION, 143
Single-collision condition, 321
Single–double–partial triple see SDpT
Single ionization, 303, 314–15, 387–9
– direct, 314, 384–5, 393
– electron-impact, 359
Single-photon DFCS, 15–22

Single photon sources, 457–9
Singular-value decomposition (SVD), 264, 268–70, 274, 282
Slater integrals, 239–40, 247, 256–62, 276
Slater matrix elements, 257, 260, 262, 283, 286
SMS, 228–30
Sodium (Na)
– alkali ion collisions on, 162–70
– electron capture from excited atoms, 159–61
– multi-charged ion collisions on, 156–9
– proton collisions on, 150–5
Sodium-like ions, 387, 391
Solitons, 63–4, 122–5
– bright, 122, 123, 125
– dark, 123
Space charge, 321, 322, 330, 349
Spatial focussing, 143
Spatial overlap, 319, 325–6
SPDC, 436, 439, 441
Specific mass shift (SMS), 228–30
Spectrometers, 30
Spectroscopy, collisional, 382–4
Spline Galerkin method, 236, 253, 256, 282
Spline orbital basis methods, 283–4
Spontaneous parametric down conversion (SPDC), 436, 439, 441
Square lattice cluster states, 450–2
Stark effect, 184
Stark shift, 19, 26, 209
Stark states, 326–7
State scattering rate, 15
Stationary conditions, 242, 243, 246, 247, 271
Stellarator, 327
STIRAP, 178, 179
Storage rings, 298, 328–9, 352–4, 381, 392
Strong field coherent control, 40–5
Structural radiation (SR), 201, 202
Sub-Doppler spectroscopy, 16
Superelastic scattering, 317, 336, 343, 357
SVD, 264, 268–70, 274, 282
Symmetric atomic mode, 424, 433
Synchrotron light, 381

T matrix
– confined, 72–3
– one-dimensional

– – even, 74
– – odd, 83
Temporal focussing, 143
Test Storage Ring (TSR), 352, 354, 382–4, 392
Θ-pinch, 327
Thomas–Fermi potential, 289
Three-body problems, 117–22
Three-body recombination, 301
Ti:sapphire amplifier, 30
Ti:sapphire laser, 17–9, 21, 27, 31, 38, 457
Time reversal, 317–8, 378–9, 381
Tokamak, 296, 327, 328
Tonks–Girardeau gas, 86–8, 98, 110
– finite-range, 88
– harmonic axial potential in, 88
Tonks–Girardeau regime, 104–7
Total cross sections, 308, 310
Transition amplitude, 15
Transition matrix element, 193, 194, 201–3, 209, 213–4
Transverse extraction, 144–6
Trapped-ion techniques, 330–5
Trapped ions, 434
Tree graphs, 435, 439–42
Trielectronic capture, 299, 300, 302, 389
Trielectronic recombination, 372–4
Triple ionization, 396
Trochoidal analyzers, 345–7

TSR, 352, 354, 382–4, 392
Two-photon absorption, 23, 29
Two-photon interference, 453–7
Two-photon process, 43
Two-photon spectroscopy, 27
Two-photon transition, 4, 14, 24, 26, 28
Two-photon transition amplitude, 14

Vacuum coefficient, 427, 429, 430
Velocity distribution, 319–20, 322
VIPA spectrometer, 30–2, 36
Virial theorem, 272
Virtually-imaged phased array (VIPA) spectrometer, 30–2, 36
VUV spectroscopy, 11, 46, 49–50

Wave-packets, 41, 42
Waveform generation, 51–2
Wiley–McLauren spectrometer, 143
Window resonances, 388

X-ray astrophysics, 294, 332, 399
XMM-Newton, 295
XUV radiation, 297
XUV spectroscopy, 46–9

Yang–Baxter relation, 91, 118

ZZ measurement gates, 445–52

CONTENTS OF VOLUMES IN THIS SERIAL

Volume 1
Molecular Orbital Theory of the Spin Properties of Conjugated Molecules, *G.G. Hall and A.T. Amos*
Electron Affinities of Atoms and Molecules, *B.L. Moiseiwitsch*
Atomic Rearrangement Collisions, *B.H. Bransden*
The Production of Rotational and Vibrational Transitions in Encounters between Molecules, *K. Takayanagi*
The Study of Intermolecular Potentials with Molecular Beams at Thermal Energies, *H. Pauly and J.P. Toennies*
High-Intensity and High-Energy Molecular Beams, *J.B. Anderson, R.P. Anders and J.B. Fen*

Volume 2
The Calculation of van der Waals Interactions, *A. Dalgarno and W.D. Davison*
Thermal Diffusion in Gases, *E.A. Mason, R.J. Munn and Francis J. Smith*
Spectroscopy in the Vacuum Ultraviolet, *W.R.S. Garton*
The Measurement of the Photoionization Cross Sections of the Atomic Gases, *James A.R. Samson*
The Theory of Electron–Atom Collisions, *R. Peterkop and V. Veldre*
Experimental Studies of Excitation in Collisions between Atomic and Ionic Systems, *F.J. de Heer*
Mass Spectrometry of Free Radicals, *S.N. Foner*

Volume 3
The Quantal Calculation of Photoionization Cross Sections, *A.L. Stewart*
Radiofrequency Spectroscopy of Stored Ions I: Storage, *H.G. Dehmelt*
Optical Pumping Methods in Atomic Spectroscopy, *B. Budick*
Energy Transfer in Organic Molecular Crystals: A Survey of Experiments, *H.C. Wolf*
Atomic and Molecular Scattering from Solid Surfaces, *Robert E. Stickney*
Quantum, Mechanics in Gas Crystal-Surface van der Waals Scattering, *E. Chanoch Beder*
Reactive Collisions between Gas and Surface Atoms, *Henry Wise and Bernard J. Wood*

Volume 4
H.S.W. Massey—A Sixtieth Birthday Tribute, *E.H.S. Burhop*
Electronic Eigenenergies of the Hydrogen Molecular Ion, *D.R. Bates and R.H.G. Reid*
Applications of Quantum Theory to the Viscosity of Dilute Gases, *R.A. Buckingham and E. Gal*
Positrons and Positronium in Gases, *P.A. Fraser*
Classical Theory of Atomic Scattering, *A. Burgess and I.C. Percival*
Born Expansions, *A.R. Holt and B.L. Moiseiwitsch*
Resonances in Electron Scattering by Atoms and Molecules, *P.G. Burke*

Relativistic Inner Shell Ionizations, *C.B.O. Mohr*

Recent Measurements on Charge Transfer, *J.B. Hasted*

Measurements of Electron Excitation Functions, *D.W.O. Heddle and R.G.W. Keesing*

Some New Experimental Methods in Collision Physics, *R.F. Stebbings*

Atomic Collision Processes in Gaseous Nebulae, *M.J. Seaton*

Collisions in the Ionosphere, *A. Dalgarno*

The Direct Study of Ionization in Space, *R.L.F. Boyd*

Volume 5

Flowing Afterglow Measurements of Ion-Neutral Reactions, *E.E. Ferguson, F.C. Fehsenfeld and A.L. Schmeltekopf*

Experiments with Merging Beams, *Roy H. Neynaber*

Radiofrequency Spectroscopy of Stored Ions II: Spectroscopy, *H.G. Dehmelt*

The Spectra of Molecular Solids, *O. Schnepp*

The Meaning of Collision Broadening of Spectral Lines: The Classical Oscillator Analog, *A. Ben-Reuven*

The Calculation of Atomic Transition Probabilities, *R.J.S. Crossley*

Tables of One- and Two-Particle Coefficients of Fractional Parentage for Configurations $s^\lambda s^{tu} p^q$, *C.D.H. Chisholm, A. Dalgarno and F.R. Innes*

Relativistic Z-Dependent Corrections to Atomic Energy Levels, *Holly Thomis Doyle*

Volume 6

Dissociative Recombination, *J.N. Bardsley and M.A. Biondi*

Analysis of the Velocity Field in Plasmas from the Doppler Broadening of Spectral Emission Lines, *A.S. Kaufman*

The Rotational Excitation of Molecules by Slow Electrons, *Kazuo Takayanagi and Yukikazu Itikawa*

The Diffusion of Atoms and Molecules, *E.A. Mason and T.R. Marrero*

Theory and Application of Sturmian Functions, *Manuel Rotenberg*

Use of Classical Mechanics in the Treatment of Collisions between Massive Systems, *D.R. Bates and A.E. Kingston*

Volume 7

Physics of the Hydrogen Maser, *C. Audoin, J.P. Schermann and P. Grivet*

Molecular Wave Functions: Calculations and Use in Atomic and Molecular Process, *J.C. Browne*

Localized Molecular Orbitals, *Harel Weinstein, Ruben Pauncz and Maurice Cohen*

General Theory of Spin-Coupled Wave Functions for Atoms and Molecules, *J. Gerratt*

Diabatic States of Molecules—Quasi-Stationary Electronic States, *Thomas F. O'Malley*

Selection Rules within Atomic Shells, *B.R. Judd*

Green's Function Technique in Atomic and Molecular Physics, *Gy. Csanak, H.S. Taylor and Robert Yaris*

A Review of Pseudo-Potentials with Emphasis on Their Application to Liquid Metals, *Nathan Wiser and A.J. Greenfield*

Volume 8

Interstellar Molecules: Their Formation and Destruction, *D. McNally*

Monte Carlo Trajectory Calculations of Atomic and Molecular Excitation in Thermal Systems, *James C. Keck*

Nonrelativistic Off-Shell Two-Body
  Coulomb Amplitudes, *Joseph C.Y. Chen
  and Augustine C. Chen*
Photoionization with Molecular Beams,
  *R.B. Cairns, Halstead Harrison and
  R.I. Schoen*
The Auger Effect, *E.H.S. Burhop and
  W.N. Asaad*

Volume 9
Correlation in Excited States of Atoms,
  A.W. Weiss
The Calculation of Electron–Atom
  Excitation Cross Section, *M.R.H. Rudge*
Collision-Induced Transitions between
  Rotational Levels, *Takeshi Oka*
The Differential Cross Section of
  Low-Energy Electron–Atom Collisions,
  D. Andrick
Molecular Beam Electric Resonance
  Spectroscopy, *Jens C. Zorn and Thomas
  C. English*
Atomic and Molecular Processes in the
  Martian Atmosphere, *Michael
  B. McElroy*

Volume 10
Relativistic Effects in the Many-Electron
  Atom, *Lloyd Armstrong Jr. and Serge
  Feneuille*
The First Born Approximation, *K.L. Bell
  and A.E. Kingston*
Photoelectron Spectroscopy, *W.C. Price*
Dye Lasers in Atomic Spectroscopy,
  W. Lange, J. Luther and A. Steudel
Recent Progress in the Classification of the
  Spectra of Highly Ionized Atoms,
  B.C. Fawcett
A Review of Jovian Ionospheric Chemistry,
  Wesley T. Huntress Jr.

Volume 11
The Theory of Collisions between Charged
  Particles and Highly Excited Atoms,
  I.C. Percival and D. Richards

Electron Impact Excitation of Positive
  Ions, *M.J. Seaton*
The R-Matrix Theory of Atomic Process,
  P.G. Burke and W.D. Robb
Role of Energy in Reactive Molecular
  Scattering: An Information-Theoretic
  Approach, *R.B. Bernstein and
  R.D. Levine*
Inner Shell Ionization by Incident Nuclei,
  Johannes M. Hansteen
Stark Broadening, *Hans R. Griem*
Chemiluminescence in Gases, *M.F. Golde
  and B.A. Thrush*

Volume 12
Nonadiabatic Transitions between Ionic
  and Covalent States, *R.K. Janev*
Recent Progress in the Theory of Atomic
  Isotope Shift, *J. Bauche and
  R.-J. Champeau*
Topics on Multiphoton Processes in Atoms,
  P. Lambropoulos
Optical Pumping of Molecules, *M. Broyer,
  G. Goudedard, J.C. Lehmann and
  J. Vigué*
Highly Ionized Ions, *Ivan A. Sellin*
Time-of-Flight Scattering Spectroscopy,
  Wilhelm Raith
Ion Chemistry in the D Region, *George
  C. Reid*

Volume 13
Atomic and Molecular Polarizabilities—
  Review of Recent Advances, *Thomas
  M. Miller and Benjamin Bederson*
Study of Collisions by Laser Spectroscopy,
  Paul R. Berman
Collision Experiments with Laser-Excited
  Atoms in Crossed Beams, *I.V. Hertel and
  W. Stoll*
Scattering Studies of Rotational and
  Vibrational Excitation of Molecules,
  Manfred Faubel and J. Peter Toennies

Low-Energy Electron Scattering by Complex Atoms: Theory and Calculations, *R.K. Nesbet*
Microwave Transitions of Interstellar Atoms and Molecules, *W.B. Somerville*

Volume 14

Resonances in Electron Atom and Molecule Scattering, *D.E. Golden*
The Accurate Calculation of Atomic Properties by Numerical Methods, *Brain C. Webster, Michael J. Jamieson and Ronald F. Stewart*
(e, 2e) Collisions, *Erich Weigold and Ian E. McCarthy*
Forbidden Transitions in One- and Two-Electron Atoms, *Richard Marrus and Peter J. Mohr*
Semiclassical Effects in Heavy-Particle Collisions, *M.S. Child*
Atomic Physics Tests of the Basic Concepts in Quantum Mechanics, *Francies M. Pipkin*
Quasi-Molecular Interference Effects in Ion–Atom Collisions, *S.V. Bobashev*
Rydberg Atoms, *S.A. Edelstein and T.F. Gallagher*
UV and X-Ray Spectroscopy in Astrophysics, *A.K. Dupree*

Volume 15

Negative Ions, *H.S.W. Massey*
Atomic Physics from Atmospheric and Astrophysical, *A. Dalgarno*
Collisions of Highly Excited Atoms, *R.F. Stebbings*
Theoretical Aspects of Positron Collisions in Gases, *J.W. Humberston*
Experimental Aspects of Positron Collisions in Gases, *T.C. Griffith*
Reactive Scattering: Recent Advances in Theory and Experiment, *Richard B. Bernstein*
Ion–Atom Charge Transfer Collisions at Low Energies, *J.B. Hasted*

Aspects of Recombination, *D.R. Bates*
The Theory of Fast Heavy Particle Collisions, *B.H. Bransden*
Atomic Collision Processes in Controlled Thermonuclear Fusion Research, *H.B. Gilbody*
Inner-Shell Ionization, *E.H.S. Burhop*
Excitation of Atoms by Electron Impact, *D.W.O. Heddle*
Coherence and Correlation in Atomic Collisions, *H. Kleinpoppen*
Theory of Low Energy Electron–Molecule Collisions, *P.O. Burke*

Volume 16

Atomic Hartree–Fock Theory, *M. Cohen and R.P. McEachran*
Experiments and Model Calculations to Determine Interatomic Potentials, *R. Düren*
Sources of Polarized Electrons, *R.J. Celotta and D.T. Pierce*
Theory of Atomic Processes in Strong Resonant Electromagnetic Fields, *S. Swain*
Spectroscopy of Laser-Produced Plasmas, *M.H. Key and R.J. Hutcheon*
Relativistic Effects in Atomic Collisions Theory, *B.L. Moiseiwitsch*
Parity Nonconservation in Atoms: Status of Theory and Experiment, *E.N. Fortson and L. Wilets*

Volume 17

Collective Effects in Photoionization of Atoms, *M.Ya. Amusia*
Nonadiabatic Charge Transfer, *D.S.F. Crothers*
Atomic Rydberg States, *Serge Feneuille and Pierre Jacquinot*
Superfluorescence, *M.F.H. Schuurmans, Q.H.F. Vrehen, D. Polder and H.M. Gibbs*
Applications of Resonance Ionization Spectroscopy in Atomic and Molecular

Physics, *M.G. Payne, C.H. Chen, G.S. Hurst and G.W. Foltz*
Inner-Shell Vacancy Production in Ion–Atom Collisions, *C.D. Lin and Patrick Richard*
Atomic Processes in the Sun, *P.L. Dufton and A.E. Kingston*

Volume 18
Theory of Electron–Atom Scattering in a Radiation Field, *Leonard Rosenberg*
Positron–Gas Scattering Experiments, *Talbert S. Stein and Walter E. Kauppla*
Nonresonant Multiphoton Ionization of Atoms, *J. Morellec, D. Normand and G. Petite*
Classical and Semiclassical Methods in Inelastic Heavy-Particle Collisions, *A.S. Dickinson and D. Richards*
Recent Computational Developments in the Use of Complex Scaling in Resonance Phenomena, *B.R. Junker*
Direct Excitation in Atomic Collisions: Studies of Quasi-One-Electron Systems, *N. Andersen and S.E. Nielsen*
Model Potentials in Atomic Structure, *A. Hibbert*
Recent Developments in the Theory of Electron Scattering by Highly Polar Molecules, *D.W. Norcross and L.A. Collins*
Quantum Electrodynamic Effects in Few-Electron Atomic Systems, *G.W.F. Drake*

Volume 19
Electron Capture in Collisions of Hydrogen Atoms with Fully Stripped Ions, *B.H. Bransden and R.K. Janev*
Interactions of Simple Ion Atom Systems, *J.T. Park*
High-Resolution Spectroscopy of Stored Ions, *D.J. Wineland, Wayne M. Itano and R.S. Van Dyck Jr.*

Spin-Dependent Phenomena in Inelastic Electron–Atom Collisions, *K. Blum and H. Kleinpoppen*
The Reduced Potential Curve Method for Diatomic Molecules and Its Applications, *F. Jenč*
The Vibrational Excitation of Molecules by Electron Impact, *D.G. Thompson*
Vibrational and Rotational Excitation in Molecular Collisions, *Manfred Faubel*
Spin Polarization of Atomic and Molecular Photoelectrons, *N.A. Cherepkov*

Volume 20
Ion–Ion Recombination in an Ambient Gas, *D.R. Bates*
Atomic Charges within Molecules, *G.G. Hall*
Experimental Studies on Cluster Ions, *T.D. Mark and A.W. Castleman Jr.*
Nuclear Reaction Effects on Atomic Inner-Shell Ionization, *W.E. Meyerhof and J.-F. Chemin*
Numerical Calculations on Electron-Impact Ionization, *Christopher Bottcher*
Electron and Ion Mobilities, *Gordon R. Freeman and David A. Armstrong*
On the Problem of Extreme UV and X-Ray Lasers, *I.I. Sobel'man and A.V. Vinogradov*
Radiative Properties of Rydberg States in Resonant Cavities, *S. Haroche and J.M. Raimond*
Rydberg Atoms: High-Resolution Spectroscopy and Radiation Interaction—Rydberg Molecules, *J.A.C. Gallas, G. Leuchs, H. Walther, and H. Figger*

Volume 21
Subnatural Linewidths in Atomic Spectroscopy, *Dennis P. O'Brien, Pierre Meystre and Herbert Walther*
Molecular Applications of Quantum Defect Theory, *Chris H. Greene and Ch. Jungen*

Theory of Dielectronic Recombination, *Yukap Hahn*
Recent Developments in Semiclassical Floquet Theories for Intense-Field Multiphoton Processes, *Shih-I Chu*
Scattering in Strong Magnetic Fields, *M.R.C. McDowell and M. Zarcone*
Pressure Ionization, Resonances and the Continuity of Bound and Free States, *R.M. More*

Volume 22
Positronium—Its Formation and Interaction with Simple Systems, *J.W. Humberston*
Experimental Aspects of Positron and Positronium Physics, *T.C. Griffith*
Doubly Excited States, Including New Classification Schemes, *C.D. Lin*
Measurements of Charge Transfer and Ionization in Collisions Involving Hydrogen Atoms, *H.B. Gilbody*
Electron Ion and Ion–Ion Collisions with Intersecting Beams, *K. Dolder and B. Peart*
Electron Capture by Simple Ions, *Edward Pollack and Yukap Hahn*
Relativistic Heavy-Ion–Atom Collisions, *R. Anholt and Harvey Gould*
Continued-Fraction Methods in Atomic Physics, *S. Swain*

Volume 23
Vacuum Ultraviolet Laser Spectroscopy of Small Molecules, *C.R. Vidal*
Foundations of the Relativistic Theory of Atomic and Molecular Structure, *Ian P. Grant and Harry M. Quiney*
Point-Charge Models for Molecules Derived from Least-Squares Fitting of the Electric Potential, *D.E. Williams and Ji-Min Yan*
Transition Arrays in the Spectra of Ionized Atoms, *J. Bauche, C. Bauche-Arnoult and M. Klapisch*

Photoionization and Collisional Ionization of Excited Atoms Using Synchrotron and Laser Radiation, *F.J. Wuilleumier, D.L. Ederer and J.L. Picqué*

Volume 24
The Selected Ion Flow Tube (SIDT): Studies of Ion-Neutral Reactions, *D. Smith and N.G. Adams*
Near-Threshold Electron–Molecule Scattering, *Michael A. Morrison*
Angular Correlation in Multiphoton Ionization of Atoms, *S.J. Smith and G. Leuchs*
Optical Pumping and Spin Exchange in Gas Cells, *R.J. Knize, Z. Wu and W. Happer*
Correlations in Electron–Atom Scattering, *A. Crowe*

Volume 25
Alexander Dalgarno: Life and Personality, *David R. Bates and George A. Victor*
Alexander Dalgarno: Contributions to Atomic and Molecular Physics, *Neal Lane*
Alexander Dalgarno: Contributions to Aeronomy, *Michael B. McElroy*
Alexander Dalgarno: Contributions to Astrophysics, *David A. Williams*
Dipole Polarizability Measurements, *Thomas M. Miller and Benjamin Bederson*
Flow Tube Studies of Ion–Molecule Reactions, *Eldon Ferguson*
Differential Scattering in He–He and He^+–He Collisions at keV Energies, *R.F. Stebbings*
Atomic Excitation in Dense Plasmas, *Jon C. Weisheit*
Pressure Broadening and Laser-Induced Spectral Line Shapes, *Kenneth M. Sando and Shih-I. Chu*
Model-Potential Methods, *C. Laughlin and G.A. Victor*

Z-Expansion Methods, *M. Cohen*
Schwinger Variational Methods, *Deborah Kay Watson*
Fine-Structure Transitions in Proton–Ion Collisions, *R.H.G. Reid*
Electron Impact Excitation, *R.J.W. Henry and A.E. Kingston*
Recent Advances in the Numerical Calculation of Ionization Amplitudes, *Christopher Bottcher*
The Numerical Solution of the Equations of Molecular Scattering, *A.C. Allison*
High Energy Charge Transfer, *B.H. Bransden and D.P. Dewangan*
Relativistic Random-Phase Approximation, *W.R. Johnson*
Relativistic Sturmian and Finite Basis Set Methods in Atomic Physics, *G.W.F. Drake and S.P. Goldman*
Dissociation Dynamics of Polyatomic Molecules, *T. Uzer*
Photodissociation Processes in Diatomic Molecules of Astrophysical Interest, *Kate P. Kirby and Ewine F. van Dishoeck*
The Abundances and Excitation of Interstellar Molecules, *John H. Black*

Volume 26
Comparisons of Positrons and Electron Scattering by Gases, *Walter E. Kauppila and Talbert S. Stein*
Electron Capture at Relativistic Energies, *B.L. Moiseiwitsch*
The Low-Energy, Heavy Particle Collisions—A Close-Coupling Treatment, *Mineo Kimura and Neal F. Lane*
Vibronic Phenomena in Collisions of Atomic and Molecular Species, *V. Sidis*
Associative Ionization: Experiments, Potentials and Dynamics, *John Weiner Françoise Masnou-Seeuws and Annick Giusti-Suzor*
On the β Decay of ^{187}Re: An Interface of Atomic and Nuclear Physics and Cosmochronology, *Zonghau Chen, Leonard Rosenberg and Larry Spruch*
Progress in Low Pressure Mercury-Rare Gas Discharge Research, *J. Maya and R. Lagushenko*

Volume 27
Negative Ions: Structure and Spectra, *David R. Bates*
Electron Polarization Phenomena in Electron–Atom Collisions, *Joachim Kessler*
Electron–Atom Scattering, *I.E. McCarthy and E. Weigold*
Electron–Atom Ionization, *I.E. McCarthy and E. Weigold*
Role of Autoionizing States in Multiphoton Ionization of Complex Atoms, *V.I. Lengyel and M.I. Haysak*
Multiphoton Ionization of Atomic Hydrogen Using Perturbation Theory, *E. Karule*

Volume 28
The Theory of Fast Ion–Atom Collisions, *J.S. Briggs and J.H. Macek*
Some Recent Developments in the Fundamental Theory of Light, *Peter W. Milonni and Surendra Singh*
Squeezed States of the Radiation Field, *Khalid Zaheer and M. Suhail Zubairy*
Cavity Quantum Electrodynamics, *E.A. Hinds*

Volume 29
Studies of Electron Excitation of Rare-Gas Atoms into and out of Metastable Levels Using Optical and Laser Techniques, *Chun C. Lin and L.W. Anderson*
Cross Sections for Direct Multiphoton Ionization of Atoms, *M.V. Ammosov, N.B. Delone, M.Yu. Ivanov, I.I. Bandar and A.V. Masalov*
Collision-Induced Coherences in Optical Physics, *G.S. Agarwal*

Muon-Catalyzed Fusion, *Johann Rafelski and Helga E. Rafelski*
Cooperative Effects in Atomic Physics, *J.P. Connerade*
Multiple Electron Excitation, Ionization, and Transfer in High-Velocity Atomic and Molecular Collisions, *J.H. McGuire*

Volume 30

Differential Cross Sections for Excitation of Helium Atoms and Helium-Like Ions by Electron Impact, *Shinobu Nakazaki*
Cross-Section Measurements for Electron Impact on Excited Atomic Species, *S. Trajmar and J.C. Nickel*
The Dissociative Ionization of Simple Molecules by Fast Ions, *Colin J. Latimer*
Theory of Collisions between Laser Cooled Atoms, *P.S. Julienne, A.M. Smith and K. Burnett*
Light-Induced Drift, *E.R. Eliel*
Continuum Distorted Wave Methods in Ion–Atom Collisions, *Derrick S.F. Crothers and Louis J. Dube*

Volume 31

Energies and Asymptotic Analysis for Helium Rydberg States, *G.W.F. Drake*
Spectroscopy of Trapped Ions, *R.C. Thompson*
Phase Transitions of Stored Laser-Cooled Ions, *H. Walther*
Selection of Electronic States in Atomic Beams with Lasers, *Jacques Baudon, Rudalf Dülren and Jacques Robert*
Atomic Physics and Non-Maxwellian Plasmas, *Michèle Lamoureux*

Volume 32

Photoionization of Atomic Oxygen and Atomic Nitrogen, *K.L. Bell and A.E. Kingston*
Positronium Formation by Positron Impact on Atoms at Intermediate Energies, *B.H. Bransden and C.J. Noble*
Electron–Atom Scattering Theory and Calculations, *P.G. Burke*
Terrestrial and Extraterrestrial H_3^+, *Alexander Dalgarno*
Indirect Ionization of Positive Atomic Ions, *K. Dolder*
Quantum Defect Theory and Analysis of High-Precision Helium Term Energies, *G.W.F. Drake*
Electron–Ion and Ion–Ion Recombination Processes, *M.R. Flannery*
Studies of State-Selective Electron Capture in Atomic Hydrogen by Translational Energy Spectroscopy, *H.B. Gilbody*
Relativistic Electronic Structure of Atoms and Molecules, *I.P. Grant*
The Chemistry of Stellar Environments, *D.A. Howe, J.M.C. Rawlings and D.A. Williams*
Positron and Positronium Scattering at Low Energies, *J.W. Humberston*
How Perfect are Complete Atomic Collision Experiments?, *H. Kleinpoppen and H. Handy*
Adiabatic Expansions and Nonadiabatic Effects, *R. McCarroll and D.S.F. Crothers*
Electron Capture to the Continuum, *B.L. Moiseiwitsch*
How Opaque Is a Star?, *M.T. Seaton*
Studies of Electron Attachment at Thermal Energies Using the Flowing Afterglow–Langmuir Technique, *David Smith and Patrik Španěl*
Exact and Approximate Rate Equations in Atom–Field Interactions, *S. Swain*
Atoms in Cavities and Traps, *H. Walther*
Some Recent Advances in Electron-Impact Excitation of $n = 3$ States of Atomic Hydrogen and Helium, *J.F. Williams and J.B. Wang*

Volume 33

Principles and Methods for Measurement of Electron Impact Excitation Cross

Sections for Atoms and Molecules by Optical Techniques, *A.R. Filippelli, Chun C. Lin, L.W. Andersen and J.W. McConkey*

Benchmark Measurements of Cross Sections for Electron Collisions: Analysis of Scattered Electrons, *S. Trajmar and J.W. McConkey*

Benchmark Measurements of Cross Sections for Electron Collisions: Electron Swarm Methods, *R.W. Crompton*

Some Benchmark Measurements of Cross Sections for Collisions of Simple Heavy Particles, *H.B. Gilbody*

The Role of Theory in the Evaluation and Interpretation of Cross-Section Data, *Barry I. Schneider*

Analytic Representation of Cross-Section Data, *Mitio Inokuti, Mineo Kimura, M.A. Dillon, Isao Shimamura*

Electron Collisions with N_2, O_2 and O: What We Do and Do Not Know, *Yukikazu Itikawa*

Need for Cross Sections in Fusion Plasma Research, *Hugh P. Summers*

Need for Cross Sections in Plasma Chemistry, *M. Capitelli, R. Celiberto and M. Cacciatore*

Guide for Users of Data Resources, *Jean W. Gallagher*

Guide to Bibliographies, Books, Reviews and Compendia of Data on Atomic Collisions, *E.W. McDaniel and E.J. Mansky*

Volume 34

Atom Interferometry, *C.S. Adams, O. Carnal and J. Mlynek*

Optical Tests of Quantum Mechanics, *R.Y. Chiao, P.G. Kwiat and A.M. Steinberg*

Classical and Quantum Chaos in Atomic Systems, *Dominique Delande and Andreas Buchleitner*

Measurements of Collisions between Laser-Cooled Atoms, *Thad Walker and Paul Feng*

The Measurement and Analysis of Electric Fields in Glow Discharge Plasmas, *J.E. Lawler and D.A. Doughty*

Polarization and Orientation Phenomena in Photoionization of Molecules, *N.A. Cherepkov*

Role of Two-Center Electron–Electron Interaction in Projectile Electron Excitation and Loss, *E.C. Montenegro, W.E. Meyerhof and J.H. McGuire*

Indirect Processes in Electron Impact Ionization of Positive Ions, *D.L. Moores and K.J. Reed*

Dissociative Recombination: Crossing and Tunneling Modes, *David R. Bates*

Volume 35

Laser Manipulation of Atoms, *K. Sengstock and W. Ertmer*

Advances in Ultracold Collisions: Experiment and Theory, *J. Weiner*

Ionization Dynamics in Strong Laser Fields, *L.F. DiMauro and P. Agostini*

Infrared Spectroscopy of Size Selected Molecular Clusters, *U. Buck*

Fermosecond Spectroscopy of Molecules and Clusters, *T. Baumer and G. Gerber*

Calculation of Electron Scattering on Hydrogenic Targets, *I. Bray and A.T. Stelbovics*

Relativistic Calculations of Transition Amplitudes in the Helium Isoelectronic Sequence, *W.R. Johnson, D.R. Plante and J. Sapirstein*

Rotational Energy Transfer in Small Polyatomic Molecules, *H.O. Everitt and F.C. De Lucia*

Volume 36

Complete Experiments in Electron–Atom Collisions, *Nils Overgaard Andersen and Klaus Bartschat*

Stimulated Rayleigh Resonances and
 Recoil-Induced Effects, *J.-Y. Courtois
 and G. Grynberg*
Precision Laser Spectroscopy Using
 Acousto-Optic Modulators, *W.A. van
 Mijngaanden*
Highly Parallel Computational Techniques
 for Electron–Molecule Collisions, *Carl
 Winstead and Vincent McKoy*
Quantum Field Theory of Atoms and
 Photons, *Maciej Lewenstein and Li You*

Volume 37

Evanescent Light-Wave Atom Mirrors,
 Resonators, Waveguides, and Traps,
 *Jonathan P. Dowling and Julio
 Gea-Banacloche*
Optical Lattices, *P.S. Jessen and
 I.H. Deutsch*
Channeling Heavy Ions through Crystalline
 Lattices, *Herbert F. Krause and Sheldon
 Datz*
Evaporative Cooling of Trapped Atoms,
 Wolfgang Ketterle and N.J. van Druten
Nonclassical States of Motion in Ion Traps,
 *J.I. Cirac, A.S. Parkins, R. Blatt and
 P. Zoller*
The Physics of Highly-Charged Heavy Ions
 Revealed by Storage/Cooler Rings,
 P.H. Mokler and Th. Stöhlker

Volume 38

Electronic Wavepackets, *Robert R. Jones
 and L.D. Noordam*
Chiral Effects in Electron Scattering by
 Molecules, *K. Blum and D.G. Thompson*
Optical and Magneto-Optical Spectroscopy
 of Point Defects in Condensed Helium,
 Serguei I. Kanorsky and Antoine Weis
Rydberg Ionization: From Field to Photon,
 G.M. Lankhuijzen and L.D. Noordam
Studies of Negative Ions in Storage Rings,
 *L.H. Andersen, T. Andersen and
 P. Hvelplund*

Single-Molecule Spectroscopy and
 Quantum Optics in Solids, *W.E. Moerner,
 R.M. Dickson and D.J. Norris*

Volume 39

Author and Subject Cumulative Index
 Volumes 1–38
Author Index
Subject Index
Appendix: Tables of Contents of Volumes
 1–38 and Supplements

Volume 40

Electric Dipole Moments of Leptons,
 Eugene D. Commins
High-Precision Calculations for the
 Ground and Excited States of the
 Lithium Atom, *Frederick W. King*
Storage Ring Laser Spectroscopy, *Thomas
 U. Kühl*
Laser Cooling of Solids, *Carl E. Mangan
 and Timothy R. Gosnell*
Optical Pattern Formation, *L.A. Lugiato,
 M. Brambilla and A. Gatti*

Volume 41

Two-Photon Entanglement and Quantum
 Reality, *Yanhua Shih*
Quantum Chaos with Cold Atoms, *Mark
 G. Raizen*
Study of the Spatial and Temporal
 Coherence of High-Order Harmonics,
 *Pascal Salières, Ann L'Huillier, Philippe
 Antoine and Maciej Lewenstein*
Atom Optics in Quantized Light Fields,
 *Matthias Freyburger, Alois
 M. Herkommer, Daniel S. Krähmer,
 Erwin Mayr and Wolfgang P. Schleich*
Atom Waveguides, *Victor I. Balykin*
Atomic Matter Wave Amplification by
 Optical Pumping, *Ulf Janicke and
 Martin Wikens*

Volume 42

Fundamental Tests of Quantum Mechanics, *Edward S. Fry and Thomas Walther*

Wave-Particle Duality in an Atom Interferometer, *Stephan Dürr and Gerhard Rempe*

Atom Holography, *Fujio Shimizu*

Optical Dipole Traps for Neutral Atoms, *Rudolf Grimm, Matthias Weidemüller and Yurii B. Ovchinnikov*

Formation of Cold ($T \leq 1$ K) Molecules, *J.T. Bahns, P.L. Gould and W.C. Stwalley*

High-Intensity Laser-Atom Physics, *C.J. Joachain, M. Dorr and N.J. Kylstra*

Coherent Control of Atomic, Molecular and Electronic Processes, *Moshe Shapiro and Paul Brumer*

Resonant Nonlinear Optics in Phase Coherent Media, *M.D. Lukin, P. Hemmer and M.O. Scully*

The Characterization of Liquid and Solid Surfaces with Metastable Helium Atoms, *H. Morgner*

Quantum Communication with Entangled Photons, *Herald Weinfurter*

Volume 43

Plasma Processing of Materials and Atomic, Molecular, and Optical Physics: An Introduction, *Hiroshi Tanaka and Mitio Inokuti*

The Boltzmann Equation and Transport Coefficients of Electrons in Weakly Ionized Plasmas, *R. Winkler*

Electron Collision Data for Plasma Chemistry Modeling, *W.L. Morgan*

Electron–Molecule Collisions in Low-Temperature Plasmas: The Role of Theory, *Carl Winstead and Vincent McKoy*

Electron Impact Ionization of Organic Silicon Compounds, *Ralf Basner, Kurt Becker, Hans Deutsch and Martin Schmidt*

Kinetic Energy Dependence of Ion–Molecule Reactions Related to Plasma Chemistry, *P.B. Armentrout*

Physicochemical Aspects of Atomic and Molecular Processes in Reactive Plasmas, *Yoshihiko Hatano*

Ion–Molecule Reactions, *Werner Lindinger, Armin Hansel and Zdenek Herman*

Uses of High-Sensitivity White-Light Absorption Spectroscopy in Chemical Vapor Deposition and Plasma Processing, *L.W. Anderson, A.N. Goyette and J.E. Lawler*

Fundamental Processes of Plasma–Surface Interactions, *Rainer Hippler*

Recent Applications of Gaseous Discharges: Dusty Plasmas and Upward-Directed Lightning, *Ara Chutjian*

Opportunities and Challenges for Atomic, Molecular and Optical Physics in Plasma Chemistry, *Kurl Becker Hans Deutsch and Mitio Inokuti*

Volume 44

Mechanisms of Electron Transport in Electrical Discharges and Electron Collision Cross Sections, *Hiroshi Tanaka and Osamu Sueoka*

Theoretical Consideration of Plasma-Processing Processes, *Mineo Kimura*

Electron Collision Data for Plasma-Processing Gases, *Loucas G. Christophorou and James K. Olthoff*

Radical Measurements in Plasma Processing, *Toshio Goto*

Radio-Frequency Plasma Modeling for Low-Temperature Processing, *Toshiaki Makabe*

Electron Interactions with Excited Atoms and Molecules, *Loucas G. Christophorou and James K. Olthoff*

Volume 45

Comparing the Antiproton and Proton, and Opening the Way to Cold Antihydrogen, *G. Gabrielse*

Medical Imaging with Laser-Polarized Noble Gases, *Timothy Chupp and Scott Swanson*

Polarization and Coherence Analysis of the Optical Two-Photon Radiation from the Metastable $2^2Si_{1/2}$ State of Atomic Hydrogen, *Alan J. Duncan, Hans Kleinpoppen and Marian O. Scully*

Laser Spectroscopy of Small Molecules, *W. Demtröder, M. Keil and H. Wenz*

Coulomb Explosion Imaging of Molecules, *Z. Vager*

Volume 46

Femtosecond Quantum Control, *T. Brixner, N.H. Damrauer and G. Gerber*

Coherent Manipulation of Atoms and Molecules by Sequential Laser Pulses, *N.V. Vitanov, M. Fleischhauer, B.W. Shore and K. Bergmann*

Slow, Ultraslow, Stored, and Frozen Light, *Andrey B. Matsko, Olga Kocharovskaya, Yuri Rostovtsev George R. Welch, Alexander S. Zibrov and Marlan O. Scully*

Longitudinal Interferometry with Atomic Beams, *S. Gupta, D.A. Kokorowski, R.A. Rubenstein, and W.W. Smith*

Volume 47

Nonlinear Optics of de Broglie Waves, *P. Meystre*

Formation of Ultracold Molecules ($T \leq 200 \mu K$) via Photoassociation in a Gas of Laser-Cooled Atoms, *Françoise Masnou-Seeuws and Pierre Pillet*

Molecular Emissions from the Atmospheres of Giant Planets and Comets: Needs for Spectroscopic and Collision Data, *Yukikazu Itikawa, Sang Joon Kim, Yong Ha Kim and Y.C. Minh*

Studies of Electron-Excited Targets Using Recoil Momentum Spectroscopy with Laser Probing of the Excited State, *Andrew James Murray and Peter Hammond*

Quantum Noise of Small Lasers, *J.P. Woerdman, N.J. van Druten and M.P. van Exter*

Volume 48

Multiple Ionization in Strong Laser Fields, *R. Dörner Th. Weber, M. Weckenbrock, A. Staudte, M. Hattass, R. Moshammer, J. Ullrich and H. Schmidt-Böcking*

Above-Threshold Ionization: From Classical Features to Quantum Effects, *W. Becker, F. Grasbon, R. Kapold, D.B. Milošević, G.G. Paulus and H. Walther*

Dark Optical Traps for Cold Atoms, *Nir Friedman, Ariel Kaplan and Nir Davidson*

Manipulation of Cold Atoms in Hollow Laser Beams, *Heung-Ryoul Noh, Xenye Xu and Wonho Jhe*

Continuous Stern–Gerlach Effect on Atomic Ions, *Günther Werth, Hartmut Haffner and Wolfgang Quint*

The Chirality of Biomolecules, *Robert N. Compton and Richard M. Pagni*

Microscopic Atom Optics: From Wires to an Atom Chip, *Ron Folman, Peter Krüger, Jörg Schmiedmayer, Johannes Denschlag and Carsten Henkel*

Methods of Measuring Electron–Atom Collision Cross Sections with an Atom Trap, *R.S. Schappe, M.L. Keeler, T.A. Zimmerman, M. Larsen, P. Feng, R.C. Nesnidal, J.B. Boffard, T.G. Walker, L.W. Anderson and C.C. Lin*

Volume 49

Applications of Optical Cavities in Modern Atomic, Molecular, and Optical Physics, *Jun Ye and Theresa W. Lynn*

Resonance and Threshold Phenomena in Low-Energy Electron Collisions with Molecules and Clusters, *H. Hotop, M.-W. Ruf, M. Allan and I.I. Fabrikant*

Coherence Analysis and Tensor Polarization Parameters of $(\gamma, e\gamma)$ Photoionization Processes in Atomic Coincidence Measurements, *B. Lohmann, B. Zimmermann, H. Kleinpoppen and U. Becker*

Quantum Measurements and New Concepts for Experiments with Trapped Ions, *Ch. Wunderlich and Ch. Balzer*

Scattering and Reaction Processes in Powerful Laser Fields, *Dejan B. Milošević and Fritz Ehlotzky*

Hot Atoms in the Terrestrial Atmosphere, *Vijay Kumar and E. Krishnakumar*

Volume 50

Assessment of the Ozone Isotope Effect, *K. Mauersberger, D. Krankowsky, C. Janssen and R. Schinke*

Atom Optics, Guided Atoms, and Atom Interferometry, *J. Arlt, G. Birkl, E. Rasel and W. Ertmet*

Atom–Wall Interaction, *D. Bloch and M. Ducloy*

Atoms Made Entirely of Antimatter: Two Methods Produce Slow Antihydrogen, *G. Gabrielse*

Ultrafast Excitation, Ionization, and Fragmentation of C_{60}, *I.V. Hertel, T. Laarmann and C.P. Schulz*

Volume 51

Introduction, *Henry H. Stroke*

Appreciation of Ben Bederson as Editor of Advances in Atomic, Molecular, and Optical Physics

Benjamin Bederson Curriculum Vitae

Research Publications of Benjamin Bederson

A Proper Homage to Our Ben, *H. Lustig*

Benjamin Bederson in the Army, World War II, *Val L. Fitch*

Physics Needs Heroes Too, *C. Duncan Rice*

Two Civic Scientists—Benjamin Bederson and the other Benjamin, *Neal Lane*

An Editor *Par Excellence*, *Eugen Merzbacher*

Ben as APS Editor, *Bernd Crasemann*

Ben Bederson: Physicist–Historian, *Roger H. Stuewer*

Pedagogical Notes on Classical Casimir Effects, *Larry Spruch*

Polarizabilities of ^{3}P Atoms and van der Waals Coefficients for Their Interaction with Helium Atoms, *X. Chu and A. Dalgarno*

The Two Electron Molecular Bonds Revisited: From Bohr Orbits to Two-Center Orbitals, *Goong Chen, Siu A. Chin, Yusheng Dou, Kishore T. Kapale, Moochan Kim, Anatoly A. Svidzinsky, Kerim Urtekin, Han Xiong and Marlan O. Scully*

Resonance Fluorescence of Two-Level Atoms, *H. Walther*

Atomic Physics with Radioactive Atoms, *Jacques Pinard and H. Henry Stroke*

Thermal Electron Attachment and Detachment in Gases, *Thomas M. Miller*

Recent Developments in the Measurement of Static Electric Dipole Polarizabilities, *Harvey Gould and Thomas M. Miller*

Trapping and Moving Atoms on Surfaces, *Robert J. Celotta and Joseph A. Stroscio*

Electron-Impact Excitation Cross Sections of Sodium, *Chun C. Lin and John B. Boffard*

Atomic and Ionic Collisions, *Edward Pollack*

Atomic Interactions in Weakly Ionized Gas: Ionizing Shock Waves in Neon, *Leposava Vušković and Svetozar Popović*

Approaches to Perfect/Complete Scattering Experiments in Atomic and Molecular Physics, *H. Kleinpoppen, B. Lohmann, A. Grum-Grzhimailo and U. Becker*

Reflections on Teaching, *Richard E. Collins*

Volume 52

Exploring Quantum Matter with Ultracold Atoms in Optical Lattices, *Immanuel Bloch and Markus Greiner*

The Kicked Rydberg Atom, *F.B. Dunning, J.C. Lancaster, C.O. Reinhold, S. Yoshida and J. Burgdörfer*

Photonic State Tomography, *J.B. Altepeter, E.R. Jeffrey and P.G. Kwiat*

Fine Structure in High-*L* Rydberg States: A Path to Properties of Positive Ions, *Stephen R. Lundeen*

A Storage Ring for Neutral Molecules, *Floris M.H. Crompvoets, Hendrick L. Bethlem and Gerard Meijer*

Nonadiabatic Alignment by Intense Pulses. Concepts, Theory, and Directions, *Tamar Seideman and Edward Hamilton*

Relativistic Nonlinear Optics, *Donald Umstadter, Scott Sepke and Shouyuan Chen*

Coupled-State Treatment of Charge Transfer, *Thomas G. Winter*

Volume 53

Non-Classical Light from Artificial Atoms, *Thomas Aichele, Matthias Scholz, Sven Ramelow and Oliver Benson*

Quantum Chaos, Transport, and Control—in Quantum Optics, *Javier Madroñero, Alexey Ponomarev, Andrí R.R. Carvalho, Sandro Wimberger, Carlos Viviescas, Andrey Kolovsky, Klaus Hornberger, Peter Schlagheck, Andreas Krug and Andreas Buchleitner*

Manipulating Single Atoms, *Dieter Meschede and Arno Rauschenbeutel*

Spatial Imaging with Wavefront Coding and Optical Coherence Tomography, *Thomas Hellmuth*

The Quantum Properties of Multimode Optical Amplifiers Revisited, *G. Leuchs, U.L. Andersen and C. Fabre*

Quantum Optics of Ultra-Cold Molecules, *D. Meiser, T. Miyakawa, H. Uys and P. Meystre*

Atom Manipulation in Optical Lattices, *Georg Raithel and Natalya Morrow*

Femtosecond Laser Interaction with Solid Surfaces: Explosive Ablation and Self-Assembly of Ordered Nanostructures, *Juergen Reif and Florenta Costache*

Characterization of Single Photons Using Two-Photon Interference, *T. Legero, T. Wilk, A. Kuhn and G. Rempe*

Fluctuations in Ideal and Interacting Bose–Einstein Condensates: From the Laser Phase Transition Analogy to Squeezed States and Bogoliubov Quasiparticles, *Vitaly V. Kocharovsky, Vladimir V. Kocharovsky, Martin Holthaus, C.H. Raymond Ooi, Anatoly Svidzinsky, Wolfgang Ketterle and Marlan O. Scully*

LIDAR-Monitoring of the Air with Femtosecond Plasma Channels, *Ludger Wöste, Steffen Frey and Jean-Pierre Wolf*

Volume 54

Experimental Realization of the BCS-BEC Crossover with a Fermi Gas of Atoms, *C.A. Regal and D.S. Jin*

Deterministic Atom–Light Quantum Interface, *Jacob Sherson, Brian Julsgaard and Eugene S. Polzik*

Cold Rydberg Atoms, *J.-H. Choi, B. Knuffman, T. Cubel Liebisch, A. Reinhard and G. Raithel*

Non-Perturbative Quantal Methods for Electron–Atom Scattering Processes, *D.C. Griffin and M.S. Pindzola*

R-Matrix Theory of Atomic, Molecular and Optical Processes, *P.G. Burke, C.J. Noble and V.M. Burke*

Electron-Impact Excitation of Rare-Gas Atoms from the Ground Level and Metastable Levels, *John B. Boffard, R.O. Jung, L.W. Anderson and C.C. Lin*

Internal Rotation in Symmetric Tops, *I. Ozier and N. Moazzen-Ahmadi*

Attosecond and Angstrom Science,
  Hiromichi Niikura and P.B. Corkum
Atomic Processing of Optically Carried RF
  Signals, *Jean-Louis Le Gouët, Fabien
  Bretenaker and Ivan Lorgeré*

Controlling Optical Chaos,
  Spatio-Temporal Dynamics, and
  Patterns, *Lucas Illing, Daniel J. Gauthier
  and Rajarshi Roy*

Supplements

Atoms in Intense Laser Fields, edited by
Mihai Gavrila (1992)

Multiphoton Ionization, *H.G. Muller, P.
  Agostini and G. Petite*
Photoionization with Ultra-Short Laser
  Pulses, *R.R. Freeman, P.H. Bucksbaum,
  W.E. Cooke, G. Gibson, T.J. McIlrath
  and L.D. van Woerkom*
Rydberg Atoms in Strong Microwave
  Fields, *T.F. Gallagher*
Muiltiphoton Ionization in Large
  Ponderomotive Potentials, *P.B. Corkum,
  N.H. Burnett and F. Brunel*
High Order Harmonic Generation in Rare
  Gases, *Anne L'Huillier, Louis-André
  Lompré, Gerard Manfrey and Claude
  Manus*
Mechanisms of Short-Wavelength
  Generation, *T.S. Luk, A. McPherson, K.
  Boyer and C.K. Rhodes*
Time-Dependent Studies of Multiphoton
  Processes, *Kenneth C. Kulander, Kenneth
  J. Schafer and Jeffrey L. Krause*
Numerical Experiments in Strong and
  Super-Strong Fields, *J.H. Eberly, R.
  Grobe, C.K. Law and Q. Su*
Resonances in Multiphoton Ionization, *P.
  Lambropoulos and X. Tang*
Nonperturbative Treatment of Multiphoton
  Ionization within the Floquet
  Framework, *R.M. Potvliege and Robin
  Shakeshaft*
Atomic Structure and Decay in High
  Frequency Fields, *Mihai Gavrila*

Cavity Quantum Electrodynamics,
edited by Paul R. Berman (1994)

Perturbative Cavity Quantum
  Electrodynamics, *E.A. Hinds*
The Micromaser: A Proving Ground for
  Quantum Physics, *Georg Raithel,
  Christian Wagner, Herbert Walther,
  Lorenzo M. Narducci and Marlan O.
  Scully*
Manipulation of Nonclassical Field States
  in a Cavity by Atom Interferometry,
  S. Haroche and J.M. Raimond
Quantum Optics of Driven Atoms in
  Colored Vacua, *Thomas W. Mossberg
  and Maciej Lewenstein*
Structure and Dynamics in Cavity
  Quantum Electrodynamics, *H.J. Kimble*
One Electron in a Cavity, *G. Gabrielse and
  J. Tan*
Spontaneous Emission by Moving Atoms,
  Pierre Meystre and Martin Wilkens
Single Atom Emission in an Optical
  Resonator, *James J. Childs, Kyungwon
  An, Ramanchandra R. Dasari and
  Michael S. Feld*
Nonperturbative Atom–Photon Interactions
  in an Optical *Cavity, H.J. Carmichael, L.
  Tian, W. Ren and P. Alsing*
New Aspects of the Casimir Effect:
  Fluctuations and Radiative Reaction, *G.
  Barton*